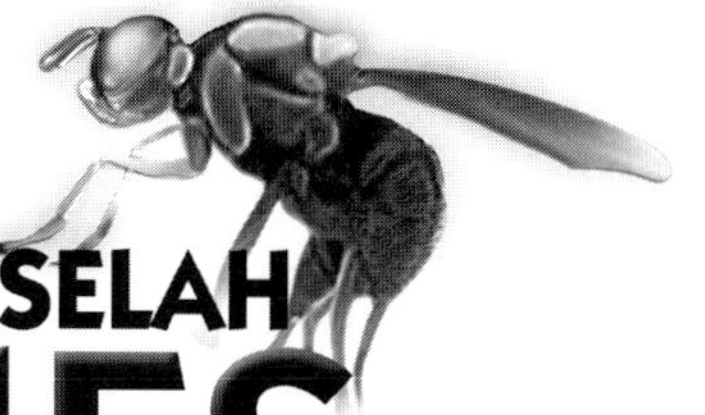

METHUSELAH FLIES

A Case Study in the Evolution of Aging

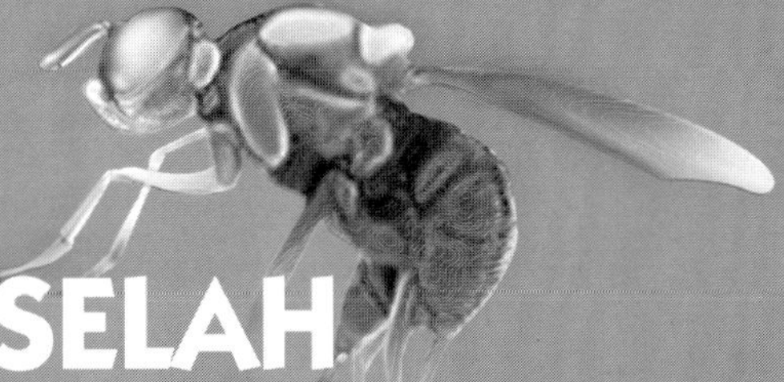

METHUSELAH FLIES

A Case Study in the Evolution of Aging

Michael R. Rose
University of California, Irvine, USA

Hardip B. Passananti
Orrick, Herrington & Sutcliffe LLP, Irvine, USA

Margarida Matos
University of Lisbon, Portugal

World Scientific

NEW JERSEY • LONDON • SINGAPORE • BEIJING • SHANGHAI • HONG KONG • TAIPEI • CHENNAI

Published by

World Scientific Publishing Co. Pte. Ltd.
5 Toh Tuck Link, Singapore 596224
USA office: Suite 202, 1060 Main Street, River Edge, NJ 07661
UK office: 57 Shelton Street, Covent Garden, London WC2H 9HE

British Library Cataloguing-in-Publication Data
A catalogue record for this book is available from the British Library.

We are grateful to the following publishers for their permission to reproduce the articles found in this volume:

Blackwell Scientific Publications (*Funct. Ecol.; J. Evol. Biol.*)
Elsevier Inc. (*Exper. Gerontol.*)
Genetics Society of America (*Genetics*)
Gerontology Society of America (*J. Gerontol. Biol. Sci.*)
Kluwer Academic Publishing (*Genetica*)
Nature Publishing Group (*Nature*)
Society for the Study of Evolution (*Evolution*)
The Company of Biologists Ltd. (*J. Exper. Biol.*)
University of Chicago Press (*Physiol. Zool.*)

METHUSELAH FLIES: A CASE STUDY IN THE EVOLUTION OF AGING

ISBN 981-238-741-2

Printed in Singapore.

CONTENTS

Introduction: Why Methuselah Flies?

Michael R. Rose[1], Hardip B. Passananti[2] and Margarida Matos[3]

[1]*Department of Ecology and Evolutionary Biology, University of California, Irvine, CA 92697-2525, USA*
mrrose@uci.edu; also to whom correspondence should be sent

[2]*Orrick, Herrington & Sutcliffe LLP, 4 Park Plaza, Suite 1600, Irvine, CA 92614*
hpassananti@orrick.com

[3]*Centro de Biologia Ambiental/Departamento de Biologia Animal, Faculdade de Ciências da Universidade de Lisboa, 1749-016 Lisboa Portugal*
mmatos@fc.ul.pt

The study of aging is one of the most challenging of all scientific endeavors. Aging is a process that can last a long time, more than 80 years in humans, even longer in some whales and trees. It is tremendously variable even within species; organisms of the same age can differ greatly in their deterioration. Most importantly, it is a process that involves numerous physiological mechanisms, and these component mechanisms vary from species to species. In mammals, aging involves increasing risks of cancer, cardiovascular disease, and reproductive incompetence. In insects, the most common animals, aging prominently involves the loss of body parts, from wing panels to legs, along with failures of metabolism and reproduction. But cancer is very rare in insects. The accumulation of pigment is widespread, as is the loss of flexibility and muscle strength, among many animal species. Making sense of the physiology of aging has been intensely difficult.

On the other hand, evolutionary biologists have developed a theory of aging that overcomes the physiological complexity of aging, revealing a profound unity underlying the diversity of aging processes. This evolutionary theory of aging can be summarized easily: aging is caused by a fall in the force of natural selection at later adult ages. [Further background for this point can be found in the earlier books of Charlesworth (1980/1994) and Rose (1991).] Natural selection gradually loses its efficiency at "filtering" less fit genotypes with age. What this means is that while natural selection actively favors the maintenance of health at early ages, it stops doing so at later ages. Since there is no other evolutionary force that can sustain health at later ages, we lose that health, as do most other species. This evolutionary theory underlies the experimental work presented in this book.

There are two genetic mechanisms for the evolution of aging: antagonistic pleiotropy and age specificity of gene action (see Rose, 1991). Antagonistic pleiotropy is based on the concept that natural selection will favor alleles that have good effects at early ages even if they have bad effects at much later adult ages. Even without such genetic interconnections between ages, genes that have effects only at late ages will accumulate deleterious mutations at a higher rate than genes expressed at young ages. Both of these genetic mechanisms thus explain the evolution of aging.

The evolutionary theory of aging, and in particular these specific genetic mechanisms, has been tested repeatedly using studies of genetic variation (past work is reviewed in Rose, 1991; see Hughes *et al.*, 2002 for a recent study). These tests are difficult to perform and interpret, because they hinge on genetic variances and covariances. These parameters are notoriously tough to measure accurately. The available evidence tends to support the evolutionary theory of aging in general terms, though there remain controversies about the specific genetic mechanisms involved.

An alternative approach to the study of aging is to create genetic stocks with slowed or delayed aging. This has been done in fruit flies, and the resulting long-lived stocks have been called "Methuselah flies" in a number of popular publications, an example being the book *Longevity, The Science of Staying Young* by Kathy Keeton (p. 33, 1992).

The evolutionary theory of aging enables scientists to make Methuselah flies using a very simple procedure: the denial of reproduction until much later ages. Such a regimen forces natural selection to act at later ages. The rationale is simple:

if aging is due to a loss of natural selection at later ages, increasing natural selection at later ages will lead to a delay of the aging processes. When it is compelled to act at later ages for many generations, natural selection progressively builds longer lived organisms. This will be explained in detail in this book.

Creating Methuselah flies is scientifically profound for several reasons. First, it supports the validity of the evolutionary theory of aging, the only general theory of aging that has much credibility at this time. Second, it shows that there are no fundamental limits to the lifespan of an organism. Change its genetics and it can live longer. Third, it suggests the possibility of using Methuselah flies to figure out the physiology of aging, and thereby move toward the postponement of aging in humans.

Several labs have created Methuselah flies, but the first lab to deliberately do so, and the lab that has created the most stocks of this type, is that of Michael Rose and his students and colleagues. Among all Methuselah stocks, the flies that he first published on in 1984 are the most completely analyzed. This book contains a sampling of work with these flies and flies closely related to them.

The book is laid out in six Parts, each of which treats a main theme of the scientific work that has been done with the Methuselah flies: 1 — creation of these flies; 2 — stress resistance and aging; 3 — reproduction and food; 4 — molecular biology of postponed aging; 5 — reverse evolution of aging; and 6 — development and crowding. Each Part begins with an introduction to the area of work covered. After the introduction, each Part reprints the key previous publications in its area. After these reprints, additional new articles are included in three of the six parts. These are previously unpublished works that have been included in this book for the sake of completeness. They were anonymously reviewed in an editorial process coordinated by Margarida Matos, who had sole authority over the publication of these additional papers. [One such work was rejected as a result of the review process.] Work on extremely late life is omitted as a topic, even though it has used the Methuselah flies discussed here, because a later book is planned in that area.

Since it will be somewhat, though not completely, neglected after this introduction, we now offer some comments on the relevance of aging research that employs other methods. The first thing to be said is that it is very difficult to make scientific progress through the study of longitudinal patterns of aging, the study of non-genetic variation in aging, and related methods. Aging varies, and that variation is important scientific material. But the study of aging is fraught with artifacts and confounds that swamp the study of mere variation in the process. What is needed is research systems that involve drastic differences in aging, differences that supply analytical power to studies of the causes and features of aging.

A longstanding method of creating differences in aging between animals is to change their immediate environment. In invertebrates that do not control their body temperatures metabolically, increasing ambient temperatures leads to increased rates of metabolism, which in turn generate earlier death. [For documentation of this and other features of the natural history of aging, see Finch (1990).] The opposite effect happens at moderately lower temperatures, above freezing. Lifespan is tuned by metabolism, higher metabolic rate shortening life, lower metabolic rate lengthening life. Data showing this effect in fruit flies was published before 1920. There are limits to this pattern, because extreme temperatures are harmful, but it is widespread. Somewhat similar effects occur with nutrition. As nutrition increases, lifespan tends to fall, unless increased nutrition relieves starvation. And conversely with decreased nutrition. Many other environmental parameters modulate rates of aging, too. But the problem with all this information is that it is difficult to pinpoint which, of the many, physiological effects of such environmental changes causally controls aging.

Another old approach is to study genetic mutants that have their lifespan altered. An early study of this kind appeared in 1922, as part of Raymond Pearl's *The Biology of Death*. Pearl studied mutants that died sooner. The problem with such mutants is that they may be dying off as a result of pathological mechanisms that have nothing in common with the causes of aging, an argument made as long ago as 1966 by Maynard Smith, if not earlier by others. This makes short-lived mutants of little value in most research on aging.

The same criticism does not apply to mutants that have increased lifespan, providing the increased lifespan is not a misleading artifact. Such mutants have been found in yeast, nematode worms, and fruit flies. Some of the details concerning their creation are given in *Ageless Quest, One Scientist's Search for Genes that Prolong Youth*, by Lenny Guarante (2003). These mutants are of great significance for our understanding of aging.

But there are still problems with these mutants. Brewer's yeast are single-celled organisms that undoubtedly age. But their aging is mixed up with their cycle of cell division. This complicates the interpretation of yeast experiments on aging.

Nematode aging mutants are in the throes of a controversy right now. Some nematode researchers, like Wayne Van Voorhies (e.g. Van Voorhies, 2002), think that nematode mutants usually, and perhaps only, live longer when they have their metabolism slowed. If that is all there is to increasing nematode life span, it is no more profound than fruit flies that live longer when they are colder. This we knew by 1920. But it is hard to measure metabolism in nematodes which can hardly be seen by the naked eye.

Fruit fly mutants might seem easier to interpret, partly because it is less difficult to measure metabolic rate in them. But it turns out that reproduction presents a problem. Fruit flies that have been castrated genetically or by irradiation live longer. Many other organisms are similar, from soybean to salmon to institutionalized men (Finch, 1990). Reproduction is costly for survival, a major theme in work on Methuselah flies. A mutant which generally reduces reproduction will probably increase lifespan. But the difficulty is that such mutants are not usually evolutionarily important, because natural selection will tend to eliminate them. Nor are they of much medical interest, because people have an interest in reproductive functions that is comparable to their interest in survival.

Another common feature in studies involving mutants is that it is seldom clear what causes an association between a specific mutant and the changed value in life-span of its bearer. Even when specific metabolic pathways are identified or targeted (e.g. using transgenic or knock-out techniques), the precise relation between the affected function and aging may be unknown. Correlation does not mean causality, and the distinction is particularly relevant when dealing with complex, pervasive biological phenomena such as aging.

An apparently promising line of research, and one with abundant empirical data, is the study of mutant mice, both spontaneous and genetically manipulated (e.g. see Finch, 1990; Migliaccio *et al.*, 1999). In spite of the usefulness of such studies, which are particularly appealing because of the high degree of evolutionary relatedness between mice and humans, they too share the aforementioned limitations, as well as the added difficulties in experimental logistics that come from working with a mammal.

For these reasons, research on mutants that affect aging has not given as many useful results as aging researchers have hoped. This is not to deny the eventual value of research on mutants. Here our purpose is to provide a background against which the value of research on Methuselah flies can be seen. These flies are uniquely important for the study of aging. No research organism is quite so easily used to find interpretable results concerning aging, as will be shown. Even though other types of animal will also provide useful insights into aging, Methuselah flies have a strong claim on the attention of all those interested in the science of aging.

We are grateful to Timothy Bradley, Adam Chippindale, and Laurence Mueller for their work on the introductory material for the book. Denise Deckert-Cruz and Adam Chippindale, and their co-workers, also contributed new research articles that helped to fill out the research. Kimberly Nguyen helped with editorial work at UC Irvine. Of course, this book would have been impossible to produce without the authors of the many experimental studies that are published here. Our indebtedness to them is profound.

At World Scientific Publishing we are grateful to Joy Quek and Deborah Heng for their contributions to the publication of this book. Our greatest debt is to Sook Cheng Lim, who has carefully nurtured the book throughout.

References

Charlesworth, B. 1980/1994. *Evolution in Age-structured Populations*, 1st/2nd edition. Cambridge University Press, Cambridge.

Finch, C.E. 1990. *Longevity, Senescence, and the Genome*. University of Chicago Press, Chicago.

Guarante, L. 2003. *Ageless Quest, One Scientist's Search for Genes that Prolong Youth*. Cold Spring Harbor Press, Cold Spring Harbor, New York.

Hughes, K.A., J.A. Alipaz, J.M. Drnevich, and R.M. Reynolds. 2002. A test of evolutionary theories of senescence. *Proceedings of the National Academy of Sciences USA* **99**: 14286–14291.

Keeton, K. 1992. *Longevity, The Science of Staying Young*. Viking Press, New York.

Maynard Smith, J. 1966. Theories of aging. In: *Topics in the Biology of Aging*, P.L. Krohn (ed.). Interscience, New York.

Migliaccio, E., M. Giorgio, S. Mele, G. Pelicci, P. Reboldl, P. P. Pandolfi, L. Lanfrancone, and P. G. Pelicci. 1999. The $p66^{shc}$ adaptor protein controls oxidative stress response and life span in mammals. *Nature* **402**: 309–313.

Pearl, R. 1922. *The Biology of Death*. J.B. Lippincott, Philadelphia.

Rose, M.R. 1991. *Evolutionary Biology of Aging*. Oxford University Press, New York.

Van Voorhies, W.A. 2002. The influence of metabolic rate on aging in the nematode *Caenorhabditis elegans*. *Aging Cell* **1**: 91–101.

I

CREATION AND LONG-TERM EVOLUTION OF METHUSELAH FLIES

The Creation of Methuselah Flies by Laboratory Evolution

Michael R. Rose[1] and Margarida Matos[2]

[1]*Department of Ecology and Evolutionary Biology,*
University of California, Irvine, CA 92697-2525, USA
mrrose@uci.edu; also to whom correspondence should be sent

[2]*Centro de Biologia Ambiental/ Departamento de Biologia Animal,*
Faculdade de Ciências da Universidade de Lisboa,
1749-016 Lisboa Portugal
mmatos@fc.ul.pt

At the heart of the research on Methuselah flies is their creation. In some respects, it is easy to create longer-lived fruit flies. It has been done at least twice inadvertently. In the 1960's, J.M. Wattiaux wanted to test the Lansing Effect. Lansing had found that cultures of rotifers reproduced using older parents exclusively quickly died out (Lansing, 1947, 1948). Wattiaux was expecting to observe a similar effect in *Drosophila pseudoobscura* and *Drosophila subobscura* in two experiments contrasting the fates of populations reproduced either at young ages or at old ages, exclusively. Unfortunately for his plan, he found the opposite pattern: the cultures reproduced using older parents over a number of generations lived longer (Wattiaux, 1968a,b). Nonetheless, he interpreted his results in a similar manner to Lansing, as some kind of non-Mendelian, cumulative, spontaneously reversible effect.

An experiment of Comfort (1953) is relevant to the non-genetic interpretations of Lansing and Wattiaux. An inbred stock of *D. subobscura* was cultured using eggs laid by females that had achieved at least 30 days of age. After eight generations, no change in survival was detected. While this was a short period for evolution to act, Comfort's result does not support a non-genetic or Lamarckian interpretation of the evolution of aging.

An entirely different interpretation of such experiments was published by Edney and Gill in 1968. They pointed out that the force of natural selection at later ages would be strengthened by reproducing laboratory cultures using older parents exclusively. However, they did not think that such experiments were very practical, because of the large amount of time required to let evolution work, given the long generation times resulting from reproduction by older parents. They proposed, instead, that experiments on aging and the force of natural selection should proceed using cultures with even younger parents. This, they predicted, should lead to more rapid aging after a number of generations of evolution.

Sokal (1970) performed an experiment of the type recommended by Edney and Gill (1978). He used flour beetles of the genus *Tribolium*, comparing a culture in which adults were only allowed three days to lay eggs with a mass culture in which a mix of young and old adults could reproduce. The young-adult culture was maintained for 40 generations. The results were mixed, but there was some tendency for cohorts derived from the young-adult culture to die sooner. Mueller (1987) also performed a young-adult experiment, using replicate cultures of *D. melanogaster*. He measured aging using the decline in age-specific fecundity. Older females from the young-adult cultures evolved reduced fecundity later in life, suggesting that they aged at a faster rate. Later data from these stocks (unpublished) also showed a longevity reduction. But a problem in both of these experiments was inbreeding depression. If selected lines become inbred, as small lines usually do, then they will show reductions in fitness-related characters, such as longevity and fecundity.

Despite the analysis of Edney and Gill (1968), the better experiment is to postpone the age of culture reproduction. Hamilton's (1966) formal analysis of the force of selection shows that natural selection on survival remains at full strength until the age of the first act of reproduction in the population. After that, the force of natural selection falls off rapidly. Natural selection should therefore act powerfully to sustain survival if that first age of reproduction is greatly postponed. While this necessarily lengthens the evolutionary process in units of time, the experimental result is less ambiguous because a successful experiment must show an enhancement in adult survival. Inbreeding depression, among other artifacts, is unlikely to produce such an enhancement in later survival.

In order to test whether older adult reproduction was likely to have such a predictable effect on aging, Rose and Charlesworth (1980; not reprinted here) set up a *D. melanogaster* culture in which eggs were collected only from females who had survived at least 21 days as adults. This culture was maintained for 12 generations, and then sampled for offspring that were reared as a cohort together with a cohort of flies that had been maintained from cultures in which the parents were no more than six days old as adults. The stock cultured using older females showed an increase in both longevity and later-life fecundity. This experiment, together with Wattiaux's inadvertent results, suggested that it should be easy to postpone aging in stocks of *Drosophila* simply by increasing the age of the parents used to create the next generation.

In the early 1980's, two laboratories set about performing properly replicated experiments in which aging would be genetically postponed by strengthening the force of natural selection at later ages. The two laboratories were headed by Rose (Rose, 1984; reprinted here) and by Luckinbill and Arking (Luckinbill *et al.*, 1984). The studies were not strictly independent of each other, because Rose visited the Luckinbill/Arking lab and discussed experimental design ideas with them. However, the experiments were then executed independently with stocks of different origins. Both laboratories proceeded by increasing the age of the females that were used to found the next generation, and this increased age was imposed over multiple generations. The Rose lab used five late-age cultures and five early-age cultures. The Luckinbill/Arking lab used fewer replicates, but they assayed their populations more often. In both cases, the age of culture reproduction was progressively increased over the course of the first dozen or so generations. In the first publications on these stocks, assays were made after no more than 13–15 generations, but the results were still statistically significant: longevity and late-life fecundity increased. These first reports appeared side-by-side in the journal *Evolution* in 1984. They essentially defined the modern era of gerontological research, because they established experimental systems in which the biology of aging could be investigated straightforwardly.

Here we reprint one of these papers, the one establishing the Methuselah fly system, Rose (1984). Not all the stocks were analyzed at first, but later studies have consistently compared the five longer-lived (O) stocks with their five (B) controls. After the generations reported in Rose (1984), the late-reproduced (O) populations have been consistently maintained using flies 70 days from their deposition as eggs, except when environmental mishaps, such as bad food, have required shorter generations. The paper by Hutchinson and Rose (1991; reprinted here) describes the Methuselah, or "O", fly stocks after 25 generations of late-reproduction. They were about 30% longer-lived as adults compared with their controls, the "B"s.

Hutchinson and Rose extensively crossed the flies within and between-types, checking for inbreeding depression. They found no evidence for it. Population sizes for each of the B and O stocks have been about 1000 at the time of reproduction. At such population sizes, inbreeding is expected to impair fitness-related characters very slowly.

One of the common criticisms of fruit fly experiments on aging is that they usually do not establish a slower rate of "actuarial aging." That is, while longevity may be increased, this might be due to a general reduction in the death rate, not a reduction in the rate of increase in the death rate during aging. Nusbaum *et al.* (1996; reprinted here) show that the longer-lived Methuselah flies do indeed have a slower rate of "actuarial aging," while other types of longer-lived fly, such as dietarily restricted flies, do not. That is, the Methuselah flies age at a slower rate.

One of the perennial issues in research on aging in fruit flies is the genetic mechanisms that underlie it, as has already been mentioned. In addition to enhanced later fecundity and longevity, the longer-lived flies of Rose and Charlesworth (1980), Luckinbill *et al.* (1984), and Rose (1984) all showed reduced early fecundity. Given the few generations required to produce this effect, and the large population sizes involved, a correlated response to selection, rather than inbreeding depression, is the more plausible explanation for this result. Antagonistic pleiotropy between later life and early fecundity is a genetic mechanism that can explain this type of correlated response (cf. Rose, 1991). Suppose that there are alleles that postpone aging, and so increase in frequency with late reproduction, but these alleles also have an antagonistic pleiotropic effect that reduces early fecundity. This would explain the common experimental result of reduced early fecundity in Methuselah flies.

But an important feature of the data in this field is that reduced early fecundity does not always evolve together with late reproduction (cf. Partridge and Fowler, 1992). One of the more perplexing findings with the Methuselah flies of Rose (1984) is that while longevity continued to increase in these flies, their early fecundity began to recover and even overtake that of the control populations during their subsequent evolution (Leroi *et al.*, 1994a; reprinted here). These results showed that the trade-off between early fecundity and later life was not absolute. What Leroi *et al.* (1994a,b) were able to show was why this trade-off, this form of antagonistic pleiotropy, had disappeared. The problem was small differences between experimental conditions and assay procedures. When the longer-lived O populations were given the B environment during the assay procedure, the O flies still showed a striking reduction in early fecundity. The earlier assays that did not show this reduction were not appropriate. This still does not establish the universality of a fecundity trade-off. Other studies in which an early fecundity reduction does not occur may be free of such trade-offs, or there may be trade-offs that involve characters other than early fecundity, perhaps developmental rate or viability. [See the discussion of these characters in the last section.]

The disparities of results among the studies mentioned in the last paragraph have been interpreted as the result of deep flaws in experimental design. Partridge and collaborators (see Partridge and Andrews, 1985; Partridge and Fowler, 1992) have suggested that selection for early *versus* late reproduction might lead to changes in survival due to changes in risks arising from acts of reproduction, and not to fixed costs of reproduction. Service (1989) showed that suppressing differences in mating activity between the B and O lines, by performing assays in virgin individuals, maintained differences in longevity between the O and B lines, indicating that the difference in longevity between these types of flies was not due to risks arising from reproduction.

Another criticism has emerged concerning the maintenance regimes of the B and O stocks (e.g. Promislow and Tatar, 1998; Linnen *et al.*, 2001). According to Promislow and collaborators, long-established laboratory populations maintained in non-overlapping generations will have a high load of mutations with late-acting effects, accumulated since lab foundation. If this evolutionary mechanism does play an important role in aging, then lab populations with discrete generations may senesce faster than populations maintained with overlapping generations. While this is undeniable, we do not agree that it poses any special problems for studies of Methuselah flies, where populations with discrete generations are compared with each other. In such experimental designs, hypotheses are tested using the relative comparison of O and B flies, regardless of putative longer-lived stocks in the wild or elsewhere that might live longer.

It has also been argued that laboratory populations selected for longer life are simply recovering their capacities of late reproduction and survival before foundation, that is their performance in the wild (Sgrò and Partridge, 2000; Hoffmann *et al.*, 2001; Linnen *et al.*, 2001). Three points deserve consideration here: (1) Not all studies of domestication have indicated that there is a drop of adult survival since introduction to the laboratory environment. In fact, the more detailed temporal studies characterizing evolutionary trajectories of fly populations adapting to the lab suggest the opposite, at least during the first generations after foundation (Matos *et al.*, 2000, 2002; cf. Hoffmann *et al.*, 2001); (2) Even if there were a consistent drop in survival as populations adapt to the laboratory environment, what does this tell us about what their performance in the wild was? As soon as populations are transferred to a novel laboratory environment, the effect of genotype-by-environment interactions preclude any interpretation as to what their attributes were in nature; (3) Finally, suppose that in fact longer-lived *Drosophila* populations are simply recovering the survival capacities of their ancestral, founder populations? Would this put at risk the research on these populations, endangering our understanding of the genetics and evolution of aging? After all, Methuselah flies are different in comparison with their control populations. Experimental evolution is concerned with differences from controls, not absolute values.

The O Methuselah Flies are examples of a range of fly stocks in which aging has been postponed using evolution in the laboratory. Among their advantages relative to other fly stocks is their five-fold replication, their large population sizes, the long laboratory domestication of their ancestors, their well-defined controls, and their long history of selection, now more than 24 years, over 100 generations. The creation of these flies illustrates principles of evolution that are now well established. The most interesting task now is the use of these flies to unravel the genetics and biology of aging.

References

Charlesworth, B. 1980. *Evolution in Age-structured Populations*. Cambridge University Press, Cambridge.

Comfort, A. 1953. Absence of a Lansing effect in *Drosophila subobscura*. *Nature* **172**: 83–84.

Edney, E.B. and R.W. Gill. 1968. Evolution of senescence and specific longevity. *Nature* **220**: 281–282.

Hamilton, W.D. 1966. The moulding of senescence by natural selection. *J Theor Biol* **12**: 12–45.

Hoffmann, A.A., R. Hallas, C. Sinclair, and L. Partridge. 2001. Rapid loss of stress resistance in *Drosophila melanogaster* under adaptation to laboratory culture. *Evolution* **55**: 436–438.

Hutchinson, E.W. and M.R. Rose. 1991. Quantitative genetics of postponed aging in *Drosophila melanogaster*. I. Analysis of outbred populations. *Genetics* **127**: 719–727.

Ives, P.T. 1970. Further studies of the South Amherst population of *Drosophila melanogaster*. *Evolution* **38**: 507–518.

Lansing, A.I. 1947. A transmissible, cumulative and reversible factor in aging. *J Gerontol* **2**: 228–239.

Lansing, A.I. 1948. Evidence for aging as a consequence of growth cessation. *Proceedings of the National Academy of Sciences USA* **34**: 304–310.

Leroi, A.M., A.K. Chippindale, and M.R. Rose. 1994a. Long-term laboratory evolution of a genetic life-history trade-off in *Drosophila melanogaster*. I. The role of genotype-by-environment interaction. *Evolution* **48**: 1244–1257.

Leroi, A.M., W.R. Chen, and M.R. Rose. 1994b. Long-term laboratory evolution of a genetic life-history trade-off in *Drosophila melanogaster*. 2. Stability of genetic correlations. *Evolution* **48**: 1258–1268.

Linnen, C., M. Tatar, and D. Promislow. 2001. Cultural artifacts: A comparison of senescence in natural, laboratory-adapted and artificially selected lines of *Drosophila melanogaster*. *Evol Ecol Res* **3**: 877–888.

Luckinbill, L.S., R. Arking, M.J. Clare, W.C. Cirocco, and S.A. Buck. 1984. Selection for delayed senescence in *Drosophila melanogaster*. *Evolution* **38**: 996–1003.

Matos, M., M.R. Rose, M.T. Rocha Pité, C. Rego, and T. Avelar. 2000. Adaptation to the laboratory environment in *Drosophila subobscura*. *J Evol Biol* **13**: 9–19.

Matos, M., T. Avelar, and M.R. Rose. 2002. Variation in the rate of convergent evolution: adaptation to a laboratory environment in *Drosophila subobscura*. *J Evol Biol* **15**: 673–682.

Mueller, L.D. 1987. Evolution of accelerated senescence in laboratory populations of *Drosophila*. *Proceedings of the National Academy of Sciences USA* **84**: 1974–1977.

Nusbaum, T.J., L.D. Mueller, and M.R. Rose. 1996. Evolutionary patterns among measures of aging. *Exper Gerontol* **31**: 507–516.

Partridge, L. and R. Andrews. 1985. The effect of reproductive activity on the longevity of male *Drosophila melanogaster* is not caused by an acceleration of ageing. *J Insect Physiol* **36**: 419–425.

Partridge, L. and A.K. Fowler. 1992. Direct and correlated responses to selection on age at reproduction in *Drosophila melanogaster*. *Evolution* **46**: 76–91.

Promislow, D. and M. Tatar. 1998. Mutation and senescence: where genetics and demography meet. *Genetica* **103**: 299–314.

Rose, M.R. 1984. Laboratory evolution of postponed senescence in *Drosophila melanogaster*. *Evolution* **38**: 1004–1010.

Rose, M.R. 1991. *Evolutionary Biology of Aging*. Oxford University Press, New York.

Rose, M.R. and B. Charlesworth. 1980. A test of evolutionary theories of senescence. *Nature* **287**: 141–142.

Service, P.M. 1989. The effect of mating status on lifespan, egg laying, and starvation resistance in *Drosophila melanogaster* in relation to selection on longevity. *J Insect Physiol* **35**: 447–452.

Sgrò, C.M. and L. Partridge. 2000. Evolutionary responses of the life history of wild-caught *Drosophila melanogaster* to two standard methods of laboratory culture. *Am Nat* **156**: 341–353.

Sokal, R.R. 1970. Senescence and genetic load: Evidence from *Tribolium*. *Science* **167**: 1733–1734.

Wattiaux, J.M. 1968a. Cumulative parental effects in *Drosophila subobscura*. *Evolution* **22**: 406–421.

Wattiaux, J.M. 1968b. Parental age effects in *Drosophila pseudoobscura*. *Exper Gerontol* **3**: 55–61.

Evolution, 38(5), 1984, pp. 1004–1010

LABORATORY EVOLUTION OF POSTPONED SENESCENCE IN *DROSOPHILA MELANOGASTER*

MICHAEL R. ROSE
Department of Biology, Dalhousie University, Halifax, Nova Scotia, Canada B3H 4J1

Received May 24, 1983. Revised December 31, 1983

Evolutionary genetics seems to have found the fundamental cause of senescence: the decline in the sensitivity of natural selection to gene effects expressed at later ages in most populations of organisms with separate somatic and germline tissue. (Here "senescence" refers to decline in age-specific fitness-components after the onset of reproductive maturity.) This idea traces back to Haldane (1941) and Medawar (1946, 1952), with considerable elaboration and elucidation since then (Williams, 1957; Hamilton, 1966; Edney and Gill, 1968; Emlen, 1970; Charlesworth and Williamson, 1975; Charlesworth, 1980; Rose, 1983*a*). While there are still clear limitations to the mathematical formulation of this theory (cf. Hamilton, 1966; Charlesworth, 1980), the basic formal analysis leads to a straightforward conclusion: the first partial derivative of fitness with respect to appropriately scaled changes in age-specific life-history characters usually declines in magnitude with the age of these changes. The force of natural selection thus declines with age.

This overall theory and its particular subsidiary variants lead to a number of empirically testable corollaries (Rose, 1983*a*, 1983*b*). Some of these corollaries are specific to the subsidiary variants of the theory (Rose and Charlesworth, 1980, 1981*a*, 1981*b*; Rose, 1983*b*), so that tests of them individually do not test the theory as a whole. Fortunately, there are two corollaries which follow from the general theory itself: the reproductive schedule of an outbred population will give rise to natural selection acting to (i) *accelerate* senescence in populations with a relatively earlier age of reproduction and (ii) *postpone* senescence in populations with a relatively *later* age of reproduction (Edney and Gill, 1968; Rose, 1983*a*). The former prediction has been corroborated by Sokal (1970) using *Tribolium castaneum*, while the latter has been corroborated by Wattiaux (1968*a*, 1968*b*) and by Rose and Charlesworth (1980, 1981*b*), using *Drosophila* species.

Once a theory has been well-developed mathematically and then empirically corroborated, attention turns to experiments in which the theory either is not clearly corroborated or is ostensibly refuted. It would be misleading to claim that all relevant experimental results directly corroborate the evolutionary theory of senescence. Sokal (1970) and Mertz (1975) using *Tribolium castaneum* and Taylor and Condra (1980) using *Drosophila pseudoobscura* found heterogeneity between lines in experiments with replication, such that some lines did not exhibit the predicted response to the imposed selective regime. Taylor and Condra (1980) also found a difference in the response of the sexes which was later attributed to the pattern of female mating preference (Taylor et al., 1981). More problematic still are the studies from the Lints laboratory, one of which failed to obtain a direct response to artificial selection for longevity (Lints et al., 1979), while another gave puzzling fluctuations in life-history attributes (Lints and Hoste, 1974, 1977). Lints (1978, 1983) has made a great deal of these problems, contending that they cast doubt on all proposed evolutionary theories of senescence.

While it can be argued that these puzzling results are due to technical artifacts such as inbreeding, genetic disequilibrium, and inadequate controls (cf. Rose and Charlesworth, 1981*b*), the only effective way to resolve such empirical ambiguities is persistent repetition of these experiments under well-defined conditions. Here I report experiments which reprise a previous experiment of this kind (Rose and Charlesworth, 1981*b*) both by wholesale repetition and by internal replication, in an attempt to remove any question about the potential reproducibility of the predicted results under conditions in which artifacts are forestalled.

Materials and Methods

The experimental population used in these experiments was an outbred laboratory population derived from the wild South Amherst, Massachusetts, *Drosophila melanogaster* population studied by Ives (1970), as outlined in Rose and Charlesworth (1981*a*). This population was intensively studied during 1977–1979, when it was found to have a great deal of additive (selectable) genetic variability for many adult life-history characters (Rose and Charlesworth, 1981*a,* 1981*b*). In particular, there were several lines of evidence indicating negative additive genetic correlation between early and late life-history characters, such as early fecundity and longevity.

In February, 1980, after more than 130 generations of laboratory culture at 25 C with unlimited food and 14 day discrete generations, ten experimental populations were derived from a single generation of the base population. Five of these were then maintained for about 50 generations in the same fashion as the base population. Here they will be referred to as the "B" populations, with subscripts 1–5 to indicate replicate number. The remaining five populations were kept under the same culture conditions as the B's, but the day on which eggs were collected for the next generation was progressively postponed. For four generations, eggs were collected from adults that were 28 days of age. Then, there were two generations of 35 days, followed by six generations of 42 days. The next two generations lasted 56 days, and finally, there was a generation of 70 days, taking the population to December, 1981. These populations were thus subject to 15 generations of laboratory evolution in which only surviving older females were able to reproduce. Here they are called "O" populations, with subscripts as before.

The culture conditions changed over the two-year period of laboratory evolution. For the first year and three months, the populations were maintained on cornmeal-molasses medium and a 12L:12D light regime, at the Laboratory of Genetics of the University of Wisconsin–Madison.

The populations were then transported to Dalhousie University, where they were maintained on banana-molasses medium and a 24L:0D light regime. Except for the period of transport, the flies were kept at 25 C. Food was always abundant in quantity, and populations experienced no severe crowding. Population sizes were in the thousands at the start of each generation, except for the O_2 population, which accidentally fell in numbers at one point. One generation of the O_5 population was accidentally terminated early. Since O_2 and O_5 were thereby aberrant, they were excluded from the life-history assays reported here. For the purpose of matching B's with O's, populations B_4 and B_5 were also dropped from the assays.

Two major life-history assays were performed from January to June, 1982. Both used adults obtained after two generations of controlled density sampling, with 14-day generations, from both B and O populations, as in Rose and Charlesworth (1981*a*). This was done to remove parental effects. The major difference in technique was that 20 eggs were harvested using a dull scalpel for each rearing tube. The first assay was of lifetime female daily fecundity and the longevity of both sexes, using the media and transfer procedures of Rose and Charlesworth (1981*a*). Complete fecundity records were obtained for 416 females, for a total of over 700,000 eggs counted. A total of 791 longevities, 424 female and 367 male, were recorded. The second assay was of

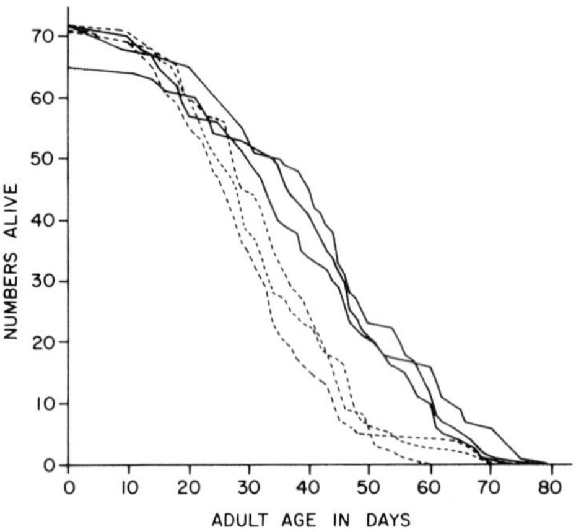

FIG. 1. Surviving numbers of females from the start of the adult life-history assay period. B population samples are shown as dashed lines, O samples as solid lines.

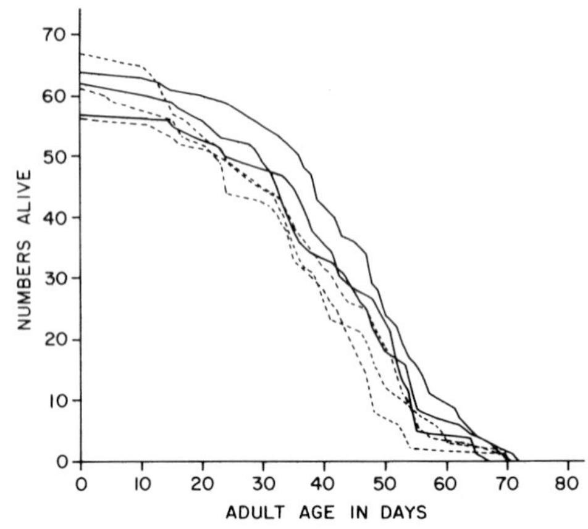

FIG. 2. Surviving number of males from the start of the adult life-history assay period. B population samples are shown as dashed lines, O samples as solid lines.

the viability of eggs laid by females over the first 8 weeks of adult life, as ascertained by the survival of 50 eggs reared together in one tube from oviposition to 14 days of age. These eggs were obtained weekly from mass layings by females kept with males in population cages. The total number of larvae assayed in this experiment was 81,250.

To reduce the likelihood of uncontrolled biases, B and O populations were handled as pairs following a consistent order. Thus, in the first assay, first B_1 and O_1 were transferred together, then B_2 and O_3, and finally B_3 and O_4. For the fecundity characters, the relative order of pair handling made a substantial difference. This was probably due to diurnal variability in laboratory temperature. These data were accordingly analysed using general mixed model maximum-likelihood analysis of variance methods, as they take into account the possibility of such design effects. For the longevity characters, there was no evidence for such an effect from mixed model analysis, so that an unbalanced, random effects, hierarchical model was used for the analysis of variance instead. All mixed model analysis proceeded using the BMDPV.3 program documented in Dixon (1983). The hierarchical analysis used the standard equations for such models (cf. Kempthorne, 1957).

RESULTS

Longevity

Figures 1 and 2 show the pattern of survivorship differences between O and B populations for females and males, respectively. The overall mean female longevities for O and B populations are 42.81 and 33.28 days, respectively, while the overall mean male longevities for O and B populations are 44.14 and 38.49 days, respectively. Using a hierarchical analysis of variance, the null hypothesis of no longevity difference between O and B treatments can be rejected ($P < .01$) for both males and females. Longevity data from this population tend to be positively skewed and leptokurtic, but not so much as to render an analysis of variance invalid.

Another test of survivorship differentiation is the comparison of mortality rates. This was done as follows: (i) Age-specific mortality rates were calculated for each 10-day interval from day 1 to day 50 of the assay. (After day 50, there were too few B individuals.) (ii) the mortality rate difference between the paired

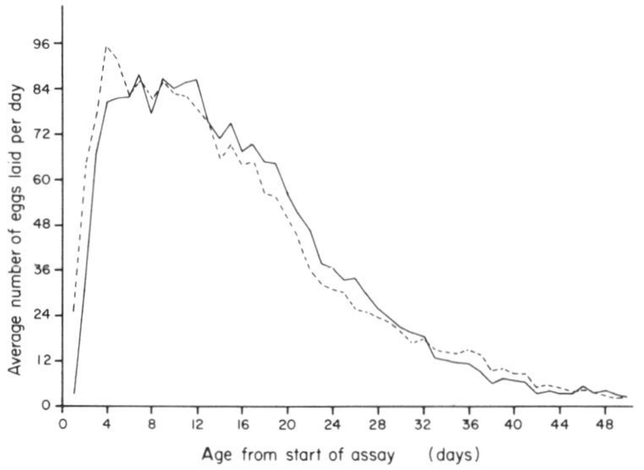

FIG. 3. Treatment means of daily fecundity plotted against days of age from the start of the adult life-history assay. The B treatment means are shown as a dashed line, the O means as a solid line.

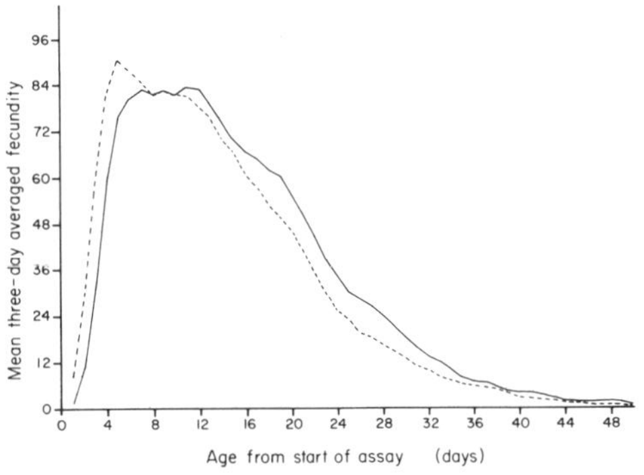

FIG. 4. Treatment means of three-day averaged daily fecundity, irrespective of survival, plotted against days of age from the start of the adult life-history assay. The B treatment means are shown as a dashed line, the O means as a solid line.

B and O populations was calculated for each of the three pairs. (iii) This difference was then divided by the total mortality rate of that pair in that age-interval.

This was done in order to compensate for the scale effect arising from the substantial increase in mortality with age. For the purpose of data analysis, the mortality rate differences of each pair are subject to serial correlation problems. This problem can be surmounted by treating the parameters of the distribution of the mortality rate differences of each pair as raw data. There are two obvious hypotheses to test concerning mortality rate differences. First, do O's have lower mortality rates than B's, on average? The mean of the average mortality rate difference over the three B-O pairs is .2821 ± .0911 (standard error) in females and .2440 ± .1181 in males. Both are significantly greater than zero ($P <$ .05), allowing us to reject the null hypothesis of no mortality rate differences. Second, is there any trend with respect to age in mortality rate differences? This was tested using least-squares linear regression through each pair's mortality rate data. The slopes were −.0617 ± .0703 and −.0700 ± .0699 for females and males, respectively. Both of these are within about one standard error of zero, giving no justification for the rejection of the null hypothesis. Thus O's were mortality-resistant compared to the B's, with no apparent age-specificity in this differentiation.

Fecundity

The overall pattern of B and O fecundity is shown in Figure 3. Initially, B's have greater fecundity, while O's have greater fecundity from about day 10 to day 30. B's again have greater fecundity from days 33 to 44, after which the pattern is unclear.

One encounters many problems when analyzing data of this kind: (a) It is known that fecundity is serially correlated among individuals (Rose and Charlesworth, 1981*a*). Thus one cannot compare fecundities day-by-day. (b) There are problems with the fecundity character of dead individuals, which bias the data analysis if such individuals are simply excluded. (c) Handling gave rise to day-to-day cycles in the fecundity of individual females. To deal with the last two problems, three-day average fecundities were calculated for each female irrespective of survival. These gave the fecundity curves of Figures 4 and 5. Differential survival plays little role for the first 20 days, the survivorship difference amounting to only 5.3% at age 20 days, while the three-day averaged fecundity difference is 20.9%. The problem of serial correlation is more severe. One way around such difficulties is to pick a pair of days, early and late, using the significance levels for those two days only to evaluate the null hypothesis of no treatment differences early or late. For many of the first six days, this hypothesis can be rejected, at the $\alpha = .05$ level. Similarly, from day 17 to day 37, this hypothesis can also be rejected, as well as sporadically thereafter, at the $\alpha = .05$ level. Ideally, one would want a single parametric test for the entire fecundity pattern, but any such parameter would have to be derived by complicated model-fitting, given the nonlinearity of the fecundity patterns. However, it seems reasonable to conclude that a pattern of postponement in reproductive output is present in the O's relative to the B's.

Mean total fecundities for the B and O treatments were 1635.01 and 1733.41, respectively. This is not a statistically significant difference at the $\alpha = .05$, or even $\alpha = .1$, level. There is certainly no evidence for an overall depression in O population fecundity. Thus the disparity in daily fecundities appears to arise from the redistribution of an approximately fixed total reproductive output.

Viability

Unlike the fecundity and longevity data, there was no apparent differentiation between treatments, with mean viabilities of .793 and .744 for B's and O's, respectively. Statistical analysis corroborates this inference. (In fact, the probability of getting this much treatment differentiation by chance alone is .879.) Nor were members of the pairs appreciably different from each other. The only statistically discernable pattern was a fall in viability with parental age.

Discussion

The overall thrust of the results is in accord with those of earlier studies of the same population (cf. Rose and Charlesworth, 1980, 1981*a*, 1981*b*). For example, Figure 5 may be compared with Figure 1 of Rose and Charlesworth (1981*b*). Essentially the same pattern of an initial

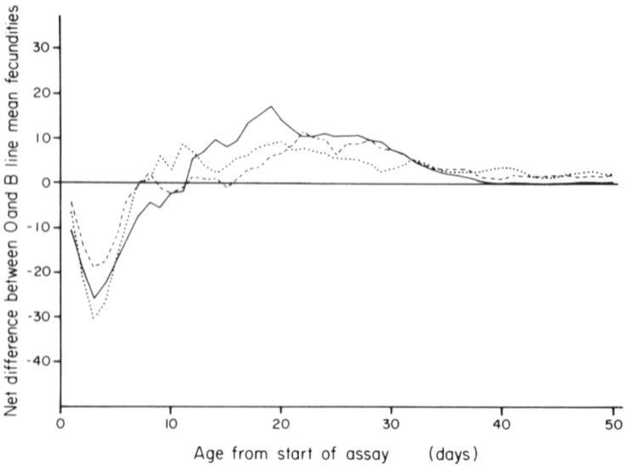

FIG. 5. Net difference between population sample means for three-day averaged daily fecundity plotted against days of age from the start of the adult life-history assay. The solid line gives the difference between B_1 and O_1. The dashed line gives the difference between B_2 and O_3. The dotted line gives the difference between B_3 and O_4.

dip below the zero-axis followed by a rise above it is present in both figures. This study thus underscores conclusions found before: (i) female longevity is increased by culturing populations at later ages, (ii) late fecundity is similarly enhanced, and (iii) these changes are associated with depression of early fecundity. In terms of life-history characters, senescence has been postponed.

The present study provides the following new conclusions. First, replication can give rise to the same *relative* changes, so that results of this kind can be reproducible (cf. Mertz, 1975). Second, the variability over replicates shows that there need be no assayable absolute changes in such characters, making paired control and selected populations during phenotypic assay essential for proper hypothesis evaluation. Third, male longevity is also increased by late culturing, although not to the same extent (cf. Taylor and Condra, 1980; Taylor et al., 1981). Fourth, there do not appear to be any large changes in viability arising from postponement of senescence. None of these new conclusions clash with those found before.

The scientific significance of these conclusions is that senescence in this population of *Drosophila melanogaster* appears to be due to antagonistic pleiotropy, such that genes which postpone senescence appear to depress early fitness-components. Put another way, these results corroborate the hypothesis of a cost to reproduction (Williams, 1957, 1966), since prolonged life seems to require reduced early reproductive output. (However, it should be pointed out that total reproductive output is not reduced with increased lifespan, so that the reproductive cost is not "absolute" or irrespective of timing.) Thus this set of interrelated hypotheses has received another experimental corroboration.

What then of other experiments which have not corroborated these hypotheses? There are a number of ways in which one could fail to get results of the present kind. First, an initially inbred population need not exhibit any appropriate response to changes in reproductive schedule, as the requisite genetic variability may not be present. Second, because small differences in assay conditions can have large effects on life-history characters, as has been shown here experimentally, careful pairing of "experimental" and control samples is important. Third, the development of inbreeding depression during an experiment of this kind could also obscure any selection response. Fourth, because these experiments take on the order of 30–50 normal generations (and thus years with insect species), they may be prematurely terminated, before sufficient opportunity has been given for selection to act. It is therefore not surprising that some of the experiments of this type should have failed to corroborate the evolutionary theory of senescence, though which technical difficulty was involved in which experiment would be difficult to ascertain.

Lints (1983) has attacked previous results which have been regarded as corroborations of the evolutionary theory of senescence. First, he has suggested that the mean longevity of the base population studied here is abnormally low. There is no force in this criticism, as this population has been deliberately selected for high initial fertility, with no selection for later survival. That it should have low longevity is therefore not problematic, but corroborative. Second, Lints draws attention to evolutionarily uninterpretable differences in *absolute* fecundities and longevities in the experiments of Rose and Charlesworth (1981*b*), but, as has been shown here, these are of no necessary significance. Third, Lints points out that experiments from his laboratory (Lints and Hoste, 1974, 1977) have shown that the absolute means of life-history characters can vary in odd ways over generations. There are two rejoinders to this criticism: again there are no clear expectations for absolute responses of the characters, and the choices of the generations for assay in this experiment and its predecessor (Rose and Charlesworth, 1981*b*) were arbitrary, yet the results were consistent. All told, Lints provides no cogent critique of the experimental evidence in support of the evolutionary theory of senescence.

Another criticism of experiments like the present one has been that they do not provide information concerning the epigenetic mechanisms underlying the postponement of senescence (Hart and Turturro, 1983). In effect, the present experiments treat the organism as an evolving black-box. This need not remain the case. The B and O populations developed here, with their different rates of senescence, provide suitable material for the analysis of the biological basis of postponed senescence.

Summary

A general corollary of population-genetic theories of senescence is that the culture of outbred laboratory populations using females of later ages should lead to the evolution of postponed senescence. This has been tested before, but the predicted results have not been consistently reproducible. An experiment of this kind was both repeated and replicated, using a previously studied *Drosophila melanogaster* population. The results uniformly corroborated the evolutionary theory of senescence. Senescence was consistently postponed. Other, anomalous, results are explained in terms of experimental artifact.

ACKNOWLEDGMENTS

I thank A. M. Coyle and M. L. Dory for technical assistance. D. Hamilton and the referees provided helpful suggestions concerning the statistical analyses. The research was supported by Natural Sciences and Engineering Research Council of Canada grant number U0178, while the author was a University Research Fellow.

LITERATURE CITED

CHARLESWORTH, B. 1980. Evolution in Age-Structured Populations. Cambridge Univ. Press, London.

CHARLESWORTH, B., AND J. A. WILLIAMSON. 1975. The probability of survival of a mutant gene in an age-structured population and implications for the evolution of life-histories. Genet. Res. 26:1–10.

DIXON, W. J., ed. 1983. BMDP Statistical Software. Univ. California Press, Berkeley.

EDNEY, E. B., AND R. W. GILL. 1968. Evolution of senescence and specific longevity. Nature 220: 281–282.

EMLEN, J. M. 1970. Age-specificity and ecological theory. Ecology 51:588–601.

HALDANE, J. B. S. 1941. New Paths in Genetics. Allen and Unwin, London.

HAMILTON, W. D. 1966. The moulding of senescence by natural selection. J. Theoret. Biol. 12: 12–45.

HART, R. W., AND A. TURTURRO. 1983. Theories of aging. Rev. Biol. Res. Aging 1:5–17.

IVES, P. T. 1970. Further studies of the South Amherst population of *Drosophila melanogaster.* Evolution 24:507–518.

KEMPTHORNE, O. 1957. An Introduction to Genetic Statistics. John Wiley and Sons, N.Y.

LINTS, F. A. 1978. Genetics and Aging. S. Karger. Basel, Switzerland.

———. 1983. Genetic influences on lifespan in *Drosophila* and related species. Rev. Biol. Res. Aging 1:51–72.

LINTS, F. A., AND C. HOSTE. 1974. The Lansing effect revisited. I. Lifespan. Exp. Geront. 9:51–69.

———. 1977. The Lansing effect revisited. II. Cumulative and spontaneously reversible parental age effects on fecundity in *Drosophila melanogaster.* Evolution 31:387–404.

LINTS, F. A., J. STOLL, G. GRUWEZ, AND C. V. LINTS. 1979. An attempt to select for increased longevity in *Drosophila melanogaster.* Gerontology 25:192–204.

MEDAWAR, P. B. 1946. Old age and natural death. Mod. Quart. 1:30–56.

———. 1952. An Unsolved Problem of Biology. H. K. Lewis, London.

MERTZ, D. B. 1975. Senescent decline in flour beetles selected for early adult fitness. Physiol. Zool. 48:1–23.

ROSE, M. R. 1983*a*. Evolution of aging. Rev. Biol. Res. Aging 1:19–24.

———. 1983*b*. Theories of life-history evolution. Amer. Zool. 23:15–23.

ROSE, M., AND B. CHARLESWORTH. 1980. A test of evolutionary theories of senescence. Nature 287:141–142.

———. 1981*a*. Genetics of life-history in *Drosophila melanogaster.* I. Sib analysis of adult females. Genetics 97:173–186.

———. 1981*b*. Genetics of life-history in *Drosophila melanogaster.* II. Exploratory selection experiments. Genetics 97:187–196.

SOKAL, R. R. 1970. Senescence and genetic load: evidence from *Tribolium.* Science 167:1733–1734.

TAYLOR, C. E., AND C. CONDRA. 1980. r- and K-selection in *Drosophila pseudoobscura.* Evolution 34:1183–1193.

TAYLOR, C. E., C. CONDRA, M. CONCONI, AND T. PROUT. 1981. Longevity and male mating success in *Drosophila pseudoobscura.* Amer. Natur. 117:1035–1039.

WATTIAUX, J. M. 1968*a*. Parental age effects in *Drosophila pseudoobscura.* Exp. Geront. 3:55–61.

———. 1968*b*. Cumulative parental age effects in *Drosophila subobscura.* Evolution 22:406–421.

WILLIAMS, G. C. 1957. Pleiotropy, natural selection, and the evolution of senescence. Evolution 11:398–411.

———. 1966. Natural selection, the costs of reproduction, and a refinement of Lack's principle. Amer. Natur. 100:687–690.

Corresponding Editor: R. E. Ricklefs

Quantitative Genetics of Postponed Aging in *Drosophila melanogaster*. I. Analysis of Outbred Populations

Edward W. Hutchinson[1] and Michael R. Rose[2]

Department of Biology, Dalhousie University, Halifax, Nova Scotia, Canada B3H 4J1, and Department of Ecology and Evolutionary Biology, School of Biological Sciences, University of California, Irvine, California 92717

Manuscript received May 15, 1989
Accepted for publication December 13, 1990

ABSTRACT

Selection has been used to create replicated outbred stocks of *Drosophila melanogaster* with increased longevity, increased later fecundity, and increased levels of physiological performance at later ages. The present study analyzed the quantitative transmission patterns of such stocks, employing extensive replication in numbers of stocks, individuals, and assayed characters. The populations used derived from five lines with postponed aging and five control lines, all created in 1980 from the same founding base population. The following characters were studied: early 24-hr fecundity, early ovary weight, early female starvation resistance, early male starvation resistance, female longevity and male longevity. Numerous crosses were performed to test for non-Mendelian inheritance, average dominance, maternal effects, sex-linkage and between-line heterogeneity. There was only slight evidence for any of these phenomena arising reproducibly in the characters studied. These findings suggest the value of this set of stocks for studies of the physiological basis of postponed aging.

DROSOPHILA melanogaster is one of only two species for which stocks having genetically postponed aging are available (ROSE 1984; LUCKINBILL *et al.* 1984), the other being *Caenorhabditis elegans* (JOHNSON and WOOD 1982; JOHNSON 1987; FRIEDMAN and JOHNSON 1988). These *D. melanogaster* stocks were not created by mutagenesis, but by selection on quantitative genetic variability. It has not proven practical to use mutagenesis to produce *Drosophila* stocks with postponed aging (*e.g.* ROBERTS and IREDALE 1985), with one possible exception (LEFFELAAR and GRIGLIATTI 1984). This limitation on genetic analysis arises because *Drosophila* spp. are subject to inbreeding depression for fitness-related characters, like longevity (CLARKE and MAYNARD SMITH 1955), making the isolation of mutant strains with postponed aging very difficult.

The use of selection is feasible for two reasons. First, there is abundant quantitative genetic variability for life-historical characters in outbred stocks of *D. melanogaster* (ROSE and CHARLESWORTH 1981a,b). Secondly, an indirect selection procedure can be used, in which natural selection is directed to act at later ages by the use of eggs laid by older females exclusively. When applied repeatedly over a number of generations, this type of selective screen leads to the evolution of postponed aging (WATTIAUX 1968a,b; ROSE and CHARLESWORTH 1981b; ROSE 1984; LUCKINBILL *et al.* 1984). This procedure also involves population sizes large enough (300–3000) that population size alone should not radically limit the gains made by selection (*cf.* YOO 1980; WEBER 1990; WEBER and DIGGINS 1990).

The *D. melanogaster* populations that have been created using these methods have been analyzed in two different ways. First, these populations have been compared with control populations, of identical origin but lacking postponed aging, with regard to morphology (ROSE *et al.* 1984; LUCKINBILL *et al.* 1988a) and physiology (SERVICE *et al.* 1985; SERVICE 1987; LUCKINBILL *et al.* 1988a; GRAVES, LUCKINBILL and NICHOLLS 1988). This research has endeavored to find mechanisms that could causally account for the postponed aging of the selected lines. Second, one of these populations has been examined genetically by means of biometrical analysis (CLARE and LUCKINBILL 1985; LUCKINBILL *et al.* 1987) and chromosomal substitution (LUCKINBILL *et al.* 1988b). The present article combines both of these avenues of research in an attempt to ascertain the quantitative genetic basis of postponed aging in *D. melanogaster*.

Of particular concern for the present work is the degree of replication in the studies of CLARE and LUCKINBILL (1985) and LUCKINBILL *et al.* (1987). Those studies used a single pair of long-lived and control populations, measuring longevity and fecundity alone among the characters that have responded to selection for postponed aging (*cf.* SERVICE *et al.* 1985; LUCKINBILL *et al.* 1988a). The present study analyzes five different stocks having postponed senescence, comparing them with five different control stocks. In addition, these stocks were not derived from

[1] Present address: Institute for Behavioral Genetics, Box 447, University of Colorado, Boulder, Colorado 80309.
[2] To whom all correspondence should be addressed at the University of California.

Genetics 127: 719–727 (April, 1991)

TABLE 1

B diallel experiments

Experiment	Character assayed	Populations assayed per character	Mean No. assayed per population	Total No. assayed
DB1	Ovary weight	5 × 5 = 25	29.6	740
	Female starvation	5 × 5 = 25	54.7	1368
DB2	Ovary weight	3 × 3 = 9	26.1	235
Total		59		2343

TABLE 2

O diallel experiments

Experiment	Character assayed	Populations assayed per character	Mean No. assayed per population	Total No. assayed
DO1	Ovary weight	5 × 5 = 25	28.0	700
	Female starvation	5 × 5 = 25	55.4	1384
DO2	Female starvation	5 × 5 = 25	52.8	1321
DO3	Ovary weight	3 × 3 = 9	26.0	234
DO4	Female starvation	5 × 5 = 25	25.6	639
	Male starvation	5 × 5 = 25	25.4	636
Total		134		4914

the stocks analyzed by CLARE and LUCKINBILL (1985) and LUCKINBILL *et al.* (1987). Results which are general to all these lines might therefore be general to the species as a whole.

The basic techniques used in the present analysis involve crosses of populations, both within the selected and control groups of lines and between the selected and control groups of lines. With some exceptions, we find that our results tend to indicate additive inheritance among populations and little differentiation between lines within a given type, postponed aging or control.

MATERIALS AND METHODS

Stocks: The postponed-aging stocks, called "O"s, that were used in the present study were derived by culture selection from the Ives population studied by ROSE and CHARLESWORTH (1981a,b). The date of founding for these stocks, and their controls, was February, 1980. The culture selection procedure involved stock maintenance using discrete generations and females of progressively greater age. Eventually, the females used were 70 days of age from the egg. For more details, consult ROSE (1984) or SERVICE *et al.* (1985). These stocks exhibit increased male and female longevities, decreased early fecundity, decreased early ovary weight, increased resistance to certain stresses, decreased early metabolic rate, increased lipid content, decreased early locomotor activity, and increased later locomotor activity (ROSE 1984; ROSE *et al.* 1984; SERVICE *et al.* 1985; SERVICE 1987), relative to the control stocks. The control stocks, called "B"s, were maintained using the same media and procedures as the O stocks, except that females of 14 days

TABLE 3

B and O crossing experiments

Experiment	Character assayed	Populations assayed per character	Mean No. assayed per population	Total No. assayed
BO1	Ovary weight	3 × 5 = 15	29.5	443
	Female starvation	4 × 5 = 20	31.2	624
	Male starvation	4 × 5 = 20	31.2	624
BO2	Ovary weight	4 × 5 = 20	30.0	600
	Fecundity	4 × 5 = 20	58.9	1,177
	Conditional fecundity	4 × 5 = 20	58.9	1,177
	Female starvation	4 × 5 = 20	54.0	1,080
	Female longevity	4 × 5 = 20	57.0	1,140
BO3	Ovary weight	4 × 3 = 12	45.2	542
	Female starvation	4 × 3 = 12	40.0	480
	Female longevity	4 × 3 = 12	49.9	599
BO4	Female starvation	4 × 3 = 12	33.7	404
BO5	Fecundity	4 × 3 = 12	76.5	918
	Conditional fecundity	4 × 3 = 12	75.3	903
	Female starvation	4 × 3 = 12	78.3	940
	Male starvation	4 × 3 = 12	78.3	940
BO6	Fecundity	4 × 3 = 12	77.4	929
	Conditional fecundity	4 × 3 = 12	75.9	911
	Female starvation	4 × 3 = 12	78.8	945
	Male starvation	4 × 3 = 12	78.8	945
	Female longevity	3 × 3 = 9	98.9	890
	Male longevity	3 × 3 = 9	98.9	886
BO7	Fecundity	4 × 3 = 12	70.1	841
	Conditional fecundity	4 × 3 = 12	69.3	831
	Female starvation	4 × 3 = 12	71.8	863
	Male starvation	4 × 3 = 12	71.8	861
	Female longevity	4 × 3 = 12	58.5	702
	Male longevity	4 × 3 = 12	57.8	694
BO8	Fecundity	4 × 1 = 4	59.0	236
	Conditional fecundity	4 × 1 = 4	58.0	232
	Female starvation	4 × 1 = 4	60.0	240
	Male starvation	4 × 1 = 4	60.0	240
	Female longevity	4 × 1 = 4	58.2	233
	Male longevity	4 × 1 = 4	57.3	229
Total		413		24,299

of age were used to start each generation. Population sizes for all stocks were of the order of 10^3.

Culture media: The medium used for culture maintenance and serial transfer of adults, when fecundity was not measured, was banana-molasses, as in ROSE (1984). Charcoal high-agar medium with yeast paste on the surface was used for fecundity counts, as in ROSE and CHARLESWORTH (1981a). The flies were kept in shell vials when they were cultured and handled, except for maintenance of the O culture adults, which were kept in cages with the medium replenished every 2–3 days.

Assays: With the exceptions indicated below, all assayed flies were reared at a density of 30 larvae per vial. Longevity was measured in groups of four to five mated pairs, transferred every four to five days. Fecundity was measured as the total number of eggs laid in 24 hr by one female, aged three to five days from pupal eclosion, kept with a single male. The term "conditional fecundity" (ROSE and CHARLESWORTH 1981a) refers to fecundity data in which zero fecundity is treated as a missing data point. Ovaries were obtained from females also aged 3–5 days, dried for 24 hr,

TABLE 4

B and O diallel heterosis effects

Character and experiment	Mean ± SEM		*t*-test		ANOVA
	Parentals	Crosses	Separate *t*	Pooled *t*	*F*
Ovary weight (mg)					
DB1	0.136 ± 0.011	0.142 ± 0.002	0.54	0.78	0.62
DB2	0.172 ± 0.008	0.163 ± 0.010	0.75	0.61	0.33
DO1	0.068 ± 0.014	0.083 ± 0.005	0.98	1.18	1.40
DO3	0.136 ± 0.021	0.136 ± 0.007	0.03	0.03	0.001
Female starvation (hr)					
DB1	37.9 ± 2.2	38.1 ± 0.7	0.09	0.12	0.01
DO1	36.1 ± 1.7	37.0 ± 0.9	0.47	0.46	0.21
DO2	65.3 ± 4.0	59.9 ± 1.5	1.27	1.53	2.28
DO4	35.5 ± 1.5	34.5 ± 0.7	0.58	0.61	0.52
Male starvation (hr)					
DO4	27.4 ± 1.0	28.4 ± 0.4	0.97	1.02	1.52

TABLE 5

B and O diallel line differentiation

Character and experiment	ANOVA		
	Mother *F*	Father *F*	Combined *F*
Ovary weight			
DB1	0.28	1.15	0.71
DB2	0.75	0.57	0.67
DO1	4.72*	1.06	2.89*
DO3	34.03**	0.06	17.10**
Female starvation			
DB1	1.38	2.63	2.04
DO1	1.86	1.60	1.74
DO2	2.44	5.27*	3.91**
DO4	0.98	1.45	1.25
Male starvation			
DO4	1.03	2.74	1.92

* $P < 0.05$, ** $P < 0.01$.

TABLE 6

B and O diallel maternal effects

Character and experiment	ANOVA	
	Method 1 *F*	Method 2 *F*
Ovary weight		
DB1	0.24	0.98
DB2	1.32	2.45
DO1	4.46	14.78*
DO3	576.84**	40.98*
Female starvation		
DB1	0.53	0.67
DO1	1.17	0.96
DO2	0.46	0.41
DO4	0.67	0.54
Male starvation		
DO4	0.38	0.57

* $P < 0.05$, ** $P < 0.01$.

and then weighed separately in the pairs that came from single females using a Cahn electronic microbalance, as in ROSE *et al.* (1984). Except as noted, starvation resistance was measured as the number of hours that a fly aged 3–5 days would respond to provocation in a vial prepared without medium, but provided with rayon saturated with water to sustain humidity, sealed with Parafilm, as in SERVICE *et al.* (1985).

Diallel analysis: A series of diallel analyses (MATHER and JINKS 1982) were performed to test for heterogeneity among lines of a given type. In a diallel design, all possible crosses are performed between a set of populations, including the distinct reciprocal crosses. The coding for these experiments involves three character positions: in the first position, *D* indicates a diallel experiment; in the second position, *B* or *O* indicates whether the populations used were B or O stocks; and the character in the third position is a numeral indicating which experiment it was, in chronological order. Thus the first experiment performed was DB1. Tables 1 and 2 give the experiment codes, the characters assayed, the populations assayed (*e.g.* 5 × 5 indicating all possible crosses of the five replicate populations of a type; 3 × 3 indicating all possible crosses of the first three populations), and the number of individuals assayed. In experiment DO2, starvation was assayed in individuals of 17–20 days of age from pupal eclosion. Each indicated experiment is entirely independent from any other; the same data are not incorporated in more than one experiment.

Transmission pattern experiments: The series of experiments on transmission of postponed aging characters is outlined in Table 3. These experiments are coded with "BO" in the first two positions, indicating crosses of B and O populations. The numerals then refer to the specific experiments. In all these experiments, but one (BO1), B and O parental populations were assayed together with both their reciprocal crosses. (This gives rise to the "4×" terms.) In most experiments, either all five replicates of each type were used or three replicates of each type were used. Crossings used the correspondingly subscripted populations (which has no significance), so that B_1 was always crossed to O_1 and so on for stocks two to five of each stock-type. In this way, each cross was independent of every other cross of the same type. In experiment BO4, starvation was assayed at 17–20 days of age. In experiment BO7, the flies for assay were reared at a density of 90 per vial. In experiment BO8, the parental B and O lines were obtained by a synthetic

TABLE 7

B and O line differentiation–1-way ANOVA

Character and experiment	ANOVA B lines	ANOVA O lines
Ovary weight		
BO1	3.03*	
BO2	5.06**	5.98**
BO3	0.30	23.18**
Fecundity		
BO2	29.45**	49.13**
BO5	12.32**	2.65
BO6	2.09	2.79
BO7	74.54**	27.98**
Conditional fecundity		
BO2	29.45**	49.13**
BO5	11.63**	5.81**
BO6	7.13**	1.20
BO7	115.67**	43.95**
Female starvation		
BO1	3.90**	1.02
BO2	7.20**	27.35**
BO3	3.12*	1.58
BO4	1.65	1.58
BO5	8.66**	4.71**
BO6	8.78**	14.34**
BO7	4.08*	5.95**
Male starvation		
BO1	2.86*	2.95*
BO5	3.72*	0.55
BO6	41.35**	18.76**
BO7	4.47*	0.90
Female longevity		
BO2	12.34**	8.06**
BO3	8.07**	8.27**
BO6	11.33**	5.45**
BO7	2.03	0.74
Male longevity		
BO6	0.24	2.77
BO7	3.12*	0.11

* $P < 0.05$, ** $P < 0.01$.

TABLE 8

B and O line differentiation–2-way ANOVA

Character	ANOVA B lines F	ANOVA O lines F
Ovary weight	0.56	0.35
Fecundity	0.66	1.86
Conditional fecundity	0.59	1.77
Female starvation	0.99	2.61
Male starvation	0.77	0.49
Female longevity	2.44	1.81
Male longevity	1.28	2.88

cross of three B and three O stocks, respectively.

General statistical procedures: All data analysis was performed using SAS (SAS Institute 1988) and SYSTAT (WILKINSON 1988) software packages, which treat data analysis with unbalanced designs, but do not interpolate data. Data were analyzed without transformation and with both WRIGHT's (1968) and Taylor power law (DOWNING 1979) transformations. Neither transformation changed the results of any hypothesis test, so only the data analysis with untransformed data is reported. Throughout the tables, a single asterisk (*) is used to indicate a result with $P < 0.05$, while a double asterisk (**) is used to indicate a result with $P < 0.01$.

RESULTS

Diallel analysis: There are three separate questions which we will address in the diallel analysis. First, is there any evidence for heterosis, or conversely inbreeding depression, in crosses between lines within either B or O stocks? The data were analyzed using both *t*-tests and a nested analysis of variance, the null hypothesis under test being no difference between the average of the two parental lines and the average of the two reciprocal crosses. The results are shown in Table 4. The two types of *t*-test in this table differ with respect to both the pooling of the sample variances in the calculation of the *t*-statistic and the number of degrees of freedom. The null hypothesis is not rejected, indicating an absence of heterosis, whatever way the data are analyzed.

Second, to what extent are the lines within a stock-type differentiated from each other? This was analyzed using a mixed analysis of variance model in which the maternal and paternal lines used in each cross appear as factorial design components. In this design, the *F*-statistic determines whether the ratio of the line-component variance to the within-line component is significantly greater than expected by chance. The results are shown in Table 5. In some cases, there were significant differences between lines, but these were not reproducible between experiments. This statistical design confounds the effects of a line with the choice of parents from that line. This can lead to the erroneous inference of a line effect when none is present. Therefore, the absence of reproducible line effects suggests that the results are ambiguous in significance. More light is shed on this question in the analysis of the B × O experiments discussed below.

Third, to what extent do maternal effects outweigh paternal effects, within stock-types? This question is addressed with greater experimental power in the transmission pattern experiments, but there is the possibility that it arises in the diallel designs. This question was examined using the *F*-statistic produced by the ratio of female parent to male parent variance components in the diallels. ANOVA methods 1 and 2 differ with respect to inclusion (method 1) or exclusion (method 2) of the uncrossed parental lines in the data analysis. The results from the two methods differed only in the analysis of experiment DO1. As shown in Table 6, significant maternal effects arise only in the cases in which there were significant line

TABLE 9

B and O differences

Character and experiment	Mean ± SEM B	Mean ± SEM O	t-test Indep. t	t-test Paired t	ANOVA F
Ovary weight (mg)					
BO2	0.109 ± 0.009	0.057 ± 0.006	4.61**	3.75*	14.09*
BO3	0.177 ± 0.003	0.112 ± 0.022	2.94*	3.14	9.65
Fecundity (eggs/24 hours)					
BO2	102.1 ± 3.8	92.3 ± 6.0	1.38	3.86*	14.74*
BO5	80.7 ± 4.3	85.0 ± 2.0	0.92	1.91	3.70
BO6	94.2 ± 1.9	93.3 ± 2.5	0.26	0.42	0.18
BO7	81.9 ± 9.7	87.3 ± 6.3	0.47	0.82	0.68
BO8	95.2 ± 4.0	98.6 ± 2.2			0.57
Conditional fecundity (eggs/24 hr)					
BO2	102.1 ± 3.8	92.3 ± 6.0	1.38	3.86*	14.74*
BO5	82.2 ± 3.6	86.2 ± 2.6	0.90	3.16	10.42
BO6	96.3 ± 2.9	95.6 ± 1.4	0.22	0.26	0.06
BO7	83.2 ± 10.1	88.9 ± 6.6	0.47	0.80	0.64
BO8	98.9 ± 3.5	98.6 ± 2.2			0.01
Female starvation (hr)					
BO1	28.7 ± 0.9	32.7 ± 0.6	3.74**	2.88*	8.25*
BO2	47.3 ± 2.7	68.2 ± 6.3	3.05*	4.10*	16.65*
BO3	44.8 ± 1.6	55.5 ± 1.3	5.19**	7.10*	50.40*
BO4	39.6 ± 2.5	47.2 ± 2.0	2.38	10.99*	116.65**
BO5	27.9 ± 1.3	35.8 ± 1.2	4.42*	7.47*	56.19*
BO6	27.3 ± 1.0	33.8 ± 2.0	2.89*	6.41*	40.92*
BO7	34.3 ± 1.0	48.1 ± 1.8	6.54*	6.62*	43.78*
BO8	27.2 ± 0.8	35.4 ± 0.9			45.51**
Male starvation (hr)					
BO1	19.9 ± 0.8	25.6 ± 0.9	5.04**	10.45**	109.94**
BO5	17.8 ± 0.6	26.7 ± 0.4	12.96**	53.07**	2799.13**
BO6	20.0 ± 1.8	31.0 ± 2.6	3.48*	13.86**	191.88**
BO7	24.3 ± 0.8	36.4 ± 0.5	12.45**	11.62**	135.08**
BO8	21.3 ± 0.8	26.1 ± 0.6			25.02**
Female longevity (days)					
BO2	50.1 ± 2.7	62.5 ± 2.7	3.26*	9.15**	76.07**
BO3	25.2 ± 2.9	48.3 ± 2.7	5.86**	5.86*	34.32*
BO6	40.4 ± 2.3	51.4 ± 2.2	3.48*	2.48	6.15
BO7	36.6 ± 1.1	48.6 ± 0.9	8.33**	8.41*	71.22*
BO8	34.9 ± 1.5	44.9 ± 2.0			16.46**
Male longevity (days)					
BO6	31.9 ± 0.3	52.0 ± 1.4	13.83**	11.92**	142.15**
BO7	31.8 ± 1.3	48.0 ± 0.4	11.87**	9.90**	98.67**
BO8	31.1 ± 1.2	45.0 ± 1.6			50.77**

* $P < 0.05$, ** $P < 0.01$.

effects (from Table 5), and those were not reproducible in any case. It would seem dubious to draw any firm conclusions from these maternal-effect results alone, without the maternal-effect tests using the transmission pattern results, below.

Transmission pattern experiments: Four features of the transmission data are of importance: (i) differences between B and O lines, within treatments; (ii) differences between B and O treatments; (iii) average dominance, as measured by the deviation of the crosses from the mid-parent value; and (iv) maternal effects, as measured by differences between the two reciprocal cross means.

Tables 7 and 8 summarize the analysis of the between-line, within-treatment, variance components. Two types of analysis were performed, one-way model II analysis of variance for each experiment and an overall two-way analysis that combined data over experiments. In the latter case, the relevant F-statistics are obtained from the ratio of the variance between replicate lines divided by the variance component for the replicates in interaction with experiments. That is, the first analysis of variance, shown in Table 7, tests for a significant deviation of particular lines of a type within each experiment, while the second analysis of variance, shown in Table 8, tests for a significant deviation of particular lines that is sustained over all experiments. Evidently, the results of these two analyses suggest that specific lines can be significantly deviant within particular experiments, but these deviations are not sustained over repeated experiments. This also makes some sense of the diallel results for

TABLE 10

B and O maternal effects

Character and experiment	Mean ± SEM		*t*-test		ANOVA
	B female × O male	O female × B male	Indep. *t*	Paired *t*	*F*
Ovary weight (mg)					
BO1	0.122 ± 0.008	0.124 ± 0.006	0.28	0.27	0.07
BO2	0.098 ± 0.009	0.090 ± 0.009	0.47	0.86	0.75
BO3	0.143 ± 0.014	0.161 ± 0.009	1.04	0.81	0.67
Fecundity (eggs/24 hr)					
BO2	97.9 ± 4.8	97.0 ± 3.6	0.15	0.47	0.22
BO5	88.0 ± 3.7	85.3 ± 5.6	0.41	1.41	2.15
BO6	91.1 ± 3.2	88.2 ± 3.0	1.67	0.160	2.60
BO7	87.7 ± 4.3	84.5 ± 3.3	0.54	10.64**	114.70**
BO8	98.0 ± 5.2	102.9 ± 3.7			0.28
Conditional fecundity (eggs/24 hr)					
BO2	97.9 ± 4.8	97.3 ± 3.6	0.10	0.33	0.11
BO5	89.0 ± 3.3	86.7 ± 4.9	0.39	1.18	1.46
BO6	91.9 ± 3.0	90.3 ± 3.8	0.31	1.43	1.98
BO7	87.7 ± 4.3	85.1 ± 4.9	0.40	3.77	14.04
BO8	99.7 ± 5.0	102.9 ± 3.7			0.13
Female starvation (hr)					
BO1	31.8 ± 1.1	30.5 ± 1.2	0.81	0.66	0.44
BO2	58.6 ± 4.8	50.5 ± 3.8	1.34	3.06	7.70
BO3	50.8 ± 1.8	52.2 ± 2.9	0.41	0.48	0.23
BO4	47.0 ± 3.6	42.8 ± 5.9	0.60	0.53	0.28
BO5	29.2 ± 1.4	32.1 ± 2.6	0.98	1.99	3.89
BO6	27.7 ± 0.9	28.7 ± 2.0	0.44	0.76	0.58
BO7	38.3 ± 0.7	40.7 ± 0.7	2.34	2.46	6.05
BO8	29.6 ± 1.0	29.7 ± 1.3			0.00
Male starvation (hr)					
BO1	22.0 ± 0.6	22.3 ± 0.8	0.30	0.50	0.25
BO5	20.9 ± 0.8	25.3 ± 0.7	4.05	17.51**	300.16**
BO6	25.1 ± 0.7	27.1 ± 2.0	0.94	1.30	1.71
BO7	31.0 ± 0.7	32.0 ± 0.1	1.52	1.40	1.97
BO8	23.6 ± 0.6	22.9 ± 1.0			0.32
Female longevity (days)					
BO2	57.7 ± 2.5	55.9 ± 2.3	0.53	0.59	0.32
BO3	36.3 ± 1.7	37.9 ± 1.5	0.73	0.68	0.47
BO7	41.5 ± 2.7	41.4 ± 2.1	0.03	0.04	0.00
BO8	41.4 ± 2.0	40.6 ± 2.2			0.07
Male longevity (days)					
BO7	35.6 ± 2.3	39.4 ± 2.3	1.17	1.66	2.75
BO8	37.1 ± 1.7	38.4 ± 1.6			0.30

* $P < 0.05$, ** $P < 0.01$.

line effects, which were not consistent over experiments (Tables 5 and 6). Such effects evidently can arise, but do not seem to reflect a durable differentiation of lines within treatments.

Table 9 gives the results of the analysis of between-treatment differences. The *t*-tests compare the means of the assayed groups either using the conventional test for differences of two group means ("Indep.") or using a test for a mean paired difference ($B_i - O_i$) greater than zero ("Paired"). The ANOVA model is mixed, with treatments taken as fixed effects and populations within treatments taken as random effects. In this model, treatment effects are tested using the *F*-statistic defined by the ratio of the treatment mean-square divided by the mean-square for the interaction of treatments and populations. Most of the known character differences have apparently been preserved, except for early fecundity. This last result is discussed further below. For longevity and starvation resistance, at least, there is consistent duplication of earlier findings.

Tables 10 and 11 give the results for dominance and maternal effects. There are only a few cases of statistical significance, some of which can be expected to arise by chance with repeated statistical tests.

As with the line-effects analysis of Table 8, the data were also analyzed using a three-way analysis of variance, with treatment differences (DIF), maternal effects (MAT), and average dominance (DOM) all modeled as fixed effects. The mean square for each of these effects ("eff") was divided by

$$MS_{EXP \times eff} + MS_{REP \times eff} - MS_{EXP \times REP \times eff}$$

TABLE 11

B and O average dominance effects

Character and experiment	Mean ± SEM		*t*-test		ANOVA
	Parentals	Crosses	Indep. *t*	Paired *t*	*F*
Ovary weight (mg)					
BO2	0.083 ± 0.004	0.094 ± 0.008	1.27	1.89	3.58
BO3	0.142 ± 0.010	0.152 ± 0.004	0.92	1.91	3.66
Fecundity (eggs/24 hr)					
BO2	97.2 ± 4.9	97.4 ± 4.1	0.04	0.18	0.03
BO5	82.7 ± 3.2	86.7 ± 4.6	0.71	1.64	2.62
BO6	93.7 ± 0.9	89.7 ± 3.1	1.08	0.82	0.67
BO7	84.7 ± 8.0	86.1 ± 4.3	0.16	0.29	0.08
BO8	96.9 ± 2.3	99.2 ± 4.0			0.30
Conditional fecundity (eggs/24 hr)					
BO2	97.2 ± 4.9	97.6 ± 4.2	0.07	0.28	0.08
BO5	84.1 ± 3.1	87.9 ± 4.1	0.73	1.73	2.91
BO6	95.9 ± 1.8	91.1 ± 3.4	1.25	1.02	1.03
BO7	86.1 ± 7.8	86.4 ± 4.6	0.04	0.07	0.01
BO8	98.7 ± 2.0	100.5 ± 3.8			0.19
Female starvation (hr)					
BO1	30.4 ± 0.3	31.1 ± 0.6	0.94	1.40	2.05
BO2	59.0 ± 4.4	53.9 ± 4.1	0.84	3.36*	11.21*
BO3	50.1 ± 1.2	51.5 ± 1.9	0.59	1.92	3.67
BO4	44.3 ± 2.4	45.1 ± 3.1	0.21	0.89	0.80
BO5	31.9 ± 1.1	30.7 ± 2.0	0.51	0.84	0.71
BO6	30.5 ± 1.5	28.3 ± 0.6	1.03	14.33**	202.54**
BO7	41.2 ± 1.1	39.5 ± 0.5	1.41	2.50	6.23
BO8	31.3 ± 0.7	29.6 ± 0.8			2.22
Male starvation (hr)					
BO1	22.3 ± 0.8	22.1 ± 0.6	0.22	0.38	0.14
BO5	22.2 ± 0.5	23.1 ± 0.7	0.99	4.28	18.12
BO6	25.5 ± 2.2	26.3 ± 1.2	0.31	0.70	0.50
BO7	30.3 ± 0.4	31.5 ± 0.3	2.28	3.21	10.29
BO8	23.7 ± 0.5	23.4 ± 0.5			0.13
Female longevity (days)					
BO2	55.9 ± 2.8	56.9 ± 1.9	0.30	0.42	0.19
BO3	36.7 ± 2.0	37.1 ± 1.0	0.16	0.24	0.06
BO6	45.9 ± 0.3	47.2 ± 0.9	1.28	1.84	3.39
BO7	42.6 ± 0.8	41.4 ± 1.9	0.57	0.60	0.36
BO8	40.0 ± 1.3	41.0 ± 1.5			0.25
Male longevity (days)					
BO6	42.0 ± 0.7	39.1 ± 0.9	2.64	3.03	9.16
BO7	40.2 ± 0.3	37.5 ± 2.0	1.33	1.53	2.35
BO8	37.8 ± 1.1	37.8 ± 1.2			0.00

* $P < 0.05$, ** $P < 0.01$.

TABLE 12

B and O differences, maternal effects, and average dominance effects–3-way ANOVA

Character	ANOVA		
	DIF *F*	MAT *F*	DOM *F*
Ovary weight	51.65**	0.32	0.33
Fecundity	0.27	1.01	0.12
Conditional fecundity	0.33	0.67	0.02
Female starvation	9.34*	0.44	1.74
Male starvation	100.78*	3.29	1.39
Female longevity	45.77*	0.01	0.03
Male longevity	57.68*		4.77

* $P < 0.05$, ** $P < 0.01$.

to obtain the *F*-statistics shown in Table 12. As may be seen, the only effects that are significant are the treatment effect differences. Of these treatment effect differences, only the two fecundity characters do not give significant results. It is not immediately clear why this should occur, particularly given that ovary weight, which essentially reflects egg content, remains significantly differentiated between B and O treatments. Our interpretation is that potential fecundity is different between B and O treatments, but a difference in actual fecundity is not always elicited in particular experiments. With respect to the general transmission pattern, we conclude that the data indicate additive inheritance among populations averaged over loci without maternal effects. This does *not* indicate an absence of dominance in the transmission patterns of the individual loci involved in postponed senescence.

DISCUSSION

Overall, the present results suggest the absence of consistent directional dominance or maternal effects in the inheritance of postponed aging in the *D. melanogaster* stocks studied. Maternal or line effects, when present, are not consistent over experiments. It cannot be said that any of the postponed-aging stocks is consistently superior to any other. On average, hybrids of postponed-aging and control stocks appear to be intermediate; there is no reproducible heterosis, inbreeding depression, or directional dominance.

Like CLARE and LUCKINBILL (1985), who studied fewer characters, fewer lines, and far fewer individuals, we find essentially additive inheritance in population crosses. Since the stocks involved are independent, the conclusions of CLARE and LUCKINBILL (1985) seem to be strongly supported.

But these conclusions may be weakened by the possibility that the populations that have been analyzed are still highly polymorphic for the alleles involved in postponed aging. In particular, crosses of highly polymorphic populations will not give clean tests of average dominance of differentiated alleles. In addition, parental lines that aren't extremely differentiated will not be as distinguishable from their F_1's, as indeed fecundity was not so differentiated in these experiments. For these reasons, we set about creating selected lines that would be more differentiated with respect to at least some of the characters involved in postponed aging, in the hope of then performing a more refined genetic analysis. The results of that study are reported in a companion paper by HUTCHINSON, SHAW and ROSE (1991) (in this issue).

However, some features of the present results are of importance as they stand. One of the outstanding problems affecting research on aging has been the persistence of disbelief that aging is made up of normal phenotypes amenable to genetic analysis and selection (*e.g.* LINTS 1978; LINTS and HOSTE 1974, 1977; LINTS *et al.* 1979). The present results together with those of Luckinbill and his colleagues (CLARE and LUCKINBILL 1985; LUCKINBILL and CLARE 1985) indicate that aging phenotypes are typical quantitative characters.

While a number of Drosophila stocks exhibit different aging patterns, upon crossing there is often extensive heterosis (CLARKE and MAYNARD SMITH 1955; GOWEN and JOHNSON 1946), indicating inbreeding depression. This made the study of *C. elegans* aging particularly attractive, because frequent self-fertilization in that species appears to prevent inbreeding depression for aging (JOHNSON and WOOD 1982). The lack of hybrid vigor in the extensive crosses performed in the present study indicates that stocks with postponed aging created using the methods of ROSE (1984) will not suffer from the problem. Thus they can be used as material for the investigation of physiological hypotheses concerning mechanisms for the postponement of normal aging (ROSE *et al.* 1984; SERVICE *et al.* 1985; SERVICE 1987; LUCKINBILL *et al.* 1988a).

The authors are grateful to P. M. SERVICE for help in the planning and execution of the reported experiments. R. E. LENSKI and L. D. MUELLER commented upon earlier drafts of the manuscript. We thank D. ARAB, H. GILLIS, G. GLAZOV, K. GRIMM, L. E. JOHNSTON, S. JOHNSON, D. M. LANE, J. JUDAH, M. D. MACKINLEY, B. MUSGRAVE, J. NELSON, S. O'KEEFE, D. PRINGLE, A. J. SHAW, B. SINGH, D. STEWART and B. TREMBLAY for technical assistance. This research was supported by Natural Sciences and Engineering Research Council of Canada grant U0178 and U.S. Public Health Service grant AG06346 to M.R.R., as well as U.S. Public Health Service grant AG08322 to THOMAS E. JOHNSON.

LITERATURE CITED

CLARE, M. J., and L. S. LUCKINBILL, 1985 The effects of gene-environment interaction on the expression of longevity. Heredity **55:** 19–29.

CLARKE, J. M., and J. MAYNARD SMITH, 1955 The genetics and cytology of *Drosophila subobscura*. XI. Hybrid vigour and longevity. J. Genet. **53:** 172–180.

DOWNING, J. A., 1979 Aggregation, transformation and the design of benthos sampling programs. J. Fish. Res. Board Can. **36:** 1454–1463.

FALCONER, D. S., 1981 *Introduction to Quantitative Genetics*, Ed. 2. Longman, New York.

FRIEDMAN, D. B., and T. E. JOHNSON, 1988 A mutation in the *age-1* gene in *Caenorhabditis elegans* lengthens life and reduces hermaphroditic fertility. Genetics **118:** 75–86.

GOWEN, J. W., and L. E. JOHNSON, 1946 On the mechanism of hete rosis. I. Metabolic capacity of different races of *Drosophila melanogaster* for egg production. Am. Nat. **80:** 149–179.

GRAVES, J. L., L. S. LUCKINBILL and A. NICHOLS, 1988 Flight duration and wing beat frequency in long- and short-lived *Drosophila melanogaster*. J. Insect Physiol. **34:** 1021–1026.

HUTCHINSON, E. W., A. J. SHAW and M. R. ROSE, 1991 Quantitative genetics of postponed aging in *Drosophila melanogaster*. II. Analysis of selected lines. Genetics **127:** 729–737.

JOHNSON, T. E., 1987 Aging can be genetically dissected into component processes using long-lived lines of *Caenorhabditis elegans*. Proc. Natl. Acad. Sci. USA **84:** 3777–3781.

JOHNSON, T. E., and W. B. WOOD, 1982 Genetic analysis of life-span in *Caenorhabditis elegans*. Proc. Natl. Acad. Sci. USA **79:** 6603–6607.

LEFFELAAR, D., and T. A. GRIGLIATTI, 1984 A mutation in Drosophila that appears to accelerate aging. Dev. Genet. **4:** 199–210.

LINTS, F. A., 1978 *Genetics and Aging, Interdisciplinary Topics in Gerontology*, Vol. 14, S. Karger, Basel.

LINTS, F. A., and C. HOSTE, 1974 The Lansing effect revi sited. I. Life-span. Exp. Gerontol. **9:** 51–69.

LINTS, F. A., and C. HOSTE, 1977 The Lansing effect revisited. II. Cumulative and spontaneously reversible parental age effects on fecundity in *Drosophila melanogaster*. Evolution **31:** 387–404.

LINTS, F. A., J. STOLL, G. GRUWEZ and C. V. LINTS, 1979 An attempt to select for increased longevity in *Drosophila melanogaster*. Genetics **25:** 192–204.

LUCKINBILL, L. S., R. ARKING, M. J. CLARE, W. C. CIROCCO and S. A. BUCK, 1984 Selection for delayed senescence in *Drosophila melanogaster*. Evolution **39:** 996–1003.

LUCKINBILL, L. S., and M. J. CLARE, 1985 Selection for life span in *Drosophila melanogaster*. Heredity **55:** 9–18.

LUCKINBILL, L. S., M. J. CLARE, W. L. KRELL, W. C. CIROCCO and P. A. RICHARDS, 1987 Estimating the number of genetic elements that defer senescence in *Drosophila*. Evol. Ecol. **1:** 37–46.

LUCKINBILL, L. S., J. L. GRAVES, A. TOMKIW and O. SOWIRKA, 1988a A qualitative analysis of life-history characters in *Drosophila melanogaster*. Evol. Ecol. **2:** 85–94.

LUCKINBILL, L. S., J. L. GRAVES, A. H. REED and S. KOETSAWANG, 1988b Localizing genes that defer senescence in *Drosophila melanogaster*. Heredity **60:** 367–374.

MATHER, K., and J. L. JINKS, 1982 *Biometrical Genetics*, Ed. 3. Chapman & Hall, London.

ROBERTS, P. A., and R. B. IREDALE, 1985 Can mutagenesis reveal major genes affecting senescence. Exp. Gerontol. **20:** 119–121.

ROSE, M. R., 1984 Laboratory evolution of postponed senescence in *Drosophila melanogaster*. Evolution **38:** 1004–1010.

ROSE, M. R., and B. CHARLESWORTH, 1981a Genetics of life history in *Drosophila melanogaster*. I. Sib analysis of adult females. Genetics **97:** 173–186.

ROSE, M. R., and B. CHARLESWORTH, 1981b Genetics of life history in *Drosophila melanogaster*. II. Exploratory selection experiments. Genetics **97:** 187–196.

ROSE, M. R., M. L. DOREY, A. M. COYLE and P. M. SERVICE, 1984 The morphology of postponed senescence in *Drosophila melanogaster*. Can. J. Zool. **62:** 1576–1580.

SAS Institute, 1988 SAS/STAT User's Guide: Release 6.03. SAS Institute Inc., Cary, N.C.

SERVICE, P. M., 1987 Physiological mechanisms of increased stress resistance in *Drosophila melanogaster* selected for postponed senescence. Physiol. Zool. **58:** 380–389.

SERVICE, P. M., E. W. HUTCHINSON, M. D. MACKINLEY and M. R. ROSE, 1985 Resistance to environmental stress in *Drosophila melanogaster* selected for postponed senescence. Physiol. Zool. **58:** 380–389.

WATTIAUX, J. M., 1968a Parental age effects in *Drosophila pseudoobscura*. Exp. Gerontol. **3:** 55–61.

WATTIAUX, J. M., 1968b Cumulative parental age effects in *Drosophila subobscura*. Evolution **22:** 406–421.

WEBER, K. E., 1990 Increased selection response in larger populations. I. Selection for wing-tip height in *Drosophila melanogaster*. Genetics **125:** 579–584.

WEBER, K. E., and L. T. DIGGINS, 1990 Increased selection response in larger populations. II. Selection for ethanol vapor resistance in *Drosophila melanogaster* at two population sizes. Genetics **125:** 585–597.

WILKINSON, L., 1988 SYSTAT: The System for Statistics. Systat Inc., Evanston, Ill.

YOO, B. H., 1980 Long-term selection for a quantitative character in large replicate populations of *Drosophila melanogaster*. I. Response to selection. Genet. Res. **35:** 1–17.

Communicating editor: A. G. CLARK

Evolution, 48(4), 1994, pp. 1244–1257

LONG-TERM LABORATORY EVOLUTION OF A GENETIC LIFE-HISTORY TRADE-OFF IN *DROSOPHILA MELANOGASTER*. 1. THE ROLE OF GENOTYPE-BY-ENVIRONMENT INTERACTION

ARMAND M. LEROI,[1] ADAM K. CHIPPINDALE, AND MICHAEL R. ROSE
Department of Ecology and Evolutionary Biology, University of California, Irvine, California 92717

Abstract.—Trade-offs among life-history traits are often thought to constrain the evolution of populations. Here we report the disappearance of a trade-off between early fecundity on the one hand, and late-life fecundity, starvation resistance, and longevity on the other, over 10 yr of laboratory selection for late-life reproduction. Whereas the selected populations showed an initial depression in early-life fecundity, they later converged upon the controls and then surpassed them. The evolutionary loss of the trade-off among life-history traits is considered attributable to the following factors: (1) the existence of differences in the culture regimes of the short- and long-generation populations *other* than the demographic differences deliberately imposed; (2) adaptation of one or both of these sets of populations to the unique aspects of their culture regimes; (3) the existence of an among-environment trade-off in the expression of early fecundity in the two culture regimes, as reflected in assays that mimic those regimes. The trade-off between early and late-life reproductive success, as manifest among divergently selected populations, is apparent or not depending on the assay environment. This demonstration that strong genotype-by-environment interactions can obscure a fundamental trade-off points to the importance of controlling all aspects of the culture regime of experimental populations and the difficulty of doing so even in the laboratory.

Key words.—Antagonistic pleiotropy, *Drosophila*, genotype × environment interaction, life history, trade-offs.

Received August 21, 1992. Accepted September 17, 1993.

The trade-off between longevity and early fecundity in outbred, laboratory-adapted populations of *Drosophila melanogaster* has been proposed as a possible example of a fundamental physiological constraint on life-history evolution (Parker and Maynard Smith 1990; Stearns 1992). This trade-off, manifest as a negative genetic correlation and antagonistic selection response between these traits, has been found repeatedly in the same population at different times (Rose and Charlesworth 1981a,b; Rose 1984). This result is also true for populations derived from different geographic localities (Luckinbill et al. 1984; Luckinbill and Clare 1985; Tucić et al. 1988), although there is at least one report of a failure to detect it in yet other populations (Partridge and Fowler 1992). Generally, however, selection for late-life reproductive success in such populations has resulted in an increase in longevity and a correlated diminution of early fecundity (Rose 1984; Luckinbill et al. 1984). Although the physiological basis of this trade-off is still unclear, it is thought to be related to patterns of lipid metabolism and storage (Service 1987; Service et al. 1985; Service et al. 1988; Luckinbill et al. 1990). Starvation resistance, partly a function of lipid reserves (Service 1987), has been found to be positively genetically correlated with longevity (Rose et al. 1992) and, like longevity, negatively genetically correlated with early fecundity (Service and Rose 1985; Service et al. 1988). A causal role for lipid reserves in the trade-off between longevity and early fecundity has been further substantiated by experiments involving environmental manipulation (Chippindale et al. 1993).

We report here the evolution of the trade-off between early fecundity, on the one hand, and longevity and starvation resistance, on the other, in an assemblage of populations on which selection has been imposed for either early or late-life reproductive success. Five of these populations, called B_1–B_5, have been maintained since their founding in 1980 on a 2-wk generation cycle. Five other populations, called O_1–O_5, have been maintained on a longer generation cycle. These populations were all derived from a common ancestral population, IV (for details see Materials and Methods, below, and Rose 1984), and

[1] Department of Molecular Genetics, Albert Einstein College of Medicine, 1300 Morris Park Avenue, Bronx, New York 10461. E-mail: leroi@aecom.yu.edu.

were first assayed in 1982, 2 yr after their founding. Consistent with the earlier finding of a negative genetic correlation between these traits in the ancestral population (Rose and Charlesworth 1981a,b), mean longevity increased and mean early fecundity decreased in the O's relative to the B's (Rose 1984). With continued selection for late-life reproductive success, the difference in longevity between the B's and O's continued to increase until last assayed in 1990 (fig. 1). This was not true, however, for the correlated response of early fecundity. Far from increasing, by 1984 the difference in early fecundity between the B's and O's had begun to diminish (Hutchinson 1989); by 1986, the two sets of populations were identical for early fecundity (Hutchinson 1989); and by 1990, the long-lived O's had a *higher* early fecundity than the short-lived B's (Chippindale et al. 1993) (fig. 1). Thus, the trade-off between longevity and early fecundity was no longer manifest among these lines. Furthermore, where Rose (1984) failed to detect any difference in total lifetime fecundity among the B's and O's in 1982, by 1990 the O's had evolved a higher total fecundity than the B's—having a greater fecundity than the B's at all times of life (fig. 2). Insofar as the O's were found in 1984 to have a greater starvation resistance than the B's (Service et al. 1985), a difference that since has been maintained or increased, the trade-off between this trait and early fecundity had disappeared as well. Because to our knowledge, high relative late-life fecundity, longevity, and starvation resistance are invariably associated among stocks, and because we do not know the physiological relationship among these traits, we will not always treat them as genetically separate traits; instead we will simply refer to a single trade-off between early fecundity and the suite of traits associated with late-life reproductive success.

This paper and its companion (Leroi et al. 1994) are concerned with the genetic and selective factors involved in evolutionary loss of the trade-off between early fecundity and the suite of traits associated with longevity. In this paper, we ask whether the disappearance of this trade-off among populations also has occurred in environments other than those in which the life-history traits in question are usually assayed. This question is of interest because of the possibility that life-history traits exhibit strong genotype-by-environment interaction, such that negative genetic correlations, absent in some environments, are present in others (Service and Rose 1985; de Jong 1990a,b; Stearns et al. 1991). Furthermore, if selection acts on such g × e variation, the mean environmental responses of populations may diverge such that interpopulation trade-offs, absent in some environments, are present in others. We argue that the B and O stocks have in fact diverged in their mean norms of reaction because of having been kept under subtly different culture conditions and that this divergence is responsible for the ostensible disappearance of the trade-off among populations. Furthermore, we investigate the historical causes for this divergence in phenotypic response.

Materials and Methods

B and O Culture Regimes

The origin and early propagation of these stocks is described in Rose (1984) and Service et al. (1985). Briefly, the B and O stocks were derived from an outbred population, IV, which had been maintained in laboratory culture at 25°C with unlimited banana-molasses medium and 14-d discrete generations for more than 130 generations. Therefore, considerable opportunity existed for the founder stock to adapt to the laboratory environment; the five replicate B populations have been maintained on this regime since their origin in February 1980. The five replicate O populations, founded at the same time, have been under selection for late-life reproductive success since their origin. This selection was accomplished by propagating them from eggs laid by old females, such that since 1982 only females that are alive and fertile 10 wk after oviposition, more than 8 wk after pupal eclosion, will contribute to the next generation.

Initially, both the B and O stocks were cultured in bottles or vials with medium. Because this culture system necessitated frequent (three times per week) changing of culture vessels for the last 8 wk of the selection treatment in the O's, after 1981 the adults of these populations were maintained in population cages. Since then, eggs have been collected from medium-filled petri plates inserted into these cages. The B populations, which have but a brief (at most 4 d) existence as adults, never have been cultured in population cages. Rather, egg collection always has been accomplished by transferring flies from their culture vials or bottles to new culture vessels filled with fresh medium and no yeast. The culture environment of the B and O stocks between egg laying and 14 d has been parallel from their initial divergence till now. In particular, larval rearing densities have not been systematically different between the B and O stocks, although they did vary between 50 and 150 larvae per vial from generation to generation.

That the O populations have been propagated from eggs collected in population cages, and the B populations from eggs collected in vials or bottles has entailed several other differences in culture regime. These differences include the following:

(*1*) *Medium Before Egg Collection.*—A natural consequence of the different culture regimes of B's and O's is that the conditioning environment prior to egg laying differs with respect to nutrition. The B's are retained in the culture vials in which they have grown and pupated and thus experience medium previously consumed ("spent") by the larvae, immediately before oviposition in new vials. Though the O's experience the same preselection treatment, they have an extended imaginal phase, during which they are exposed to only fresh food. Two d before egg collection, this food is supplemented with live yeast paste. In sum, the B's are normally conditioned for egg laying on "spent" medium, whereas the O's are provided rich, fresh food for egg laying.

(*2*) *Adult Density.*—The number of adults per cage is around 2000 in O populations at egg collection; the number of adults per vial is between 50 and 150 in B populations at egg collection. Given the different dimensions of their enclosures, the density (flies/cm^3) of adults B's is approximately 10 times greater than that of O's at egg laying.

(*3*) *Time Span of Egg Collection.*—Whereas the aged O populations are allowed 24 h to produce the eggs required for propagation, the B populations provide enough in about 1 h, at which point egg collection is ceased.

(*4*) *Use of CO_2 Anesthesia Prior to Egg Collection.*—The B populations are gassed to permit mixing of flies from different culture vials within each population; the O populations are not.

Evolution of Relative B and O Fecundity, Longevity, and Ovary Weight

Preassay Handling.—For two generations prior to all assays, experimental populations were cultured on a common 2-wk generation schedule identical to the routine B culture to standardize parental environment and age as well as synchronize the emergence of experimental adults.

Standard Fecundity Assay: Fresh Medium + Yeast.—A standard technique was used to assay relative fecundity in the B and O stocks from 1980 to 1990; the fecundities reported in figures 1 and 2 were collected using this technique. In 1982, the early fecundity of 416 females from six populations: B_{1-3} and $O_{1,3,4}$ (Rose 1984) were assayed. In 1984, 589 females from all 10 populations were assayed, and in 1986, 1724 females from six populations were assayed in three blocks (for details, see Hutchinson 1989). In 1990, 348 females from all 10 populations were assayed (Chippindale et al. 1993).

In this assay procedure, flies from B and O populations that eclosed within the same 24–36-h period are transferred in pairs to assay vials. These vials contained sucrose-yeast medium with charcoal powder added for coloring (vid. Rose 1984). The laying surface is covered with a thin yeast paste to which dilute acetic acid has been added and which is allowed to dry overnight. Flies are transferred to fresh vials at 24-h intervals; B and O populations were handled in pairs: that is, B_1 and O_1 were transferred simultaneously. Early fecundity is assayed on day 4 after eclosion. All eggs laid on the medium surface or on the sides of the laying tubes are counted. All flies are handled at room temperature, using CO_2 anesthesia and kept in the same incubator at 25°C under constant illumination.

This assay method served, with some variation, as a control for all new fecundity assays reported here. New assays included the following variations in treatment: (1) CO_2 anesthesia, its presence or absence in the daily transfer of flies to fresh assay vials; (2) adult density, four pairs per vial instead of one pair per vial during the conditioning on saturation yeast levels; (3) yeast feeding, providing flies with sufficient yeast for the whole 4-d conditioning period instead of transferring them daily to fresh conditioning vials; (4) larval densities, 60–80 per vial in some experiments, 120–175 per vial (which may more closely approximate actual culture densities) in others. These changes in procedure were used to test the influence of a specific environmental variable on relative fecundity. We take them to serve as adequate control treatments if O mean early fecundity exceeds B mean early fecundity, as it does in the "standard" assay environment defined above. These control treatments all entail conditioning on a background of fresh, carbohydrate-rich food with an excess of dietary live yeast at low adult densities in an 8-dram vial.

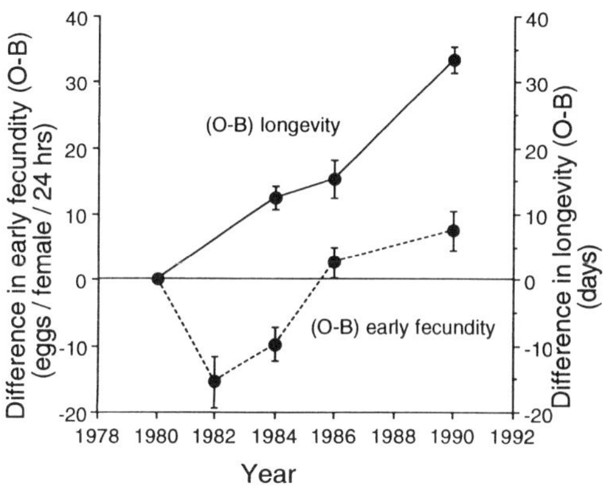

FIG. 1. Paradoxical evolution of relative B and O fecundity and longevity from 1982 to 1992. The solid line represents the progress of the difference (O − B) mean longevity: as the O's, which are selected for late-life reproductive success, increase in mean longevities, the difference becomes increasingly positive. The broken line represents the progress of the difference (O − B) mean early-life fecundity. Initially in the divergence of the stocks, the difference is negative as the O's decline, an apparent trade-off. Later, the direction of evolution of relative fecundity reverses itself, such that by 1992 the O's have a *greater* early fecundity than the B's. Longevity was assayed on unyeasted banana-molasses medium; early fecundity was assayed on yeasted banana or charcoal food medium, that is, "standard" conditions. The 1982 data are from Rose (1984); 1984 and 1986 data from Hutchinson (1989). Means (and standard errors of the means) based on the difference (O − B) over five pairs of replicate populations.

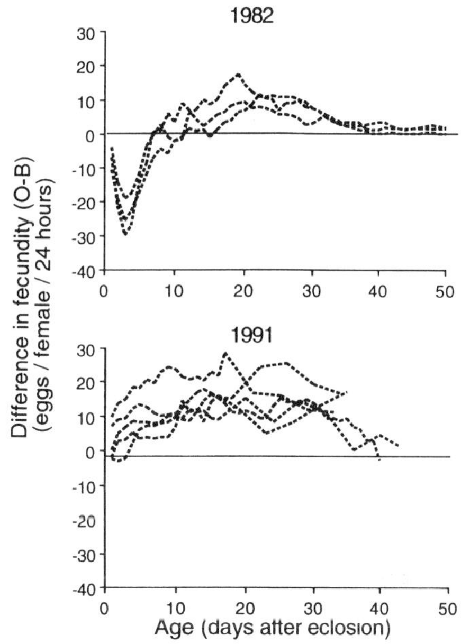

FIG. 2. The evolution of relative B and O lifetime fecundities. Above, in 1982, the O's have a lower early fecundity than the B's (days 3–7 after eclosion) and a greater fecundity later in life; data from Rose (1984). Below, in 1992, the O's have equal or greater fecundity at all ages; data from Chippindale et al. (1993). Both assays in "standard" conditions.

Standard Ovary Weight Assay.—In the standard ovary weight assay, flies were handled exactly as in a standard fecundity assay described above, except that at peak fecundity—days 4–5 after eclosion—females were anesthetized with CO_2 and frozen. The ovaries of these flies were subsequently dissected out of their bodies; both bodies and ovaries were dried to stable weight (48 h at 50°C) and then weighed on a Cahn electronic microbalance.

In 1983, the early ovary weights of 1350 females from all 10 populations were assayed over two experiments (Rose et al. 1984). In 1985 and 1986, respectively, 300 and 251 females from six populations, B_{1-3} and O_{1-3} were assayed (Hutchinson 1989). In 1992, 400 females from all 10 populations, B_{1-5} and O_{1-5}, were assayed.

Longevity Assays.—In the standard longevity assay, flies from paired B and O populations that had eclosed within the same 24–36-h period were transferred in pairs to assay vials. These vials contained banana-molasses medium, but no additional live surface yeast. Flies were transferred to fresh food three times weekly, and a daily record of mortality was kept. Flies whose mates had died were paired with another survivor of the opposite sex from the same population as soon as one became available. In 1984, 416 females from all 10 populations were assayed. In 1985, 1814 females from populations B_{1-3} and O_{1-3} were assayed (for details see Hutchinson and Rose 1991). In 1991, 300 females from all 10 populations were assayed.

Longevity was also assayed twice by a slightly different method. Here, the assay vials differed from those above in that the medium was covered with additional live surface yeast + acetic acid, an environment that corresponds to the "standard" fecundity assay environment. Using this method, the longevities of 416 flies from six populations, B_{1-3} and O_{1-3}, were assayed in 1982 (Rose 1984), as well as that of 348 females from all 10 populations in 1990 (Chippindale et al. 1993).

Influence of Environment on the Expression of the Trade-Off

Experiment 1: The Influence of Culture Regime on Relative B and O Early Fecundity.—In this experiment, we examined relative B and O fecundity in an environment that approximates, as closely as possible, the normal propagation environment of the B's. This environment we call the "B-type" assay environment; it consists of maintaining flies in their culture vials until day 14 after oviposition and then assaying fecundity. As an appropriate control, we chose a "standard" fecundity assay involving conditioning on fresh medium with additional live yeast and daily transfer to fresh food as discussed above.

After 4 d of conditioning in either the "B-type" or control treatments, fecundity was assayed by transferring flies, with the aid of CO_2 anesthesia, to a charcoal food + yeast vial. In the case of the "B-type" treatment, all 100–150 flies were transferred to such an assay vial, and the females were permitted to lay eggs for 1 h $\pm$ 2 min; female fecundity was then estimated by dividing the number of eggs in each vial by the number of females in that vial. In the case of the "standard" assay environment, each mated pair of flies was transferred to a "standard" assay vial and the female was permitted to lay eggs for 24 h. The "B-type" and the "standard" conditioning treatments differ, then, in numerous environmental variables of which a nonexhaustive list is as follows: the freshness of the banana-molasses medium, the presence of excess live yeast, adult density, and the time permitted for oviposition. In this experiment, all 10 B and O populations were assayed. The culture vials from each population were assigned randomly to the "standard" and "B-type" treatments; larval density was 100–150 per vial in both treatments. In the "standard" treatment, 18–20 females were assayed per population; in the "B-type" treatment 20 vials of flies—each containing 50–75 females per vial—were assayed per population.

Experiment 2: The Influence of CO_2 Anesthesia on Relative B and O Early Fecundity.—This experiment consisted of two treatments: a "CO_2" and a "no-CO_2" treatment. Both treatments were based on the "standard" fecundity assay (with a 24-h oviposition period), but where flies of the "CO_2" treatment were transferred daily to fresh vials with the aid of CO_2, the "no-CO_2" treatment flies were transferred by gently tapping the flies from one vial to another. Altogether, 669 females from all 10 B and O populations were assayed.

Experiment 3: The Influence of Live Dietary Yeast on Relative B and O Fecundity.—This experiment consisted of two treatments: "yeast" and "no yeast." Both treatments were based on the "standard" fecundity assay (with a 24-h oviposition period), but whereas flies of the "yeast" treatment were given the usual supplement of live yeast + acetic acid daily, the "no-yeast" flies were not. Flies in both treatments were transferred without CO_2 anesthesia. Altogether, 759 females from all 10 B and O populations were assayed.

Experiment 4: The Influence of Adult Density on Relative B and O Fecundity.—In addition to examining the influence of adult density on relative fecundity, this experiment was also designed to determine the effect of density per se rather than food deprivation caused by high density. For this reason, the experiment consisted of two factors crossed to give four treatments. The first factor was adult density per conditioning vial: "low-density" vials had 1 pair of flies per vial, whereas "high-density" vials had 32 pairs of flies per vial. The second factor was yeast level: "low-yeast" vials had a drop of dried yeast solution in each vial as described in the "standard" fecundity assay; the "high-yeast" vials had an excess of a thick yeast paste. This latter treatment was designed to ensure an excess of dietary yeast even at high densities. Both the "low-density, low-yeast" and "low-density, high-yeast" treatments correspond to the "standard" fecundity assay, insofar that in each case the flies are provided with a considerable excess of yeast. Daily transfer was without CO_2 anesthesia. Fecundity was assayed on day 4 after eclosion, and the females in the "high-density" vials were counted again to take account of mortality in the conditioning phase. Larval density was the same for all treatments. Each of the "low-density" treatments had 20 replicate vials per treatment; each of the "high-density" treatments had four replicate vials per population.

Experiment 5: The Influence of Oviposition Time on Relative B and O Fecundity.—We attempted to separate the effects of oviposition time on relative fecundity from the physiological effects of the conditioning environment. This experiment is identical to experiment 1 except that we allowed females to oviposit for 1, 6, or 12 h.

All 10 B and O populations were assayed. The culture vials for each population were assigned randomly to the "standard" and "B-type" treatments. Flies from each culture vial were then assigned randomly to one of the three oviposition time treatments. In the "standard" treatments, 18–20 females were assayed per population for each of the three time intervals; in the "B-type" treatments, 20 vials of flies—each containing 50–75 females per vial—were assayed per population for each of the three time intervals.

Experiment 6: The Influence of Oviposition Time and Medium Quality on Relative B and O Fecundity.—In this experiment, we evaluated whether two properties of the "B-type" environment interact to influence relative fecundity. These factors are oviposition time and the medium found in "B-type" culture vials. This medium is a banana-molasses one in which flies have undergone development from egg to adult at typical culture densities (120–170 per 5-ml medium), hence the name, "spent medium." This medium is presumed to contain high levels of waste, dead flies, and is probably low in nutritional content. To separate the effects of spent medium from the other factors that distinguish the "B-type" and "standard" assay environments, we collected large quantities of it from culture vials, stored it at −4°C until use, and then thawed and redistributed it to clean 8 dram vials in quantities of ~ 3 mL per vial.

The assay conditions for both the experimental "spent medium" and control flies were identical except for the conditioning medium itself. In this case, the appropriate control was "standard" assay conditions where conditioning took place on fresh banana-molasses food with excess yeast for 4 d and no daily transfer. At day 14 after oviposition, after 4 d of conditioning, fecundity was assayed by transferring each pair, with the aid of CO_2 anesthesia, to a charcoal food + yeast fecundity assay vial in which the female was permitted to lay eggs for either 6, 12, 18, or 24 h. Females from all treatments and populations were assayed on the same day. Altogether, 1685 females from all 10 B and O populations were assayed.

Experiment 7: The Influence of Culture Regime on Relative B and O Early Ovary Weight.—The conditioning phase of this experiment was identical to that of experiment 1 (above), except that at day 14 after oviposition, flies from both the control and "B-type" treatments were frozen and assayed for ovary weight, instead of fecundity (see above). Altogether, 1000 females from all 10 B and O populations were assayed.

Experiment 8: The Influence of Culture Regime on Relative B and O Starvation Resistance.—The conditioning phase of this experiment was identical to that of experiment 1 (above), except that at day 14 after oviposition, flies from both the control and "B-type" treatments were assayed for starvation resistance instead of fecundity. Altogether, 400 females from all 10 B and O populations were assayed.

Experiment 9: The Influence of Medium Quality on Relative B and O Starvation Time.—The conditioning phase of this experiment was identical to that of experiment 6 (above), except that at day 14 after oviposition, flies from both the control and "spent" treatments were assayed for starvation resistance instead of fecundity. Altogether, 1411 females from all 10 B and O populations were assayed.

Statistical Analysis and Experimental Design

All hypothesis testing was done using the means of replicate populations as observed variates. All means and standard errors are based on three or five replicate populations assayed. Given this procedure, the distribution of population means is not tested for normality. The distribution of population means cannot be tested usefully because of the few populations, typically five. All statistical work was carried out using STATVIEW on Macintosh computers.

RESULTS

Evolution of Relative B and O Fecundity, Longevity, and Ovary Weight.

Longevity 1982–1992.—Rose (1984) first assayed longevity in 1982 on medium supplemented with surface live yeast + acetic acid as in our "standard" fecundity assay. He found that B and O female longevities were 33.28 (± 1.00) and 42.81 (± 1.67) d, respectively [means (standard errors); $P < 0.01$ two-tailed probability estimated using the t-distribution with $N - 1 = 4$ df]. Using the same assay in 1991, B and O mean longevities were 29.32 (± 1.15) and 44.44 (± 1.54) d, respectively [$P < 0.0005$, two-tailed probability estimated using the t-distribution with $N - 2 = 8$ df] (Chippindale et al. 1993). This suggests that from 1982 to 1991, the O's have increased from being approximately 28% longer lived than the B's to being 51% longer lived in this environment.

Further evidence for a continued increase in

TABLE 1. The influence of CO_2 anaesthesia, dietary yeast, and adult density upon relative B and O early fecundity. Means and standard errors estimated using replicate populations as observed variates. All treatments allowed 24 h for oviposition.

	B mean (± SE)	O mean (± SE)
Experiment 2		
CO_2	65.03 (± 2.98)	74.54 (± 3.39)
No CO_2	70.76 (± 1.87)	80.07 (± 3.36)
Experiment 3		
Yeast	78.55 (± 3.41)	88.02 (± 4.61)
No yeast	5.21 (± 0.53)	6.91 (± 0.79)
Experiment 4		
Low density, low yeast	75.88 (± 0.92)	88.36 (± 4.12)
High density, low yeast	3.07 (± 0.23)	2.71 (± 0.26)
Low density, high yeast	78.61 (± 2.05)	102.65 (± 2.93)
High density, high yeast	30.39 (± 1.95)	36.16 (± 1.83)

Experiment 2. ANOVA: stock $P < 0.01$; use of CO_2 $0.1 > P > 0.05$; stock-by-use of CO_2 interaction $P > 0.1$.

Experiment 3. ANOVA: stock $0.1 > P > 0.05$; presence of dietary yeast $P < 0.0005$; stock-by-presence of dietary yeast $P > 0.1$.

Experiment 4. Low-yeast treatments only. ANOVA: stock $P < 0.0005$; adult density $P < 0.0005$; stock-by-density $P < 0.0005$. High-yeast treatments only. ANOVA stock $P < 0.0005$; adult density $P < 0.0005$; stock-by-density $P < 0.0005$.

O longevity over B comes from a series of assays done on banana-molasses medium without yeast (fig. 1). In a 1984 experiment, Hutchinson and Rose (1991) found that O mean longevity was greater than B mean longevity [B: 50.08 (± 2.71) d; O: 62.52 (± 2.69) d; $P < 0.05$ two-tailed probability estimated using the t-distribution with $N - 2 = 8$ df]. In three 1985 experiments, they found that O longevity continued to be greater than B longevity [means over experiments, B: 34.03 (± 1.47) d; O: 49.41 (± 1.91) d]. A two-way ANOVA with stock and experiment as cross-classified fixed effects shows a significant effect of stock ($P < 0.0005$), a significant effect of experiment ($P < 0.005$), and a significant stock × experiment interaction ($P < 0.05$). A 1991 assay also shows greater O mean longevity than B [B: 40.97 (± 1.76) d; O: 74.34 (± 1.63); $P < 0.0005$, two-tailed probability estimated using the t-distribution with $N - 1 = 8$ df]. Thus, over the three sets of assays performed in this environment from 1984 to 1991, the O's are successively 25%, 45%, and 81% longer lived than the B's.

Early Fecundity 1982–1992.—Rose (1984) defined early fecundity as fecundity on days 3, 4, and 5 after eclosion. Reanalyzing his 1982 data (fig. 2), a two-way ANOVA with day and stock as cross-classified fixed effects, shows a significant stock effect ($P < 0.01$) with B greater than O, no significant day effect ($0.5 > P > 0.1$), and no significant stock-by-day effect ($P > 0.5$). Because daily fecundity was estimated over the entire life span of individual females, the 1991 data (fig. 2) are directly comparable to the 1982 data. A similar two-way ANOVA on the 1991 data shows a significant stock effect ($P < 0.05$) with O greater than B, a significant day effect ($P < 0.005$), and no stock-by-day effect ($P > 0.5$).

Further evidence for an evolutionary reversal in relative early fecundity comes from additional experiments performed in the mid-1980s and early 1990s, which assayed peak fecundity, day 4 after eclosion, alone. In a 1984 experiment, Hutchinson and Rose (1991) found that peak B fecundity was greater than peak O fecundity ($P < 0.05$; two-tailed probability estimated using the t-distribution with $N - 1 = 4$ df). In three 1985 experiments, however, B early fecundity was not significantly different from O: a two-way ANOVA with stock and experiment as cross-classified fixed effects shows no effect of stock ($P > 0.5$), no effect of experiment ($0.5 > P > 0.1$), and no stock-by-experiment interaction ($P > 0.5$) (data from Hutchinson and Rose 1991). Finally, 10 assays were carried out between 1990 and 1991 using the "standard" procedure or some close variant of it. Considering these 1990–1991 data together, an ANOVA with stock and experiment as fixed cross-classified effects shows a significant effect of stock ($P < 0.0005$) with O early fecundity greater than B early fecundity, a significant effect of experiment ($P < 0.0005$), but no significant stock × experiment interaction ($0.5 > P > 0.1$), thus confirming the results of the 1991 lifetime fecundity assay, above. Estimated in this way, mean B fecundity is, in 1992, 67.82 eggs per day (± 0.47), and mean O fecundity is 80.57 eggs per day (± 0.79) (data partially shown in tables 1 and 2).

The progress of relative fecundity at day 4 after eclosion (fig. 1) shows a pattern of gradual reversal in relative fecundity not easily attributable to any secular change in the environment. Where absolute *Drosophila* fecundity is a trait notoriously subject to block effects, the rank order of early fecundity of long- and short-lived stocks, prior to this report, has been highly replicable among laboratories, workers, and even stocks (e.g., Rose 1984; Luckinbill et al. 1984). Since

TABLE 2. The influence of oviposition time upon relative B and O fecundity when females are conditioned on "standard" medium.

	Time permitted for oviposition (h)								
	Experiment 1		Experiment 5			Experiment 6			
	1	24	1	6	12	6	12	18	24
B1	3.18	49.50	3.67	17.56	34.28	21.35	38.65	57.20	73.11
B2	4.19	50.56	3.92	15.21	29.00	21.05	37.35	58.30	71.20
B3	2.06	50.67	2.78	21.00	32.32	13.60	44.10	58.55	65.25
B4	1.42	48.16	2.51	20.76	36.30	22.10	46.05	57.50	63.16
B5	1.76	58.82	3.39	14.05	36.82	16.55	40.00	60.15	67.68
Mean	2.52	51.54	3.25	17.72	33.74	18.93	41.23	58.34	68.08
SE	0.51	1.87	0.27	1.41	1.43	1.65	1.65	0.52	1.84
O1	1.12	71.06	1.34	10.53	45.16	25.80	49.83	79.00	84.05
O2	0.61	61.56	0.54	9.22	36.72	0.70	40.55	52.00	62.05
O3	0.30	56.05	1.39	3.70	40.28	14.30	40.70	69.85	84.20
O4	0.22	69.33	0.49	6.45	24.84	7.75	48.90	58.60	72.50
O5	0.82	56.71	0.78	7.89	35.05	12.65	38.85	67.40	71.68
Mean	0.61	62.94	0.91	7.56	36.41	12.24	43.77	65.37	74.90
SE	0.17	3.12	0.19	1.18	3.37	4.13	2.31	4.66	4.19

Experiment 1. ANOVA: stock $P < 0.05$; oviposition time $P < 0.0005$; stock-by-oviposition time $P < 0.005$.
Experiment 5. ANOVA: stock $P < 0.05$; oviposition time $P < 0.0005$; stock-by-oviposition time $P < 0.005$.
Experiment 6. ANOVA: stock $P > 0.1$; oviposition time $P < 0.0005$; stock-by-oviposition time $0.1 > P > 0.05$.

the 1982, 1984, 1986, and 1990–1992 assays were performed by different workers in different laboratories (albeit using a standardized assay method), we can only infer that the rank order had indeed changed. We cannot tell whether, in this environment, O flies have increased their early fecundity, or whether B flies have decreased their early fecundity.

Early Ovary Weight 1984–1992.—Hutchinson and Rose (1991) assayed early-life ovary weight twice, in 1984 and 1985. Both experiments showed that females from B populations had significantly larger ovaries than those from O populations—corresponding, presumably, to their then higher fecundities—although the difference was of somewhat smaller magnitude in the second experiment (table 4). In 1992, B and O ovary weights were not different in the "standard" environment (table 4).

Influence of Environment on the Expression of the Trade-Off

Early Fecundity.—In experiments 1 through 6, we tested whether the loss of the trade-off found in the "standard" fecundity assay holds true for other environments. Given the difference in oviposition environment that the B populations and the O populations have experienced in the course of their evolution (see above), we compared B and O fecundities in the "standard" assay environment (which entails conditioning at low adult density and fresh medium with a yeast supplement followed by a 24-h oviposition period), with relative fecundity in an environment that approximates the normal propagation environment of the B's: the "B-type" environment (which entails conditioning at high densities and "spent" medium followed by a 1-h oviposition period).

The fecundities of both B and O were very low in the "B-type" environment compared with the "standard" environment (tables 2, 3, experiment 1). These low fecundities are most obviously because of the difference in the time course of the assays (1 versus 24 h), but the crowding and absence of high levels of live yeast in the former may also contribute. More interestingly, *relative* B and O fecundity in the "B-type" environment is the reverse of what it is in the "standard" environment (fig. 3). In the "B-type" treatment, each B female laid a mean of 1.7 eggs, whereas each O female laid a mean of only 0.33 eggs (table 3); in the "standard" treatment, each B female laid a mean of 51.54 eggs, whereas each O female laid a mean of 62.94 eggs (table 2). In the following experiments, we tried to determine which of the several differences between the "B-type" and "standard" assay environments cause this environmental dependence of relative fecundity.

Table 1 shows the results from three experiments in which the effects of CO_2 anesthesia, live dietary yeast, and adult density (that is, density

TABLE 3. The influence of oviposition time upon relative B and O fecundity when females are conditioned in the "B-type" environment or on "spent" medium.

	Time permitted for oviposition (h)							
	Experiment 1 (B-type)	Experiment 5 (B-type)			Experiment 6 (Spent medium)			
	1	1	6	12	6	12	18	24
B1	2.01	2.79	5.83	7.75	5.60	11.60	15.90	16.10
B2	1.42	3.82	4.56	3.44	11.55	8.47	14.80	17.85
B3	1.93	1.50	1.30	3.30	9.50	18.45	17.25	26.25
B4	1.37	4.58	7.20	8.25	8.00	10.45	21.85	16.65
B5	1.75	3.98	8.40	3.60	6.70	15.90	15.80	16.74
Mean	1.70	3.33	5.46	5.27	8.27	12.97	17.12	18.72
SE	0.13	0.54	1.22	1.12	1.05	1.83	1.25	1.90
O1	0.24	1.85	2.63	2.50	3.15	8.79	17.80	17.55
O2	0.17	0.74	1.15	1.75	1.10	2.35	8.45	9.40
O3	0.41	1.64	0.80	2.15	3.45	10.00	15.30	17.37
O4	0.25	0.48	0.45	1.89	1.75	10.70	16.65	15.35
O5	0.57	0.38	2.79	3.30	1.60	10.10	20.85	17.89
Mean	0.33	1.02	1.56	2.32	2.21	8.39	15.81	15.51
SE	0.07	0.30	0.48	0.28	0.46	1.54	2.06	1.59

Experiment 1. ANOVA: stock $P < 0.0005$.
Experiment 5. ANOVA: stock $P < 0.0005$; oviposition time $0.1 > P > 0.05$; stock-by-oviposition time $P > 0.1$.
Experiment 6. ANOVA: stock $P < 0.005$; oviposition time $P < 0.0005$; stock-by-oviposition time $P > 0.1$.

at time of oviposition) on relative fecundity were investigated. The results of experiment 2 show that the use of CO_2 anesthesia has no detectable effect on relative B and O fecundity. Similarly, the results of experiment 3 show that the presence of live dietary yeast has no detectable effect on relative B and O fecundity. In experiment 4,

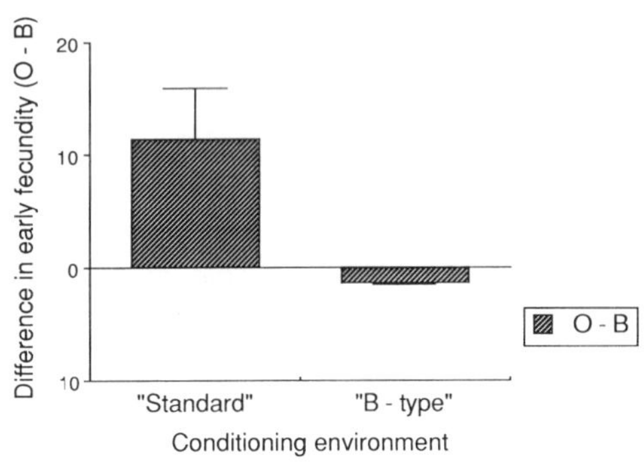

FIG. 3. Relative B and O early fecundity when females are conditioned in "standard" and "B-type" environments. Early fecundity is the number of eggs laid by a female in a 24-h period at day 4 after eclosion. Again, in the "standard" environment the O's have a higher fecundity than the B's. In the "B-type" environment, however, the B's have a higher fecundity than the O's. An ostensible trade-off exists among environments. Means and standard errors of the means based on the difference (O − B) over five pairs of replicate populations.

the effect of adult density on relative fecundity was examined in two dietary environments: low yeast and excess yeast. Considering the former treatment alone, there was a significant stock-by-density interaction. Initial examination of the means (table 1, experiment 4) suggests a reversal of relative B and O fecundity across density treatments; however, B and O are not significantly different under low-yeast, high-density conditions ($P > 0.1$; two-tailed probability estimated using the t-distribution with $N - 2 = 8$ df). Considering the high-yeast treatment alone, a significant stock × density interaction term is also found. However, examination of the means (table 1, experiment 4) shows that the significant interaction term does not reflect a reversal of relative B and O fecundity under high densities. To sum up, it appears that neither CO_2 anesthesia, live dietary yeast, nor high adult densities are solely responsible for the change in sign of relative fecundity between the "standard" and "B-type" fecundity environments.

The influence of time permitted for oviposition on relative fecundity was evaluated in experiments 1, 5, and 6 (table 2). Experiment 1 included a treatment in which flies were conditioned in a manner identical to those in the "standard treatment" (that is, low density, excess dietary yeast against a background of fresh carbohydrate-rich food), but were permitted to

TABLE 4. B and O dry ovary weight (mg): 1983 to 1992. Females are conditioned in one of two environments: "B-type" or "standard" for 3–4 d after eclosion. This age corresponds to that at which "early-life" fecundity is assayed (tables 1, 2).

	B mean (± SE)	O mean (± SE)	Reference
"Standard" conditioning environment			
1984	0.109 (± 0.009)	0.057 (± 0.006)	Hutchinson 1989
1985	0.177 (± 0.003)	0.112 (± 0.022)	Hutchinson 1989
1992	0.120 (± 0.010)	0.139 (± 0.013)	
"B-type" conditioning environment			
1983	0.139 (± 0.006)	0.072 (± 0.009)	Rose et al. 1984
1992	0.142 (± 0.014)	0.105 (± 0.011)	

"Standard": in 1984, $P < 0.005$, df = 8; in 1985, $P < 0.05$, df = 4; in 1992, $P > 0.1$, df = 8.
"B-type": in 1983, $P < 0.01$, df = 8; in 1992, $0.1 > P > 0.05$, df = 8. All probabilities are two-tailed estimated using the t-distribution. All means and standard errors are estimated using independent populations as replicates.

lay eggs for only 1 h—as in the "B-type" assay—against the 24 h typically permitted in the "standard" assay. The B's had a higher fecundity than the O's after 1 h of egg laying ($P < 0.05$; two-tailed probability estimated using the t-distribution with $N - 2 = 8$ df) where, in the same experiment, the O's had a higher fecundity after 24 h of egg laying (table 2, experiment 1). The time dependence of relative fecundity when flies are conditioned in the "standard" environment was confirmed by experiments 5 and 6; the first shows a significant effect, the second a marginally significant effect of stock × oviposition time interaction (table 2). From all three experiments, it appears that the B's lay more eggs for about the first 6 h after being transferred to a vial of fresh medium, whereas after about 12 h the O's lay eggs more rapidly.

Is the time permitted for oviposition, the only difference between the "standard" and "B-type" environments that is causally responsible for the reversal in relative fecundity? It appears not. The results of experiment 5 (table 3) show the influence of time permitted for oviposition when flies are conditioned in the "B-type" environment. In contrast to conditioning in the "standard" environment, rank fecundity seems independent of oviposition time. It appears that the B's lay more eggs than the O's early on, and then flies from both stocks largely cease to lay. What is it about conditioning in the "B-type" environment that prevents the expression of maximal fecundity in both the B's or O's and, presumably, the expression of O superiority? In experiment 6, the influence of conditioning flies on the "spent" medium found in the "B-type" environment and its interaction with oviposition time was investigated. It appears that of the many factors that differ between the "standard" and "B-type" conditioning environments "spent medium" contributes substantially to the low fecundities of both the B and O stocks in the "B-type" environment, and the relative superiority of the B populations (table 3).

Ovary Weight—Rose et al. (1984) first assayed the ovary weight of flies conditioned in the "B-type" environment in 1983. As in the "standard" environment, the B's were found to have significantly heavier ovaries than the O's (table 4). It is unclear whether this is still the case: B females conditioned in the "B-type" environment in 1992 still seem to have larger ovaries than O females, but the difference is not significant (table 4). Whereas in 1983 the B's ovaries were 93% heavier than those of the O's, by 1992 this difference had diminished to 35%.

If the "standard" and "B-type" ovary weight data from 1992 are considered in a two-way ANOVA in which stock and conditioning environment are fixed, cross-classified effects, then no effects of stock ($P > 0.1$) or environment ($P > 0.1$) are detected, but a significant stock × environment effect ($P < 0.05$) exists. In 1992, just as relative B and O fecundity reverses among environments, so too, does relative ovary weight.

Starvation Resistance.—In experiment 8, we tested the effects of the "standard" and "B-type" conditioning environments on female starvation resistance. A two-way ANOVA in which stock and conditioning environment are fixed, cross-classified effects shows a significant effect of stock (O > B; $P < 0.0005$); a significant effect of environment ("B-type" > "standard"; $P < 0.0005$); and a significant stock × environment interaction term ($P < 0.05$). In both environments, the O's had a higher starvation resistance than the

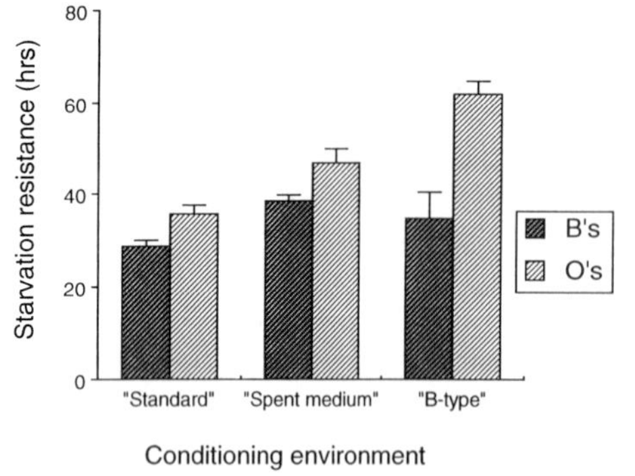

FIG. 4. B and O starvation resistance when females are conditioned in "standard," "B-type" or "spent medium" environments. In each case, the O's have a significantly greater starvation resistance than the B's. Means (and standard errors of the means) based on five replicate populations.

B's (fig. 4). Similarly, in experiment 9, we asked what the effect of one component of the "B-type" environment, "spent medium," is on female starvation resistance. The "standard" environment was used again here as a control. A two-way ANOVA in which stock and medium are fixed, cross-classified effects shows a significant effect of stock (O > B; $P < 0.005$); a significant effect of environment ("spent medium" > "standard"; $P < 0.0005$); but no significant stock × environment interaction term ($P > 0.1$). Again, the O's had a higher starvation resistance than the B's regardless of conditioning medium (fig. 4). The stock means and standard errors shown in figure 4 for the "standard" environment are calculated across both experiments 8 and 9.

DISCUSSION

The Interpopulation Trade-Off between Early Fecundity and Starvation Resistance is Manifest in Some Environments but not in Others.—In the "standard" environment, the long-lived O's now have a superior early fecundity relative to the short-lived B's and continue to maintain their historically greater late fecundity, starvation resistance, and mean longevity. Thus, it is clear that the trade-off among life-history traits, once present in the comparison of these stocks under these conditions, has now disappeared. However, relative B and O early fecundity is strongly dependent on assay environment. Given the presumably close functional relationship between fecundity and starvation resistance (Service and Rose 1985; Service et al. 1988; Hutchinson et al. 1991; Chippindale et al. 1993), we might expect that as relative fecundity changes sign among the "standard" and "B-type" environments so too would starvation resistance such that the sign of their relationship over populations was always constant. This is not the case. Instead, the O's have a higher starvation resistance in *all* environments, although their superiority is more pronounced in some environments than others (fig. 4). Thus, it appears that the trade-off between starvation resistance and fecundity, as inferred from a comparison of the B's and O's, exists in some environments (e.g., "B-type") but not in others (e.g., "standard").

The Disappearance of the Trade-Off is Probably not Caused by Either Drift or Inbreeding.—The fivefold replication of the B's and O's and the use of each replicate population as a single observation when hypothesis testing makes a role for drift in causing the loss of the trade-off unlikely. Furthermore, as discussed below, these populations have census sizes that are sufficiently large that even after 55 (O) and 270 (B) generations, the cumulative effects of drift are not expected to be great.

Whenever the value of a fitness trait declines unexpectedly in the course of a selection experiment—as has early-life B fecundity relative to that of the O's—then inbreeding depression must be suspected. An inbreeding depression explanation for the disappearance of the trade-off must entail the short generation B's being inbred relative to the O's, for it is the B's that now have lower fecundity in the "standard" environment. We believe, for several reasons, that inbreeding depression is not responsible for the loss of the trade-off. First, it is unlikely that the B's are inbred relative to the O's, for where the census population sizes for populations of both stocks are on the order of 10^3 at the beginning of each generation, the O's reproduce very late in life—and for this reason alone should have a lower N_e than the B's, which reproduce early—all other things being equal. Second, although B early fecundity is relatively low in some environments (the "standard" environment), it is clear that they lay more eggs than the O's in others (e.g., the "B-type" environment). That is, each stock has the higher fecundity in the environment in which it has been propagated and to which it has presumably adapted. Although inbreeding depression can be environment specific, it is implausible that its effects should mimic the likely outcome of environment-specific adaptation so exactly. Third, the egg-to-adult development time, which should be greater in inbred stocks, is in fact less in the B's than in the O's (Chippindale et al. unpubl data). Fourth, B populations have been crossed in an attempt to detect the F_1 heterosis that might be indicative of inbreeding. This was done for fecundity in 1985–1986 (Hutchinson and Rose 1991), but no sign of inbreeding depression was detected. Fifth, no sign exists that the variance for fecundity in the "standard" environment is greater among the five replicate B populations than among the five replicate O populations (see standard-error estimates above), as might be expected if the former were relatively inbred. Thus, it appears that the disappearance of the trade-off in the "standard" environment between early fecundity on the one hand, and longevity and starvation resistance on the other, is best attributed to the action of selection. The question remains, however, to what type of selection is this disappearance due?

The Disappearance of the Trade-Off is Caused by the Adaptation of the B and O Stocks to Differences in Culture Regime Rather than Differences in Demographic Regime.—The major difference between the B's and O's is their demographic regime: The two-wk generation cycle of the B's against the 10-wk generation cycle of the O's. But the finding that the O's have a superior fecundity to the B's *except* in that environment in which the B's are actually cultured suggests immediately that adaptation to subtle differences in culture regime, rather than these imposed differences in demography, has caused the ostensible loss of the trade-off (fig 3). This conclusion is bolstered by further experiments in which the components of the "B-type" environment were dissected, showing that (1) B's have an initially higher oviposition rate than the O's on being placed in a fresh culture vial, and (2) lay relatively more eggs after conditioning on "spent" medium (tables, 2, 3). An ability to lay eggs rapidly after having lived for a few days on "spent medium" would be of considerable selective benefit to a B female in her own culture regime but of no use to an O female. Indeed, between the effects of conditioning environment and oviposition time, the average B female now lays about five times as many eggs of an average O female in the 1 h "B-type" assay environment. Conversely, the ability to lay many eggs over 24 h after being maintained for days on fresh medium with ample live dietary yeast (that is, a high "standard" environment fecundity), is precisely that which would be of selective value to an O female being maintained in her own culture regime. In this environment, at day four after eclosion, the O's lay about 18% more eggs than do the B's; later in life this difference would be even greater.

The reversal of relative B and O fecundity among the "B-type" and "standard" environments (fig. 3) suggests the existence of a trade-off, presumably evolving as a negative genetic correlation, between fecundity in the two environments. Several recent workers have inferred negative genetic correlations among fitness components assayed in different environments (e.g., Scheiner et al. 1989; Fry 1990; Robertson 1965), but the generality of such trade-offs is still unclear. The generic relationship between fecundity in the "standard" and "B type" environments is further investigated in a companion paper (Leroi et al. 1994).

The Disappearance of the Trade-Off is Probably Caused by Evolution of Increased Acquisitive Ability in the O Populations in the Cage Environment.—Trade-offs among life-history traits are often thought to be caused by "structural pleiotropies" (Stearns et al. 1991), that is, caused by underlying physiological relations that are largely invariant among environments. Among the simplest conceivable physiological relations that can give rise to antagonistic pleiotropy is a Y-shaped one, in which two life-history traits share some resource represented by the common branch (van Noordwijk and de Jong 1986). Evidence exists from surgical, mutational, and environmental manipulations that fecundity and starvation resistance may be viewed as end points of such a Y-shaped pathway in which the shared common resource is metabolic triglycerides (Butterworth and Bodenstein 1968; Doane 1960a,b,c). If we assume that loci affect either the allocation or acquisition of resources in this pathway, but not both, then we might imagine that at selective equilibrium the bulk of alleles that remain segregating in a population are those that influence allocation. Those variants that influence acquisition would be already fixed or eliminated, allowing for a constant environment and negligible mutation (cf. Houle 1991).

In this view, the original negative genetic correlation between early fecundity and starvation resistance detected in the ancestral IV population was caused by alleles segregating at loci that influenced the traffic of resources into either fecundity or starvation resistance. In reconstructing the paradoxical evolution of the B's and O's since then, no strong statement can be made about which of the stocks is evolving—indeed both probably are. But it should be noted that the "B-type" environment is the ancestral one: that in which IV was propagated between 1975 and 1980 for more than 130 generations. The O culture regime with its conditioning in fresh medium and yeast, however, is relatively novel. For this reason, we suspect that the relative superiority of the B's in the "B-type" environment is caused less by their having adapted to it, than to the O's having lost fitness in the "B-type" environment as they have gained fitness in their own environment. In other words, alleles for resource-acquisition favored in O populations have had the pleiotropic consequence of depressing fecundity in the B's culture regime.

The progress of relative and absolute B and O ovary weight in the "standard" environment is consistent with the idea that the O's are better at acquiring resources in this environment. Where the O's originally had very small ovaries in the "standard" environment, their ovaries are now at least as large as those of the B's, which show no real change (table 4). Less consistent with the idea of environment-specific evolution of resource acquisition, is the observation that O ovary weight *also* has increased in the "B-type" environment. Although it is true that their ovaries are still smaller than the B's in this environment (table 4), they are larger than might be expected from the O's very low relative fecundities in the "B-type" environment (table 3). We do not know enough about the physiology and quantitative genetics of oviposition, egg production, egg storage, and nutrient uptake in *Drosophila* to explain these observations.

Implications for the Inference of Evolutionary Trade-Offs from Laboratory and Natural Populations.—The existence of strong genotype-by-environment interaction for life-history traits permits populations to evolve widely varying responses to the environment. We have argued that such evolved differences in environmental responses can complicate the expression of trade-offs among populations. We find it noteworthy that these complications arise in a carefully controlled long-term selection experiment. These results also point to the necessity of having assay environments correspond closely to the selection environments if selection responses (and hence trade-offs) are to be accurately identified. Such lack of correspondence may account for the failure of Partridge and Fowler (1992) to detect a decline in fecundity in stocks selected for late-life reproductive success.

The complex relationship between environment and the expression of life-history trade-offs demonstrated here raises several questions. For example, what is the genetic structure of B and O populations, as they are now, with respect to life-history traits? A negative genetic correlation existed originally in the ancestral IV population between early fecundity on the one hand, and longevity and starvation resistance on the other (Rose and Charlesworth 1981a; Service and Rose 1985). Now that this correlation is no longer apparent from the comparison of the B's and O's, does it still exist *within* either or both of these sets of populations? To put it another way, is the trade-off truly gone in the "standard" environment (the genetic correlation being now positive), or is it merely obscured, for example, obscured by fixation of beneficial alleles at acquisition loci in the O populations. In a companion paper (Leroi et al. 1994), we investigate this and other questions using additional selection lines derived recently from the B's and O's.

Acknowledgments

We thank A. R. Bennett, A. Gibbs, M. Matos, L. D. Mueller, D. Reznick, J. Shiotsugu, and two anonymous reviewers for their comments on the manuscript. We are furthermore grateful to L. D. Mueller for sage advice throughout the course of the study, T. J. Bradley for advice on weighing ovaries, as well as for the use of laboratory space and equipment, and E. W. Hutchinson for providing helpful information concerning his experiments. We thank R. Chen, P. Campos, and H. Saing for assisting with the fecundity, starvation-resistance, and ovary-weight assays, respectively, and S. B. Kim and J. Shiotsugu for assistance throughout the course of the work. This research was supported by United States–Public Health Services grants AG06346 and AG09970 to M.R.R. from the National Institute on Aging.

Literature Cited

Butterworth, F. M., and D. Bodenstein. 1968. Adipose tissue of *Drosophila*. III. The effect of the ovary on cell growth and the storage of lipid and glycogen in the adult-tissue. Journal of Experimental Zoology 167:207–218.

Chippindale, A. K., A. M. Leroi, S. B. Kim, and M. R. Rose. 1993. Phenotypic plasticity and selection in *Drosophila* history evolution. 1. Nutrition and the cost of reproduction. Journal of Evolutionary Biology 6:171–193.

Doane, W. W. 1960a. Developmental physiology of the mutant female *sterile* (*2*) *adipose* of *Drosophila melanogaster*. I. Adult morphology, longevity, egg production, and egg lethality. Journal of Experimental Zoology 145:1–22.

———. 1960b. Developmental physiology of the mutant female *sterile* (*2*) *adipose* of *Drosophila melanogaster*. II. Effects of altered environment and residual genome on its expression. Journal of Experimental Zoology 145:23–42.

———. 1960c. Developmental physiology of the mutant female *sterile* (*2*) *adipose* of *Drosophila melanogaster*. III. Corpus allatum—complex and ovarian transplantation. Journal of Experimental Zoology 145:275–298.

Fry, J. 1990. Trade-offs in fitness on different hosts: evidence from a selection experiment with a phytophagous mite. American Naturalist 136:569–580.

Houle, D. 1991. Genetic covariance of fitness correlates: what genetic correlations are made of and why it matters. Evolution 45:630–645.

Hutchinson, E. W. 1989. Quantitative genetics of postponed senescence, Ph.D. diss. Dalhousie University, Halifax.

Hutchinson, E. W., and M. R. Rose. 1991. Quantitative genetics of postponed aging in *Drosophila melanogaster*. I. Analysis of outbred populations. Genetics 127:719–727.

Hutchinson, E. W., A. J. Shaw, and M. R. Rose. 1991. Quantitative genetics of postponed aging in *Drosophila melanogaster*. II. Analysis of selected lines. Genetics 127:729–737.

de Jong, G. 1990a. Genotype-by-environment interaction and the genetic covariance between environments: multilocus genetics. Genetica 81:171–177.

———. 1990b. Quantitative genetics of reaction norms. Journal of Evolutionary Biology 3:447–468.

Leroi, A. M., W. R. Chen, and M. R. Rose. 1994. Long-term laboratory evolution of a life-history trade-off in *Drosophila melanogaster*. 2. Stability of genetic correlations. Evolution 48:1258–1268.

Luckinbill, L. S., M. J. Clare. 1985. Selection for life span in *Drosophila melanogaster*. Heredity 55:9–18.

Luckinbill, L. S., M. J. Arking, M. J. Clare, W. C. Cirocco, and S. Buck. 1984. Selection for delayed senescence in *Drosophila melanogaster*. Evolution 38:996–1003.

Luckinbill, L. S., V. Riha, S. Rhine, and T. A. Grundzien. 1990. The role of glucose-6-phosphate dehydrogenase in the evolution of longevity in *Drosophila melanogaster*. Heredity 65:29–38.

van Noordwijk, A., and G. de Jong. 1986. Acquisition and allocation of resources: their influence on variation in life history tactics. American Naturalist 128.137–142.

Parker, G. A., and J. Maynard Smith. 1990. Optimality theory in evolutionary biology. Nature 348: 27–33.

Partridge, L., and K. Fowler. 1992. Direct and correlated responses to selection on age at reproduction in *Drosophila melanogaster*. Evolution 46:76–91.

Robertson, F. W. 1965. The analysis and interpretation of population differences. Pp. 95–115 *in* H. G. Baker and G. L. Stebbins, eds. The genetics of colonizing species. Academic Press, New York.

Rose, M. R. 1984. Laboratory evolution of postponed senescence in *Drosophila melanogaster*. Evolution 38:1004–1010.

Rose, M. R., and B. Charlesworth. 1981a. Genetics of life-history in *Drosophila melanogaster*. I. Sib analysis of adult females. Genetics 97:173–186.

———. 1981b. Genetics of life-history in *Drosophila melanogaster*. II. Exploratory selection experiments. Genetics 97:187–196.

Rose, M. R., M. L. Corey, A. M. Coyle, and P. M. Service. 1984. The morphology of postponed senescence in *Drosophila melanogaster*. Canadian Journal of Zoology 62:1576–1580.

Rose, M. R., L. N. Vu, S. U. Park, and J. L. Graves. 1992. Selection on stress resistance increases longevity in *Drosophila melanogaster*. Experimental Gerontology 27:241–250.

Scheiner, S. M., R. L. Caplan, and R. F. Lyman. 1989. A search for trade-offs among life-history traits in *Drosophila melanogaster*. Evolutionary Ecology 3:51–63.

Service, P. M. 1987. Physiological mechanisms of increased stress resistance in *Drosophila melanogaster* selected for postponed senescence. Physiological Zoology 60:321–326.

Service, P. M., and M. R. Rose. 1985. Genetic covariation among life-history components: the effects of novel environments. Evolution 39:943–945.

Service, P. M., E. W. Hutchinson, M. D. MacKinley, and M. R. Rose. 1985. Resistance to environmental stress in *Drosophila melanogaster* selected for postponed senescence. Physiological Zoology 58: 380–389.

Service, P. M., E. W. Hutchinson, and M. R. Rose. 1988. Multiple genetic mechanisms for the evolution of senescence in *Drosophila melanogaster*. Evolution 42:708–716.

Stearns, S. C. 1992. The evolution of life histories. Oxford University Press, New York.

Stearns, S. C., G. de Jong, and G. Newman. 1991. The effects of phenotypic plasticity on genetic correlations. Trends in Ecology and Evolution 6:122–126.

Tucić, N., D. Cvetković, and D. Milanović. 1988. The genetic variation and covariation among fitness components in *Drosophila melanogaster* females and males. Heredity 60:55–60.

Corresponding Editor: D. Reznick

Experimental Gerontology, Vol. 31, No. 4, pp. 507–516, 1996

0531-5565/96 $15.00 + .00

EVOLUTIONARY PATTERNS AMONG MEASURES OF AGING

THEODORE J. NUSBAUM, LAURENCE D. MUELLER and MICHAEL R. ROSE

Department of Ecology and Evolutionary Biology, University of California, Irvine, California 92717

Abstract—Maximum lifespan has been one of the most common aging measures in comparative studies, while the Gompertz model has recently attracted both proponents and critics of its capacity to adequately describe the acceleration of mortality in the oldest age classes. The Gompertz demographic model describes age-dependent mortality rate acceleration and age-independent mortality using the parameters α and A, respectively. Evolutionary biologists have predominantly used average longevity in studies of aging. Little is known about the evolutionary relationships of these measures on the microevolutionary time scale. We have simultaneously compared Gompertz parameters, average longevity, and maximum longevity in 50 related populations of *Drosophila melanogaster*, many of which have been selected for postponed aging. Overall, these populations have differentiated significantly for the A and α parameter of the Gompertz equation, as well as average and maximum longevity. These indices of aging appear to measure the same genetic changes in aging. However, in some specific population comparisons, the relationships among these measures are more complex. In a second experiment, environmental manipulation of longevity had substantially different effects from genetic differentiation, with the A parameter accounting for changes in overall mortality. The adequacy of the maximum lifespan and the Gompertz equation as indices of aging in evolutionary studies is discussed.

Key Words: Gompertz equation, measures of aging, evolution of aging

INTRODUCTION

RECENT WORK in the evolutionary biology of aging has proceeded largely with the use of such characters as average lifespan and average late fertility (Rose, 1991). However, other studies of aging have been based on maximum lifespan (Comfort, 1979) or on fitted mortality models (e.g., Finch, 1990; Finch *et al.*, 1990; Carey *et al.*, 1992; Curtsinger *et al.*, 1992; Brooks *et al.*, 1994). Many of the latter studies have measured what they consider rates of aging using the α parameter of the Gompertz equation, where this equation gives the dependence of age-specific mortality on age as follows:

Correspondence to: T.J. Nusbaum
(*Received* 28 *May* 1995; *Accepted* 5 *January* 1996)

$$m(x) = A \exp(\alpha x)$$

where $m(x)$ is the mortality rate at age x, A is the initial mortality rate, and α is the exponent term for the acceleration in mortality rate as a function of age. However, the rationales for both maximum lifespan and the Gompertz parameters as measures of aging are somewhat obscure. Maximum lifespan has the advantage that it is not as dependent on environmental vagaries (Comfort, 1979; Rose, 1991), but it is inherently sample size dependent. As for the Gompertz model and its variants, even their proponents remark that, "The Gompertz model is empirical and is not based on any principle or law" (Finch, 1990, p. 16).

This situation may be contrasted with that of the evolutionary theory of aging, which explains patterns of aging by deduction from genetic principles, but makes no use of the Gompertz model (Charlesworth, 1980; Rose, 1991). It may be asked whether maximum lifespan and Gompertz parameters have a heritable genetic basis and, thus, might be subject to selection. But even if they have heritable underpinnings, what would the pattern of selection be for these characters? It is impossible to make deductively derived predictions where these models are concerned, absent ad hoc assumptions connecting them to evolutionary theory (cf. Promislow, 1993). Another issue is whether maximum lifespan and Gompertz parameters reflect similar or different underlying processes or determinants of aging, compared to characters like age-specific survival probability or fecundity. While all these characters may be related to one another algebraically, they are by no means identical. For example, in principle average longevity can be changed by manipulating A, or α, or both. Is there any need to consider the maximum longevity and Gompertz parameters separately? An additional issue is whether or not the demography of aging is essentially a fixed characteristic of each species, as argued by Fries (1980), and, therefore, impossible to change. More particularly, are average longevity, maximum longevity, and the parameters of the Gompertz equation unresponsive to selection or environmental manipulation? Given the effects of nutrition in modulating life histories (Stearns, 1992), one might expect such environmental intervention to influence the expression of these measures of aging. If these measures are responsive, does selection cause different patterns in their relationship than does environmental manipulation?

In the present study, we attempt to address questions of this kind using a set of survival data collected from 50 *Drosophila melanogaster* populations that are closely related, but have, nonetheless, diverged considerably with respect to average lifespan. These data are analyzed for each population to obtain all the basic types of aging measures that have been of interest. Because these populations have been subjected to strong selection for some time, it is unlikely that their mortality schedules are changing rapidly, if at all. Consequently, these measures of aging are likely to be stable through time. We further analyze the data to compare patterns among populations for these indices. We find that there seem to be fundamental underlying survival changes that affect various measures in similar ways in analyses of all populations. However, in certain population comparisons, more complex patterns of association among these indices are apparent. We also find evidence for differences in the patterns of association among these measures of aging caused by selection and environmental manipulation.

Although the Gompertz equation has been in use for about 170 years (Gompertz, 1825), to our knowledge this is the first large-scale test of the Gompertz parameters against other commonly used indices of aging across many replicated, outbred populations of *Drosophila melanogaster*, all derived from a common ancestral population. In addition, this comparison was conducted using populations for which the genetic and physiological basis of aging is somewhat understood. Similarly, maximum lifespan has not been studied using replicated, outbred populations within a species that are highly differentiated for average lifespan. By using many populations, derived from a single common ancestral population and displaying great ranges in longevity, we search for patterns in these indices of aging that have biological interpretation.

MATERIALS AND METHODS

Experimental populations

Each of these populations had been derived from a common ancestral stock, the IVs, which has been maintained in the laboratory since 1975 (Rose, 1984; Service *et al.*, 1985). The selection histories of these stocks are presented in Fig. 1. Each of these stocks are fivefold replicated and have been maintained at large populations sizes (at least 2000 flies/replicate population).

Demographically selected stocks, the Bs and Os, are derived directly from the IVs. The Bs are selected for early-life fertility, while the Os are selected for late-life fertility (Rose, 1984; Service *et al.*, 1985). The other stocks are selected for resistance to either starvation (SOs and their controls the COs, both derived from the Os, and the SBs and their controls the CBs, both derived from the Bs) or desiccation (Ds and their controls the Cs both derived from the Os) (Rose *et al.*, 1992). As controls for starvation-selected stocks, the CO and CB populations are

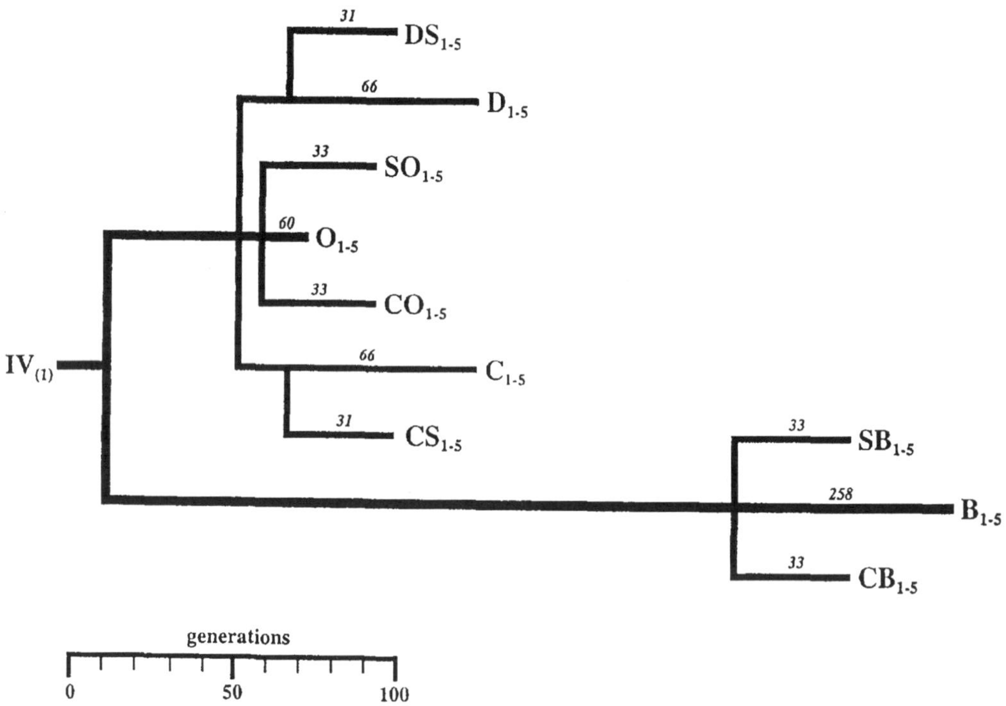

FIG. 1. Selection history of stocks employed in these studies. All stocks are ultimately derived from the unreplicated IV population of flies. All other stocks are fivefold replicated. Subscripts represent replicate numbers. Numbers above lines represent generations since derivation at the time of the longevity assays in this paper. The IV, B, O, CO, and CB populations are selected for demographic properties. All other populations are stress selected, either for starvation resistance (SO, SB, C, CS, DS) or desiccation resistance (D).

demographically selected. As controls for the Ds, the Cs are starvation selected, because the D selection protocol involves desiccation and starvation. DS and their CS controls were derived from the Ds and Cs, respectively, after the first 20 generations of selection on the latter groups of populations. Subsequently, DS and CS have both been starvation selected as in Rose *et al.* (1992). In the demographically selected stocks generation times vary from two weeks in the Bs to 10 weeks in the Os. For the stress-selected stocks typical generation times range from about 19 days in the Ds and Cs to about four weeks in the SO, CO, CS, and DS populations.

Experimental procedures

In 1990 we obtained data from mixed-sex cohorts of 86–114 *D. melanogaster* from each of the 50 replicates of the outbred populations discussed above. All flies were assayed in the same experiment except those in the nutritional manipulation assay (see below). Representative samples of flies from each replicate population were taken out of their normal selection treatment and handled identically for two generations to control for environment-specific effects. To control for density, rearing vials were standardized at 60–80 eggs. Longevity assays proceeded as in recent experiments from our laboratory (Rose *et al.*, 1992), adult flies being held uncrowded, in vials, until dead of natural causes. Longevity was measured by placing four male–female pairs in a 25 × 95 mm shell vial provisioned with standard molasses-banana medium. On average, 12 such vials were set up for each of the 50 populations. Flies were assayed for survival daily and transferred to new vials three times weekly. In addition to the assay of survival patterns among stocks, in a second experiment two selection treatment groups, B and O, were assayed at two levels of nutrition each: 1 mL and 10 mL of our standard yeast solution spread on the surface of the medium, as in Chippindale *et al.* (1993). In all other respects, these flies were handled identically to those of the first experiment. Although the exact nutritional content of the two yeast solutions is unknown, these solutions are known to differentially affect fecundity (Chippindale *et al.*, 1993), so we reasonably expect effects on survival schedules also.

Statistical methodology

The parameters of the Gompertz model were estimated by the maximum likelihood technique using the exact Gompertz model (Mueller *et al.*, 1995) directly on the survival data, without prior transformation. The rationale for this procedure is that it enabled us to fit the model without reorganizing the age classes. Earlier work has shown that this technique is superior to the typical least-squares linear regression approach used to estimate Gompertz parameters. The least-squares linear regression approach is biased and produces poorer estimates of the model parameters as a function of the duration of the age-intervals chosen and the number of individuals within the age classes (Mueller *et al.*, 1995).

The following analyses allowed us to investigate the relationships among indices of aging in these stocks. Analysis of variance (ANOVA) was performed on A, α, average longevity, and maximum longevity separately for each sex with stock type as the independent variable. Because each selection treatment type (e.g., B, O) has five independent replicates, there are five estimates of each aging measure for each selection treatment type. These provide the error estimates in the ANOVA. Paired or unpaired t-tests were used for specific comparisons between experimental populations and their controls. How populations were derived from ancestors determined whether paired or unpaired t-tests were employed in a particular case. All replicates of the B and O populations were derived from the same single ancestral population, so unpaired t-tests were used for comparison. All other populations were compared by paired t-tests because each experimental treatment replicate and its like-numbered control replicate is derived from the same unique ancestral replicate population. Data from the second experiment was analyzed by ANOVA with stock type (B or O), yeast level, and sex as independent effects. This analysis was carried out to determine the influence of nutrition on each of the four indices of aging.

RESULTS

Table 1 shows the estimates of the means for A and α from the Gompertz equation, the average longevity, and the maximum longevity for each group of fivefold replicated populations, with the sexes separated. Table 2 summarizes the statistical analyses that we have performed on these data. Table 3 displays the estimates of the four measures of aging in each of the B and O replicate populations by sex.

TABLE 1. SUMMARY OF AGING MEASURES

Stock	$A\ (\times 10^4)$	*SE* $(\times 10^4)$	α	*SE*	*Average longevity*	*(SE)*	*Maximum longevity*	*(SE)*
				Males				
B	9.13	3.78	.123	.019	49.6	0.4	66.8	2.5
O	4.75	2.09	.059	.010	86.8	2.3	128.0	2.6
SB	26.90	5.12	.046	.007	60.9	1.4	109.4	4.8
CB	3.06	0.11	.110	.003	48.9	1.5	74.2	3.2
SO	5.07	0.17	.043	.002	94.3	3.7	137.6	2.3
CO	12.60	0.60	.058	.002	63.2	0.8	96.6	4.5
D	9.25	0.29	.060	.001	82.2	1.4	124.0	2.1
C	15.90	0.50	.047	.002	69.9	0.6	71.7	0.5
DS	11.90	1.10	.040	.007	85.9	4.8	130.4	1.6
CS	26.10	1.50	.041	.003	85.8	4.1	125.8	3.1
B low	27.80	1.28	.072	.006	43.4	1.1	68.2	5.2
B high	47.70	2.53	.080	.005	34.0	1.4	57.2	2.6
O low	8.73	0.30	.065	.003	63.0	1.7	91.0	2.5
O high	35.10	1.03	.046	.003	55.9	1.7	86.4	4.7
				Females				
B	18.06	7.66	.099	.013	41.0	1.8	63.4	5.9
O	20.43	6.78	.051	.006	74.3	1.6	116.4	4.6
SB	14.16	0.60	.061	.003	58.0	2.5	91.6	3.7
CB	4.21	0.46	.090	.005	45.8	2.0	68.4	4.8
SO	7.53	0.25	.047	.002	82.0	2.5	124.6	3.7
CO	17.30	0.92	.056	.002	58.6	0.7	90.4	4.0
D	17.40	2.04	.045	.005	66.0	2.5	110.2	5.3
C	17.43	1.68	.051	.005	62.6	2.0	103.4	4.5
DS	17.09	1.12	.036	.002	67.0	3.6	120.6	4.4
CS	19.07	1.10	.039	.003	72.7	6.9	115.2	3.9
B low	15.68	0.81	.102	.005	36.2	0.7	56.6	3.0
B high	53.78	9.74	.062	.016	29.1	1.1	51.2	4.9
O low	18.37	0.38	.058	.002	57.8	1.2	86.3	4.8
O high	31.32	1.50	.064	.003	44.5	1.5	73.4	1.5

Each of the 10 stocks employed in these studies is fivefold replicated. Each entry is the mean of the five replicates. SE represents the standard error of the mean. Measures of aging studied include A (day^{-1}) and α (day^{-1}) from the Gompertz model, average longevity (days), and maximum longevity (days). The B and O stocks were also measured in two nutritional environments, high yeast (=high) and low yeast (=low), in a second experiment.

The most obvious conclusion is that these stocks are differentiated from each other with respect to all indices: *A*, α, average longevity, and maximum longevity, in both males and females (Table 2a). This differentiation indicates heritable genetic variation among stocks for these characters, because the stocks are derived from a common ancestor, the IVs. These results also corroborate the findings of Hughes and Charlesworth (1994), who detected heritable genetic variation in the IV stock. Selection on the fecundity schedules or on stress-resistance characters of these populations caused the indirect evolution of A, α, average longevity, and maximum longevity. While the evolutionary theory of aging predicts that changes in demographic selection will lead to changes in the pattern of aging within a population, the indirect

TABLE 2. SUMMARY OF STATISTICAL ANALYSES

	A	α	*Ave. longevity*	*Max. longevity*
(a) Univariate Analyses				
One-way ANOVA with stock as the independent variable				
		Males		
All Stocks	<0.026	<0.001	<0.001	<0.001
B vs. O	NS	0.018	<0.001	<0.001
		Females		
All Stocks	0.002	0.001	<0.001	<0.001
B vs. O	NS	0.011	<0.001	<0.001
t-Tests comparing pairs of experimental and control treatment groups				
		Males		
B-O	NS	0.03	<0.001	<0.001
C-D	<0.001	NS	0.002	<0.001
CB-SB	<0.01	0.0004	0.004	0.003
CO-SO	<0.0003	<0.004	0.001	0.002
CS-DS	NS	NS	NS	NS
O high-O low	<0.00002	0.01	0.004	NS
B high-B low	0.004	NS	0.009	NS
		Females		
B-O	NS	0.02	<0.001	<0.001
C-D	NS	NS	NS	NS
CB-SB	0.0002	<0.007	0.031	0.044
CO-SO	<0.0006	<0.03	<0.001	0.002
CS-DS	<0.001	NS	NS	NS
O high-O low	0.0007	NS	<0.001	NS
B high-B low	<0.02	NS	0.005	NS

	A	α	*Ave. longevity*	*Max. longevity*
(b) Multivariate Analysis				
Three-way ANOVA (stock yeast, sex) on B and O data from second experiment				
Stock	<0.001	<0.001	<0.001	<0.001
Yeast	<0.001	NS	<0.001	0.005
Sex	NS	0.006	<0.001	0.003

Each of the 10 stocks employed in these studies is fivefold replicated. The unit of observation is the estimated value for a replicate. Indices of aging studied are A (day^{-1}), and α (day^{-1}), from the Gompertz model, average longevity (days), and maximum longevity (days). Column entries are *p*-values unless otherwise indicated. The B and O stocks were also measured in two nutritional environments, high yeast (= high) and low yeast (= low), in a second experiment.
All interactions NS except stock · yeast · sex for *A* ($p = 0.005$) and α ($p = 0.044$).

TABLE 3. B AND O DATA

Stock	*Sex*	A $(\times 10^4)$	α	*Average longevity*	*Maximum longevity*
B1	M	3.89	0.103	49.8	68
B2	M	7.29	0.094	48.0	71
B3	M	0.11	0.182	50.8	62
B4	M	1.25	0.081	49.8	73
B5	M	2.19	0.156	49.6	60
O1	M	1.05	0.041	85.0	129
O2	M	4.00	0.057	80.3	119
O3	M	8.55	0.040	88.0	133
O4	M	0.07	0.092	86.6	126
O5	M	0.59	0.064	94.3	133
B1	F	9.37	0.096	43.3	64
B2	F	18.7	0.112	34.1	49
B3	F	12.2	0.093	43.7	64
B4	F	3.07	0.137	42.2	56
B5	F	47.0	0.055	41.5	84
O1	F	31.6	0.060	71.4	120
O2	F	38.9	0.041	72.2	120
O3	F	20.7	0.036	74.6	123
O4	F	7.01	0.049	80.5	121
O5	F	3.91	0.066	73.0	98

Values of the Gompertz parameters A (day^{-1}) and α (day^{-1}), average longevity (days), and maximum longevity (days) for both sexes of each replicate of the B and O stocks from Experiment 1.

response of these measures to selection for stress-resistance characters is less straightforward. However, these indirect responses can be explained by the causal relationship between stress-resistance and longevity in these populations (Rose *et al.*, 1992). Moreover, the finding of indirect responses among measures of aging to selection on demography and stress-resistance immediately indicates that the idea of a fixed species-specific longevity is not universally valid, as perusal of Table 1 also suggests.

One of the most important specific comparisons of stocks is that of the B and O treatment groups (Table 3). These populations are well known for their contrasting aging patterns, from survival to physiology (Rose, 1984; Service *et al.*, 1985). Previously, it had not been shown that they were differentiated with respect to α. The present study now shows that they are differentiated in the manner expected, the short-lived Bs showing a faster acceleration of mortality rate with adult age than the long-lived Os (Table 1, Table 2a). It may also be noted that the B-O contrast has three indices of aging moving together, average longevity and maximum longevity increasing in Os as the acceleration of mortality rate, α, decreases. Although it is generally true that all measures differentiate across all stocks, concordant changes of α, average longevity, and maximum longevity are not evident in each comparison of selection treatment groups or nutritional treatments, as shown in Table 2 for the C-D, CS-DS, O high-O low, and B high-B low comparisons.

Moreover, in the nutritional experiments, B males and B and O females show variation in average longevity with no accompanying significant variation in maximum longevity or α (Table 2a). In addition, nutritional manipulation of longevity appears to affect the A parameter, with two- to threefold increases in A when a given population is reared on high yeast (Table 2b).

Environmental manipulation, as opposed to genetic differentiation, influences mortality schedules through the age-independent parameter, not the rate of aging (Table 2b). Only selection, as is the case with the SOs and SBs, has produced the fully concordant changes among all measures of aging.

DISCUSSION

The Gompertz equation is not based on any principle or law (Finch, 1990), and there has been controversy over the demographic theory it embodies (Carey *et al.*, 1993; Curtsinger *et al.*, 1993; but see Nusbaum *et al.*, 1993; Hughes and Charlesworth, 1994). Nevertheless, our results suggest that it can accurately summarize environmental and genetic alterations of longevity in a manner more or less consistent with other indices of aging. Thus, we have a general "postponed aging" syndrome, involving A, α, average longevity, and maximum longevity. However, not all stocks exhibit the full syndrome.

Johnson (1987, 1990) and colleagues observed differences in α, average longevity, and maximum longevity, but not A, among standard, recombinant-inbred, and mutant strains of *Caenorhabditis elegans.* This is a direct parallel to the B-O outcome, except that the differentiation of strains is not produced by selection. Finch (1990, see his Table 10.1, p. 508) surveyed studies of dietarily restricted rats and observed differences in both mortality rate doubling time (a statistic proportional to α) and maximum longevity. No consistent differences in the A parameter (called IMR) were evident. However, protocols differed among studies and no statistical analyses were conducted. In any case, the closest studies to our own (Johnson, 1987, 1990) produced analogous results, despite the use of inbred and mutant strains of *C. elegans* and our use of laboratory-selected outbred populations of *Drosophila.* Because of the analogous results between populations that differ in population genetic structure, we hypothesize that redundancy among these measures of aging has some credibility.

Here, we have shown that environmentally induced changes in aging largely affect A, the age-independent parameter of the Gompertz equation. Is it likely to be the case that environmental factors will normally alter A, as opposed to α, and, conversely, for genetic factors? Perhaps the dichotomy is one between environmental modulation of lifespan and genetic modulation of lifespan in which fitnesses are always reasonably high, the former changing A, the latter changing α. Testing this idea would require the assembly of a broad range of data, covering diverse genetic and environmental effects upon lifespan. On the other hand, because there is no profound theoretical basis for an association between genetic variation and the α parameter, or an association between environmental effects and the A parameter, any such study might prove largely vacuous.

Despite the heritable genetic basis for and the empirical relationships among these four indices of aging in these 50 populations, we do not argue that the Gompertz model is the best or even necessarily a very good descriptor of aging in populations. There are reasons, grounded in first principles of age-specific population genetics theory and age-specific patterns of gene expression (Charlesworth, 1980), to expect that rates of aging may not continue to increase exponentially in the oldest age classes, as required by the model. In the evolutionary theory of aging, the force of selection falls with chronological age after reproductive maturity (Hamilton, 1966; Charlesworth, 1980; Rose, 1991). This fall may give rise to an accelerating increase in age-specific mortality rate, for a time. However, there is no mathematical necessity to any particular second-order form for the initial decline in the force of natural selection. The form will instead be determined by the age-specific mortality and fecundity schedules of the population. A still more important point is that the fall in the force of natural selection eventually slows in rate, because that force cannot go below zero. It reaches zero, and then remains at that value thereafter. From that point on, if we assume complete age specificity in the evolution of life-history characters, the expectation is for a reduction in the upward acceleration of the age-specific mortality rate, all other things being equal. In other words, we expect a failure of the Gompertz model to fit at extremely late ages. This, of course, is the pattern that has been adduced by Gavrilov and Gavrilova (1991), Carey *et al.* (1992) and Curtsinger *et al.* (1992). Thus, the Gompertz model has little credibility as a generally appropriate model, if perfect age specificity of life-history evolution is assumed.

A considerable complication to this analysis is the question of pleiotropy between age classes. Any discussion that directly connects the force of natural selection to the shaping of age-specific fitness components assumes the effective absence of pleiotropy. But the theoretical (Charlesworth, 1980; Rose, 1985) and experimental (Rose, 1991) importance of pleiotropy in the evolution of aging is well established, even though it is not a universal phenomenon (Mueller, 1987; Service *et al.*, 1988). Therefore, the preceding argument is necessarily flawed. We cannot really make strong statements connecting the force of natural selection to demographic patterns, except for the broadest qualitative assertions that are embodied in the evolutionary theory of aging in general. Natural selection does not directly shape the timing of aging, even though it is the ultimate cause of that aging. Complex patterns of pleiotropy intervene. For example, the continued survival of some individuals in the laboratory, long after reproduction and, thus, long after force of natural selection has hit zero, indicates some kind of "pleiotropic echo," in which earlier adult adaptations for survival are keeping much older individuals alive. Thus, we cannot plausibly invoke the quantitative force of natural selection to guide the interpretation of life-history schedules, at least not exactly.

Finally, there is the purely practical question: should the Gompertz parameters or maximum lifespan be used in studies of the evolution of aging, given their redundancy? To this point, characters like average longevity and fertility have been used preferentially within evolutionary biology. They do not require fitting to a population's mortality pattern. They can be studied in terms of correlation between relatives and during artificial selection (cf. Falconer, 1981). Moreover, there are no profound theoretical motivations to use the maximum lifespan or the Gompertz parameters in evolutionary studies of aging, and, indeed, there are the theoretical and experimental problems with them discussed above. We conclude that the utility of studying these parameters is now an open question that requires further experimental investigation. This is not to question, for example, the utility of estimating life tables in studies of the evolutionary biology of aging, especially because the evolutionary theory of aging is an age-specific theory of survival and fertility. However, there may always be confusions generated by continuing attempts to explain demographic patterns in terms of ad hoc theory (Promislow, 1991, 1993), rather than the evolutionary theory developed for life history (e.g., Charlesworth, 1980).

Acknowledgments—This research was supported, in part, by NIH Grant #AG09970.

REFERENCES

BROOKS, A., LITHGOW, G.J., and JOHNSON, T.E. Mortality rates in a genetically heterogeneous population of *Caenorhabditis elegans*. *Science* **263,** 668–671, 1994.

CAREY, J.R., LIEDO, P., OROZCO, D., and VAUPEL, J.W. Slowing of mortality rates at older ages in large medfly cohorts. *Science* **258,** 457–461, 1992.

CHARLESWORTH, B. *Evolution in Age-Structured Populations*, Cambridge University Press, London, 1980.

CHIPPINDALE, A.K., LEROI, A. M., KIM, S.B., and ROSE, M.R. Phenotypic plasticity and selection in *Drosophila* life history evolution. I. Nutrition and the cost of reproduction. *J. Evolut. Biol.* **6,** 171–193, 1993.

COMFORT, A. *The Biology of Senescence,* 3rd ed., Churchill Livingstone, Edinburgh, 1979.

CURTSINGER, J.W., FUKUI, H.H., TOWNSEND, D.R., and VAUPEL, J.W. Demography of genotypes: Failure of the limited life-span paradigm in *Drosophila melanogaster. Science* **258,** 461–463, 1992.

FALCONER, D.S. *Introduction to Quantitative Genetics,* 2nd ed., Longman, London, 1981.

FINCH, C.E., PIKE, M.C., and WITTEN, M. Slow increases of the Gompertz mortality rate during aging in certain animals approximate that of humans. *Science* **249,** 902–905, 1990.

FINCH, C.E. *Longevity, Senescence, and the Genome,* University of Chicago Press, Chicago, 1990.

FRIES, J.F. Aging, natural death, and the compression of morbidity. *N. Eng. J. Med.* **303,** 130–135, 1980.

GAVRILOV, L.A. and GAVRILOVA, N.S. *The Biology of Life Span: A Quantitative Approach,* Harwood Academic Publishers, New York, 1991.

GOMPERTZ, B. On the nature of the function expressive of the law of human mortality, and on a new mode of determining the value of life contingencies. *Philos. Trans. R. Soc. Lond. Series A* **115,** 513–585, 1825.

HAMILTON, W.D. The moulding of senescence by natural selection. *J. Theor. Biol.* **12,** 12–45, 1966.

HUGHES, K.A., and CHARLESWORTH, B. A genetic analysis of senescence in *Drosophila. Nature* **367,** 64–66, 1994.

JOHNSON, T.E. Aging can be genetically dissected into component processes using long-lived lines of *Caenorhabditis elegans. Proc. Natl. Acad. Sci. USA* **84,** 3777–3781, 1987.

JOHNSON, T.E. Age-1 mutants of *Caenorhabditis elegans* prolong life by modifying the Gompertz rate of aging. *Science* **249,** 908–912, 1990.

MUELLER, L.D. The evolution of accelerated senescence in laboratory populations of *Drosophila. Proc. Natl. Acad. Sci. USA* **84,** 1974–1977, 1987.

MUELLER, L.D., NUSBAUM, T.J., and ROSE, M.R. The Gompertz equation as a predictive tool in demography. *Exp. Gerontol.* **30,** 553–569, 1995.

NUSBAUM, T.J., GRAVES, J.L., MUELLER, L.D., and ROSE, M.R. Letter. *Science* **260,** 1567, 1993.

PROMISLOW, D.E.L. Senescence in natural populations of mammals: A comparative study. *Evolution* **45,** 1869–1877, 1991.

PROMISLOW, D.E.L. On size and survival: Progress and pitfalls in the allometry of lifespan. *J. Gerontol. Biol. Sci.* **48,** B115–B123, 1993.

ROSE, M.R. Laboratory evolution of postponed senescence in *Drosophila melanogaster. Evolution* **38,** 1004–1010, 1984.

ROSE, M.R. Life history evolution with antagonistic pleiotropy and overlapping generations. *Theor. Popul. Biol.* **28,** 342–358, 1985.

ROSE, M.R. *Evolutionary Biology of Aging,* Oxford University Press, New York, 1991.

ROSE, M.R., VU, L.N., PARK, S.U., and GRAVES, J.L., JR. Selection on stress resistance increases longevity in *Drosophila melanogaster. Exp. Gerontol.* **27,** 241–250, 1992.

SERVICE, P.M., HUTCHINSON, E.W., MAC KINLEY, M.D., and ROSE, M.R. Resistance to environmental stress in *Drosophila melanogster* selected for postponed senescence. *Physiol. Zool.* **58,** 380–389, 1985.

SERVICE, P.M., HUTCHINSON, E.W., and ROSE, M.R. Multiple genetic mechanisms for the evolution of senescence in *Drosophila melanogaster. Evolution* **42,** 708–716, 1988.

STEARNS, S.C. *Evolution of Life-Histories,* Oxford University Press, Oxford, 1992.

II

STRESS RESISTANCE, PHYSIOLOGY, AND AGING

Increasing Stress Resistance Often Postpones Aging

Timothy J. Bradley[1] and Michael R. Rose[1,2]

[1]*Department of Ecology and Evolutionary Biology, University of California, Irvine, CA 92697-2525, USA*

[2]*mrrose@uci.edu; to whom correspondence should be sent*

Soon after the Methuselah flies were characterized for their longevity, the question of *how* they lived longer was raised. One of the first clues to their postponed, or slower, aging was that they seemed to be more durable under normal laboratory conditions, and under conditions in which various forms of stress were imposed on them. This finding led to a series of experiments from 1983 to 1985, primarily performed by Philip M. Service, then a post-doctoral student in the Rose laboratory. The philosophy behind these experiments was that the O flies must be able to survive events (either environmental or internal in origin) that lead to death in the B flies. Enhanced longevity in the Methuselah flies therefore might reflect identifiable physiological mechanisms associated with the evolution of postponed aging. This hypothesis was the starting point for screening physiological mechanisms of postponed aging in B and O fruit flies.

It has been established in fruit fly research that the so-called "heat shock response," in which proteins are chaperoned to prevent degradation due to heat, is a generalized stress response, of value in surviving other forms of stress, as well as heat. If that is true, and the O flies lived longer because of stress resistance, then Service should have found a consistent improvement in survival at temperatures that produce heat shock. As shown in Service *et al.* (1985; reprinted here), no such heat shock response difference arose between O and B flies. The response was close to statistical significance, but not quite there. Later molecular work showed that there were molecular differences in the heat shock gene expression of B and O flies (Kurapati *et al.*, 2000; reprinted here). Those data will be presented in the fourth Section, "Genetics and Molecular Biology."

Service was successful, however, in demonstrating statistically significant differences in the Methuselah flies for two other forms of stress resistance, those involving starvation and desiccation. Starvation resistance was operationally defined as the time it took a fly to die under conditions of complete food deprivation in the presence of high humidity. Desiccation resistance was operationally defined as the time it took a fly to die under conditions of complete food deprivation and in the presence of air dried by a desiccant. Resistance to both of these stresses was consistently and significantly enhanced in the male and female O flies compared with the B flies (Service *et al.*, 1985). This finding stimulated a new line of research on the Methuselah flies.

Service noted that starvation resistance showed an interesting dependence on age and gender. Females exhibited an increasing starvation resistance with age, but males did not. He looked for another physiological variable that would show a corresponding age, gender, and stock dependence. Metabolic rate and activity level did not show a decline as starvation resistance increased. Fat content did, however, show a positive correlation with starvation resistance (Service, 1987). The O flies had more fat than the B flies, and females accumulated with fat with age, while males did not.

This physiological work was followed up primarily in the laboratories of Bradley and Gibbs, working with Rose and his students, from 1990 to 2000. In a massive study of many genetic stocks of fruit fly, testing for predictors of starvation resistance, Djawdan *et al.* (1998; reprinted here) found that the total amount of stored calories, whether fat or carbohydrate, in the fly body predicted the time to death from starvation with very high precision. This broad study, examining flies from a variety of demographic and stress selection regimes found no differences in metabolic rate expressed on a per fly basis. Starvation resistance could be predicted entirely by knowing the caloric value of lipid and carbohydrate stores. For some reason, longer-lived flies stored more calories.

Another way to test for correlations between starvation and desiccation resistance and longer life is to select for increased stress resistance and observe the indirect response of longevity. Longevity should increase with such selection, if stress resistance indeed underlies longevity. This was done for both desiccation resistance and starvation resistance in the Rose laboratory (Rose *et al.*, 1992, reprinted here), starting from both the five O populations and the five B populations for starvation resistance, and from the five O populations only for desiccation resistance. The direct response to selection for both types of stress resistance was strong, both increasing a great deal. Longevity also increased in both types of selection line, even though it had not been directly selected for. This additional line of evidence strongly suggested that increased stress resistance was a causal contributor to increased longevity and postponed aging.

A later series of experiments was performed in the Rose laboratory on stress resistance and longevity in a variety of stocks related to the original Methuselah flies. The first of these experiments compared the longevities and stress resistances of these lines (Phelan *et al.*, 2003). In lines exhibiting very high levels of stress resistance, longevity was not increased and in some cases even fell. The second series of experiments followed lines selected at very high levels of stress, tracking both stress resistance and longevity (Archer *et al.*, 2003). At very high levels of stress resistance, longevity either did not continue to increase or it fell. These results suggest that *moderate* increases in stress resistance increase longevity, but extreme increases in the characters underlying stress resistance might decrease longevity.

While the physiological basis of starvation resistance was found to be tied to the accumulation of energy in the form of stored lipid and carbohydrates, these variables were not found to be critical for desiccation resistance. Djawdan *et al.* (1996, reprinted here, next section) found that flies highly selected for increased desiccation resistance had identical caloric stores as their control populations. It had been proposed (Hoffman and Parsons, 1991) that reduced metabolic rate fosters stress resistance, but both Service (1987) and Djawdan *et al.* (1996, reprinted here, next section; 1997) found no general correlation between increased stress resistance and reduced metabolic rate in the Methuselah stocks. Detailed work on the physiological mechanisms of desiccation resistance in the Bradley and Gibbs laboratories showed that physiological characters regulating water balance were the best predictors of desiccation resistance (Gibbs *et al.*, 1997). Reduced rates of transcuticular water loss, and increased water storage (Gibbs *et al*, 1997; Williams and Bradley, 1998), specifically in the hemolymph (Folk *et al.*, 2001), are the critical elements in resistance to desiccation. Nghiem *et al.* (2000, reprinted here) examined these variables specifically in the Methuselah flies. They found that the longer-lived flies had evolved reduced cuticular water permeability and increased whole-body water content, but no change in water content at death.

This was a satisfying set of results. The storage and retention of calories and water predict resistances to starvation and to desiccation respectively, which in turn predict survival rates to later ages. But this conclusion is valid only *within a limited range* of stress resistances. Moderately enhanced storage of lipid, carbohydrate, and water can increase the stress resistance of flies and promote longevity. At extremely high levels of stress resistance, the pattern breaks down. Put another way, moderately increased stress resistance enhances aging, while greatly increased stress resistance does not.

Following the original 1985 publication of Service *et al.*, other gerontological researchers discovered that stress resistance was connected to aging. Jazwinski (1996) reviewed related stress resistance results in yeast, nematodes, mice, and fruit flies. It has now been widely shown that increased stress resistance confers increased longevity, and the connection between stress resistance and aging seems to be general to a variety of organisms. While research with Methuselah flies was the first to demonstrate this connection, it was also the first to demonstrate the limitations of the correlation.

References

Archer, M.A., J.P. Phelan, K.A. Beckman, and M.R. Rose. 2003. Breakdown in correlations during laboratory evolution. II. Selection on stress resistance in *Drosophila* populations. *Evolution* **57**: 536–543.

Djawdan, M., A.K. Chippindale, M.R. Rose, and T.J. Bradley. 1998. Metabolic reserves and evolved stress resistance in *Drosophila melanogaster. Physiol Zool* **71**: 584–594.

Djawdan, M., M.R. Rose, and T.J. Bradley. 1997. Does selection for stress resistance lower metabolic rate? *Ecology* **78**: 828–837.

Djawdan, M., T. Sugiyama, L. Schlaeger, T.J. Bradley, and M.R. Rose. 1996. Metabolic aspects of the trade-off between fecundity and longevity in *Drosophila melanogaster. Physiol Zool* **69**: 1175–1195.

Folk, D.G; C. Han, and T.J. Bradley. 2001. Water acquisition and partitioning in *Drosophila melanogaster*: Effects of selection for desiccation resistance. *J Exper Biol* **204**: 3323–3331.

Gibbs, A.G., A.K. Chippindale, and M.R. Rose. 1997. Physiological mechanisms of evolved desiccation resistance in *Drosophila melanogaster. J Exper Biol* **200**: 1821–1832.

Hoffmann, A.A. and P.A. Parsons. 1991. *Evolutionary Genetics and Environmental Stress.* Oxford University Press. New York.

Jazwinski, S.M. 1996. Longevity, genes, and aging [Review]. *Science* **273**: 54–59.

Kurapati, R., H.B. Passananti, M.R. Rose, and J. Tower. 2000. Increased *hsp22* RNA Levels in *Drosophila* lines genetically selected for increased longevity. *J Gerontol Series A: Biol Sci Med Sci* **55**: B552–B559.

Nghiem, D., A.G. Gibbs, M.R. Rose, and T.J. Bradley. 2000. Postponed aging and desiccation resistance in *Drosophila melanogaster. Exper Gerontol* **35**: 957–969.

Phelan, J.P., M.A. Archer, K.A. Beckman, A.K. Chippindale, T.J. Nusbaum, and M.R. Rose. 2003. Breakdown in correlations during laboratory evolution. I. Comparative analyses of *Drosophila* populations. *Evolution* **57**: 527–535.

Rose, M.R., L.N. Vu, S.U. Park, and J.L. Graves. 1992. Selection for stress resistance increases longevity in *Drosophila melanogaster*. *Exper Gerontol* **27**: 241–250.

Service, P.M. 1987. Physiological mechanisms of increased stress resistance in Drosophila melanogaster selected for postponed senescence. *Physiol Zool* **60**: 321–326.

Service, P.M., E.W. Hutchinson, M.D. MacKinley, and M.R. Rose. 1985. Resistance to environmental stress in *Drosophila melanogaster* selected for postponed senescence. *Physiol Zool* **58**: 380–389.

Williams, A.E. and T.J. Bradley. 1998. The effect of respiratory pattern on water loss in desiccation resistant *Drosophila melanogaster*. *J Exper Biol* **201**: 2953–2959.

RESISTANCE TO ENVIRONMENTAL STRESS IN DROSOPHILA MELANOGASTER SELECTED FOR POSTPONED SENESCENCE[1]

P. M. SERVICE, E. W. HUTCHINSON, M. D. MACKINLEY, AND M. R. ROSE[2]

Department of Biology, Dalhousie University, Halifax, Nova Scotia, Canada B3H 4J1

(Accepted 12/20/84)

Adult *Drosophila melanogaster* from populations that exhibit postponed senescence were compared to flies from control populations. Male and female flies from populations displaying postponed senescence were significantly more resistant than control flies to desiccation, starvation, and the vapor of a 15% ethanol solution. In the presence of desiccant or when humidity was not controlled, flies from postponed-senescence populations were more resistant than controls to heat stress in the range 37–39 C. Differences in heat tolerance disappeared when experiments were conducted at high humidity. Tolerance to starvation increased with age for females but remained approximately constant for males. Tolerance to desiccation decreased with age for both sexes. Differences between control and postponed-senescence populations with respect to starvation resistance and desiccation resistance were obtained at all ages examined. Females from postponed-senescence populations had lower proportional water content than did females from control populations. Differences in water content cannot explain the greater desiccation resistance of flies from postponed-senescence populations. Genetically increased life span in these *D. melanogaster* populations appears to result proximally from greater resistance to mortality arising from environmental stress.

INTRODUCTION

The longevity of insects can be altered in the laboratory by age-specific selection on fecundity. Sokal (1970) and Mertz (1975) selected for high fecundity at early ages in *Tribolium*. They obtained a correlated reduction in adult life span, although results were not consistent across replicates in either experiment. Conversely, selection for high fecundity at *later* ages results in *increased* longevity (Wattiaux 1968*a*, 1968*b*; Rose and Charlesworth 1981; Luckinbill et al. 1984; Rose 1984*c*). Increased mean life span is associated with reduced fecundity of young adult females (Rose and Charlesworth 1981; Luckinbill et al. 1984; Rose 1984*c*). This pattern of postponed senescence and concomitant reduction in early-life fitness is implicit in the "antagonistic pleiotropy" hypothesis for the evolution of senescence (Williams 1957; Rose 1983): genes favored by natural selection because of early beneficial effects have later, pleiotroic, deleterious effects that engender senescence. This is an inherently physiological hypothesis, which calls for elucidation of the underlying physiological mechanisms that give rise to antagonism between early-life and late-life fitness characters.

Rose (1984*c*) produced by selection five replicate populations of *Drosophila melanogaster* that exhibit increased mean longevity relative to five control populations. These populations have several advantages for physiological studies of life history variation: (1) they are derived from a common base population, (2) differences between long-lived and short-lived flies that are consistent across replicates must arise from the basic biological requirements of increased longevity, and (3) a great deal is already known about the biology of *D. melanogaster*.

Our goal is a mechanistic understanding of the evolutionary modification of life history. One avenue of approach to that goal is the comparative physiology of long-lived and short-lived flies, and this paper reports our initial physiological investigations. The specific question motivating the

[1] We thank K. Nishikawa for assistance with experiments. This research was supported by operating grant U0178 to M.R.R. from the Natural Sciences and Engineering Research Council of Canada.

[2] To whom reprint requests should be addressed.

Physiol. Zool. 58(4):380–389. 1985.

 0031-935X/85/5804-8480$02.00

present experiments was whether or not increased longevity is the result of greater resistance to environmental causes of mortality, such as excessive heat, low humidity, or poor nutrition, in particular or in general.

MATERIAL AND METHODS

GENERAL

The experimental flies were derived from five control populations (referred to as B_1–B_5) and five populations (O_1–O_5) exhibiting postponed senescence. The history of these populations is given in Rose (1984*c*); additional details about culture methods are given here. The B stocks are maintained in discrete generations of 14-day length, and the O stocks are maintained in 70-day-long discrete generations. Larval, pupal, and early-adult development of each O generation takes place in 25 × 95-mm shell vials without replacement of food. At 1–4 days of adult age, O flies are transferred to population cages. Food, in 100 × 15-mm plastic petri dishes, is replaced in the O population cages three times weekly. B flies spend their entire 14-day lives in shell vials, without replacement of food. All stocks are maintained on banana-agar medium that contains corn and malt syrups and killed baker's yeast. Live yeast is *not* added to the medium. At the time of the present experiments, there had been approximately 95 B generations and 24 O generations.

Samples of 1,200–1,500 eggs were obtained from each of the 10 stock populations and transferred at a density of 30 eggs per vial to 25 × 95-mm shell vials containing the usual banana-agar medium. Subsequently, each generation of *experimental* flies was used to produce the next experimental generation at approximately 2-wk intervals. Forty to 50 virgins (per population) were collected from each generation and arbitrarily paired with males from different rearing tubes. Mated females were allowed to feed and oviposit individually in tubes containing yeasted charcoal medium (Rose 1979) for 48 h and were then discarded. Thirty eggs were collected from each laying tube and transferred to a fresh rearing tube containing banana-agar medium. Thus, each population of experimental flies was maintained over several generations at a controlled density with controlled mating. Experimental flies were taken from the second and subsequent generations. Effective population sizes were 60–80 per generation after the initial sampling from stocks. Flies from the same rearing tube constituted sibships, and in each experiment only one female and one male were used from each tube.

The 2-wk interval between experimental generations amounted to a relaxation of selection for the O populations. In addition, the relatively small population sizes in the experimental lines would have given rise to inbreeding artifacts (cf. Rose 1984*a*, 1984*b*) if they had been maintained for many generations. (It should, however, be pointed out that these artifacts would give rise to a failure to detect systematic differentiation rather than to spurious inferences.) Therefore, after seven experimental generations we decided to derive a second series of experimental generations from the stock populations.

Stocks and experimental flies were maintained at 25 C under a 24L:0D regime. Food was always abundant in the rearing tubes used for experimental flies.

HEAT STRESS

Foam-plastic test tube stoppers (35 mm long) were used to confine adult flies to the lower 15 mm of weighted, empty 25 × 95-mm shell vials. The flies were subjected to thermal stress in a water bath consisting of a 20-liter glass aquarium equipped with an immersion circulator/heater. The space available to the flies was kept completely submerged, with the tops of the vials protruding above the water surface. Flies were tested at 36, 37, 38, and 39 C, the upper and lower limits of the range being determined by practical necessity. A common temperature for induction of heat-shock puffing in polytene chromosomes is 37 C (Ashburner and Bonner 1979). Males and females were tested separately.

There were four flies in each vial. Flies tested simultaneously constituted experimental blocks, a complete block consisting of one vial per population for each of 10 populations. Thus, 40 flies were tested simultaneously and an experiment comprised as many as 10 blocks. Water temperature variation was less than 0.1 C throughout an experiment. Each vial was examined at regular intervals (1–10 min depending on the temperature used), and the number of immobile flies was recorded. In this manner, it was possible to estimate the time to complete immobility of each fly. Lack of mobility is not equated with death. In general, however, immobile flies virtually never recovered after the termination of heat stress.

In our initial series of experiments, the tops of the vials used for stressing flies were open to the atmosphere. Thus, humidity was not controlled. In order to distinguish the effects of heat from possible effects of desiccation, we conducted additional experiments at 38 C. In one experiment, ~0.7 g of absorbent rayon and 5 ml of distilled water were added to each shell vial in the space above a short (17 mm) foam-plastic plug used to confine the flies. The tops of the vials were covered with Parafilm M laboratory film, and a small hole was made in the film in order to relieve pressure. These flies were thereby maintained at or near 100% relative humidity during the heat stress. In a second experiment, the rayon and water were replaced by 3 g of calcium sulfate–indicator desiccant (size, 8 mesh).

STARVATION

Short foam-plastic plugs were used, as before, to confine five adult flies to the lower 15 mm of 25 × 95-mm shell vials. Approximately 0.7 g of absorbent rayon and 5 ml of distilled water were added above the plugs, and the open ends of the vials were covered with Parafilm. Experimental vials were organized into blocks; however, all blocks were run concurrently, rather than consecutively as in the heat stress tests. Males and females were tested simultaneously. Vials were examined every 12 h, and the number of dead flies was recorded. Tests were conducted at room temperature (24–25 C).

DESICCATION

Flies were sacrificed using the same methods as those used in the starvation tests, with the following changes: (1) 3 g desiccant was used in place of the absorbent rayon and water, and (2) the observation interval was 2 h.

AGE-SPECIFIC STARVATION AND DESICCATION

Starvation and desiccation tests were conducted repeatedly on one experimental generation. Starvation tests were conducted during the first, second, third, fourth, and sixth weeks of adult life and desiccation tests during the first, second, third, and fifth weeks. Procedures were identical to those outlined above, except that 3 ml rather than 5 ml of water per vial was used to maintain humidity in the starvation tests.

Rearing-tube cohorts were transferred intact to fresh food tubes every 3–5 days. Only one female and one male were used per cohort per test per week. However, cohorts were sampled over several weeks, so tests conducted in different weeks are not strictly independent. During weeks 2–4, temperatures in the laboratory declined ~0.5 C.

ETHANOL VAPOR

Using methods similar to those of the starvation tests, adult flies were exposed to ethanol vapor at room temperature. In place of water, 5 ml of 45%, 30%, or 15% ethanol solution in distilled water was added to the vials. The observation interval was 10–30 min depending on ethanol concentration. Males and females were tested consecutively rather than concurrently. Flies were also subjected to heat and ethanol vapor stresses simultaneously. Procedures were identical to those outlined above for the heat stress tests with distilled water, except that 15% ethanol solution was used. Tests were conducted at 38 C only.

FLY WEIGHTS

Adult flies <1-wk-old from eclosion were sampled from rearing tubes. As usual, only one female and one male were taken from each tube. Flies were frozen individually in microcentrifuge tubes at −55 C. Wet and dry weights of whole flies were determined with a Cahn electronic microbalance, flies being desiccated at 55–60 C for 24 h.

STATISTICAL METHODOLOGY

The comparisons of interest are between flies from populations that exhibit postponed senescence and those from populations that do not. We were not interested in the causes or patterns of within-population variation. Accordingly, we have used the five B population means and five O population means in each experiment as the basic data for analysis. Throughout, *t*-tests with 8 degrees of freedom are used, the null hypotheses being that there are no differences between B and O treatment means. We present two-tailed probabilities for the *t*-statistics, although one-tailed tests could be justified based on our a priori expectation of greater stress resistance in O flies. Differences among populations within treatments did not display consistent patterns. We have chosen, therefore, to present only the overall treatment means and their associated standard errors (each based on five population averages).

RESULTS

HEAT STRESS

Under conditions of uncontrolled humidity, O males are more resistant than B males to heat stress (fig. 1). Treatment means were not significantly different at 36 C ($P > .05$), although the direction of difference was consistent with the results of tests at higher temperatures. Females showed a significant difference ($P < .02$) at 38 C only (fig. 2). However, the direction of difference is consistent over all temperatures, and a combined probability test (Sokal and Rohlf 1981, p. 779) indicates that O population females are significantly more resistant than B females to heat stress under conditions of uncontrolled humidity ($P < .01$). When humidity is maintained at a high level, there is no significant difference between B and O population means at 38 C for either sex (table 1). For females, the direction of difference is opposite to that in tests with uncontrolled humidity. In the presence of desiccant, O males are significantly more resistant than B males to heat stress at 38 C; there is no significant difference for females (table 1).

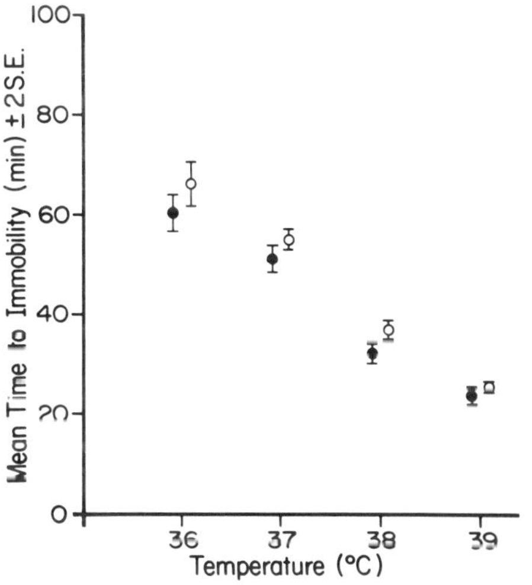

FIG. 1.—Temperature and time to immobility for male flies. Open circles represent O populations; solid circles represent B populations. Data points have been offset in order to separate error bars. Flies were 11–15 days old from eclosion for test at 36 C, 8–14 days old from eclosion for test at 37 C, 6–10 days old from eclosion for test at 38 C, and 3–8 days old from eclosion for test at 36 C. Flies used for tests at 36 C and 38 C were from the same experimental generation, as were flies tested at 37 C and 39 C.

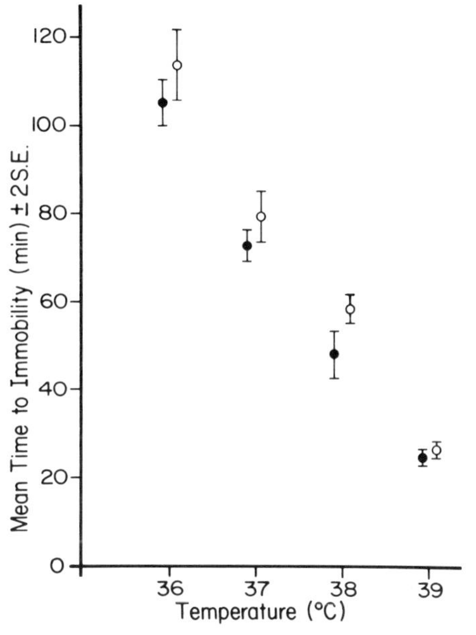

FIG. 2.—Temperature and time to immobility for female flies. Open circles represent O populations; solid circles represent B populations. Data points have been offset in order to separate error bars. Flies were 9–13 days old from eclosion for test at 36 C, 7–13 days old from eclosion for test at 37 C, 2–6 days old from eclosion for test at 38 C, and 2–7 days old from eclosion for test at 39 C. Flies used for tests at 36 C and 38 C were from the same experimental generation, as were flies tested at 37 C and 39 C.

TABLE 1

TIME (min) TO IMMOBILITY FOR FLIES SUBJECTED TO 38 C HEAT STRESS WITH CONTROLLED HUMIDITY

Environment and Results	Males		Females	
	B	O	B	O
A. High humidity:				
Treatment $\bar{X}$	48.8	52.6	56.4	52.8
SE	2.27	1.12	4.32	2.77
t	1.52		.71	
P	N.S.		N.S.	
B. With desiccation:				
Treatment $\bar{X}$	28.9	31.0	44.5	45.1
SE	.40	.45	.83	1.52
t	3.53		.35	
P	<.01		N.S.	

NOTE.—Males were 11–14 days old from eclosion; females were 7–12 days old from eclosion for tests at high humidity and 10–13 days old from eclosion for tests with desiccation. Average sample size for each population was 38. N.S. = not significant.

STARVATION AND DESICCATION

Young adult flies from O populations are significantly more resistant than flies from B populations to starvation (table 2) and desiccation (table 3). The results for starvation were obtained twice, using two different generations of flies.

AGE-SPECIFIC STARVATION AND DESICCATION

The relatively greater resistance of O flies to starvation and desiccation at room temperature (tables 2 and 3) is consistent over all ages examined (figs. 3–6): the differences are statistically significant ($P < .05$) at all ages for both sexes for both stresses, except for females subjected to desiccation at 33 days of adult age. Resistance to desiccation declines with age for both sexes (figs. 5 and 6). For females, there is a striking increase with age in resistance to starvation (fig. 4), especially during the first 2 wk after eclosion.

TABLE 2

SURVIVAL TIME (h) FOR FLIES SUBJECTED TO STARVATION

Group and Results	Males		Females	
	B	O	B	O
A. Experimental generation 6:				
Treatment $\bar{X}$	22.0	27.9	46.9	61.2
SE	.88	1.33	1.56	4.57
t	3.71		2.96	
P	<.01		<.02	
B. Experimental generation 7:				
Treatment $\bar{X}$	23.1	29.8	39.9	55.7
SE	.67	.84	2.87	.86
t	6.21		5.29	
P	<.001		<.001	

NOTE.—Generation 6 flies were 16 days old from eclosion; generation 7 flies were 13 days old from eclosion. Average sample size for each population was 34 in generation 6, and 32 in generation 7.

TABLE 3

SURVIVAL TIME (h) FOR FLIES SUBJECTED TO DESSICATION

	MALES		FEMALES	
RESULTS	B	O	B	O
Treatment $\bar{X}$	7.6	9.6	13.4	15.6
SE	.19	.23	.19	.15
t	6.70		9.08	
P	<.001		<.001	

NOTE.—Flies were 3–4 days old from eclosion. Average sample size for each population was 39.

ETHANOL VAPOR

The results for exposure to ethanol vapor at room temperature are presented in table 4. Differentiation between B and O populations was obtained at 15% ethanol concentration only, with O flies surviving longer. There are no significant differences between the B and O populations when flies are subjected to 38 C in the presence of ethanol vapor (table 5).

WATER CONTENT

Absolute and relative water contents of adult flies are presented in table 6. B females contain slightly more water than do O females, the difference being statistically significant in the case of proportional water content.

DISCUSSION

Flies from populations that exhibit postponed senescence are consistently more resistant to starvation and desiccation than are flies from control populations. These results have been obtained repeatedly in the experiments reported here and, for starvation, in an independent series of

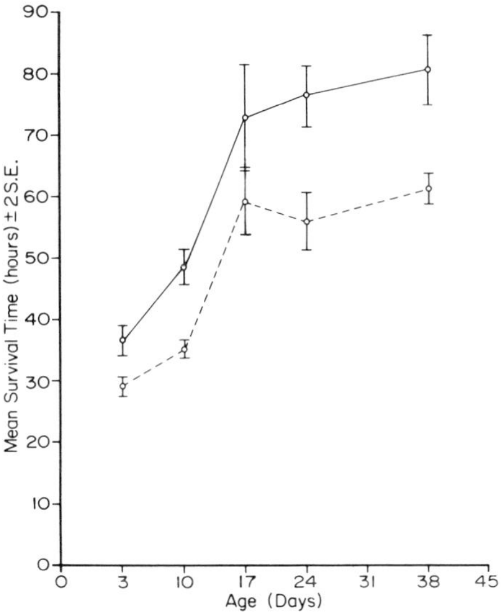

FIG. 4.—Survival time and age since pupal eclosion of female flies subjected to starvation. The solid line represents O populations; the broken line represents B populations.

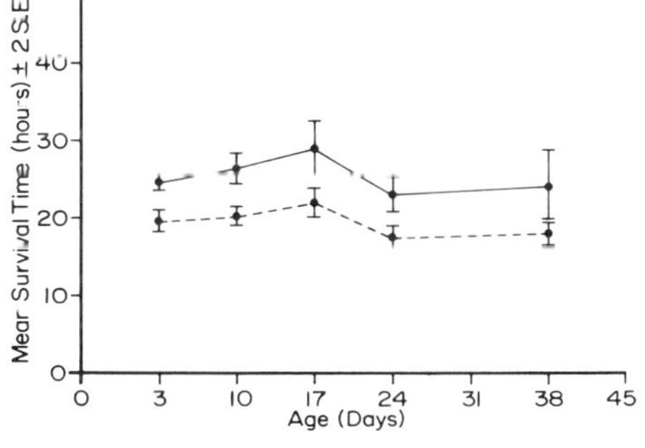

FIG. 3.—Survival time and age since pupal eclosion of male flies subjected to starvation. The solid line represents O populations; the broken line represents B populations.

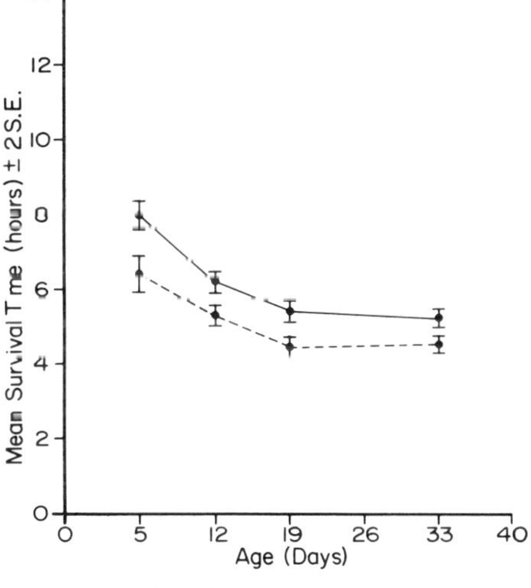

FIG. 5.—Survival time and age since pupal eclosion of male flies subjected to desiccation. The solid line represents O populations; the broken line represents B populations.

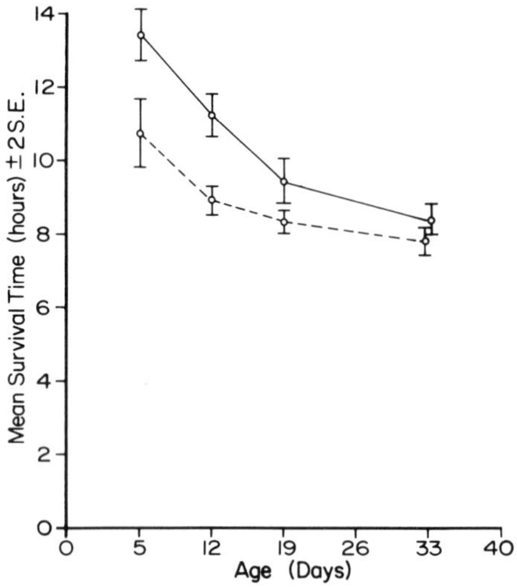

FIG. 6.—Survival time and age since pupal eclosion of female flies subjected to desiccation. The solid line represents O populations; the broken line represents B populations. Data points have been offset at age 33 days from eclosion in order to separate error bars.

experiments (E.W.H. and M.R.R., unpublished data). In addition, long-lived flies survive longer than control flies when subjected to the vapor of a 15% ethanol solution. These results apply to both sexes and are evidently genetically based.

Total ether-extractable lipid content increases with age in female *Drosophila melanogaster* (Fairbanks and Burch 1970), and starved flies are known to deplete lipid reserves (Wigglesworth 1949). It seems likely, therefore, that the increase with age in survival time of starved females (fig. 4) is the result of increased lipid content. This immediately suggests a plausible basis for the difference between B and O flies in resistance to starvation, a point that will be addressed below. In *D. melanogaster,* rate of water loss in dry air increases with age (Fairbanks and Burch 1970), and greater rates of water loss by older flies may account for their decreased survival time under desiccating conditions (figs. 5 and 6).

The results of the heat stress tests in which humidity was controlled (table 1), together with the results of the desiccation experiments (table 3, figs. 5 and 6), suggest that the greater resistance of O flies to heat stress under conditions of uncontrolled humidity (figs. 1 and 2) is due to their greater resistance to desiccation. A similar relationship between desiccation resistance and heat tolerance has been proposed before (Parsons 1970; Lamb and McDonald 1973). Furthermore, there is a reduction in heat tolerance with age (Lamb and McDonald 1973), a trend that mirrors increased rates of water loss in older flies (Fairbanks and Burch 1970).

TABLE 4

TIME TO DEATH (h) FOR FLIES IN ETHANOL VAPOR

	Ethanol Concentration (%)								
	15			30			45		
Group and Results	B		O	B		O	B		O
A. Males:									
Treatment $\bar{X}$	9.7		12.9	3.04		3.02	1.48		1.44
SE	.84		.77	.17		.09	.03		.04
t		2.86			.10			.78	
P		<.05			N.S.			N.S.	
B. Females:									
Treatment $\bar{X}$	13.9		19.3	3.27		3.12	1.39		1.35
SE	1.11		.95	.20		.06	.05		.04
t		3.68			.78			.64	
P		<.01			N.S.			N.S.	

NOTE.—Flies were ~12 days old from eclosion in tests with 15% ethanol, 7 days old from eclosion in tests with 30% ethanol, and 10 days old from eclosion in tests with 45% ethanol. Average sample size for each population was 31. N.S. = not significant.

TABLE 5

TIME (min) TO IMMOBILITY FOR FLIES SUBJECTED TO 38 C HEAT STRESS WITH 15% ETHANOL

	MALES		FEMALES	
RESULTS	B	O	B	O
Treatment $\bar{X}$	26.9	25.7	23.7	21.9
SE	.41	.58	.74	1.12
t	1.71		1.34	
P	N.S.		N.S.	

NOTE.—Males were 9–12 days old from eclosion; females were 6–9 days old from eclosion. Average sample size for each population was 34. N.S. = not significant.

Flies that display postponed senescence survive longer than controls in three separate stress tests: starvation, desiccation, and 15% ethanol vapor. This differentiation may result from either a single physiological mechanism or several such mechanisms. For example, reduced respiration rates or reduced activity levels in O flies could account for greater resistance to starvation (all other things being equal) and for reduced sensitivity to desiccation and ethanol because of lesser requirements for gas exchange.

Alternatively, greater starvation resistance might be due to increased lipid and glycogen storage in fat bodies of O flies. Doane (1960*b*) found that mutant flies with excess lipid had greater resistance to starvation than did wild-type flies and that heterozygous wild-type/mutant flies (but not homozygous mutants) also had greater resistance to desiccation than did wild-type flies. (Interestingly, mutants with excess lipid did not show increased longevity [Doane 1960*a*]. However, the mutation is so severe—resulting in sterility of homozygotes—that the longevity data are difficult to interpret in this context.) Doane (1961) and Butterworth and Bodenstein (1968) have shown that excess lipid accumulation in the fat bodies of some *D. melanogaster* mutants can be attributed to abnormal ovary function. Furthermore, ovaries of young O females (adult age ≤1 wk) have dry weights one-third to one-half those of comparable B females (Rose et al. 1984), reflecting the reduced early fecundity of O females (Rose 1984*c*). These facts, together with known relationships between ovary and fat-body function demonstrated in endocrinological studies (Engelmann 1983; Hagedorn 1983), strongly suggest that selection for postponed senescence has acted on mechanisms that regulate allocation of resources between reproductive and somatic functions and has directly resulted in increased resistance to starvation (if not to desiccation and ethanol vapor). At least two factors complicate this hypothesis. First, O females are more resistant than B females to starvation at ages >1 wk (fig. 2),

TABLE 6

WATER CONTENT OF FLIES

	MALES		FEMALES	
VARIABLE AND RESULTS	B	O	B	O
A. Total water content (mg/fly):				
Treatment $\bar{X}$	.521	.522	.801	.792
SE	.0054	.0087	.0095	.0092
t	.10		.68	
P	N.S.		N.S.	
B. Water as proportion of wet weight:				
Treatment $\bar{X}$	.704	.700	.704	.697
SE	.0019	.0011	.0015	.0019
t	1.87		2.96	
P	N.S.		<.02	

NOTE.—Flies were 3–4 days old from eclosion. Average sample size for each population was 32. N.S. = not significant.

ages at which B females usually do not have significantly greater ovary weight (Rose et al. 1984). Second, O males are also more resistant than B males to starvation. O males have greater mean longevity than B males, but Rose et al. (1984) were unable to detect any morphological differences between B and O males comparable to the ovary weight differences between B and O females. Of course, an equivalent redistribution of somatic and reproductive investments might proceed via changes in male mating speed and frequency.

Parsons (1970) and McKenzie and Parsons (1974) showed that there is genetic variation for resistance to desiccation in wild populations of *D. melanogaster*. Parsons (1970) obtained a positive correlation between resistance to desiccation and wet weight. The greater resistance of O flies to desiccation in the present experiments cannot, however, be explained on the basis of greater absolute or proportional water content (table 6).

Survival of adult *D. melanogaster* is enhanced if flies are subjected to low concentrations of ethanol vapor (Parsons, Stanley, and Spence 1979). The threshold concentration, above which ethanol becomes a stress, has been estimated to be ~12% (Parsons 1980). Parsons and Spence (1981) showed that the threshold concentration was much lower for an *Adh*-null mutant of *D. melanogaster* than for a strain derived from a wild population. It is possible, therefore, that the variation in resistance to ethanol observed in our populations is a function of *Adh* genotype.

We have shown that adult flies selected for postponed senescence (O flies) have relatively greater resistance to several environmental stresses. This is true of young as well as of older adults and conforms to the observation that, in longevity assays, both young and old O adults have lower mortalities than B adults (Rose 1984*c*). It is possible that the increased resistance to environmental stress observed in O flies is merely a coincidental result of the culture methods employed during selection for greater longevity and that selection for longevity in a different environment would not have resulted in similar patterns of stress resistance. However, we have argued above that increased starvation resistance, at least, may be a direct result of evolutionary increase in life span coupled with reduced early fecundity.

Our findings constitute a partial unraveling of the physiological basis of postponed senescence. Much remains to be uncovered, especially the cellular basis of the physiological antagonism between ovary weight and stress resistance, the two factors that appear to underlie antagonistic pleiotropy between reproduction and longevity.

LITERATURE CITED

Ashburner, M., and J. J. Bonner. 1979. The induction of gene activity in *Drosophila* by heat shock. Cell **17**:241–254.

Butterworth, F. M., and D. Bodenstein. 1968. Adipose tissue of *Drosophila melanogaster*. III. The effect of the ovary on cell growth and the storage of lipid and glycogen in the adult tissue. J. Exp. Zool. **167**:207–218.

Doane, W. W. 1960*a*. Developmental physiology of the mutant *female sterile(2)adipose* of *Drosophila melanogaster*. I. Adult morphology, longevity, egg production, and egg lethality. J. Exp. Zool. **145**: 1–21.

———. 1960*b*. Developmental physiology of the mutant *female sterile(2)adipose* of *Drosophila melanogaster*. II. Effects of altered environment and residual genome on its expression. J. Exp. Zool. **145**:23–41.

———. 1961. Developmental physiology of the mutant *female sterile(2)adipose* of *Drosophila melanogaster*. III. Corpus allatum-complex and ovarian transplantations. J. Exp. Zool. **146**:275–298.

Engelmann, F. 1983. Vitellogenesis controlled by juvenile hormone. Pages 259–270 *in* R. G. H. Downer and H. Laufer, eds. Endocrinology of insects. Liss, New York.

Fairbanks, L. D., and G. E. Burch. 1970. Rate of water loss and water and fat content of adult *Drosophila melanogaster* of different ages. J. Insect Physiol. **16**:1429–1436.

Hagedorn, H. H. 1983. The role of ecdysteroids in the adult insect. Pages 271–304 *in* R. G. H. Downer and H. Laufer, eds. Endocrinology of insects. Liss, New York.

Lamb, M. J., and R. P. McDonald. 1973. Heat tolerance changes with age in normal and irradiated *Drosophila melanogaster*. Exp. Gerontol. **8**: 207–217.

Luckinbill, L. S., R. Arking, M. J. Clare, W. C. Cirocco, and S. A. Buck. 1984. Selection for delayed senescence in *Drosophila melanogaster*. Evolution **38**:996–1003.

McKenzie, J. A., and P. A. Parsons. 1974. The genetic architecture of resistance to desiccation in populations of *Drosophila melanogaster* and *D. simulans*. Aust. J. Biol. Sci. **27**:441–456.

MERTZ, D. B. 1975. Senescent decline in flour beetle strains selected for early adult fitness. Physiol. Zool. **48**:1–23.

PARSONS, P. A. 1970. Genetic heterogeneity in natural populations of *Drosophila melanogaster* for ability to withstand desiccation. Theor. Appl. Genet. **40**: 261–266.

———. 1980. Ethanol utilization: threshold differences among six closely related species of *Drosophila*. Aust. J. Zool. **28**:535–541.

PARSONS, P. A., and G. E. SPENCE. 1981. Longevity, resource utilization and larval preferences in *Drosophila*: inter- and intraspecific variation. Aust. J. Zool. **29**:671–678.

PARSONS, P. A., S. M. STANLEY, and G. E. SPENCE. 1979. Environmental ethanol at low concentrations: longevity and development in the sibling species *Drosophila melanogaster* and *D. simulans*. Aust. J. Zool. **27**:747–754.

ROSE, M. R. 1979. Quantitative genetics of adult female life-history in *Drosophila melanogaster*. D.Phil. thesis, University of Sussex, Brighton, England.

———. 1983. Evolution of aging. Rev. Biol. Res. Aging **1**:19–24.

———. 1984*a*. Artificial selection on a fitness component in *Drosophila melanogaster*. Evolution **38**:516–526.

———. 1984*b*. Genetic covariation in *Drosophila* life history: untangling the data. Am. Natur. **123**: 565–569.

———. 1984*c*. Laboratory evolution of postponed senescence in *Drosophila melanogaster*. Evolution **38**:1004–1010.

ROSE, M. R., and B. CHARLESWORTH. 1981. Genetics of life history in *Drosophila melanogaster*. II. Exploratory selection experiments. Genetics **97**: 187–196.

ROSE, M. R., M. L. DOREY, A. M. COYLE, and P. M. SERVICE. 1984. The morphology of postponed senescence in *Drosophila melanogaster*. Can. J. Zool. **62**:1576–1580.

SOKAL, R. R. 1970. Senescence and genetic load: evidence from *Tribolium*. Science **167**:1733–1734.

SOKAL, R. R., and F. J. ROHLF. 1981. Biometry. 2d ed. W. H. Freeman, San Francisco.

WATTIAUX, J. M. 1968*a*. Cumulative parental age effects in *Drosophila subobscura*. Evolution **22**: 406–421.

———. 1968*b*. Parental age effects in *Drosophila pseudoobscura*. Exp. Gerontol. **3**:55–61.

WIGGLESWORTH, V. B. 1949. The utilisation of reserve substances in *Drosophila* during flight. J. Exp. Biol. **26**:150–163.

WILLIAMS, G. C. 1957. Pleiotropy, natural selection, and the evolution of senescence. Evolution **11**: 398–411.

Experimental Gerontology, Vol. 27, pp. 241–250, 1992
0531-5565/92 $5.00 + .00

SELECTION ON STRESS RESISTANCE INCREASES LONGEVITY IN *DROSOPHILA MELANOGASTER*

MICHAEL R. ROSE, LINH N. VU, SUNG U. PARK and JOSEPH L. GRAVES, JR.

Department of Ecology and Evolutionary Biology, School of Biological Sciences, University of California, Irvine, California 92717

Abstract — Tests for the causal involvement of specific physiological mechanisms in the control of aging require evidence that these mechanisms can be used to increase longevity or reproductive lifespan. Selection for later reproduction in *Drosophila* has been shown to lead to increased longevity, as well as increased resistance to starvation and desiccation stresses. Selection for increased resistance to starvation and desiccation in *Drosophila melanogaster* is here shown to lead to increased longevity, indicating that alleles that increase stress resistance also may increase longevity. The responses of desiccation and starvation resistance to selection are partly independent of each other, indicating a multiplicity of physiological mechanisms involved in selectively postponed aging, and thus aging in general.

Key Words: stress, selection, *Drosophila*, longevity

INTRODUCTION

ONE OF the central problems of experimental gerontology is the delineation of the physiological mechanisms that causally determine rates and patterns of aging. There are three obvious methods of approaching this problem: (1) study of the temporal correlates of aging in defined cohorts, (2) comparison of organisms undergoing accelerated aging with normal organisms, and (3) comparison of organisms undergoing postponed or slowed aging with normal organisms. Some decades of research using the first two approaches have not led to much in the way of strong inferences concerning the physiological mechanisms controlling aging (Comfort, 1979; Finch, 1990). There are self-evident reasons for this failure. Temporal correlates of aging need not be causally significant, and the problem of delineating such causality requires other experimental methodologies. Organisms with shortened lifespan could be dying because of novel pathologies unrelated to normal aging. For these reasons, it has been argued that the study of organisms undergoing postponed aging is likely to be the most successful approach (Hutchinson and Rose, 1987; Johnson, 1987; Rose, 1991, chap. 8).

Correspondence to: M.R. Rose.
(*Received* 9 *May* 1991; *Accepted* 1 *August* 1991)

The obvious experimental design for the study of organisms whose aging is genetically postponed is to compare the physiological properties of such organisms with those from control populations (e.g., Rose *et al.*, 1984; Service *et al.*, 1985; Service, 1987; Luckinbill *et al.*, 1988). An analogous strategy can be used with organisms that have undergone physiological postponed aging as a result of an environmental treatment, such as dietary restriction (reviewed by Masoro, 1988; Weindruch and Walford, 1988). But there are several problems with either approach. With environmental manipulations that postpone aging, it remains uncertain whether a specific induced physiological difference *itself* causally postpones aging, or is merely a secondary effect of either the treatment or a mechanism that is causally involved in postponing aging. With genetically differentiated stocks, it is analogously unclear whether the physiological difference is a result of accidentally fixed alleles, or alleles changed in frequency because of linkage disequilibrium with mutant or selectively favored alleles. Either of these genetic difficulties could produce a spurious correlation between postponed aging and physiological differentiation.

To some extent, these problems with genetically postponed aging can be alleviated using replication, large population sizes, or further genetic analysis. When there are only a few stocks with postponed and normal aging, the problem of accidental genetic differentiation during mutagenesis or selection is profound. Differentiation of experimental populations due to genetic sampling effects is one of the best studied phenomena of population genetics (e.g., Wright, 1977). Given enough generations, each population becomes a unique population-genetic entity. For this reason, a relative comparison of two such populations has zero degrees of freedom for statistical inference; it is only when such comparisons are replicated over distinct pairs of populations that there is any statistical power for inference. Thus, a number of studies of genetically postponed aging (e.g., Pretzlaff and Arking, 1989) with one postponed aging population and one control population amount to anecdotal information at the genetic level. In such studies, the populations examined may indeed be differentiated as inferred but repetitions of the experiment involving different populations selected in the same manner may not give consistent results. This arises because of between-population, within-treatment variance, a variance component that such experiments systematically neglect. However, the solution to this problem is straightforward: multiple replication of selected and control populations from the outset of the experimental program (see Luckinbill *et al.*, 1984; Rose, 1984; Service *et al.*, 1985; Hutchinson and Rose, 1991).

Linkage disequilibrium is a somewhat related problem, and replication within and between populations can alleviate it to some extent. The reason for this is simply that the more generations allowed for recombination to breakdown a linkage association, the greater the approach to linkage equilibrium. Similarly, the more generations of selection in a selection experiment, the greater the opportunity for recombination to break down an association. In a genetic experiment where the position of the allele is known, the greater the number of crosses that are performed, the greater the opportunity for separating the relevant allele from others.

In this article, we present a method for forcing a breakdown in artifactual associations between physiological mechanisms and postponed aging. The method is selection upon the correlated physiological character(s) in replicated lines that are paired with control lines. Selection would be applied in the same direction that the character is differentiated in outbred stocks that have genetically postponed aging, with the goal of further postponing aging. The advantages of this procedure are that additional lines can be created, more generations may be generated in which recombination can break down linkage disequilibrium, and the genetic correlation between a given character and aging can be immediately tested. In the case of wholly accidental associations between the test character and postponed aging, this type of test should be definitive. With respect to linkage disequilibrium, this type of test can only quantitatively increase confidence in the mechanistic association between the character and postponed aging. However, any such increase is nonetheless worthwhile. The two characters that are tested in the present report are starvation resistance and desiccation resistance, both of which have been found to be associated with selectively postponed aging in *Drosophila melanogaster* (Service *et al.*, 1985, 1988).

MATERIALS AND METHODS

Stocks

All selection experiments used populations derived originally from a South Amherst, Massachusetts, *Drosophila melanogaster* population studied by P.T. Ives (1970). Specifically, the experimental stocks started from a set of 10 populations derived from this Ives stock in February 1980, with five kept under an early reproduction (at 2 weeks of total age) regime (B) and five kept with later culture reproduction, at 10 weeks of total age, since 1981 (O) (Rose, 1984; Hutchinson and Rose, 1991). The "Os" have greater longevities than the "Bs." (In 1990, mean adult male longevity over five B (about generation 270) populations ± standard error was 39.3 ± 1.0 days; for O (about generation 60) populations (±SE), it was 72.4 ± 1.9 days, $p < 0.01$ [t test].) These B and O stocks also have large differences in age-specific fecundity (Rose, 1984), morphology (Rose *et al.*, 1984), stress resistance (Service *et al.*, 1985; Service, 1987; Service *et al.*, 1988), metabolic rate (Service, 1987), locomotion (Service, 1987; Graves and Rose, 1990), and caloric reserves (Service, 1987; Graves *et al.*, in press.).

Selection procedures

In 1988, two populations were derived from each of the O populations, O_1–O_5. One population of each pair was selected for desiccation resistance by placing the young adults (both sexes) of each generation in sealed cages containing desiccant and no food for enough time to kill most (70–90%) of the adults. Surviving adults were then allowed to feed and reproduce. These selected populations were designated D_x, the subscript giving the O population of origin. The other populations were used as controls and handled like the D_x populations, except that nonnutritive agar rather than desiccant was supplied. This procedure killed very few of these control (C_x) flies. The first six generations of the response to such selection are discussed elsewhere (Rose *et al.*, 1990).

The second experiment, begun in 1989, was analogous to the first in design, except that (1) both sets of B and O populations were used to found pairs of selected and control lines, giving 20 new populations, and (2) selected flies (both sexes) were subjected to starvation rather than desiccation. Control flies had access to food medium throughout life. The control populations are designated CB_x and CO_x according to the population of origin, while the selected are SB_x and SO_x, analogously.

In addition, four desiccation-selected lines were derived from the Ives stocks in 1990 in order to test for effects of stock background on the response to selection for desiccation resistance.

The complete set of stocks of interest are outlined in Table 1.

TABLE 1. A LIST OF ALL THE DIFFERENT TYPES OF LABORATORY *D. MELANOGASTER* POPULATIONS RELEVANT TO THE PRESENT STUDY

Name	*Source*	*Year founded*	*Number*	*Selected for*
IV	Wild	1975	1	2-week net fertility
B	IV	1980	5	2-week net fertility
O	IV	1980	5	10-week net fertility
C	O	1988	5	3-week net fertility
D	O	1988	5	Resistance to desiccation
ID	IV	1990	4	Resistance to desiccation
CB	B	1989	5	4-week net fertility
SB	B	1989	5	Resistance to starvation
CO	O	1989	5	4-week net fertility
SO	O	1989	5	Resistance to starvation

Assay methods

Starvation resistance was assayed in vials of 4–5 flies provided with a humid atmosphere only, following Service *et al.* (1985). Desiccation resistance was assayed using vials containing desiccant (Service *et al.*, 1985). Longevity assays were made using vials of culture medium, three males and three females per vial, with transfers every two to three days. Dead flies were not replaced.

Statistical methods

Data were analyzed using comparisons of population means and their linear regressions on generation number, as in previous studies (e.g., Service *et al.*, 1988). The population mean is the appropriate test statistic in selection studies since quantitative genetics indicates it is the parameter that performs predictably (Falconer, 1981).

RESULTS

The response to selection for starvation resistance is indicated in Figs. 1 and 2. Starvation resistance increased substantially in the SB and SO populations over the course of selection, but not in the controls. Figure 2 shows the indirect response of longevity (with adequate nutrition) to selection for starvation resistance. Both SB and SO populations increased substantially in mean adult longevity, with total increases of about 12 and 17 days, respectively. CO populations did not change significantly in mean longevity, while CB populations did. However, even if the statistical test for significant response of longevity to starvation selection is made relative to these controls, it is still significant in both SO and SB treatments. Figure 3 gives survivorship curves for generation 19 of this experiment, males and females plotted separately.

The cause of the CB longevity increase may be the change in demographic regime imposed during selection. In response to selection, the SB and SO populations evolved greatly increased starvation resistance (Fig. 1), forcing an increase in the generation time

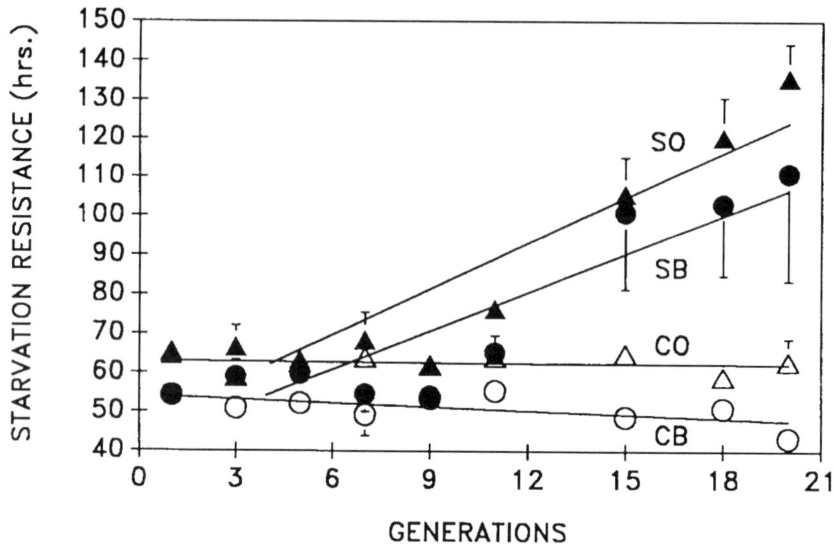

FIG. 1. The direct response to selection for increased starvation resistance. The solid symbols indicate the average of the lines undergoing selection. The hollow symbols indicate the average of lines not undergoing additional selection. The triangles indicate the lines derived from longer-lived populations created by selection in the laboratory. The circles indicate the lines derived from short-lived controls. The error bars indicate 95% confidence intervals for each treatment in each generation; where none is indicated, the confidence interval is narrower than the width of the symbol. Only the SB and SO treatments have average slopes that are significantly different from zero. For SB, the average slope over treatments ($\pm$ SE) is 3.40 ± 0.75. For SO, the average slope is 3.84 ± 0.20. These slopes are not significantly different from each other.

(from 14 days to about 28 days) that must have led to selection for a shift to later reproduction and thus postponed aging (e.g., Rose, 1984) in the CB and SB populations. The O populations had already been selected for later reproduction, so this is not expected to arise with CO and SO treatments.

There was also a significant response of desiccation resistance to selection for starvation resistance. In generation 18 of selection for starvation resistance, the desiccation resistances (mean $\pm$ SE) were as follows: CB—9.1 ± 0.1 h; SB—12.6 ± 0.2 h; CO—11.3 ± 0.4 h; SO—15.1 ± 0.2 h (CB vs. SB, $p < 0.01$, paired t test; CO vs. SO, $p < 0.01$, paired t test). In both cases, the starvation-selected lines have significantly increased desiccation resistance.

Table 2 shows the response to selection for desiccation resistance at generations 19, 30 or 31, and 45. Desiccation resistance increased substantially. Again, male and female longevities increased over the course of selection. These results, together with those of Fig. 2, indicate that longevity can be increased by selection for characters that are positively correlated with longevity in demographic selection experiments, like stress resistance. However, starvation resistance did not significantly increase.

By contrast, in the four desiccation-selected lines derived from the Ives stocks there was a statistically detectable increase in starvation resistance (average net response $\pm$ SE: $+16.5 \pm 2.5$ h, $p < 0.01$).

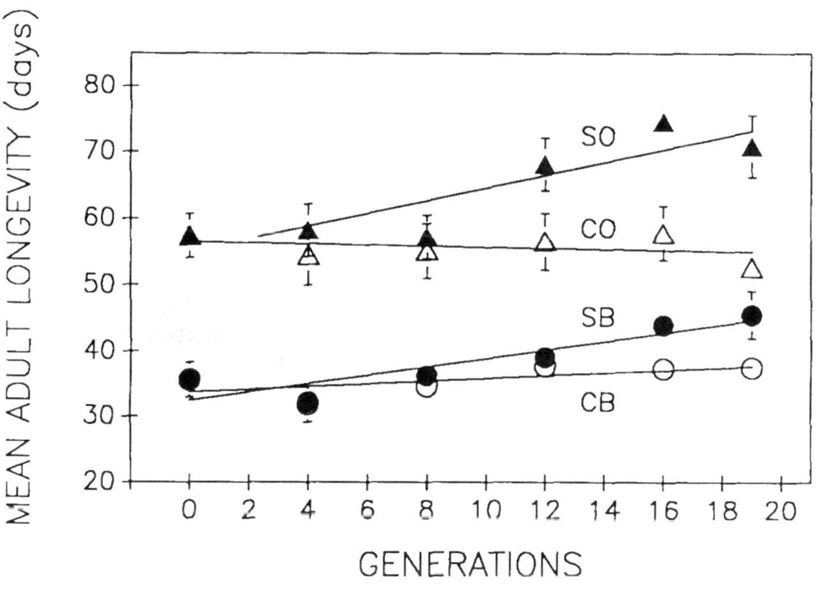

FIG. 2. The indirect response of longevity to selection for increased starvation resistance. These data average the males and females together. Analysis of the sexes separately gives similar results, with one exception noted below. The symbols indicate the same treatments as in Fig. 1. The CB, SB, and SO treatments have average slopes that are significantly different from zero: CB—0.21 ± 0.07; SB—0.66 ± 0.07; SO—0.94 ± 0.03. The SB and SO slopes are significantly greater than the CB and CO slopes, respectively. The SO slopes are significantly greater than the SB slopes for the average of the two sexes and for males, but not for females.

DISCUSSION

The present results show fairly clearly that indirect selection upon longevity-related characters can further increase longevity, and thus postpone aging. A number of additional questions remain. First, the response of longevity was significantly slower in SB males compared with SO males, though not in females. This suggests the possible importance of genetic background. However, it would be unwise to conclude much from this result, because selection differentials were not monitored. Without them, genetic differences cannot be inferred from the pattern of selection response with certainty. On the other hand, the fact that the *direct* response to selection was similar in SB and SO populations while the indirect response was not, suggests that at least the direct response to selection was similar in the two types of line. That leaves only inadvertent indirect selection on the line types as a source of artifactual differences in the indirect response of longevity, other than a difference in initial genetic structure. Note that derivation from the longer-lived O populations did not impose any limitation on the further increase of longevity under selection, relative to the shorter-lived B populations.

Second, are desiccation resistance and starvation resistance genetically correlated to such an extent that the present selection experiments are acting upon the same loci? Table 2 indicates that selection for increased desiccation resistance does not significantly increase starvation resistance, as was found in an earlier study (Rose *et al.*, 1990). How-

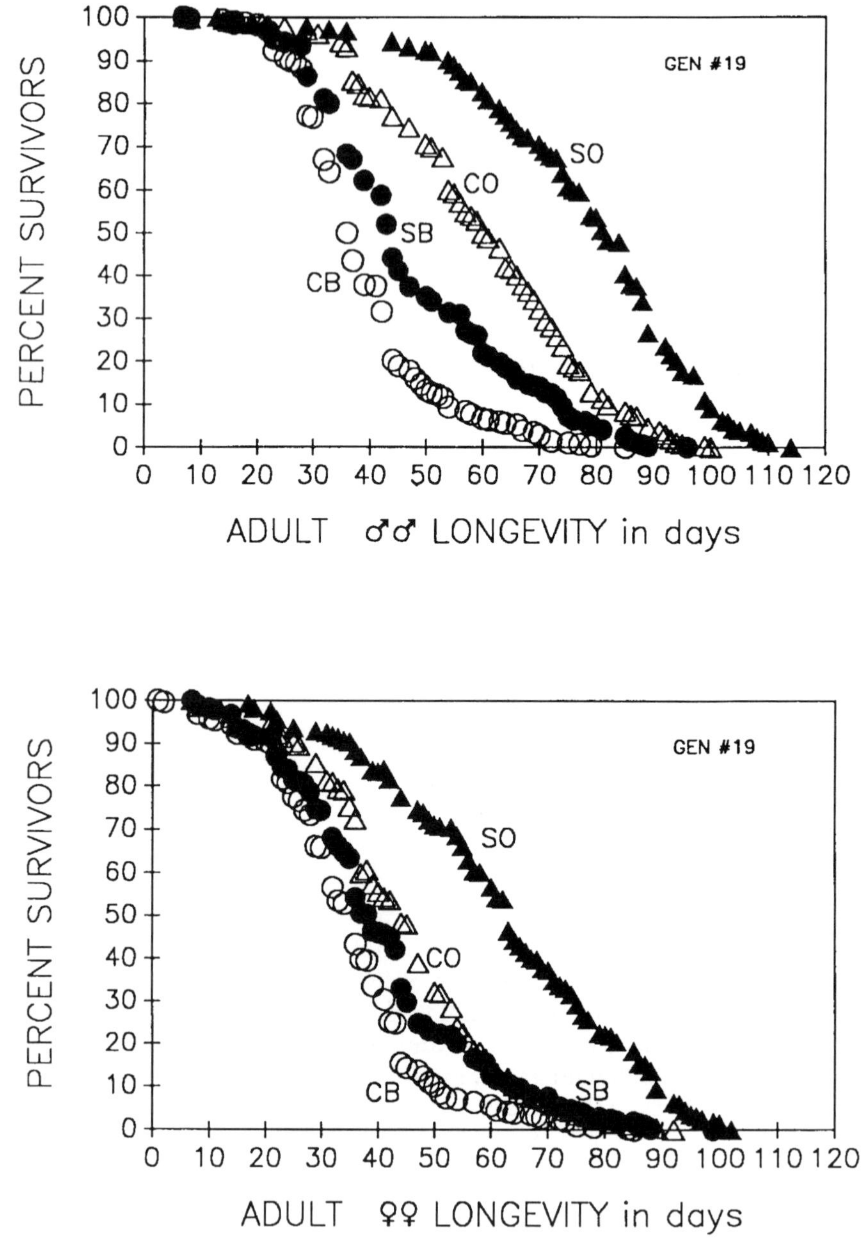

FIG. 3. Survivorship curves for generation 19 of the starvation selection experiment. Top: males; bottom: females. The solid symbols indicate the average of the lines undergoing selection. The hollow symbols indicate the average of the lines not undergoing additional selection. The triangles indicate the lines derived from longer-lived populations created by selection in the laboratory. The circles indicate the lines derived from short-lived controls.

TABLE 2. SELECTION RESPONSE WITH DESICCATION SELECTION. AVERAGE RESPONSE OF D_x STOCKS AND C_x STOCKS

	Generation					
	19		*30/31*		*45*	
Character	*D*	*C*	*D*	*C*	*D*	*C*
Desiccation resistance (h)	30.3**	19.3	29.5*	16.5	34.4*	14.7
Ethanol resistance (h)	53.0**	37.9	67.2*	35.2	77.9*	31.3
Starvation resistance (h)	64.7	60.4	64.9	54.6	69.5	77.9
Male longevity (days)	—	—	54.3**	39.4	61.2*	49.6
Female longevity (days)	—	—	64.7**	56.5	66.6**	60.4

All hypothesis tests are two tailed; all use pairwise comparison comparisons of D and C populations according to their O population of origin. All consistently significant single-generation effects are also significant when analyzed over all generations, but not otherwise (* $p < 0.05$; ** $p < 0.01$). Longevity was not measured in generation 19. Some of the stress resistance characters were measured in generation 30, and the rest in generation 31.

ever, selection for starvation resistance gave a significant increase in desiccation resistance, totalling about 3–4 h. Earlier work on the physiology of starvation resistance indicated that lipid levels were a major factor in the determination of starvation resistance (Service, 1987). In addition, our unpublished studies of glycogen levels in these flies indicate that a major determinant of desiccation resistance is glycogen level (Graves *et al.*, in press). However, glycogen is also an energy source, like lipid. O flies have increased levels of lipid relative to B flies (Service, 1987). If we assume that starvation resistance selection acts to increase both lipid and glycogen levels, with the former quantitatively more important, while desiccation selection primarily acts upon glycogen, then the present results could arise from a proportionally slight effect of increased glycogen levels upon starvation resistance in the D flies, which are derived from high-lipid O flies. The greater effect of starvation selection upon desiccation resistance found in the SO and SB populations could then be interpreted in terms of large increases in both lipid and glycogen level with sustained starvation selection. A critical test of this hypothesis is that the effect of desiccation selection upon starvation resistance should be more detectable when lipid levels are low, as in the B and Ives stocks. This is exactly the result observed when desiccation-selection was imposed on derivatives of the Ives stock.

Third, it should be noted that ethanol resistance, another character associated with increased longevity in previous studies (Service *et al.*, 1985), increases in parallel with desiccation resistance (Table 2), as found in earlier work (Rose *et al.*, 1990). There may be considerable genetic redundancy between ethanol resistance and desiccation resistance.

The present selection results can also be used to calculate roughly the contribution of starvation and desiccation resistance to postponed aging in the original O stocks, providing complete additivity of effects is assumed. (Evidence for average additivity of starvation resistance alleles has been provided elsewhere [Hutchinson and Rose, 1991].) The average total increase in starvation resistance in the SB and SO lines is about 53 h, which was associated with an average total increase in longevity of about 14.5 days. The average total increase in desiccation resistance in the D lines is about 20 h, while the average total increase in longevity is about 9 days in the D lines. If we assume that all of the increase in longevity in the D lines is due to increases in desiccation resistance, then the effect upon longevity is +0.45 days for each hour of increased desiccation resistance. In the original O stocks in 1984, desiccation resistance was increased by about 3 h (Rose, 1984), suggesting that about 1.35 days of the longevity difference of 12 days between O and B stocks was due to increased desiccation resistance, or about 11% of the total longevity increase. The average increase in desiccation resistance in the SB and SO lines was about 3.7 h, which (again assuming additivity) would give rise to an increase of about 1.6 days in longevity. Thus, the residual impact attributable to the physiological effect of increased starvation resistance alone would be about 12.9 days of increased longevity. The corrected proportionality between starvation resistance alone and longevity is about 0.24 days increased longevity for each hour of increased starvation resistance. The original difference between B and O in starvation resistance was about 11 h, which would then produce about 2.7 days of increased longevity, approximately 22% of the total longevity difference. To perhaps an order of magnitude, then, the loci determining starvation resistance and desiccation resistance are responsible for a total of about 33% of the increased longevity of the O populations. This indicates that changes in just two physiological mechanisms can substantially postpone aging. It also indicates that additional mechanisms are involved, mechanisms that might serve as further targets of selection to postpone aging.

Acknowledgments — We are grateful to Q. Dang, K.K. Grimm, J.E. Nelson, and M.C. Nguyen for technical assistance. We also thank F.J. Ayala, M. Djawdan, C.E. Finch, R.E. Lenski, and L.D. Mueller for their comments upon an earlier draft. This research was supported by US-PHS grant AG06346 to MRR.

REFERENCES

COMFORT, A. *The Biology of Senescence.* 3rd ed. Churchill Livingstone, Edinburgh, 1979.

FALCONER, D.S. *Introduction to Quantitative Genetics.* 2nd ed. Longman, London, 1981.

FINCH, C.E. *Longevity, Senescence, and the Genome.* University of Chicago Press, Chicago, IL, 1990.

GRAVES, J.L. and ROSE, M.R. Flight duration in *Drosophila melanogaster* selected for postponed senescence. In: *Genetic Effects on Aging II,* Harrison, D.E. (Editor), pp. 57–63, Telford, Caldwell, NJ, 1990.

GRAVES, J.L., TOOLSON, E.C., JEONG, C., VU, L.N., and ROSE, M.R. Desiccation, flight, glycogen and postponed senescence in *Drosophila melanogaster. Physiol. Zool.,* in press.

HUTCHINSON, E.W. and ROSE, M.R. Genetics of aging in insects. *Rev. Biol. Res. Aging* **3,** 62–70, 1987.

HUTCHINSON, E.W. and ROSE, M.R. Quantitative genetic analysis of postponed aging in *Drosophila melanogaster.* In: *Genetic Effects on Aging II,* Harrison, D.E. (Editor), pp. 66–87, Telford, Caldwell, NJ, 1990.

HUTCHINSON, E.W. and ROSE, M.R. Quantitative genetics of postponed aging in *Drosophila melanogaster.* I. Analysis of outbred populations. *Genetics* **127,** 719–727, 1991.

IVES, P.T. Further studies of the South Amherst population of *Drosophila melanogaster. Evolution* **24,** 507–518, 1970.

JOHNSON, T.E. Aging can be genetically dissected into conponent processes using long-lived lines of *Caenorhabditis elegans. Proc. Natl. Acad. Sci. U.S.A.* **84,** 3777–3781, 1987.

LUCKINBILL, L.S., ARKING, R., CLARE, M.J., CIROCCO, W.C., and BUCK, S.A. Selection for delayed senescence in *Drosophila melanogaster. Evolution* **38,** 996–1003, 1984.

LUCKINBILL, L.S., GRAVES, J.L., TOMKIW, A., and SOWIRKA, O. A qualitative analysis of some life-history correlates of longevity in *Drosophila melanogaster. Evol. Ecol.* **2,** 85–94, 1988.

MASORO, E.J. Food restriction in rodents: An evaluation of its role in the study of aging. *J. Gerontol.* **43,** B59–B64, 1988.

PRETZLAFF, R. and ARKING, R. Patterns of amino acid incorporation in long-lived genetic strains of *Drosophila melanogaster. Exp. Gerontol.* **24,** 67–81, 1989.

ROSE, M.R. Laboratory evolution of postponed senescence in *Drosophila melanogaster. Evolution* **38,** 1004–1010, 1984.

ROSE, M.R. *Evolutionary Biology of Aging.* Oxford University Press, New York, NY, 1991.

ROSE, M.R. and CHARLESWORTH, B. Genetics of life-history in *Drosophila melanogaster.* II. Exploratory selection experiments. *Genetics* **97,** 187–196, 1981.

ROSE, M.R., DOREY, M.L., COYLE, A.M., and SERVICE, P.M. The morphology of postponed senescence in *Drosophila melanogaster. Can. J. Zool.* **62,** 1576–1580, 1984.

ROSE, M.R., GRAVES, J.L., and HUTCHINSON, E.W. The use of selection to probe patterns of pleiotropy in fitness characters. In: *Insect Life Cycles: Genetics, Evolution and Coordination,* Gilbert, F. (Editor), pp. 29–42, Springer-Verlag, London, 1990.

SERVICE, P.M. Physiological mechanisms of increased stress resistance in *Drosophila melanogaster* selected for postponed senescence. *Physiol. Zool.* **60,** 321–326, 1987.

SERVICE, P.M., HUTCHINSON, E.W., MacKINLEY, M.D., and ROSE, M.R. Resistance to environmental stress in *Drosophila melanogaster* selected for postponed senescence. *Physiol. Zool.* **58,** 380–389, 1985.

SERVICE, P.M., HUTCHINSON, E.W., and ROSE, M.R. Multiple genetic mechanisms for the evolution of senescence in *Drosophila melanogaster. Evolution* **42,** 708–716, 1988.

WEINDRUCH, R. and WALFORD, R.L. *The Retardation of Aging and Disease by Dietary Restriction.* Charles C. Thomas, Springfield, IL, 1988.

WRIGHT, S. *Evolution and the Genetics of Populations. Vol. 3, Experimental Results and Evolutionary Deductions.* University of Chicago Press, Chicago, IL, 1977.

584

Metabolic Reserves and Evolved Stress Resistance in *Drosophila melanogaster*

Minou Djawdan*
Adam K. Chippindale†
Michael R. Rose
Timothy J. Bradley‡
Department of Ecology and Evolutionary Biology, University of California, Irvine, Irvine, California 92697-2525

Accepted 5/1/98

ABSTRACT

We have examined starvation and desiccation resistance in 43 outbred populations of *Drosophila melanogaster* that have diverged from a common ancestral population as a result of a variety of defined selection protocols. The populations differ up to 8.5-fold in desiccation resistance and up to 10-fold in starvation resistance. We used these populations to search for evolved physiological changes that might explain the differences in stress resistance. We examined two hypotheses for increased stress resistance that had been proposed previously in the literature: (1) that increments in starvation resistance are principally the result of differential lipid accumulation, and (2) that changes in glycogen accumulation play a role in evolved increases in resistance to desiccation stress. By quantifying desiccation resistance, starvation resistance, lipid content, and carbohydrate content in each of our populations of flies, we were able to demonstrate strong correlations between the capacity of the flies to resist starvation and the quantity of lipid or carbohydrate that the flies had stored. The strongest correlation ($R^2 = 0.99$) was observed when the total energy content of both the lipid and carbohydrate stores was regressed against starvation resistance. These results demonstrate that the flies responded to selection for starvation resistance through a genetically determined increase in both lipid and carbohydrate storage. Similar analyses of the correlation between lipid storage or total energy storage and desiccation resistance revealed no significant correlations. Carbohydrate storage was significantly correlated with desiccation resistance in female but not in male flies. These results suggest that different forms of stress are resisted with distinct physiological mechanisms and that the evolutionary response of the flies to stress selection is specific to the stress imposed.

*Current address: Department of Biology, San Diego City College, 1313 12th Street, San Diego, California 92101.
†Current address: Department of Biology @ Natural Sciences IV, University of California, Santa Cruz, Santa Cruz, California 95064.
‡To whom correspondence should be addressed; E-mail: tbradley@uci.edu.

Physiological Zoology 71(5):584–594. 1998. 0031-935X/98/7105-9745$03.00

Introduction

Stress has been defined by Koehn and Bayne (1989, p. 158) as "any environmental change that acts to reduce the fitness of an organism." There is a considerable literature on stress and stress resistance in the fields of evolutionary biology, ecology, and comparative physiology, because environmental stresses are thought to influence the survival, distribution, fitness, and evolutionary trajectories of organisms (Hoffmann and Parsons 1991). Physiologists and physiological ecologists have often sought to determine specific physiological mechanisms used by organisms to resist stress since these can have substantial adaptive significance.

Clearly, a number of mechanisms can influence survival in specific stressful regimes. Thick fur promotes survival of mammals in cold climates, and chloride cells promote salt secretion across the gills of seawater fish (Prosser 1991; Randall et al. 1997). In a search for more universal mechanisms of stress resistance, however, physiologists and ecologists have sought mechanisms of stress resistance that promote resistance to multiple forms of stress or are shared across taxonomic groups.

Comparative studies, while pointing out broad patterns of correlations, are often confounded by phylogenetic associations among the organisms examined and uncertainty about the precise organismal characteristics under selection when species are compared across habitats (Garland and Adolph 1994; Leroi et al. 1994; Rose and Lauder 1996). Desert organisms, for example, may experience several types of environmentally correlated sources of selection (heat, drought, low primary productivity, etc.).

It is valuable, therefore, to combine comparative studies with other approaches, such as laboratory selection studies. The mechanistic aspects of the short-term evolution of stress resistance can and have been pursued in laboratory selection experiments. Hoffmann and Parsons (1989*a*, 1989*b*) selected for increases in desiccation resistance in derivatives of isofemale lines of *Drosophila melanogaster*. They observed decreases in mass-specific metabolic rate in selected lines compared to controls (Hoffmann and Parsons 1989*a*, 1989*b*). Likewise, we have imposed selection for resistance to starvation as well as desiccation on *D. melanogaster* (Rose et al. 1990, 1992). In our physiological investigations using *D. melanogaster* populations, we found that although mass-specific metabolic rate is reduced in stress-selected populations, this reduction can be accounted for by increases in weight arising from lipid and carbohydrate accumulation (Djawdan et al. 1997). Stores of lipid and carbohydrate do not themselves contribute to metabolic rate. They do contribute to mass and therefore lower mass-specific metabolic rate. The evolutionary response to stress selection in these flies is not a change in the metabolic rate of individual flies but, rather, an increased accumulation of metabolic reserve substances, leading to the reduction in mass-specific metabolic rate observed.

On the basis of these results, Djawdan et al. (1997) suggested that accumulation of metabolic reserves may promote increased resistance to a variety of forms of stress. Increased storage of lipid has been associated with resistance to starvation stress in *D. melanogaster* (Service 1987; Chippindale et al. 1996). Glycogen has been shown to be increased in populations of *D. melanogaster* selected for increased desiccation resistance (Graves et al. 1992).

In the present study, we tested the hypothesis proposed by Service (1987), Graves et al. (1992), and Djawdan et al. (1997) that the accumulation of metabolic reserves, that is, lipid and carbohydrate, play an important role in resisting desiccation and starvation. We used nine selection treatments and 43 populations of *D. melanogaster* for which survival during desiccation and starvation, along with total lipid and carbohydrate content, was measured. We restricted our analyses to two forms of stress (starvation and desiccation) and two metabolic reserve substances (lipid and carbohydrate) because specific models and hypotheses are present in the literature regarding these variables.

Material and Methods

Selection Treatments

All 43 *Drosophila melanogaster* populations employed in this study were descended from the Massachusetts population studied by Ives (1970; the IV population). This population was derived from the wild in 1975 and maintained in the laboratory as an outbred population on discrete generations of 2 wk for about 120 generations before the creation of replicate lines and the imposition of different forms of selection in the laboratory. Since 1981, the standard conditions have been: 24L : 0D light schedule, moderate densities (50–150 organisms per vial), and abundant banana-molasses food. Vials were 2.3 cm in diameter and 9.5 cm tall. In all of the selection treatments discussed here, except for ACO, a standard 2 wk period from oviposition has been allowed for development of the flies before selection. The ACO treatment differed because of direct and intentional selection for rapid development (Table 1).

B and O. In February 1980, 10 populations were created from the IV stock. Five were selected for early reproduction (B_{1-5}), and five were selected for late-life reproduction (O_{1-5}; Rose 1984). The five early-reproduction populations were maintained in the same manner as the ancestral IV population. The five late-reproduction populations were selected for late-life fertility by steadily increasing the age of reproduction to 10 wk from oviposition. Since 1982, the O's have been maintained on a 10-wk (egg to egg) selection regime.

C and D. In 1988, five lines were selected for increased resistance to desiccation (D_{1-5}) and five control lines (C_{1-5}) were

Table 1: List of populations of *Drosophila melanogaster* used in the experiments, their ancestral population, selection regime, generation length, and number of generation under selection

Population	Ancestral Population	Selection Treatment	Generation Length (Egg to Egg)	Number of Generations of Selection
O	Ives	Postponed senescence	10 wk	68
B	Ives	Control for O populations	2 wk	346
D	O	Desiccation resistance	Approximately 3 wk	130
C	O	Control for D populations	Approximately 3 wk	130
SO	O	Starvation resistance	Approximately 4 wk	76
CO	O	Control for SO populations	Approximately 4 wk	76
DS	D	Starvation resistance	Approximately 4 wk	78
CS	C	Starvation resistance	Approximately 4 wk	78
ACO	CO	Accelerated development	8.5 d	95

Table 2: Survival during desiccation, survival during starvation, and total lipid and carbohydrate content in females

Selection Treatment	Survival during Desiccation (h)	Survival during Starvation (h)	Total Lipid Content (mg/fly)	Total Carbohydrate Content (mg/fly)	Total Energy Content (J/fly)
ACO	7.79 ± .82	18.52 ± 2.22	.029 ± .003	.027 ± .002	2.05 ± .226
B	8.57 ± .23	17.80 ± .96	.031 ± .002	.027 ± .003	2.14 ± .126
O	12.25 ± .78	31.34 ± 2.07	.049 ± .002	.036 ± .003	3.17 ± .174
C	15.70 ± 1.01	67.38 ± 4.25	.113 ± .005	.044 ± .004	5.99 ± .281
D	66.62 ± .96	84.33 ± 2.99	.083 ± .008	.118 ± .020	7.27 ± .442
CO	11.41 ± .73	26.07 ± 1.04	.044 ± .005	.038 ± .002	3.03 ± .215
SO	17.48 ± .96	158.78 ± 15.67	.232 ± .016	.087 ± .010	12.12 ± .970
CS	21.10 ± 1.04	183.16 ± 9.99	.215 ± .044	.133 ± .026	13.02 ± 2.593
DS	28.94 ± .68	175.47 ± 8.48	.212 ± .01	.121 ± .017	12.47 ± 1.774

Note. Values presented are the means ± standard errors of the means for the five replicates within each selection treatment.

initiated from the O treatment as described in Rose et al. (1990, 1992). Two weeks after egg collection, D populations were transferred from culture vials to sealed cages containing a cloth bag of Drierite (150–175 g). Flies were desiccated until the initial population size of 5,000–10,000 flies was reduced by 80%–90% because of mortality. Surviving flies were almost all females. Since desiccation selection consisted of at least two direct stresses, food and water deprivation, the proper control for the effects of desiccation is flies that are watered but not fed. The C treatment was such a control, maintained in exact parallel to the D-selected populations and given nonnutritional agar throughout the D selection period. The C treatment thus constitutes a control for both the demographic and starvation components of desiccation selection. C populations experienced weak selection for starvation tolerance relative to other starved treatments. At the time of these experiments (fall 1994), it took approximately 4 d to reduce the population sizes to the target value by desiccation. Selection was followed by 3 d of recovery time with fresh food and yeast paste. The total generation time of the D and C populations was therefore about 3 wk.

CO and SO. In 1989, two sets of starvation-selected populations and controls were derived from each of the B and O populations, as described in Rose et al. (1992). Here only the O-derived populations were used. The five replicate starvation-selected populations (SO_{1-5}) and five control populations (CO_{1-5}) were handled as follows: after the 2-wk development period, populations subjected to starvation selection were starved in cages with access to water from pure agar plates until the initial population of 7,000–15,000 flies was reduced to about 1,000 flies, almost all females. Controls (CO's) were maintained in parallel to the selected populations but were given food throughout the selection period.

CS and DS. In 1989, an experiment was initiated to investigate the effects of selection history on the rate of subsequent response to selection. The D and C lines, described above, having

Table 3: Survival during desiccation, survival during starvation, and total lipid and carbohydrate content in males

Selection Treatment	Survival during Desiccation (h)	Survival during Starvation (h)	Total Lipid Content (mg/fly)	Total Carbohydrate Content (mg/fly)	Total Energy Content (J/fly)
ACO	5.35 ± .03	17.03 ± 1.95	.022 ± .003	.020 ± .005	1.56 ± .111
B	5.44 ± .19	15.48 ± .37	.027 ± .002	.015 ± .001	1.60 ± .080
O	8.64 ± .55	38.76 ± 1.58	.047 ± .002	.027 ± .003	2.76 ± .134
C	9.44 ± .25	78.99 ± 1.93	.096 ± .006	.030 ± .003	4.84 ± .296
D	38.48 ± 8.55	67.79 ± 8.11	.083 ± .008	.051 ± .011	5.02 ± .564
CO	7.45 ± .34	33.53 ± 1.47	.096 ± .006	.030 ± .003	2.90 ± .170
SO	10.55 ± 1.21	146.03 ± 17.27	.151 ± .017	.055 ± .007	7.84 ± .90
CS	13.58 ± 1.26	184.68 ± 6.90	.161 ± .044	.090 ± .026	9.43 ± 2.60
DS	19.65 ± 2.21	156.77 ± 4.85	.140 ± .036	.084 ± .021	8.36 ± 2.12

Note. Values presented are the means ± standard errors of the means for the five replicates within each selection treatment.

been selected for different stresses for 21 generations, were used to found two starvation-selected treatments: DS_{1-4} and CS_{1-4}. Both of these selection treatments were then placed in selection for starvation resistance. Populations were not paired by replicate number in their handling, and no unstressed controls were maintained. The rate of response of each treatment to selection of equivalent intensity was the important feature of this experiment.

ACO. In 1991, selection on development time and extremely early fertility was initiated on derivatives of the CO populations (see above and Chippindale [1994]). The ACO_{1-5} populations were selected by transferring only the first 15% of emerging adults to population cages for reproduction. The adults were then allowed the shortest possible interval for egg laying.

Preexperiment Standardization

Before experimental assays were performed, samples of 1,500–2,000 flies of all populations from the various selection treatments were kept for two full generations under identical conditions. A 3-wk culture regimen was used to synchronize selection treatments and to remove residual nongenetic (i.e., parental and grand-parental) effects. The experimental generation was then derived from eggs collected over a 2- or 3-h period from young adults that had been given yeast-supplemented food plates for 2 d before collection.

Eggs for the experimental generation were collected at a density of exactly 60 eggs per vial, with 5 mL (±0.5 mL) of medium for growth. Fifteen vials, or 900 eggs, were prepared for each of the 43 experimental populations, for a total of 34,200 eggs. These vials were incubated in an interspersed fashion and rotated daily in the incubator at 25°C. Two weeks were allowed for preadult development and adult maturation before flies were sampled for assay. Replicate populations of the same number were handled in parallel at all stages. On day 14 from egg collection, the adults were assigned at random to starvation- or desiccation-resistance vial assays or to lipid or carbohydrate assays.

Stress Resistance

Starvation and desiccation resistance assays were conducted following the general methods of Service et al. (1985). In both starvation- and desiccation-resistance assays, four flies of the same sex were sealed into the base of cylindrical glass vials (2.3 cm in diameter and 9.5 cm tall) and capped with a foam stopper approximately 5 mm thick. About 1 cm^3 of space was allowed for the flies to move around in the closed end of the vial. For the starvation-resistance assay, rayon balls were added to the open end of the vial. This absorbent material was soaked with distilled water to humidify the vial. For the desiccation-resistance assay, Drierite was added to the vial. In each type of assay, the open end of the vial was sealed with Parafilm. The frequency of mortality checks varied by assay type and treatment group. In the starvation-resistance assays, checks were made every 6 h (0100, 0700, 1300, and 1900 hours) on nonstarvation-selected treatments (B, O, CO, ACO) and weakly starvation-selected treatments (C, D). Strongly starvation-selected populations will survive for up to 20 d without food, so checks on mortality were accordingly scaled at 12-h intervals. In the desiccation-resistance assay, checks of nondesiccation-selected treatments (every treatment except the D's) were made every hour around the clock, until all flies were dead; the D's were checked every 6 h. Forty flies of each sex were assayed for each population for each of the two types of stress, for a total of 6,080 flies.

Lipid Measurements

Lipid measurements were carried out by separating genders and drying the flies at 60°C for 45 min. For each population, six groups of 10 of each gender were dried and weighed on a Cahn electrobalance (Cerritos, Calif.). Each group of 10 flies was placed in a 10 × 50 mm Whatman thimble. Thimbles

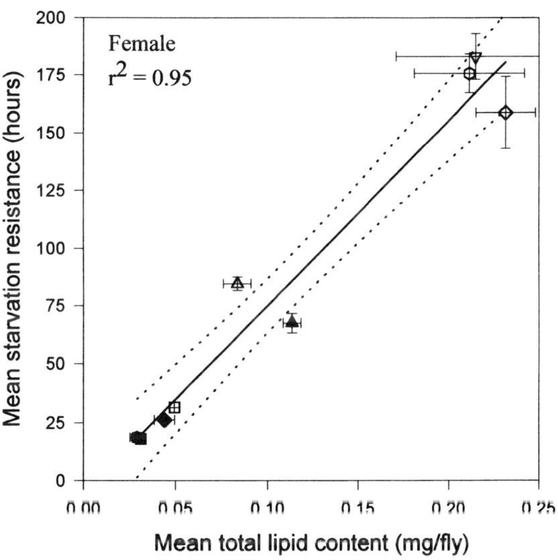

Figure 1. The relationship between mean (±SE) starvation resistance and mean (±SE) total lipid content in females of *Drosophila melanogaster* from nine distinct selection regimes: B populations (■), O populations (□), C populations (▲), D populations (△), CO populations (♦), SO populations (◇), CS populations (▽), DS populations (○), ACO populations (●). Each point represents the mean value from five replicate populations within a selection regime, except for CS and DS, where four replicate populations were available. A least squares regression is fitted to the values as a solid line. The 95% confidence intervals around each regression are shown as dotted lines.

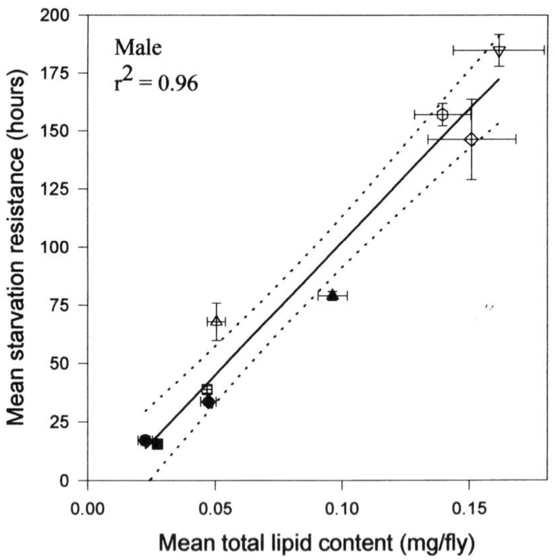

Figure 2. The relationship between mean (±SE) starvation resistance and mean (±SE) total lipid content in males of *Drosophila melanogaster* from nine distinct selection regimes: B populations (■), O populations (□), C populations (▲), D populations (△), CO populations (♦), SO populations (◇), CS populations (▽), DS populations (○), ACO populations (●). Each point represents the mean value from five replicate populations within a selection regime, except for CS and DS, where four replicate populations were available. A least squares regression is fitted to the values as a solid line. The 95% confidence intervals around each regression are shown as dotted lines.

were then stapled shut and placed in the extractor of a Soxhlet apparatus with petroleum ether as the solvent. Petroleum ether was used as the extraction solvent because the extract contains fewer nonlipids than other solvents (Dobush et al. 1985). Flies were extracted for at least 20 h, at which point they were removed from the extractor and thimble and placed in an oven to dry at 60°C for 30 min. Flies were then weighed a second time on the Cahn electrobalance. The difference between the initial dry weight and the postextraction dry weight was considered the total lipid content of flies. Males and females of a single population were analyzed separately.

Carbohydrate Measurements

In order to measure the carbohydrate content of flies, the colorimetric method of carbohydrate extraction of Van Handel (1965) was modified as described below. Flies were separated by gender and dried in an oven for 45 min at 60°C. For each population, five groups of five females each were weighed and placed in 1.7-mL microcentrifuge tubes. Five hundred microliters of water was added to each tube, and flies were ground using a handheld battery-operated grinder. Tubes were placed in boiling water for 5 min. One hundred microliters from each tube was transferred to a 13 × 100 mm test tube. Three milliliters of anthrone reagent (150 mg anthrone per 100 mL of 72% sulfuric acid) was then added to each tube, and tubes were incubated in a water bath at 90°C. Absorbance was measured using a Perkin Elmer Lambda 3B (Norwalk, Conn.) spectrophotometer at a wavelength of 620 nm. For each assay, two duplicate standard curves were prepared as well as three samples of a known amount of glycogen. These were assayed in parallel with the flies. Although the standard curves did not differ significantly, the standard curve that best predicted the known concentration was used for calculating the carbohydrate content of the flies.

Total Energy Content

Total energy content was obtained by converting the amount of lipid and carbohydrate per fly to joules using values of 17.6 J/mg of carbohydrate and 39.3 J/mg of lipid. The energy content of the carbohydrate and lipid stores were added and averaged to obtain the means for each selection treatment.

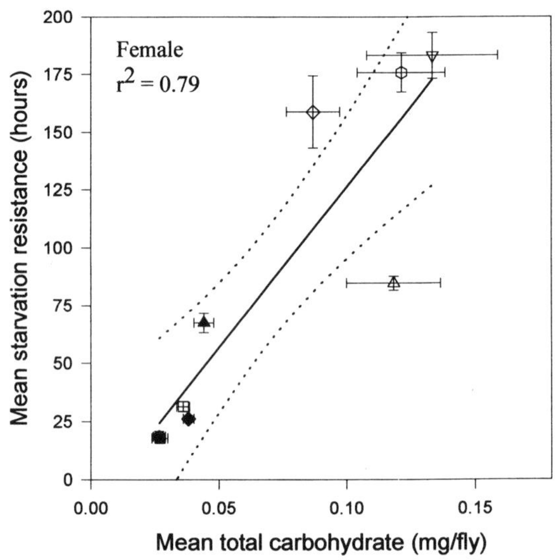

Figure 3. The relationship between mean (±SE) starvation resistance and mean (±SE) total carbohydrate content in females of *Drosophila melanogaster* from nine distinct selection regimes: B populations (■), O populations (□), C populations (▲), D populations (△), CO populations (♦), SO populations (◇), CS populations (▽), DS populations (○), ACO populations (●). Each point represents the mean value from five replicate populations within a selection regime, except for CS and DS, where four replicate populations were available. A least squares regression is fitted to the values as a solid line. The 95% confidence intervals around each regression are shown as dotted lines.

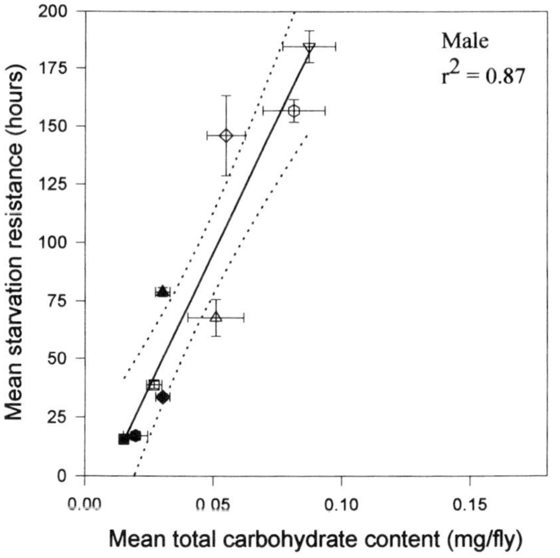

Figure 4. The relationship between mean (±SE) starvation resistance and mean (±SE) total carbohydrate content in males of *Drosophila melanogaster* from nine distinct selection regimes: B populations (■), O populations (□), C populations (▲), D populations (△), CO populations (◆), SO populations (◇), CS populations (▽), DS populations (O), ACO populations (⬢). Each point represents the mean value from five replicate populations within a selection regime, except for CS and DS, where four replicate populations were available. A least squares regression is fitted to the values as a solid line. The 95% confidence intervals around each regression are shown as dotted lines.

Statistical Analysis

A model II regression analysis was performed, using the statistical software package SAS, to test the significance of the relationship between metabolic resources and stress resistance. Genders were analyzed separately. Carbohydrate content and desiccation resistance of female flies in the D, DS, CS, and SO populations were examined for statistical similarity using Scheffé's multiple range test on SAS. A Wilcoxon paired-sample test (Sokal and Rohlf 1969) was used to test the significance of differences in the patterns of lipid and carbohydrate changes in starvation- and desiccation-selected lines.

Results

Tables 2 and 3 present the means and standard error of desiccation and starvation survival in hours, carbohydrate and lipid storage in milligrams, and energy storage in joules for females and males of each selection treatment. The relationship between starvation resistance and total lipid content for males and females is presented in Figures 1 and 2. For both males and females, the relationship between total lipid content and survival during starvation stress was positive, with correlation coefficients of 0.95 and 0.96, respectively. The regression between total lipid content and survival during starvation was significant for both females ($y = 801.16x - 4.92$, $P < 0.0001$) and males ($y = 1{,}139.51x - 11.72$, $P < 0.001$). Figures 3 and 4 present the relationship between starvation resistance and total carbohydrate content per fly. Starvation resistance increased with carbohydrate content for both females and males, with a correlation coefficient of 0.79 for females and 0.88 for males. The regression between total carbohydrate content and survival during starvation was significant for both females ($y = 1{,}391.43x - 12.95$, $P < 0.0014$) and males ($y = 2{,}226.63x - 17.57$, $P < 0.0002$). The relationship between total energy content per fly and starvation resistance is presented in Figures 5 and 6 for females and males, respectively. The relationship between total energy content and starvation resistance was positive and had a correlation coefficient of 0.99 for females and 0.97 for males. The regression for both females ($y = 15x - 17.36$, $P < 0.001$) and males ($y = 21.41x - 23.28$, $P < 0.001$) was highly significant.

Figures 7–12 present the relationship between desiccation

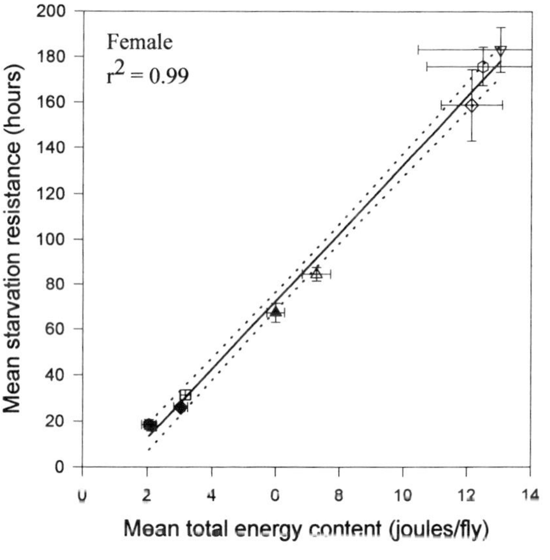

Figure 5. The relationship between mean (±SE) starvation resistance and mean (±SE) total energy content in females of *Drosophila melanogaster* from nine distinct selection regimes: B populations (■), O populations (□), C populations (▲), D populations (△), CO populations (◆), SO populations (◇), CS populations (▽), DS populations (O), ACO populations (⬢). Each point represents the mean value from five replicate populations within a selection regime, except for CS and DS, where four replicate populations were available. A least squares regression is fitted to the values as a solid line. The 95% confidence intervals around each regression are shown as dotted lines.

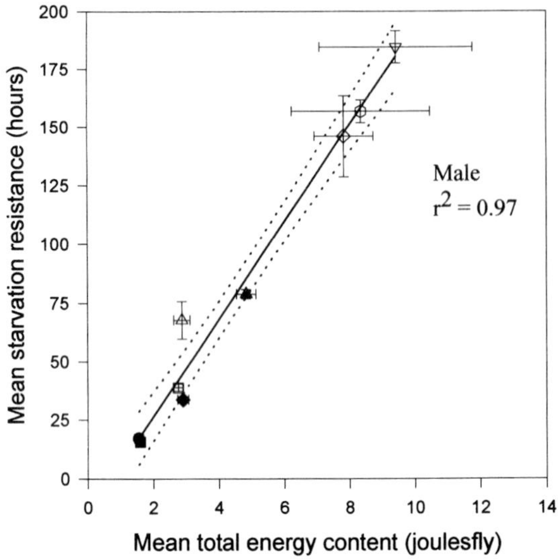

Figure 6. The relationship between mean (±SE) starvation resistance and mean (±SE) total energy content in males of *Drosophila melanogaster* from nine distinct selection regimes: B populations (■), O populations (□), C populations (▲), D populations (△), CO populations (♦), SO populations (◇), CS populations (▽), DS populations (○), ACO populations (●). Each point represents the mean value from five replicate populations within a selection regime, except for CS and DS, where four replicate populations were available. A least squares regression is fitted to the values as a solid line. The 95% confidence intervals around each regression are shown as dotted lines.

resistance and lipid content, carbohydrate content, and total energy content, respectively, for females and males. The regression between desiccation resistance and lipid content was not significant for either females ($P = 0.62$) or males ($P = 0.73$). The regression between desiccation resistance and total energy content was also not significant for either females ($P = 0.35$) or males ($P = 0.33$). The regression between carbohydrate content and desiccation resistance was significant in females ($y = 278.34x + 1.55$, $P < 0.046$), with a correlation coefficient of 0.45, but not in males ($P = 0.22$).

Discussion

The Evolution of Starvation Resistance

We have examined starvation resistance in a large number of outbred populations of *Drosophila melanogaster* that have diverged from a common ancestral population (the IV's) as a result of specific and highly defined selection protocols. These populations differ more than 10-fold in starvation resistance, providing a unique opportunity for exploring differences in physiological performance in a single species.

In examining the relationship between metabolic stores and starvation, our results are in agreement with those of Service (1987) and Chippindale et al. (1996), indicating that lipids play an important role in resistance to starvation. Regressions examining the relationship between lipid content and starvation resistance for all the lines examined (Figs. 1, 2) demonstrate an excellent fit. These results confirm the importance of lipid stores in resisting starvation in these organisms. We also observed a strong and positive correlation between carbohydrate stores and starvation resistance (Figs. 3, 4). When we calculated the total amount of energy stored in both lipids and carbohydrates (Figs. 5, 6), we found that this value provides the best predictor of starvation resistance ($r^2 = 0.99$ for females and 0.97 for males). The observation that carbohydrate as well as lipid stores contribute to starvation resistance expands our previous understanding of the physiological mechanisms for starvation resistance presented by Service (1987) and Chippindale et al. (1996).

The novel physiological findings from this study are not that lipid and carbohydrate contributed to starvation resistance; that has long been known. Instead, the results demonstrate

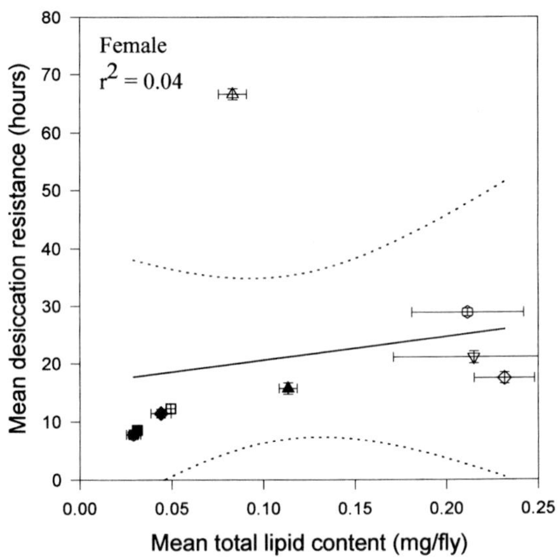

Figure 7. The relationship between mean (±SE) desiccation resistance and mean (±SE) total lipid content in females of *Drosophila melanogaster* from nine distinct selection regimes: B populations (■), O populations (□), C populations (▲), D populations (△), CO populations (♦), SO populations (◇), CS populations (▽), DS populations (○), ACO populations (●). Each point represents the mean value from five replicate populations within a selection regime, except for CS and DS, where four replicate populations were available. A least squares regression is fitted to the values as a solid line. The 95% confidence intervals around each regression are shown as dotted lines.

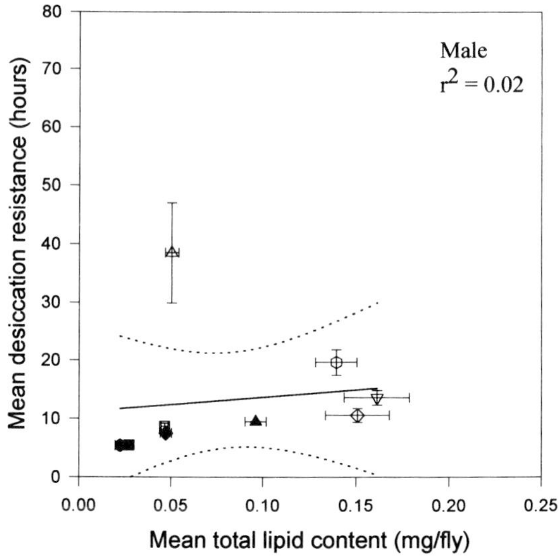

Figure 8. The relationship between mean (±SE) desiccation resistance and mean (±SE) total lipid content in males of *Drosophila melanogaster* from nine distinct selection regimes: B populations (■), O populations (□), C populations (▲), D populations (△), CO populations (♦), SO populations (◇), CS populations (▽), DS populations (○), ACO populations (●). Each point represents the mean value from five replicate populations within a selection regime, except for CS and DS, where four replicate populations were available. A least squares regression is fitted to the values as a solid line. The 95% confidence intervals around each regression are shown as dotted lines.

that populations subjected to selection for starvation resistance responded with specific changes in the storage of both carbohydrate and lipid, and these changes explain 99% of the evolved changes in stress resistance. Because these results were obtained in flies from different populations reared under identical conditions, it is clear that the differences are genetic in origin.

The Evolution of Desiccation Resistance

Using the B, O, C, and D lines, Graves et al. (1992) found a correlation between levels of carbohydrate storage and the desiccation resistance of these populations. By examining additional lines, we show here that the pattern of response to stress selection was much more complicated. Neither lipid nor total energy was significantly correlated with desiccation resistance (Figs. 7–12) over the full spectrum of populations we examined. Carbohydrate levels were positively correlated with desiccation resistance in females but not males. Female DS, CS, and SO flies did not differ from the D females with regard to the level of carbohydrate storage ($P > 0.05$; Scheffé's multiple range test), but they were significantly less resistant to desiccation stress ($P < 0.05$; Scheffé's multiple range test). It would seem that the level of carbohydrate storage was a covariate of desiccation resistance but was not sufficient to explain the extreme desiccation resistance observed in the D populations. Instead, water storage and physiological mechanisms of reducing water loss to the atmosphere appeared to play a major role in the evolution of desiccation resistance in the D flies. Gibbs et al. (1997) have shown in separate studies that flies in the D populations have increased stores of free water and reduced water loss rates compared with their controls. Williams et al. (1997) have shown that the D flies show genetically based differences in respiratory pattern that may result in reductions in rates of respiratory water loss.

We have shown that evolved differences in starvation resistance can be fully explained by observed differentiation in the levels of metabolic storage compounds. The evolution of desiccation resistance is physiologically more complex. We conclude that organisms resist stress by mechanisms specific to that stress. These mechanisms may be genetically linked, but well-developed examples of a general stress-resistance mechanism remain to be found.

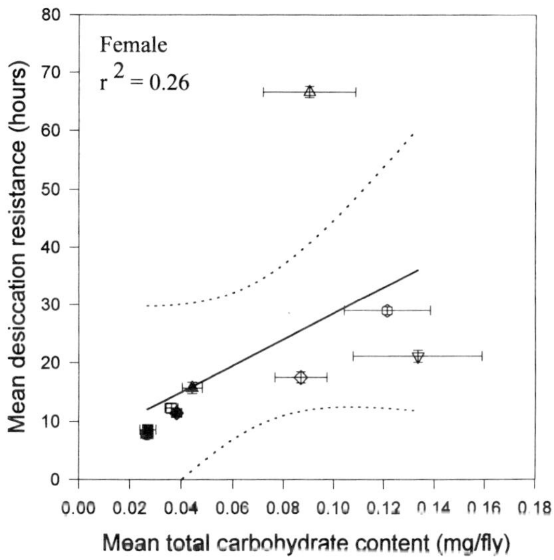

Figure 9. The relationship between mean (±SE) desiccation resistance and mean (±SE) total carbohydrate content in females of *Drosophila melanogaster* from nine distinct selection regimes: B populations (■), O populations (□), C populations (▲), D populations (△), CO populations (♦), SO populations (◇), CS populations (▽), DS populations (○), ACO populations (●). Each point represents the mean value from five replicate populations within a selection regime, except for CS and DS, where four replicate populations were available. A least squares regression is fitted to the values as a solid line. The 95% confidence intervals around each regression are shown as dotted lines.

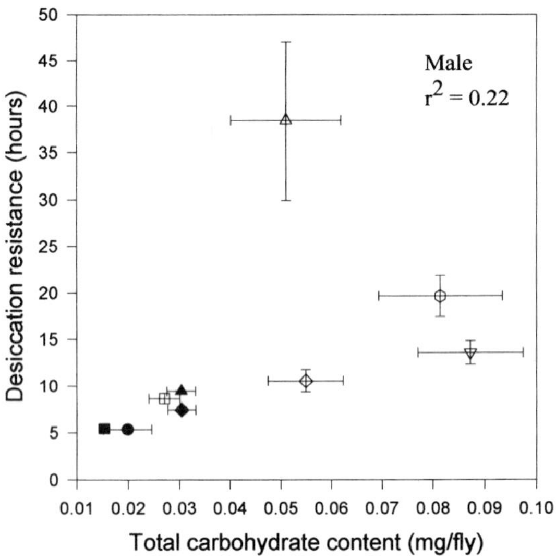

Figure 10. The relationship between mean (±SE) desiccation resistance and mean (±SE) total carbohydrate content in males of *Drosophila melanogaster* from nine distinct selection regimes: B populations (■), O populations (□), C populations (▲), D populations (△), CO populations (♦), SO populations (◇), CS populations (▽), DS populations (○), ACO populations (●). Each point represents the mean value from five replicate populations within a selection regime, except for CS and DS, where four replicate populations were available. A least squares regression is fitted to the values as a solid line. The 95% confidence intervals around each regression are shown as dotted lines.

The Search for a General Mechanism of Stress Resistance

Selection studies previously conducted on *D. melanogaster* demonstrate that flies subjected to selection for resistance to one form of stress often have improved resistance both to the stress under selection and to a variety of other forms of stress. Service et al. (1985), for example, demonstrated that flies selected for reproduction late in life show postponed senescence, as well as resistance to starvation, desiccation, and ethanol vapors. Similarly, Hoffmann and Parsons (1989*a*, 1989*b*) observed that flies selected for resistance to desiccation show increased tolerance of starvation, heat, ethanol vapors, acetic acid vapors, and gamma radiation. Rose et al. (1990) showed that selection for desiccation resistance increases resistance to ambient alcohol. On the basis of these results, researchers have suggested the existence of one or more general mechanisms of stress resistance that can protect against multiple forms of stress. For example, Hoffmann and Parsons (1991) proposed that a reduction in metabolic rate might serve as a general mechanism of resistance to a variety of stresses. Djawdan et al. (1997) found that the flies did not have a reduced overall metabolic rate, but merely a reduced metabolic rate expressed per milligram. The flies in stress-selected lines have increased metabolic reserves in the form of lipid or carbohydrate and thus have a greater body weight than their controls. Djawdan et al. (1997) suggested that the metabolic reserves observed in stress-selected lines play a role in resisting a variety of stresses. This hypothesis has been confirmed in the present study with regard to starvation resistance but not desiccation resistance.

Why Do the D Flies Have Elevated Levels of Stored Carbohydrate?

Our results indicate a strong correlation between the starvation resistance of a population of flies and the total energy stored by those flies. The form in which the energy is stored can differ substantially between populations, however. If carbohydrate content is not correlated with desiccation resistance over all the lines, why do the D populations accumulate glycogen while the C populations accumulate lipid? Unlike lipid accumulation, carbohydrate storage is compatible with abundant storage of free water. In fact, the amount of water associated with glycogen in the form of water of hydration exceeds the weight of the glycogen by three- to fivefold (Schmidt-Nielsen 1990). The

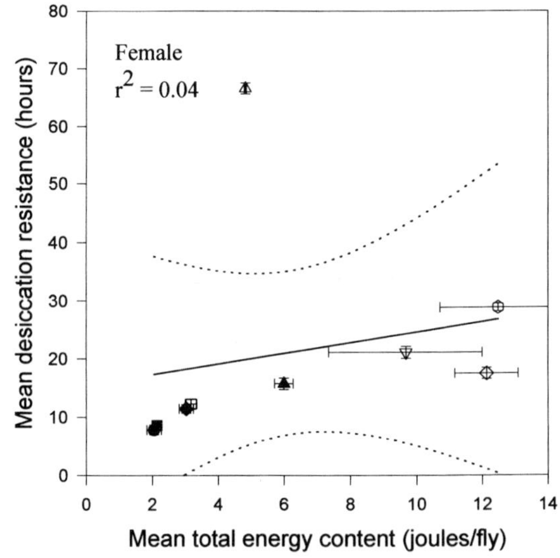

Figure 11. The relationship between mean (±SE) desiccation resistance and mean (±SE) total energy content in females of *Drosophila melanogaster* from nine distinct selection regimes: B populations (■), O populations (□), C populations (▲), D populations (△), CO populations (♦), SO populations (◇), CS populations (▽), DS populations (○), ACO populations (●). Each point represents the mean value from five replicate populations within a selection regime, except for CS and DS, where four replicate populations were available. A least squares regression is fitted to the values as a solid line. The 95% confidence intervals around each regression are shown as dotted lines.

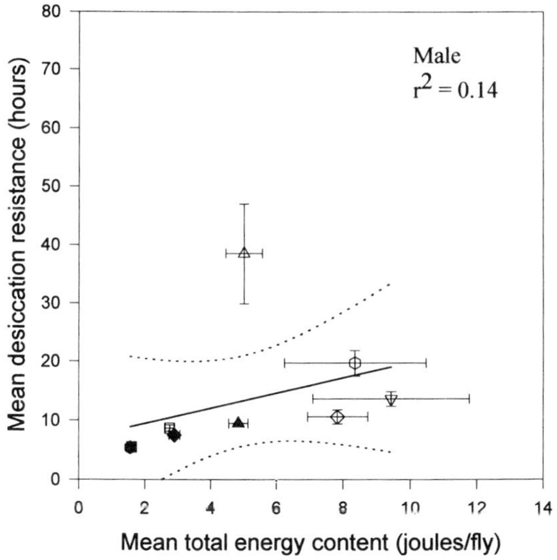

Figure 12. The relationship between mean (±SE) desiccation resistance and mean (±SE) total energy content in males of *Drosophila melanogaster* from nine distinct selection regimes: B populations (■), O populations (□), C populations (▲), D populations (△), CO populations (♦), SO populations (◇), CS populations (▽), DS populations (○), ACO populations (●). Each point represents the mean value from five replicate populations within a selection regime, except for CS and DS, where four replicate populations were available. A least squares regression is fitted to the values as a solid line. The 95% confidence intervals around each regression are shown as dotted lines.

D flies may therefore show statistically significantly higher levels of carbohydrate storage than do the C populations because this form of energy storage does not interfere with water storage. This extra water may play a critical role in the enhanced desiccation resistance observed in the D flies. Central to such an explanation is the assumption that the water of hydration associated with carbohydrate is available to the flies when the carbohydrate is metabolized.

Can We Predict Directions of Evolutionary Change?

We have already shown that the responses to different forms of stress selection are physiologically distinct. It is also of interest to know if selection studies can provide insights into whether evolutionary trajectories are predictable.

We have addressed this question by examining changes in lipid and carbohydrate storage in populations subjected to selection for starvation and desiccation resistance (Fig. 13). Figure 13 shows the response of five lines (the D populations) to strong selection for desiccation resistance, the response of five lines (the C populations) to mild selection for starvation resistance, and the response of five lines (the SO populations) to strong selection for starvation resistance.

Figure 13 shows the mean and standard deviations of the responses of the populations. The D_1, C_1, and SO_1 populations were each derived from the O_1 population, the D_2, C_2, and SO_2 populations were all derived from O_2, and so forth for each of the five O populations. The experimental design therefore employs three distinct selection treatments on three populations derived from the same ancestral population. This protocol was replicated five times. The evolutionary pathway taken by these populations as they diverged from the O populations can, as a first approximation, be estimated by the slope of the line in Figure 13. Each of the five populations yields an independent estimation of these changes. The variance in the slopes within a selection treatment varies greatly, requiring the use of a nonparametric test. A Wilcoxon paired-sample test reveals that the D populations followed a different evolutionary pathway with regard to these characters than did the C and SO populations ($P = 0.05$ for both comparisons), but the SO and C populations did not differ in their response ($P = 0.1$). Note that levels of lipid and carbohydrate storage do not evolve in parallel. Selection appeared to be able to shape these variables with at least some degree of evolutionary independence. The changes in the relative amounts of lipid and carbohydrate stored are predictable, however, for desiccation or starvation selection treatments.

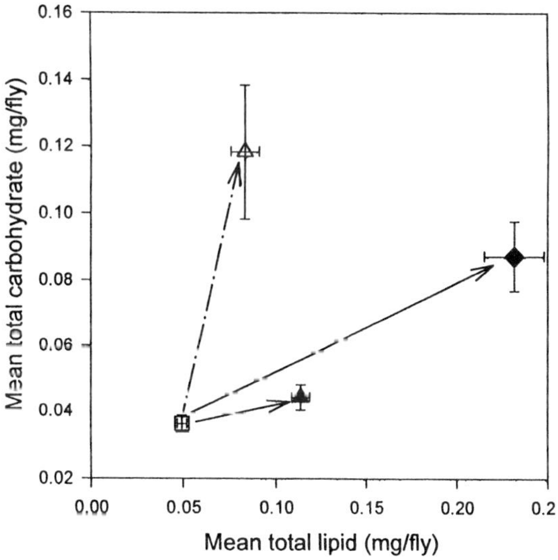

Figure 13. The relationship between mean (±SE) total carbohydrate content and mean (±SE) total lipid content in females from populations selected for starvation (C, ▲; and SO, ♦), desiccation (D, △), and their common ancestral populations (O, □).

Acknowledgments

We would like to thank Lisa K. Schlaeger and Cristina Zua Zua for their help with the lipid and carbohydrate assays and numerous other undergraduate students who helped with stock maintenance and starvation and desiccation assays. We would also like to thank our colleagues Larry Mueller, Ted Nusbaum, and Amitab Joshi, who helped with the statistical analyses; Allen Gibbs, who generously shared his space, resources, and ideas; and Al Bennett, for his always helpful insights. Funding for this research was provided by grant AGO-9970 from the U.S. Public Health Service and grants IBN-9507435 and IBN-9723404 from the National Science Foundation.

Literature Cited

Chippindale A.K. 1994. The Evolution of Life History Trade-Offs in *Drosophila melanogaster.* PhD diss. University of California, Irvine.

Chippindale A.K., T.J.F. Chu, and M.R. Rose. 1996. Complex trade-offs and the evolution of starvation resistance in *Drosophila melanogaster.* Evolution 50:753–766.

Djawdan M., M.R. Rose, and T.J. Bradley 1997. Does selection for stress resistance lower metabolic rate? Ecology 78:828–837.

Dobush G.R., D. Davidson Ankney, and D.G. Krementz. 1985. The effect of apparatus, extraction time and solvent type on lipid extraction of snow geese. Can. J. Zool. 63:1917–1920.

Garland T. and S.C. Adolph. 1994. Why not to do two-species comparisons: limitations on inferring adaptation. Physiol. Zool. 67:797–828.

Gibbs A.G., A.K. Chippindale, and M.R. Rose. 1997. Physiological mechanisms of evolved desiccation resistance in *Drosophila melanogaster.* J. Exp. Biol. 200:1821–1832.

Graves J.L., E.C. Toolson, C. Jeong, L.N. Vu, and M.R. Rose. 1992. Desiccation, flight, glycogen, and postponed senescence in *Drosophila melanogaster.* Physiol. Zool. 65:268–286.

Hoffmann A.A. and P.A. Parsons. 1989*a.* An integrated approach to environmental stress tolerance and life history variation: desiccation tolerance in *Drosophila.* Biol. J. Linn. Soc. 37:117–136.

———. 1989*b.* Selection for increased desiccation resistance in *Drosophila melanogaster:* additive genetic control and correlated responses for other stresses. Genetics 122:837–845.

———. 1991. Evolutionary Genetics and Environmental Stress. Oxford University Press, New York.

Ives P.T. 1970. Further studies of the South Amherst population of *Drosophila melanogaster.* Evolution 24:507–518.

Koehn R.K. and B.L. Bayne. 1989. Towards a physiological and genetical understanding of the energetics of stress response. Biol. J. Linn. Soc. 37:157–171.

Leroi A.M., M.R. Rose, and G.V. Lauder. 1994. What does the comparative method reveal about adaptation? Am. Nat. 143:381–402.

Prosser C.L. 1991. Environmental and Metabolic Animal Physiology. Wiley, New York.

Randall D., W. Burggren, and K. French. 1997. Animal Physiology. W. H. Freeman, New York.

Rose M.R. 1984. Laboratory evolution of postponed senescence in *Drosophila melanogaster.* Evolution 38:1004–1010.

Rose M.R., J.L. Graves, Jr., and E.W. Hutchinson. 1990. The role of selection to probe patterns of pleiotropy in fitness characters. Pp. 29–42 in Francis Gilbert, ed. Insect Life Cycles. Springer, Berlin.

Rose M.R. and G.V. Lauder. 1996. Adaptation. Academic Press, San Diego, Calif.

Rose M.R., L.N. Vu, S.U. Park, and J.L. Graves, Jr. 1992. Selection on stress resistance increases longevity in *Drosophila melanogaster.* Exp. Gerontol. 27:241–250.

Schmidt-Nielsen K. 1990. Animal Physiology. Cambridge University Press, Cambridge.

Service P.M. 1987. Physiological mechanisms of increased stress resistance in *Drosophila melanogaster* selected for postponed senescence. Physiol. Zool. 60:321–326.

Service P.M., E.W. Hutchinson, M.D. Mackinley, and M.R. Rose. 1985. Resistance to environmental stress in *Drosophila melanogaster* selected for postponed senescence. Physiol. Zool. 58:380–389.

Sokal R.R. and F.J. Rohlf 1969. Biometry. W. H. Freeman, San Francisco.

Van Handel E. 1965. Microseparation of glycogen, simple sugars, and lipids. Ann. Biochem. 11:266–271.

Williams A.E., M.R. Rose, and T.J. Bradley. 1997. CO_2 release patterns in *Drosophila melanogaster:* the effect of selection for desiccation resistance. J. Exp. Biol. 200:615–624.

The Journal of Experimental Biology 200, 1821–1832 (1997)
Printed in Great Britain © The Company of Biologists Limited 1997
JEB0798

PHYSIOLOGICAL MECHANISMS OF EVOLVED DESICCATION RESISTANCE IN *DROSOPHILA MELANOGASTER*

ALLEN G. GIBBS*, ADAM K. CHIPPINDALE† AND MICHAEL R. ROSE

Department of Ecology and Evolutionary Biology, University of California, Irvine, CA 92697, USA

Accepted 9 April 1997

Summary

We investigated physiological characters associated with water balance in laboratory populations of *Drosophila melanogaster* selected for resistance to desiccating conditions for over 100 generations. Five replicate, outbred, desiccation-selected (D) populations were compared with their control (C) populations. Water loss rates of female D flies were approximately 40 % lower than those of C females. Although excretory water loss was reduced in desiccation-selected flies, it comprised less than 10 % of total water loss, indicating that the D populations have evolved reduced cuticular and/or respiratory water loss rates. Total surface lipid amounts did not differ between the C and D flies. Cuticular hydrocarbons from D flies were longer than those from C flies and melted at slightly higher temperatures, possibly contributing to reduced water loss rates. Desiccation-selected flies contained approximately 30 % more bulk water than controls, as well as more glycogen. However, total metabolic water stores did not differ between the stocks owing to higher lipid levels in the C populations. The ability to tolerate water loss, as measured by water content at the time of death, did not differ between D and C flies. Thus, evolution of increased desiccation resistance has occurred by multiple physiological mechanisms, but some potential adaptive differences have not evolved.

Key words: desiccation, fruit fly, *Drosophila melanogaster*, water balance, evolution, cuticular lipid, laboratory selection.

Introduction

One of the most important environmental stresses facing insects in nature is desiccation. The small size (and therefore large ratio of surface area to volume) of terrestrial arthropods makes them particularly susceptible to dehydration. Numerous comparative studies have shown that insects from warmer, drier environments exhibit a variety of apparently adaptive differences in water balance, e.g. reduced cuticular permeabilities, reduced excretory water loss and differences in the quantity and composition of cuticular lipids (reviewed by Hadley, 1994*a*).

Comparative studies of adaptation can be strengthened by phylogenetically based analyses of the pattern of evolution (Harvey and Pagel, 1991; Garland and Adolph, 1994). However, reliable information about the phylogenetic relationships of organisms is often unavailable. Even if this information is available, several other factors (e.g. genetic correlations, linkage disequilibrium, pleiotropy, parental effects, genotype-by-environment interactions) can undermine one's confidence in the adaptive significance of a given trait (Lauder *et al.* 1993; Leroi *et al.* 1994). In cases in which a trait does have current adaptive value (as indicated, for example, by field studies of selection; see Endler, 1986), it is impossible to determine precisely which selective factors influenced the evolution of this trait. Did it arise as a result of the same selective forces now operating in nature, has a trait evolved for a different purpose been recruited for a new function, or is it a product of stochastic processes? If a trait has not evolved in response to selection, why not? Is its evolution constrained by the pleiotropic effects of genetic correlations with other characters, or has there simply been insufficient genetic variation or selection to affect the trait?

Comparative studies are inherently correlative, attempting to reconstruct the history of selection from its endpoints. A complementary approach to the comparative method is to study the evolutionary process in the laboratory. The laboratory evolution approach avoids many of the potential pitfalls of interspecific comparative studies, since the selective regime is known and can be manipulated by the investigator (Rose *et al.* 1990, 1996; Garland and Carter, 1994). Numerous researchers have shown that organisms respond to selection for increased resistance to environmental stresses, e.g. temperature (Bennett *et al.* 1990; Huey *et al.* 1991; Lenski and Bennett, 1993), starvation (Rose *et al.* 1992) and urea (Gupta *et al.* 1995; Joshi *et al.* 1996). However, the genetic or physiological basis of evolved stress resistance is frequently unclear. Polygenic and oligogenic variation have each been observed in response to

*e-mail: agibbs@uci.edu.
†Present address: Department of Biology, NS IV, University of California, Santa Cruz, CA 95064, USA.

selection for pesticide resistance (Crow, 1957; McKenzie and Clarke, 1988). Hoffmann and Parsons (1989, 1991) have hypothesized that reduced metabolic rates constitute a generalized response to stress selection, although this proposal has been disputed by Djawdan *et al.* (1997).

Several investigators have shown that the desiccation resistance of *Drosophila melanogaster* can be increased by selection (Hoffmann and Parsons, 1989; Rose *et al.* 1990, 1992; Blows and Hoffmann, 1993). Typical selection protocols involve exposing flies to dry air and using the longest-surviving individuals to propagate the next generation. From an organismal standpoint, there are only three mechanisms by which resistance to such an acute desiccation stress can be enhanced: rates of water loss can be reduced, flies can contain more water (as bulk water or metabolic water stores) or flies can tolerate the loss of greater amounts of water. Extreme examples of the last possibility are anhydrobiotic organisms, such as baker's yeast, brine shrimp embryos or seeds, which can survive for months or years in a dry state (Crowe *et al.* 1992). These mechanisms of desiccation resistance are not exclusive of one another.

Physiological studies of evolved mechanisms of desiccation resistance in *Drosophila* spp. have been limited and sometimes contradictory. Graves *et al.* (1992) found that desiccation-selected populations of *D. melanogaster* contained more glycogen than controls, whereas Blows and Hoffmann (1993) found no correlation with potential metabolic water stores (glycogen and lipid) in selected lines of *D. serrata*. Instead Blows and Hoffmann (1993) observed that selection had resulted in reduced metabolic rates under desiccating conditions. Djawdan *et al.* (1997), using the same populations as Graves *et al.* (1992), found no differences in metabolic rates between desiccation-selected and control populations, when corrected for differences in metabolic reserves.

The conflicting results of selection experiments may be due to differences in species, selection protocols, the number of generations of selection, etc. In addition, each of these studies examined only one or a few components of water balance. We describe here experiments investigating the mechanistic basis of evolved desiccation resistance in five outbred populations of *D. melanogaster* selected for more than 100 generations. We measured the rate at which flies lost water. We examined several aspects of cuticular lipids: quantity, composition and physical properties. We determined how much water (both bulk and metabolic water) the flies contained, and we determined how much loss of overall body water the flies could tolerate. By constructing a complete water budget for flies from these populations, we were able to assess the relative contributions of these characters to overall desiccation resistance.

Materials and methods

Drosophila melanogaster *populations*

The origins and maintenance of the populations have been described previously (Rose *et al.* 1990, 1992). In short, five O-stocks (O_1–O_5) were derived in 1980 from a large, outbred laboratory population, originally collected in Amherst, MA, USA, in 1975 (Rose, 1984). These were selected for postponed age of reproduction, resulting in greatly increased lifespan in these populations relative to five control (B_1–B_5) populations. The O-stocks have been maintained as outbred populations on a 10-week generation cycle since 1981. All stocks described here are maintained with constant light at 25 °C on banana–molasses medium.

Service *et al.* (1985) found that the O populations exhibited greater resistance to a variety of environmental stresses, including desiccation, relative to five control populations maintained on the 2-week life cycle of the stock that was ancestral to both O populations and their controls. Service *et al.* (1988) then derived reverse selection lines from the O populations, these lines being handled in the same manner as the controls, with 2-week generations. In the course of more than 20 generations, starvation resistance and early fecundity responded to reverse selection, but desiccation resistance did not. In order to determine whether this lack of response to reverse selection was due to an exhaustion of genetic variation for desiccation resistance in the O populations, Rose *et al.* (1990) initiated selection for increased desiccation resistance in derivatives of O populations (designated D_1–D_5, one from each O population). A control (C) population was also founded from each O population at the same time. The five C populations have been maintained identically to the D population, except for the absence of desiccation selection. All populations have been maintained as large outbred populations (approximately 1000 individuals per generation), in order to minimize the effects of inbreeding and genetic drift.

Since being founded in 1988, the C and D flies have been reared in vials for 2 weeks (including all pre-adult stages and 3–4 days as adults; Fig. 1). At 14 days from the egg stage, flies are transferred to population cages. The D cages contain approximately 150 g of fresh Drierite desiccant and are sealed with plastic wrap to maintain low humidity. The D flies are therefore subjected to both desiccation and the absence of food. In order to control for the effects of starvation, the C populations are reared in the same manner as the D population, but their cages are provided with a non-nutritive agar plate instead of desiccant. These flies are not subjected to desiccation selection, but do, however, undergo mild selection for starvation resistance. For a given D population, when approximately 80 % of the flies have died, the desiccant is removed, and the flies are provided with food. Each C-stock is paired with its corresponding D-stock during selection; when a given stock D_i (i=1–5) is given food, so is its corresponding C_i population.

Our experiments were designed to provide as realistic a picture as possible of the water relationships of these flies at the time they entered selection. Thus, assays were performed when the flies were 14 days from the egg stage (3–4 days post-eclosion), the age at which the D populations were subjected to desiccation selection (and the C populations were starved). For all assays, flies were reared for two generations on a

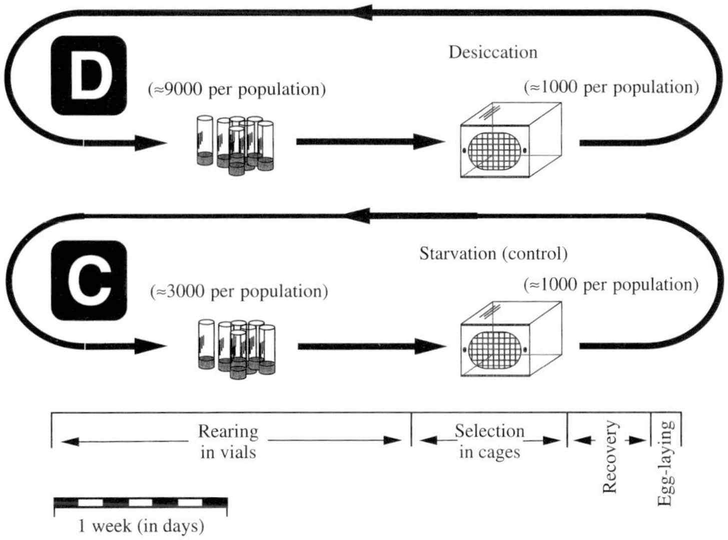

Fig. 1. Schematic illustration of the life cycles for the C and D populations. The mild starvation stress experienced by the C flies serves as a control treatment for the combined starvation and desiccation selection experienced by the D flies.

common 2-week stock cycle with relaxed selection in order to avoid potential effects of the parental environment. Because of the paired nature of the evolutionary history of the flies and the maintenance schedule, assays were performed as paired measurements, i.e. each D population was assayed for water loss measurements or sampled for other analyses along with its corresponding C control.

Assays for desiccation resistance

At 4 days post-eclosion, flies were lightly anesthetized with CO_2. Ten flies of each sex, from each of the ten D and C populations, were placed in individual 30 ml vials. A polyethylene sponge was used to keep the flies in the lower half of the vial. Fresh Drierite desiccant (approximately 3 g) was added above the sponge, and the vial was sealed with Parafilm. Flies were checked at hourly intervals for death, as indicated by failure to right themselves or to move their legs when their vials were tapped or inverted. Upon death, flies were immediately weighed for determination of tolerance to dehydration (details below).

Water loss rates

We used a Sable Systems (Henderson, NV, USA) flow-through respirometry system to measure water loss rates. Dry air was passed at a rate of 25 ml min^{-1} through 5 ml glass and aluminum chambers containing groups of 20 female flies. Up to eight chambers (including empty control chambers) could be assayed sequentially. The sensor (a modified Li-Cor LI6262 infrared humidity detector) was calibrated by injecting water drops of known volume (0.5–5 µl) into the flow stream. The instrument response was linear over this range, with an intercept statistically indistinguishable from zero. Sham injection of air gave no response. Thus, although our measurements were in the 0–0.5 µl range, we felt confident in extrapolating from our calibration curve to this range. The entire apparatus was contained in a walk-in temperature-controlled room at 25 °C.

Owing to the limitations of our respirometry system, only six sets of 20 flies could be assayed on a given day. Thus, each C_i–D_i stock pair was measured, in triplicate, on a different day. Egg collection was staggered so that all flies were 14 days from the egg stage on the day of measurement. For a given set of measurements, eight respirometry chambers were flushed with dry air overnight at 25 ml min^{-1}. Six chambers were used for flies, and two empty chambers served as controls. Twenty female flies per chamber were used in all water loss measurements. Preliminary experiments indicated that water loss rates were relatively high in the first 2 h following placement in the chambers, then stabilized for the next several hours. We therefore staggered placement of the flies in the respirometer, so that they had been in their chambers for 3 h before a measurement began. An empty chamber was read first, followed by three chambers containing flies, the second empty chamber, three more experimental chambers, and a repeated measurement of the first empty chamber. D and C flies were alternated during the day, and their overall order (C flies first or D flies first) was alternated from one day to the next. Air flow rates during measurements were 25 ml min^{-1}. Because the chambers were flushed continuously at approximately the same rate before measurement, little water vapor accumulated in the chambers, and washout times were usually short. Even so, we only used data covering approximately the last 30 min of a 1 h run.

Cuticular lipid analyses

Two-week-old flies (approximately 4 days post-eclosion)

were frozen and stored at −20 °C for cuticular lipid analyses. Epicuticular hydrocarbons were isolated as described by Toolson (1982). Ten flies were placed on a silica gel column in a Pasteur pipette, and hydrocarbons were eluted with 6 ml of HPLC-grade hexane. For quantitative determination of cuticular lipid amounts, 2.5 μg of *n*-docosane in a small volume of hexane was added to the column along with the flies as an internal standard.

We analyzed cuticular lipids in two ways. Lipid amounts were quantified by capillary gas chromatography of samples containing the *n*-docosane standard, using a 30 m×0.32 μm DB-1 column (J&W Scientific, Sacramento, CA, USA) in a Hewlett-Packard 5890A gas chromatograph, GC. The chromatograms were also used to investigate differences in cuticular hydrocarbon composition. Hydrocarbons were classified by carbon number (chain length) according to their retention times compared with *n*-alkane standards. Mean chain lengths were computed as the weighted average of the abundance of each component.

Functional differences in the water-proofing abilities of cuticular hydrocarbons depend in part upon their physical properties (Beament, 1945; Toolson *et al.* 1979; A. G. Gibbs, in preparation). Lipid melting temperatures (T_m) were determined using Fourier transform infrared (FTIR) spectroscopy, as described by Gibbs and Crowe (1991). This technique is based upon the lipid phase-dependent shift in the infrared absorbance maximum of $-CH_2-$ symmetrical stretching vibrations. Hydrocarbons extracted from groups of ten flies were placed in a temperature-controlled cell holder in a Perkin-Elmer Systems 2000 FTIR spectrometer. The sample temperature was increased in 2–3 °C increments from 5–10 °C to 50–60 °C. As indicated by FTIR, hydrocarbons began melting at 20–25 °C and had finished melting by 45–50 °C. Plots of $-CH_2-$ symmetrical stretching frequency *versus* temperature were sigmoidal. The midpoint of the phase transition (T_m) was calculated as the midpoint of a fitted logistic curve.

Water content assays

Bulk water contents were determined for six groups of five flies from a given sex and population. Flies were frozen in microcentrifuge tubes on dry ice. Within 24 h, each group was weighed on tared pieces of aluminum foil to a precision of 1 μg using a Cahn microbalance. The flies were dried overnight at 55 °C and re-weighed. Water content was calculated as the difference between the wet and dry mass. Preliminary experiments indicated that wet mass did not change for up to 48 h on dry ice.

Within a few generations of the water content assays, flies were assayed for glycogen and lipid contents, as part of a study of stress and metabolism in these and other *Drosophila* populations (M. Djawdan, A. K. Chippindale, M. R. Rose and T. J. Bradley, in preparation). Because both energetic substrates can serve as sources of metabolic water (Schmidt-Nielsen, 1990), we calculated the total amount of metabolic water that could be obtained if these substrates were completely aerobically metabolized (i.e. maximum available metabolic water). Our conversion factors were 0.56 mg H_2O mg^{-1} glycogen and 1.07 mg H_2O mg^{-1} lipid (Schmidt-Nielsen, 1990). Water bound to glycogen was included in our measurement of bulk water. Glycogen levels were determined using a modified anthrone reaction, and lipid contents were measured by extraction with a Soxhlet apparatus, as in our earlier work (Chippindale *et al.* 1996).

Dehydration tolerance

To investigate whether the C and D flies differed in their abilities to tolerate water loss, we attempted to measure the water contents of flies as close to the time of death from dehydration as possible. If a fly appeared dead during an hourly check for the desiccation resistance assay (above), it was removed from its vial, and its wet mass was measured immediately. It was then dried overnight at 55 °C, and the water content was determined as described above.

Statistical analyses

Data were analyzed by mixed-model analyses of variance (ANOVAs) using Systat software. We treated sex and treatment (i.e. C or D selection regime) as fixed effects, with population number (1–5) as a random effect. Populations were treated as D_i–C_i pairs during selection and assays, so population effects in statistical analyses are equivalent to block effects. The fundamental unit of replication in these stocks is the population, so graphical data are depicted as the means (±S.E.M.) of N=5 populations. Different numbers of flies were used in different assays (e.g. individual flies in the dehydration tolerance assay, groups of 20 flies for water loss measurements). For ease of comparison, all measurements have been converted to single fly units, and error estimates have been scaled accordingly.

Results

Desiccation resistance

Flies from desiccation-selected populations survived over twice as long in dry air as their C controls (P<0.001; Fig. 2), with one D female surviving for 84 h. Thus, selection for desiccation resistance has been quite successful, indicating that the original founding O populations contained substantial genetic variation upon which selection could act. Females survived longer than males (P<0.001). A significant gender-by-treatment effect was observed, due to the much greater desiccation resistance of D females relative to other gender–treatment combinations.

Water loss rates

Fig. 3 depicts a typical recording for water loss from a C and D stock pair. For all populations, water loss was characterized by a relatively constant background rate, with intermittent bursts of water loss. We calculated three parameters: (1) total water loss rate, calculated by integrating the area under the curve; (2) 'sporadic' water loss (area of the

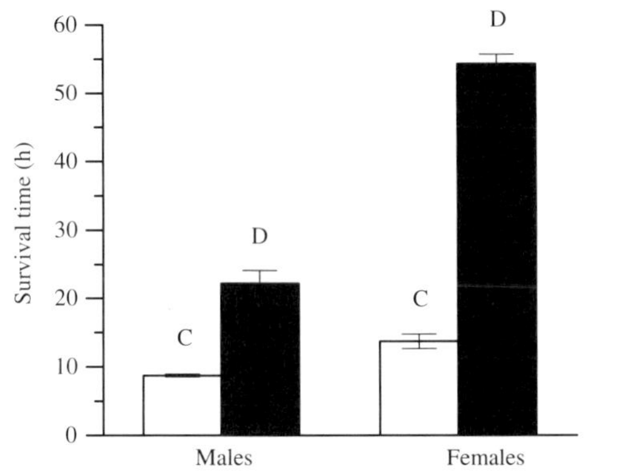

Fig. 2. Mean survival times (desiccation resistance) for individual flies in dry air. For each population (C and D), ten flies of each gender were assayed. Error bars depict standard errors for the means of *N*=5 population means.

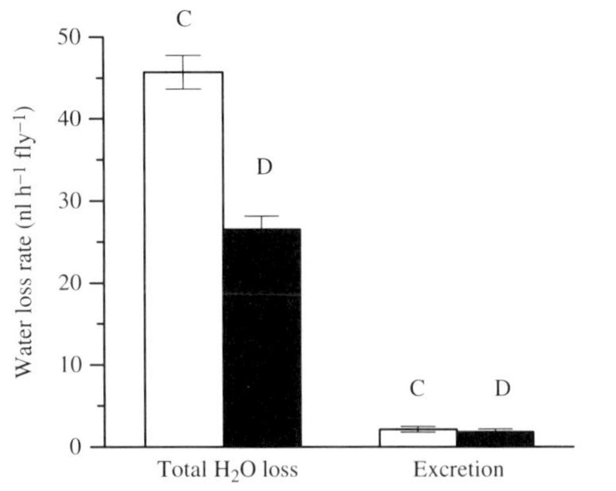

Fig. 4. Water loss rates for control (C) and desiccation-selected (D) female flies. Excretory water loss was calculated from the integrated area of the bursts of water loss shown in Fig. 3. Means and standard errors for *N*=5 populations are provided.

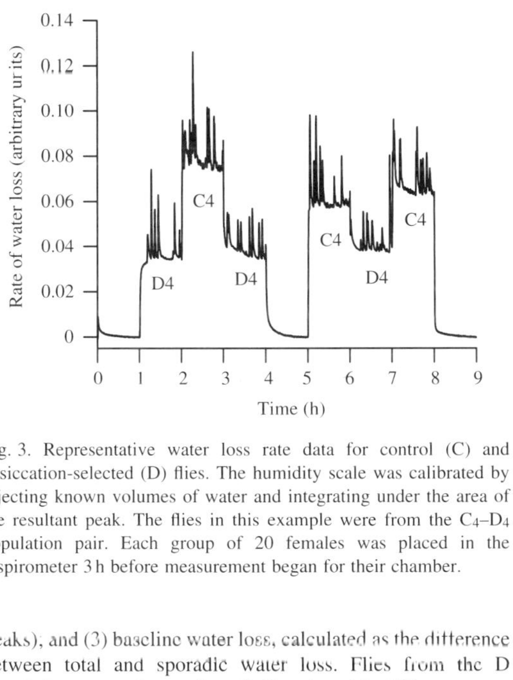

Fig. 3. Representative water loss rate data for control (C) and desiccation-selected (D) flies. The humidity scale was calibrated by injecting known volumes of water and integrating under the area of the resultant peak. The flies in this example were from the C_4–D_4 population pair. Each group of 20 females was placed in the respirometer 3 h before measurement began for their chamber.

peaks), and (3) baseline water loss, calculated as the difference between total and sporadic water loss. Flies from the D populations were larger than C flies, but this difference was due to their greater water content (see below). Because the D and C treatments did not differ in dry mass (0.444±0.031 mg and 0.399±0.024 mg (mean ± S.E.M.) per fly for D and C females, respectively), we have expressed our results on a per fly basis.

Total water loss rates differed significantly between the D and C stocks. D females lost water at approximately 60 % of the rate for C females (Fig. 4; P<0.001). The difference between the stocks was mainly due to differences in the baseline rates of water loss. Rates of 'sporadic' water loss were slightly lower in the D populations (P<0.05), but still accounted for less than 10 % of total water loss in any given population. On a percentage basis, the D populations lost relatively more water during these events (6.7 % of total water loss in the D flies, 4.6 % in the C flies; P<0.025).

Cuticular lipid analyses

Eighteen hydrocarbons accounted for approximately 97 % of the total cuticular lipid complement. On the basis of previous analyses of cuticular hydrocarbons from *D. melanogaster* (Antony and Jallon, 1982; Scott and Jackson, 1988; Graves *et al.* 1992), these peaks primarily corresponded to alkenes and alkadienes containing 21–31 carbon atoms. Although we did not perform a detailed chemical analysis of cuticular lipid structure, we note that no novel peaks were observed in either C or D flies. Thus, any differences in cuticular hydrocarbons reflected differences in the distribution of compounds common to both sets of populations, rather than the synthesis of new components.

Flies averaged 1.6 µg of total hydrocarbon per fly, and lipid quantity did not differ between genders or between the C and D populations (Table 1). Lipid melting temperatures (T_m) differed significantly between genders and treatments (Table 2). Mean T_m values (±S.E.M. for N=5 populations) were 36.6±0.6 °C and 35.7±0.5 °C for D and C females, respectively, and 33.9±0.7 °C and 30.2±0.4 °C for D and C males, respectively. An ANOVA of mean hydrocarbon chain length revealed significant gender and treatment differences consistent with the T_m results (Table 3). Surface lipids from desiccation-selected flies were longer, on average, than those of C flies. When hydrocarbons were grouped by number of carbon atoms (21–31 carbons, odd numbers only), ANOVAs performed on each group separately revealed significant

Table 1. *Analysis of variance results for cuticular hydrocarbon quantity in C and D flies*

Source	Sum of squares	d.f.	Mean square	*F*-ratio	*P*
Gender	0.0303	1	0.0303	0.2793	0.75
Treatment	0.6240	1	0.6240	1.1109	0.5
Population	1.4815	4	0.3704	2.4938	0.076
Gender × Treatment	0.8168	1	0.8168	1.1744	0.5
Gender × Population	0.4341	4	0.1085	0.7307	0.58
Treatment × Population	2.2469	4	0.5617	3.7823	0.019
Gender × Treatment × Population	0.4131	4	0.1033	0.6955	0.6
Error	2.9703	20	0.1485		

In this and the following tables, gender and treatment (D or C) were treated as fixed effects and population as a random effect. d.f., degrees of freedom.

Table 2. *Analysis of variance results for cuticular lipid melting temperature* (T_m)

Source	Sum of squares	d.f.	Mean square	*F*-ratio	*P*
Gender	260.2456	1	260.2456	79.8912	0.001
Treatment	81.9250	1	81.9250	48.4047	0.005
Population	33.9155	4	8.4789	1.8301	0.14
Gender × Treatment	30.6170	1	30.6170	4.2775	0.25
Gender × Population	13.0299	4	3.2575	0.7031	0.59
Treatment × Population	6.7701	4	1.6925	0.3653	0.83
Gender × Treatment × Population	28.6309	4	7.1577	1.5450	0.25
Error	203.8488	44	4.6329		

d.f., degrees of freedom.

Table 3. *Analysis of variance results for mean cuticular hydrocarbon chain length*

Source	Sum of squares	d.f.	Mean square	*F*-ratio	*P*
Gender	21.7436	1	21.7436	259.4079	0.001
Treatment	1.5031	1	1.5031	16.6858	0.025
Population	0.7928	4	0.1982	2.4474	0.08
Gender × Treatment	0.0221	1	0.0221	0.1447	0.75
Gender × Population	0.3353	4	0.0838	1.0351	0.41
Treatment × Population	0.3603	4	0.0901	1.1124	0.38
Gender × Treatment × Population	0.6111	4	0.1528	1.8864	0.15
Error	1.6196	20	0.0810		

d.f., degrees of freedom.

gender differences for each chain length. Males had relatively greater amounts of 21-, 23- and 25-carbon hydrocarbons, whereas females had relatively higher levels of the 27-, 29- and 31-carbon species (Fig. 5; $P<0.05$). Four lipid groups exhibited significant differences between the D and C selection treatments (C>D for 21- and 23-carbon lipids, D>C for 25- and 29-carbon species; $P<0.05$). A significant gender-by-treatment interaction was observed for hydrocarbons containing 25 carbon atoms.

Water content: bulk water and metabolic water

On a wet mass basis, female flies weighed more than males, and D flies weighed more than C flies ($P<0.01$; Table 4). The selection-related difference in mass was due to the higher water content of the desiccation-selected populations, since dry mass did not differ significantly between D and C populations. Individual female D flies averaged 241 µg more water than C females (Fig. 6). D males also contained over 20 % more water than C males (results not shown).

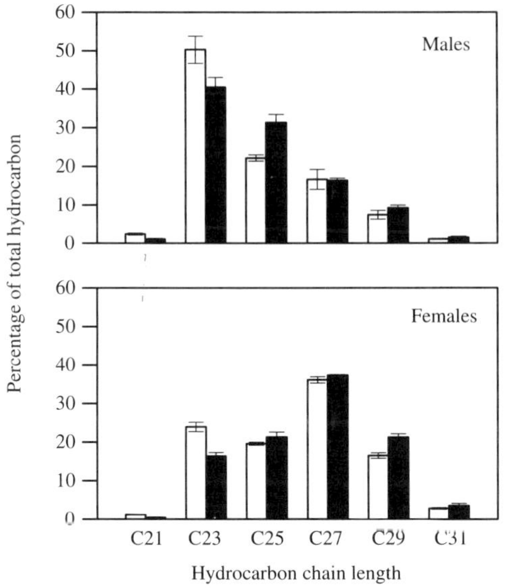

Fig. 5. Effects of selection treatment and gender upon chain lengths of cuticular hydrocarbons. Data for C23–C31 species include 2–5 different GC peaks. Open bars, C populations; filled bars, D populations. Values are means ± S.E.M. for N=5 populations.

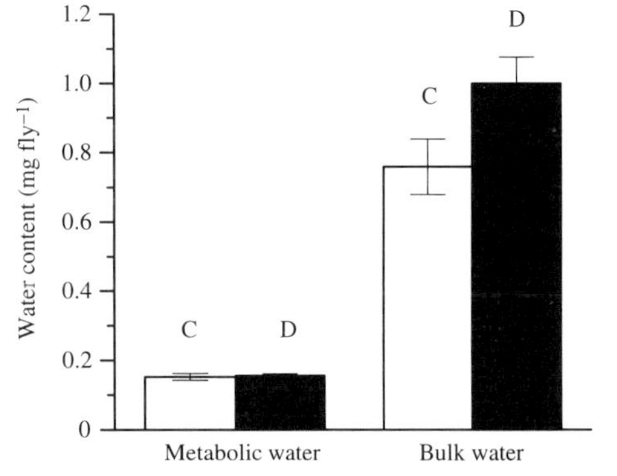

Fig. 6. Water reserves for control (C) and desiccation-selected (D) females. Bulk water was calculated from the difference between wet and dry mass, and metabolic water was calculated from data in M. Djawdan and others (in preparation). Values are means ± S.E.M. for N=5 populations.

Six generations after these measurements, glycogen levels and total lipid contents were determined by M. Djawdan, A. K. Chippindale, M. R. Rose and T. J. Bradley (in preparation). Because these substances can be used as a source of metabolic water, we used their data to calculate the potential metabolic water that could be produced from each substrate. As observed previously (Graves *et al.* 1992), the D flies contained significantly more glycogen than the C flies and, therefore, could obtain more metabolic water from glycogen metabolism. However, the C flies contained significantly more lipid. When total available metabolic water from glycogen and lipid was combined, both D and C flies had nearly identical amounts (approximately 154 µg per female fly, Fig. 6).

Table 1. *Body sizes of desiccation-selected and control populations*

Population	Wet mass per fly (mg)	Dry mass per fly (mg)
D females	1.444±0.102	0.444±0.031
D males	0.851±0.070	0.253±0.016
C females	1.158±0.019	0.399±0.024
C males	0.731±0.013	0.248±0.009

Data are means ± S.E.M. for N=5 populations, with six groups of five flies each measured per population.

Dehydration tolerance

Dehydration tolerance, as indicated by water content at death, can be quantified in two ways: either as absolute water content or as percentage water. On a percentage basis, C flies actually appeared to tolerate greater dehydration (45 % body water at death *versus* 56 % for D flies, Fig. 7A). This counterintuitive result was due to the fact that the C flies had a greater dry mass at death (P<0.001, Fig. 7B), but the D and C flies did not differ from each other in water content at time of death (P>0.25, Fig. 7C). Females contained more water when they died than males (P<0.001) owing to their larger overall size. Since absolute water contents were the same, but the C flies had a greater dry mass, percentage water content at death was greater in D flies.

Discussion

The D and C populations were founded in 1988 from the O-stocks, which at that time survived desiccation for approximately 15 h (for females; males survived approximately 9 h). The D flies have since diverged dramatically from their O ancestors and their unselected controls; desiccation survival times for D females now average over 2 days (Fig. 2). At the organismal level, there are only three possible, non-exclusive mechanisms for increased desiccation resistance: flies can contain more water, they can restrict its loss or they can tolerate loss of a greater amount of water. Below we consider the impact of each of these three mechanisms.

Water loss rates

Comparative studies have shown that *Drosophila* species and populations from apparently xeric habitats lose water more slowly and survive desiccation stress longer than those from mesic environments (Eckstrand and Richardson, 1980, 1981;

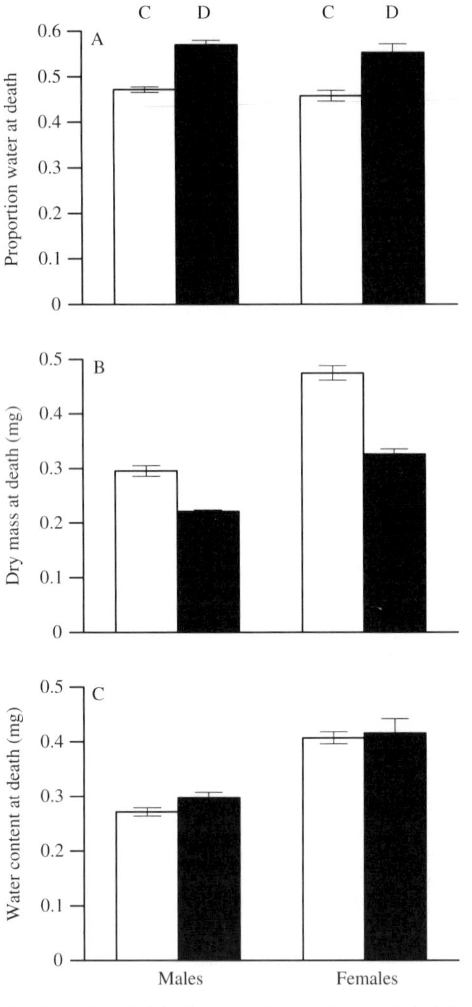

Fig. 7. Analysis of dehydration tolerance for control (C) and desiccation-selected (D) flies. Wet and dry mass for individual flies were determined immediately after their apparent death. The proportion composed of water was calculated as the water content divided by the wet mass. Means and standard errors for *N*=5 populations are shown.

Hoffmann, 1991; A. G. Gibbs and L. M. Matzkin, unpublished observations). In our laboratory-evolved populations, we found that desiccation-selected female flies lost water approximately 60 % as rapidly as their controls. Several mechanisms can be envisioned to account for these differences. Cuticular permeability could be reduced, respiratory water loss could be lowered by decreasing metabolic rate or regulating spiracular opening more tightly, or excretory water loss could be reduced.

Both C and D flies exhibited bursts of water loss at irregular intervals. On the basis of measurements of CO_2 release from individual flies (Williams *et al.* 1997), we do not believe these represented respiratory water loss associated with spiracular opening during discontinuous ventilation (Hadley 1994*a*,*b*). Instead, the bursts probably reflected excretory events, such as defecation. The D females lost slightly less water overall during these events than the C females. However, water loss by this route accounted for less than 10 % of total loss in any given population and, on a percentage basis, was higher in the D populations. Thus, it appears that reduction of excretory water loss is not a major contributor to desiccation resistance in the D flies.

Since presumed excretory water loss accounted for less than 10 % of the total, over 90 % of water loss could be attributed to differences in baseline rates. This difference included both respiratory and cuticular components, and our techniques were unable to distinguish between these routes. As noted above, ventilatory patterns may differ between D and C flies, although overall metabolic rates do not differ and are not affected by dehydration stress (Djawdan *et al.* 1997). Thus, it is unclear to what extent differences in respiratory water loss contribute to increased desiccation resistance. In several other insect species, cuticular water loss has been shown to comprise over 90 % of total water loss (Hadley, 1994*a*,*b*), equivalent to the basal contribution in our flies. Other insects that have been studied have been much larger than *Drosophila* species, and it is not known how differences in body size or taxon can affect the relative importance of respiratory water loss. However, available data suggest that lower water loss rates in the D flies may result largely from reduced cuticular permeability.

If this is case, it is reasonable that cuticular lipids might be responsible, since these form the primary passive barrier to evaporative water loss. Two aspects of cuticular lipids may be important: lipid amount and lipid composition. Graves *et al.* (1992) found no differences in cuticular hydrocarbon quantities or composition between the aging-selected O populations and their short-lived controls, despite differences in desiccation resistance and in water loss rates of dead flies. We found no significant differences between C and D flies in the quantity of cuticular hydrocarbons, indicating that desiccation-selected flies have not reduced water loss rates by increasing the thickness of the waterproofing layer. Instead, changes in surface lipid composition seem to have played a greater role in the evolution of desiccation resistance.

The mechanistic link between lipid composition and water loss rates is believed to be the physical properties of the cuticular lipids (Beament, 1945; Noble-Nesbitt, 1991). Hydrocarbons with longer chain-lengths would be expected to melt at higher temperatures and therefore provide a better barrier to water loss. We found that D flies had longer chain-length cuticular hydrocarbons than control populations, and that lipid melting temperatures were higher than those of C flies. The difference in mean hydrocarbon chain length, approximately 0.5 carbons per molecule, was consistent with the observed increase in T_m values of approximately 1 °C (Gibbs and Pomonis, 1995). We note that the lipid melting points we measured (>35 °C for females) were above the long-term (>15 years, >300 generations) thermal regime experienced by the flies (25 °C) since their ancestors were originally collected. However, most lipid samples began to melt below 25 °C, so that detectable changes in phase behavior were apparent at the flies' normal temperature. It is not known how partial lipid melting would affect cuticular permeability.

Water content

Besides reductions in water loss rates, desiccation resistance can be enhanced by having more water to start with. Graves *et al.* (1992) found that the D populations contained higher levels of glycogen than the C populations and suggested that glycogen serves as a source of metabolic water during desiccation. However, a rough estimate of the potential contribution of glycogen to desiccation resistance in the D flies indicates the situation is not so simple. M. Djawdan, A. K. Chippindale, M. R. Rose and T. J. Bradley, in preparation), in experiments performed shortly after ours, found that D females averaged 0.118 mg glycogen per individual, and we measured water loss rates of 0.0265 $\mu l\,h^{-1}\,fly^{-1}$ for D females. Assuming that metabolism of 1 g of glycogen results in the production of 0.56 g of water (Schmidt-Nielsen, 1990), metabolic water production from glycogen could offset water loss during desiccation stress for only 0.56×0.118/0.0265=2.5 h. An additional complication is that the C flies have greater lipid stores than the D flies, sufficient to give both sets of populations equivalent total metabolic water stores (Fig. 6). We note that the C populations are subjected to mild starvation selection, and that flies subjected to strong starvation selection have even greater lipid stores (Chippindale *et al.* 1996).

The largest store of water in the body is water itself. We found that D flies contained approximately 30 % more water than the C flies (Fig. 6). In fact, the *difference* in water content between C and D flies was greater than the combined available metabolic water from both energetic substrates. This finding further demonstrates that glycogen is at best a minor source of water in the D flies. Thus, higher glycogen levels in the D flies must be due to some other factor. One possibility is that glycogen serves as a metabolic fuel during desiccation stress. In accordance with this idea, Djawdan *et al.* (1997) observed that *D. melanogaster* subjected to desiccation stress primarily metabolize carbohydrates, whereas starvation-stressed flies use lipids. It should also be noted that glycogen may bind 3–5 times its mass in water (Schmidt-Nielsen, 1990). If glycogen binds 4 g water g^{-1}, carbohydrate stores in the D females could bind approximately 47 % of the total water content. Thus, glycogen may act as a 'sponge' to store water. This water would become available only as glycogen was metabolized; thus, shifting of metabolic fuels during desiccation to glycogen (Djawdan *et al.* 1997) would facilitate the release of water of hydration.

Blows and Hoffmann (1993), using desiccation-selected populations of *D. serrata*, found no differences in glycogen or lipid content, and none was evident for total water content (calculated from Table 5 in Blows and Hoffmann, 1993). Instead, they observed that metabolic rates increased at low humidity, but to a lesser extent for desiccation-selected lines. Blows and Hoffmann (1993) suggested that reduced respiratory water loss was responsible for increased desiccation resistance. For the C and D populations of *D. melanogaster* used in this study, Djawdan *et al.* (1997) found instead that metabolic rates were unaffected by desiccation stress and did not differ between selected and control flies.

The conflicting results obtained in these two studies might be explained by differences in species or number of generations of selection [14 for Blows and Hoffmann (1993); approximately 120 for Djawdan *et al.* (1997) and this study]. In addition, we feel that the differences in metabolic rates seen by Blows and Hoffmann (1993) may have been an experimental artefact, and we offer an alternative explanation. Blows and Hoffmann (1993) measured metabolic rates under humid conditions in a Gilson respirometer, by using 2.5 mol l^{-1} NaOH to absorb CO_2. 'Dry' conditions were provided by using solid NaOH instead. Sodium hydroxide is very hygroscopic, so that water lost by the flies under dry conditions could have been absorbed by the NaOH pellets and interpreted as CO_2 production. Thus, increased metabolic rates at low humidity could actually reflect increased net water loss under these conditions (Cooper, 1983), and the apparent difference in metabolic rates in dry air could have been partly due to reduced water loss in desiccation-selected lines. We suggest that the experiments by Blows and Hoffmann (1993) could not distinguish between reduced metabolic rates or water loss rates in desiccation-selected flies, and their results and those observed for the Rose laboratory populations may not be inconsistent.

Dehydration tolerance

The third physiological component of desiccation resistance is the ability to tolerate water loss. A survey of the literature on insect dehydration tolerance by Hadley (1994*a*) found widely varying estimates for measures such as percentage water content at death and percentage of total body water which an insect can lose. We found that the absolute water content at death did not differ between the C and D flies (Fig. 7C). However, owing to their continued metabolism over a much longer period of desiccation stress, the D flies had a significantly lower dry mass at death. (As noted above, dry mass did not differ between hydrated C and D flies.) This result led to the counterintuitive finding that C flies could tolerate greater dehydration, as measured by percentage body water at time of death (Fig. 7A).

The lack of differentiation in dehydration tolerance indicates that the D populations have not taken advantage of all possible mechanisms of desiccation resistance. Several possible explanations are available. The first is that little or no genetic variation in dehydration tolerance existed in the original founding populations, and that no new mutations having significant effects on dehydration tolerance have arisen and spread in the more than 100 generations since the C and D populations were founded. Another possibility is that the selection differential involving dehydration tolerance has been too small for significant differences to have appeared yet. A third alternative is that selection has acted upon the dehydration threshold, but that genetic correlations with other characters or pleiotropic effects at other life stages have acted in the opposite direction, so that no net change has occurred. A physiological explanation is that there may be an absolute limit to how little water a fly can have and remain viable. Insects may need a certain amount of hemolymph for nutrient and hormone transport, and cells may need a certain amount of water to remain viable. Dehydration tolerance may represent a physiological constraint which cannot be improved without evolving novel traits such as those exhibited by anhydrobiotes (Crowe *et al.* 1992). Our experiments were not designed to distinguish among these alternatives.

Contributions of traits to evolved desiccation resistance

Under desiccating conditions in vials (i.e. still air), D females survived for approximately 52 h, whereas C females survived for only 15 h (Fig. 2). Since we have a complete water budget, at the organismal level, for the D and C populations, it is possible to assess the relative importance of different mechanisms for desiccation resistance. We can do so by quantifying how many hours of additional desiccation resistance are conferred by reduced water loss, increased water content or differences in dehydration tolerance. Only the case for female flies will be considered, because females survive desiccation stress much longer than males (Fig. 2) and we have more complete data for females. In addition, female flies appear to be under stronger selection for survival of desiccation, because few males survive the selection procedure, as discussed below.

Since differences in dehydration tolerance (as indicated by water content at death) have not appeared during the evolution of desiccation resistance (Fig. 7), reduction of water loss and increased water content must be sufficient to explain the greater desiccation resistance of the D flies. If one assumes that water loss rates measured for groups of 20 flies after 3 h in a flow-through respirometer are an accurate index of water loss rates for individual flies during desiccation in still air, then the reduced water loss rates in the D flies (approximately 60 % of the rate for the C flies) will result in survival times of 15 h/0.6=25 h. Thus, reduction of water loss rates alone has increased desiccation resistance of the D flies by 10 h.

The greater water content in D females, which will be lost at a lower rate, can account for 0.241 mg H2O/0.0265 mg H2O lost per hour = 9.1 h. Metabolic water contents of the D and C flies were not significantly different (Fig. 6). However, the greater desiccation resistance of the D flies implies (in the absence of differences in metabolic rates; Djawdan et al. 1997) that the D flies will generate more metabolic water than the C flies before death. D females contained energy stores sufficient to generate 0.156 mg of metabolic water. This sets an upper limit of 0.156/0.0265=5.9 h of desiccation resistance available from metabolic water.

On the basis of these calculations, reduced water loss rates and increased water content could account for approximately 25 of the 37 h of increased desiccation resistance in D females. Several factors may contribute to the discrepancy between these values. Water loss rates may be lower under the conditions of still air found during desiccation assays, so these calculations could underestimate the contributions of both reduced water loss rates and increased water content. In addition, water loss rates progressively decline by up to 20 % during longer periods in the respirometer (A. G. Gibbs, unpublished results), which will further extend desiccation resistance. Finally, although every effort was taken to standardize conditions, environmental differences between experimental generations could also have affected our results.

An interesting aspect of the gender-specific differences in desiccation resistance is that males rarely survive the selection treatment. It appears that desiccation selection has had differing effects upon males and females, with D males being selected for early adult mating success rather than desiccation resistance (A. K. Chippindale, A. G. Gibbs and M. R. Rose, in preparation). Since no mating is observed during or after selection (A. K. Chippindale, unpublished observations), female D flies must mate early, survive desiccation and lay eggs afterwards. Despite the differing selection pressures upon males and females, both sexes exhibited increased resistance to desiccation, using similar physiological mechanisms (e.g. greater water content), presumably because they share most of their genetic material. Similar sex-specific responses to selection have also been observed in starvation-selected populations (Chippindale *et al.* 1996).

Conclusions

The D and C populations provide an especially straightforward example of the utility of experimental evolution studies as a complementary approach to comparative studies. The selective force is known: desiccation survival at 14 days past the egg stage. Each D population is paired with a control C population, which is treated in as identical a manner as possible, except for the lack of water. Large population sizes are maintained, minimizing artifacts associated with genetic drift or founder effects. The replicated structure of the populations provides statistical power unattainable in two-species comparisons (Garland and Adolph, 1994). These populations constitute a fivefold replicated experimental equivalent of the two-species comparison, with the additional benefit of known selection histories. Finally, insect water balance has been widely studied (Hadley, 1994*a*), so that the prospect exists for understanding the evolution of this physiological system at whatever level of organization and depth of detail one wants.

The D populations have responded to desiccation selection by reducing water loss, possibly *via* differences in cuticular lipid properties. They have a higher water content, but overall metabolic water stores do not differ between D and C populations, and contribute less than 6 h to desiccation resistance. Dehydration tolerance does not differ between D and C populations, in spite of more than 100 generations under conditions which would appear to favor differentiation. Thus, the D populations have evolved increased desiccation resistance by only some of the possible physiological mechanisms. It remains to be seen whether the lack of differentiation in metabolic water content or dehydration tolerance is due to the lack of genetic variation for these characters, to genetic or physiological constraints, or to other factors.

We thank A. Tabatabaee for assistance with water loss measurements, R. Begino for help with cuticular lipid analyses, D. Nghiem for help with desiccation resistance and dehydration tolerance assays, M. Djawdan and A. E. Williams for technical advice and access to unpublished data, J. B. Lighton for assistance with trouble-shooting the respirometer, T. J. Nusbaum and L. D. Mueller for statistical advice, and V. A. Pierce, A. F. Bennett and T. J. Bradley for helpful discussions and comments on the manuscript. The D and C populations were maintained by L. N. Vu and M. Sheik. This research was funded by a Medical Education and Research Society Young Investigator Award and NSF grant IBN-9317471 to A.G.G. and NIH grant AG09970 to M.R.R.

References

ANTONY, C. AND JALLON, J. (1982). The chemical basis for sex recognition in *Drosophila melanogaster*. *J. Insect Physiol.* **28**, 873–880.

BEAMENT, J. W. L. (1945). The cuticular lipoids of insects. *J. exp. Biol.* **21**, 115–131.

BENNETT, A. F., DAO, K. M. AND LENSKI, R. E. (1990). Rapid evolution in response to high-temperature selection. *Nature* **346**, 79–81.

BLOWS, M. W. AND HOFFMANN, A. A. (1993). The genetics of central and marginal populations of *Drosophila serrata*. I. Genetic variation for stress resistance and species borders. *Evolution* **47**, 1255–1270.

CHIPPINDALE, A. K., CHU, T. J. AND ROSE, M. R. (1996). Complex trade-offs and the evolution of starvation resistance in *Drosophila melanogaster*. *Evolution* **50**, 753–766.

COOPER, P. D. (1983). Components of evaporative water loss in the desert tenebrionid beetles *Eleodes armata* and *Cryptoglossa verrucosa*. *Physiol. Zool.* **56**, 47–55.

CROW, J. F. (1957). Genetics of insect resistance to chemicals. *A. Rev. Ent.* **2**, 227–246.

CROWE, J. H., HOEKSTRA, F. A. AND CROWE, L. M. (1992). Anhydrobiosis. *A. Rev. Physiol.* **54**, 579–599.

DJAWDAN, M., ROSE, M. R. AND BRADLEY, T. J. (1997). Does selection for stress resistance lower metabolic rate? *Ecology* (in press).

ECKSTRAND, I. A. AND RICHARDSON, R. H. (1980). Comparison of some water balance characteristics in several *Drosophila* species which differ in habitat. *Env. Ent.* **9**, 716–720.

ECKSTRAND, I. A. AND RICHARDSON, R. H. (1981). Relationships between water balance properties and habitat characteristics in the sibling Hawaiian drosophilids, *D. mimica* and *D. kambysellisi*. *Oecologia* **50**, 337–341.

ENDLER, J. A. (1986). *Natural Selection in the Wild.* Princeton, NJ: Princeton University Press.

GARLAND, T. AND ADOLPH, S. (1994). Why not to do two-species comparative studies: limitations on inferring adaptation. *Physiol. Zool.* **67**, 797–828.

GARLAND, T. AND CARTER, P. A. (1994). Evolutionary physiology. *A. Rev. Physiol.* **56**, 579–621.

GIBBS, A. AND CROWE, J. H. (1991). Intra-individual variation in cuticular lipids studied using Fourier transform infrared spectroscopy. *J. Insect Physiol.* **37**, 743–748.

GIBBS, A. AND POMONIS, J. G. (1995). Physical properties of insect cuticular hydrocarbons: the effects of chain length, methyl-branching and unsaturation. *Comp. Biochem. Physiol.* B **112**, 243–249.

GRAVES, J. L., TOOLSON, E. C., JEONG, C., VU, L. N. AND ROSE, M. R. (1992). Desiccation, flight, glycogen and postponed senescence in *Drosophila melanogaster*. *Physiol. Zool.* **65**, 268–286.

GUPTA, K., LEE, Y. AND YIN, J. (1995). Extremo-phage: *In vitro* selection of tolerance to a hostile environment. *J. molec. Evol.* **41**, 113–114.

HADLEY, N. F. (1994*a*). *Water Relations of Terrestrial Arthropods*. San Diego: Academic Press.

HADLEY, N. F. (1994*b*). Ventilatory patterns and respiratory transpiration in adult terrestrial insects. *Physiol. Zool.* **67**, 150–189.

HARVEY, P. H. AND PAGEL, M. D. (1991). *The Comparative Method in Evolutionary Biology*. Oxford: Oxford University Press. 239pp.

HOFFMANN, A. A. (1991). Acclimation for desiccation resistance in *Drosophila*: Species and population comparisons. *J. Insect Physiol.* **37**, 757–762.

HOFFMANN, A. A. AND PARSONS, P. A. (1989). An integrated approach to environmental stress tolerance and life history variation: desiccation tolerance in *Drosophila*. *Biol. J. Linn. Soc.* **37**, 117–136.

HOFFMANN, A. A. AND PARSONS, P. A. (1991). *Evolutionary Genetics and Environmental Stress.* Oxford: Oxford University Press. 284pp.

HUEY, R. B., PARTRIDGE, L. AND FOWLER, K. (1991). Thermal sensitivity of *Drosophila melanogaster* responds rapidly to laboratory natural selection. *Evolution* **45**, 751–756.

JOSHI, A., KNIGHT, C. D. AND MUELLER, L. D. (1996). Genetics of larval urea tolerance in *Drosophila melanogaster*. *Heredity* **77**, 33–39.

LAUDER, G. V., LEROI, A. M. AND ROSE, M. R. (1993). Adaptations and history. *Trends Ecol. Evol.* **8**, 294–297.

LENSKI, R. E. AND BENNETT, A. F. (1993). Evolutionary response of *Escherichia coli* to thermal stress. *Am. Nat.* **142**, S47–S64.

LEROI, A. M., ROSE, M. R. AND LAUDER, G. V. (1994). What does the comparative method reveal about adaptation? *Am. Nat.* **143**, 381–402.

MCKENZIE, J. A. AND CLARKE, G. M. (1988). Diazonin resistance, fluctuating asymmetry and fitness in the Australian sheep blow fly, *Lucilia cuprina*. *Genetics* **120**, 213–220.

NOBLE-NESBITT, J. (1991). Cuticular permeability and its control. In *Physiology of the Insect Epidermis* (ed. K. Binnington and A. Retnakaran), pp. 252–283. East Melbourne, Victoria, Australia: CSIRO Publications.

ROSE, M. R. (1984). Laboratory evolution of postponed senescence in *Drosophila melanogaster*. *Evolution* **38**, 1004–1010.

ROSE, M. R., GRAVES, J. L. AND HUTCHISON, E. W. (1990). The use of selection to probe patterns of pleiotropy in fitness characters. In *Insect Life Cycles: Genetics, Evolution and Co-ordination* (ed. F. Gilbert), pp. 29–42. New York: Springer-Verlag.

ROSE, M. R., NUSBAUM, T. J. AND CHIPPINDALE, A. K. (1996). Laboratory evolution: the experimental wonderland and the Cheshire cat syndrome. In *Adaptation* (ed. M. R. Rose and G. V. Lauder), pp. 221–241. San Diego, CA: Academic Press.

ROSE, M. R., VU, L. N., PARK, S. U. AND GRAVES, J. L. (1992).

Selection on stress resistance increases longevity in *Drosophila melanogaster*. *Exp. Gerontol.* **27**, 241–250.

SCHMIDT-NIELSEN, K. (1990). *Animal Physiology: Adaptation and Environment.* 4th edn. Cambridge: Cambridge University Press. 602pp.

SCOTT, D. AND JACKSON, L. L. (1988). Interstrain comparison of male-predominant antiaphrodisiacs in *Drosophila melanogaster*. *J. Insect Physiol.* **34**, 863–871.

SERVICE, P. M., HUTCHINSON, E. W., MACKINLEY, M. D. AND ROSE, M. R. (1985). Resistance to environmental stress in *Drosophila melanogaster* selected for postponed senescence. *Physiol. Zool.* **58**, 380–389.

SERVICE, P. M., HUTCHISON, E. W. AND ROSE, M. R. (1988). Multiple genetic mechanisms for the evolution of senescence in *Drosophila melanogaster*. *Evolution* **42**, 708–716.

TOOLSON, E. C. (1982). Effects of rearing temperature on cuticle permeability and epicuticular lipid composition in *Drosophila pseudoobscura*. *J. exp. Zool.* **222**, 249–253.

TOOLSON, E. C., WHITE, T. R. AND GLAUNSINGER, W. S. (1979). Electron paramagnetic resonance spectroscopy of spin-labelled cuticle of *Centruroides sculpturatus* (Scorpiones: Buthidae): correlation with thermal effects on cuticular permeability. *J. Insect Physiol.* **25**, 271–275.

WILLIAMS, A. E., ROSE, M. R. AND BRADLEY, T. J. (1997). CO_2 release patterns in *Drosophila melanogaster*: the effect of selection for desiccation resistance. *J. exp. Biol.* **200**, 615–624.

Experimental Gerontology

Experimental Gerontology 35 (2000) 957–969

www.elsevier.nl/locate/expgero

Postponed aging and desiccation resistance in *Drosophila melanogaster*

D. Nghiem[a], A.G. Gibbs[b], M.R. Rose[c], T.J. Bradley[c,*]

[a]*M.D. Anderson Cancer Center, 1515 Holcombe, Houston, TX 77030-4095, USA*
[b]*Department of Ecology and Evolutionary Biology, University of Arizona, Tucson, AZ, USA*
[c]*Department of Ecology and Evolutionary Biology, University of California, Irvine, CA 92697, USA*

Accepted 28 June 2000

Abstract

Studies with the fruit fly, *Drosophila melanogaster*, have repeatedly shown that selection for postponed reproduction leads to increases in mean life span and increased stress resistance; including increased resistance to desiccation, starvation and ethanol vapors. We show that desiccation resistance declines with age in both short- and long-lived flies suggesting that desiccation resistance may serve as a useful biomarker for aging-related declines in physiological performance. We examined the physical basis of desiccation resistance in five replicate populations selected for postponed reproduction and five replicate control populations. The variables examined were water content, rates of water loss during desiccation, and water content at time of death due to desiccation. In the absence of desiccation stress, both the flies exhibiting postponed senescence and their controls maintained constant water content throughout their lifetimes. In the presence of desiccation stress, the short-lived flies showed significantly higher rates of water loss at all ages than did the long-lived flies. Flies from the two treatments did not differ in water content at death. Our results indicate that water loss rates are the major determinant of desiccation resistance. Water loss rates are under genetic control and covary with age in populations with genetically-determined postponed senescence. © 2000 Published by Elsevier Science Inc.

Keywords: *Drosophila*; Biomarker; Stress resistance; Desiccation; Water loss

1. Introduction

Laboratory selection for postponed reproduction in *Drosophila melanogaster* has been shown to produce populations which exhibit increased longevity and postponed senescence compared to control populations (Rose, 1984; Luckinbill et al., 1984).

* Corresponding author. Tel.: 1-949-824-84-83.
E-mail address: tbradley@uci.edu (T.J. Bradley).

PII: S0531-5565(00)00163-7

These populations have served as valuable systems for the study of demography, as well as the physiological and genetic bases of aging.

Populations of *D. melanogaster* exhibiting postponed senescence have frequently also shown increased resistance to a variety of environmental stresses including desiccation, starvation and ethanol vapors (Service et al., 1985). Although the correlation between genetically-determined postponed aging and stress resistance in *Drosophila* has been known for a number of years, the physiological basis of this resistance has only recently begun to be elucidated.

In a series of studies aimed at examining selection for stress resistance, Hoffman and Parsons (1989a,b, 1991) suggested that increased resistance to a variety of stresses might be functionally correlated with a reduced metabolic rate. They proposed that reduced metabolic rate, and concomitantly reduced respiration, might confer increased starvation resistance, reduced respiratory water loss, and decreased exposure via the respiratory system to aerial toxins. This hypothesis is also attractive as a possible explanation for increased longevity in populations exhibiting increased stress resistance, in view of theories of aging which attribute the adverse affects of aging to aerobic activity and oxidative damage (Sohal and Orr, 1994).

Djawdan et al. (1996) examined this hypothesis using five replicate long-lived populations (O populations) and five short-lived control populations (B populations) developed by Rose (1984). Contrary to the hypothesis of metabolic change, they found no significant difference in the age-specific metabolic rates of the populations.

Djawdan et al. (1998) conducted investigations to determine the physiological basis of increased starvation resistance not just in the B and O lines but in a variety of demographically and stress selected populations. They found that they could explain 99% of the variability in starvation resistance in these populations by quantifying the energy content of lipid and carbohydrate stores in these flies. Metabolic rate did not change with selection for enhanced starvation resistance.

Gibbs et al. (1997) using populations of *D. melanogaster* selected for desiccation resistance and their control populations, demonstrated that a variety of physiological traits had evolved in flies exhibiting increased desiccation resistance, particularly water content and cuticular permeability. Archer (1998) demonstrated that these two parameters were sufficient to explain 99% percent of the variability in desiccation resistance in a variety of lines. She also verified that these same traits differentiated in a second set of lines selected for enhanced desiccation resistance (Archer, 1998).

In the present paper, we are specifically interested in the physiological basis for differential stress resistance in populations that are genetically differentiated with regard to longevity. While several forms of increased stress resistance are correlated with increased longevity in populations of *D. melanogaster*, these do not necessarily show similar age-specific changes. Starvation resistance, for example, increases with age in adult female *D. melanogaster* as the females accumulate metabolic storage compounds. Desiccation resistance, by contrast, declines with adult age in both short-lived and long-lived populations of flies. Desiccation resistance may therefore serve as a physiological biomarker (similar to cardiac function or muscle strength in humans) which is correlated in an individual both temporally and physiologically with the progression of the aging process. Adult desiccation resistance is of particular interest in *D. melanogaster* because it: (1) has been shown to be under genetic control; (2) varies between short- and long-lived populations; (3) has been shown to decline with age; and (4) is a functional character which is likely to be of significance for survival in wild habitats.

At the organismal level, desiccation resistance can be described as an issue of water balance, which involves three variables: (1) the initial water content of the flies; (2) the rate of water loss during desiccation; and (3) the threshold at which water loss becomes lethal. All the forms of physiological variability between the selected populations (e.g. cuticular, metabolic or hormonal) must affect one of these three variables if they are to influence desiccation resistance. In the present paper, we have undertaken an analysis of these three variables, and their relation to desiccation resistance during aging, in flies that are genetically differentiated with regard to longevity. We examine each of these variables as physiological characters that may decline with age.

2. Materials and methods

2.1. Stocks

The populations of *D. melanogaster* used for this study were derived from a South Amherst, Massachusetts population established by P.T. Ives in 1975. Ten lines were derived from this population by M.R. Rose in 1980. Five replicate lines (O lines) were selected for postponed age of reproduction (postponed senescence) over several generations, until they could be maintained on discrete, 10-week generations (Rose, 1984). Five other replicate lines (B lines) were maintained in the same fashion as the ancestral Ives population, on a 2-week cycle of discrete generations, to serve as controls for the O populations. For experiments, adults from both the B and O populations were maintained in cages at population densities of 10^3–10^4 flies per cage. All populations were reared and maintained at 25°C on a 24-hour light cycle, on standard banana medium.

In order to eliminate parental and grandparental effects, and thereby ensure that all differences between the B and O lines were genetically based, the experimental populations were reared through two generations of common conditions. For the experimental generation, larvae were reared at densities of 60–80 per 10-ml vial. Fourteen days after oviposition, when emerging adults were ~4 days old, flies were dumped into cages. For most experiments, all five replicate lines for the B and O treatment groups were collected on the same day. Because of the limited capacity of our respirometer, egg collections for flies used for water-loss measurements were staggered over a 5-day period, so that pairs of B and O lines could be assayed at the same age. All assays were performed on female flies.

2.2. Desiccation resistance

Desiccation resistance was measured every 7 days, beginning 4 days after eclosion. For each population, ten female flies were placed individually in 30-ml vials. A polyethylene sponge restricted them to the lower half of the vial, and 5 g silica gel desiccant was added above the sponge. The vial was sealed with Parafilm, and flies were checked hourly. When a fly was deemed dead, as indicated by the inability to right itself or move its limbs when agitated, time of death was noted and the fly was immediately removed for determination of water content (see below).

2.3. Water content

Water contents of unstressed flies were measured every 4 days, beginning with the day of eclosion. Flies were anesthetized with CO_2, placed in previously weighed micro-centrifuge tubes, and immediately weighed on a Cahn microbalance to a precision of 1 μg. After drying at 55°C for 1 h, the flies were reweighed to give the dry weight. Water contents were calculated as the difference between wet and dry weights. Five samples of ten flies per replicate line were used to determine a mean value for each population.

The water content at the time of death (dehydration tolerance) was measured in a similar manner to initial water content. As each fly died in the assay for desiccation resistance (above), it was immediately weighed on a tared piece of aluminum foil, dried at 55°C, and reweighed to estimate the water remaining when the fly died.

2.4. Water-loss rates

Rates of water loss were measured using a TR-3 flow-through respirometry system (Sable Systems, Henderson, Nevada) with a Li-Cor LI-6262 water-vapor detector (Gibbs et al., 1997). The entire system was contained in a 25°C room, and the air-flow rate was 25 ml/min. Water-loss rates were measured every 7 days, beginning with day 4 post-eclosion. Four groups of 20 female flies were assayed from each population, at each age. Because empty respirometry chambers exhibited small but measurable differences in water-vapor signal, baselines were recorded for each chamber and subtracted from water-loss rates measured with flies present. Placement of flies in the chambers was staggered so that all flies had been in the respirometry chambers for 3 h at the start of measurement. In order to minimize wash-out artefacts, dry air was flushed through the chambers not being recorded at 25 ml/min, the same flow rate as during recording.

We calibrated the respirometer by injecting known volumes (0.5–5 μl) of water into the air stream, while the system was running under identical conditions to our assays. Rates of water loss from groups of *Drosophila* were calculated from our calibration regression curve, and loss rates per fly were calculated by dividing by the number of flies in the chamber (20).

2.5. Cuticular lipids

Surface lipids of B and O flies were analysed in a separate series of experiments. Flies were reared for two generations of selection and otherwise handled in the same manner as for experiments described above. At 4-days post-eclosion, adult females were frozen at −70°C. Cuticular hydrocarbons were isolated from groups of ten flies by silica gel chromatography (Toolson, 1982). Lipids were analysed by two methods: gas chromatography (GC) and Fourier transform infrared spectroscopy (FTIR). For GC analyses, lipid amounts were quantified by comparison to a peak corresponding to 2.5 μg of *n*-docosane, which was added as an internal standard (Graves et al., 1992). The distribution of hydrocarbon chain lengths was determined by comparison to known *n*-alkane standards. Melting temperatures of hydrocarbon mixtures were determined using FTIR, as described previously (Gibbs and Crowe, 1991; Gibbs et al., 1997).

2.6. Statistical analyses

In the present study, we compare two treatment groups, each containing five populations subjected to differing life-history selection. Because we are interested in differences resulting from the selection regime, statistical analyses were performed using the population means, for $n = 5$ populations per treatment, using Tool Pack software in Microsoft Excel and SYSTAT 5.0. Because the B flies lived a much shorter time than the Os, we could not sample both sets of populations at later ages. We continued to assay the O population as long as flies were available, but direct comparisons between the B and O populations (e.g. repeated measures ANOVAS) were only performed for ages in which all populations were alive.

3. Results

In the present study, we were seeking genetically-determined differences, at the population level, between flies which have undergone selection for short generation times (the B populations) and those which have undergone selection for postponed reproduction (the O populations). The sample size for each of the points shown on the graphs is therefore five populations in each treatment.

3.1. Effects of age on desiccation resistance

Patterns of changing desiccation resistance with age are very interesting in the B and O populations. The O flies are more resistant on day 4 of adult life, the first day we tested, and this statistically significant difference persists throughout life ($P < 0.001$) (Fig. 1). A second very important aspect of the desiccation resistance is that it declines with adult age in both the B and O populations.

3.2. Effects of age on water content

Neither the B nor O populations exhibited significant changes with age in initial water content, as indicated by least-squares regression slopes that were not significantly different from zero ($P > 0.15$ for both B and O populations) (Fig. 2). A repeated measures ANOVA indicated that the percent water content of the B populations was slightly but significantly higher than that of the O populations ($P < 0.001$). This difference is in the opposite direction, however, from that expected to provide greater desiccation resistance to the O flies.

3.3. Effects of age on water content at death from dehydration

Least-squares regressions of age versus water content at the time of death indicated no statistically significant changes with age in either the B or O populations in their tolerance

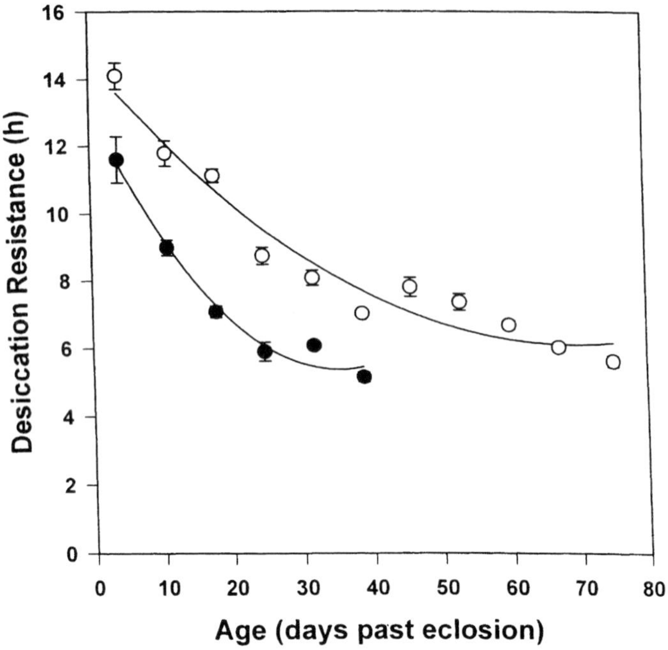

Fig. 1. The desiccation resistance in hours of flies as a function of adult age post-eclosion. The values presented are the means of the average values obtained from the five B populations (●) and the five O population (○) (mean ± S.E). The lines through the points were fit as a second-order regression.

to dehydration ($P > 0.1$ for both B and O populations) (Fig. 3). A repeated measures ANOVA revealed a slight but statistically significant difference between B and O flies ($P = 0.01$), with the O flies exhibiting greater tolerance.

3.4. Effects of age on rates of water loss

The mean rate of water loss of the B populations was significantly higher than that of the O populations as tested by repeated measures ANOVA ($P = 0.01$) (Fig. 4). Given that the populations were grown under identical conditions for two generations and tested under identical conditions, this demonstrates that the B and O populations are genetically differentiated with regard to the rates of water loss in dry air. Rates of water loss increased significantly with age for both the B ($P = 0.002$) and O ($P < 0.001$) populations. Comparison by *t*-test of the slopes of water loss versus age indicated that the O flies showed a slower rate of increase in water loss with age ($P < 0.001$). As a result, the difference in rates of water loss between B and O flies increased with age.

3.5. Cuticular lipids of B and O flies

Differences in either the amount or physical properties of surface lipids may affect rates of water loss (Gibbs, 1998; Rourke and Gibbs, 1999). In 4-day old flies, lipid amounts did not differ between B and O females (Table 1). No novel cuticular hydrocarbons were

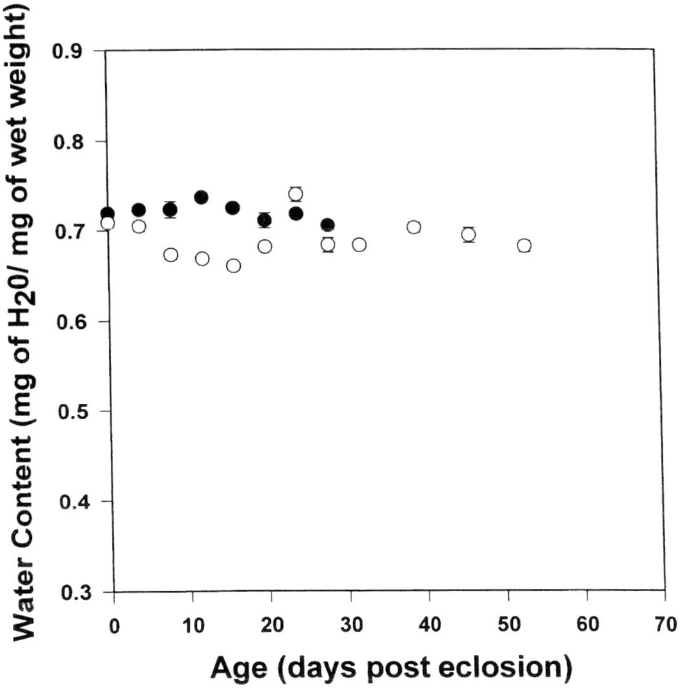

Fig. 2. Whole body water content of the B and O flies in the absence of desiccation stress, i.e. immediately after removal from the colony, as a function of adult age post-eclosion. The values presented are the means of the average values obtained from the five B populations (●) and the five O populations (○) (mean ± S.E).

detected in either set of populations, and the distribution of chain lengths was similar. Hydrocarbons ranged from 21 to 31 carbons in length, with 27-carbon species being the most abundant. Although hydrocarbons from O females were slightly longer on average than those of B females, lipid melting temperatures did not differ between selection treatments.

4. Discussion

Drosophila populations selected for postponed aging are useful models for the study of aging, because they exhibit genetically-determined differences in life span. A functional understanding of these genetic differences depends on the elucidation of age-specific changes in physiological characteristics of these populations. In particular, physiological systems that exhibit both genetic differentiation and age-specific decline in function are candidates for systems subject to late-life selection. Because reduced stress resistance may play a role in age-specific increases in mortality, a knowledge of which physiological systems decline with age may provide insight not only into the mechanisms of aging but mortality as well.

Our results in this study corroborate those obtained by Service et al. (1985), when the O

964 D. Nghiem et al. / Experimental Gerontology 35 (2000) 957–969

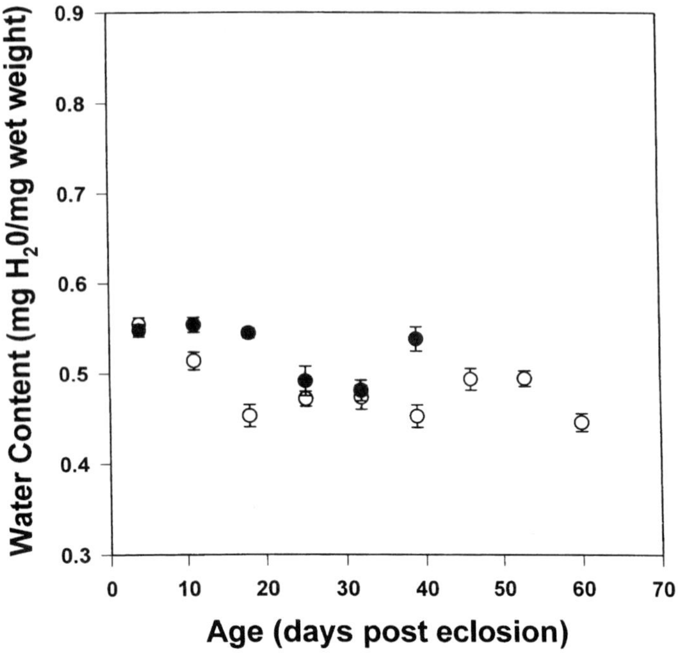

Fig. 3. Whole body water content of the B and O flies immediately following death by desiccation, as a function of adult age post-eclosion. The values presented are the means of the average values obtained from the five B populations (●) and the five O populations (○) (mean ± S.E).

populations had undergone approximately 24 generations of selection. After >100 generations of selection in the Os (and >450 generations of the B flies), the previously observed differences in desiccation resistance are still present, and the O flies remain more resistant at all ages (Fig. 1). What are the physiological mechanisms responsible for these differences? At the organismal level and under conditions where ingestion is precluded, differences in desiccation resistance at a given age can be explained by one or more of three variables: (1) the initial water content; (2) the water content at which death occurs; and (3) the rate of water loss. Age-related declines in desiccation resistance could be due to age-specific changes in any or all of these factors.

The B and O flies exhibited no age-specific changes in water content (Fig. 2). In contrast to expectations, the B flies actually had a higher percent water content than the Os. This may be due to increased lipid content in the O females, which would increase the dry mass relative to water content. We did not measure lipid contents in these experiments, but previous studies of these populations have documented higher lipid contents in O females (Service, 1987; Chippendale et al., 1998; Djawdan et al., 1998). Our results suggest that selection- and age-related differences in desiccation resistance are not caused by differences in initial water contents.

The ability of flies to tolerate loss of water, as indicated by water content at death, did not change with age (Fig. 3). This character cannot, therefore, explain age-dependent changes in desiccation resistance. The B and O flies did differ slightly in their water

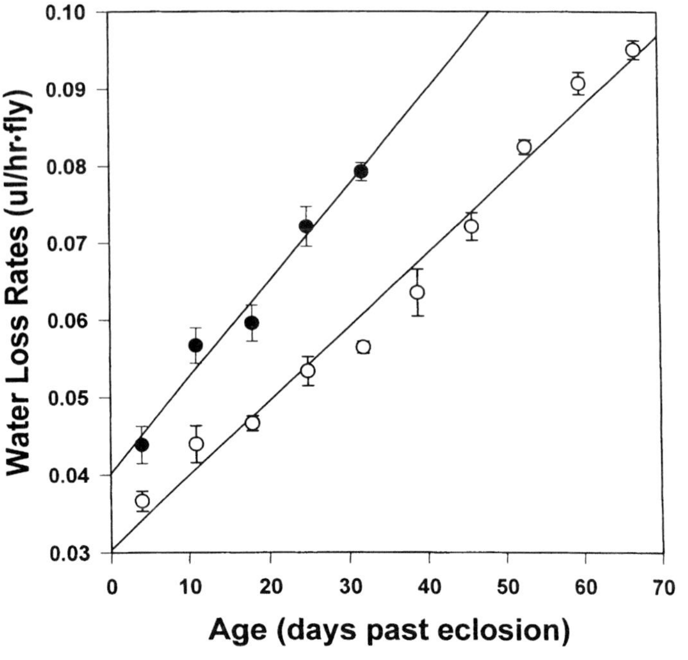

Fig. 4. Water loss rates measured by the release of water vapor into a dry stream of air, as a function of adult age post-eclosion. The values presented are the means of the average values obtained from the five B populations (●) and the five O populations (○) (mean ± S.E). The lines through the points were fit as a linear regression.

contents at death, but we conclude that these differences play only a minor role in explaining differences in desiccation resistance between the B and O populations (see further discussion of this issue below).

Because neither initial nor final water contents differ substantially as a function of age or selection treatment, we are left with water loss as the only remaining cause for changes in desiccation resistance. Beginning with our initial measurements on day 4 of adult life, the O flies exhibited consistently lower rates of water loss than the B flies (Fig. 4). In both selection treatments, water-loss rates increased with age, but the rate of increase was slower in the O populations. Water loss rates were the character that changed most with age, and that differed most between selection treatments (B versus O). The quantitative correlation between changes in the rate of water loss and changes in desiccation resistance

Table 1
Cuticular lipids of B and O females, at 4-days post-eclosion (data are means (±SE) for $n = 5$ populations per selection treatment. P-values are for unpaired t-tests ($n = 5$)

Lipid property	B flies	O flies	P-value
Amount (μg fly^{-1})	1.19 (0.12)	1.06 (0.08)	0.41
Mean hydrocarbon chain length	25.74 (0.30)	25.98 (0.09)	0.04
Melting temperature (°C)	34.45 (0.77)	33.19 (0.69)	0.26

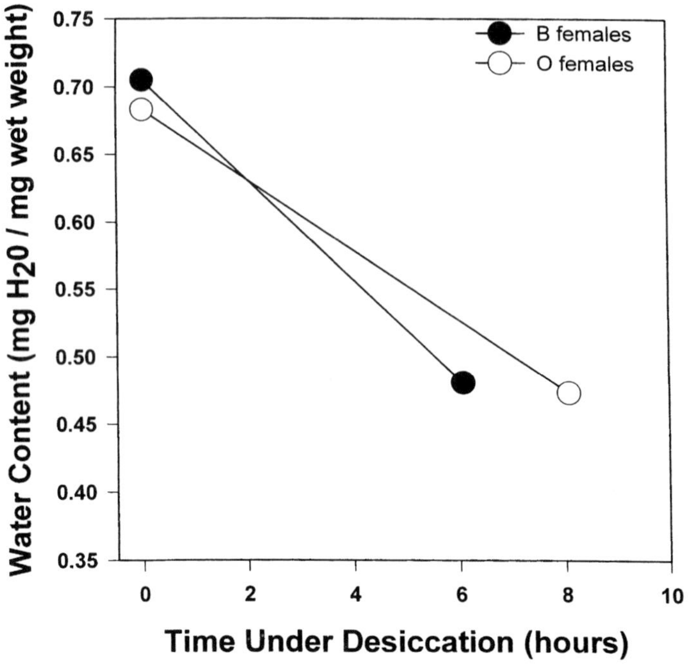

Fig. 5. A visual depiction of water loss during a desiccation bout in adult flies 32-days post-eclosion. The points at time zero indicate the water content of the B and O flies at the initiation of desiccation. With time, the B flies lose water faster than the O flies. As a result, water content declines to a lethal level sooner in the B than in the O flies. Since both groups have the same tolerance for water loss (a percent water content around 48%) the O flies resist desiccation longer than the B flies solely due to a slower rate of water loss.

is demonstrated by the observation that the rate of water loss increased 1.87-fold from day 4 to day 32 in the B populations while desiccation resistance decreased 1.90-fold. Similarly, water loss rates in the O flies populations increased 1.61-fold over this same time period and desiccation resistance decreased 1.60-fold. The physiological character that quantitatively dominates in determining changes in desiccation resistance is therefore water loss rate.

This point is well illustrated in Fig. 5, which shows the changes in water content during desiccation in flies aged 32 days post-eclosion. It can be seen that the B and O populations begin with a very similar water content, although the B flies contain slightly more. As desiccation proceeds, the flies lose water. The line depicting water loss for the B flies has a more negative slope, illustrating a more rapid loss of body water. The B flies die following 6 h of desiccation, while the O flies can survive for 8 h. Since the water content at death is essentially identical in the B and O flies, the difference in desiccation tolerance is solely due to the difference in water loss rate.

We have thus identified a change in the rate of water loss as the physiological basis for both selection- and age-related changes in desiccation resistance. Clearly, it would be interesting to gain a more detailed understanding of the specific mechanisms responsible. Insects lose water through three distinct routes: cuticular, respiratory and excretory. Our experiments were not designed to distinguish which of these pathways has been modified in the B and O flies. Pathways of water loss have been investigated in detail in *D. melanogaster*, however, particularly in populations selected for desiccation resistance and their controls (see reviews by Bradley et al., 1999; and Gibbs, 1999).

Excretory events in our flow-through respirometry system can be identified as peaks of water vapor appearing at the humidity sensor (Gibbs et al., 1997). Excretory water loss has previously been quantified by integration of these peaks, in two sets of populations descended from the O flies. These include the desiccation-selected D populations, and their control C populations. In these O-derived stocks, excretory water loss accounted for $<5\%$ of total water loss (Gibbs et al., 1997). We did not perform a detailed analysis of excretory water loss in our study, but inspection of water-loss recordings revealed no evidence that excretion accounted for a significant fraction of total water loss, at any age in any of the B or O populations. Although excretory water loss may differ with age, it was not sufficient to account for the two-fold increase in total losses as flies aged.

Insects can modulate respiratory loss from the tracheal system by changing the patterns with which the external spiracles are opened and closed (Lighton, 1994). Williams et al. (1997) examined B and O flies and found no differences in respiratory patterns, whereas the desiccation-selected D populations had differentiated patterns of spiracular control compared to their control (C) populations. Subsequent analyses of simultaneous CO_2 and H_2O release patterns revealed that these changes actually had no effect on overall water loss (Williams and Bradley, 1998; Williams et al., 1998). Because the B and O flies exhibit no differences in respiratory pattern at day 4 (when overall water loss differed by 1.22-fold; Fig. 4), and highly differentiated patterns of respiratory control have no effect on water loss, it is unlikely that respiratory losses play an important role in the differences in water-loss rates between B and O flies.

Of the three routes of water loss from insects, only the cuticular pathway remains. Can we further narrow our focus to a particular aspect of cuticular structure? An obvious candidate is the epicuticular lipids, which provide the primary barrier to transpiration through the cuticle (Gibbs, 1998). Our analyses of surface lipids extracted from 4-day old flies revealed no significant differences in lipid amounts or physical properties. In accordance with previous work by Graves et al. (1992), gender was the primary factor affecting surface lipid composition (data not shown). A similar lack of differentiation was obtained in comparisons of the D and C populations, which exhibit even greater differences in rates of water loss (Gibbs et al., 1997). These results suggest that we can exclude cuticular lipids as the cause of differentiation between water-loss rates of young flies. Other components of the cuticle (e.g. chitin, cuticular proteins), or hormonal changes affecting cuticular permeability, may be involved.

A limitation of this analysis is that detailed studies of respiration, cuticular lipids, and excretion in these populations have been performed only on young flies. Further mechanistic studies of the effects of aging on pathways of water loss in *Drosophila* clearly are in order. The above studies indicate that the techniques for conducting such studies are available.

In summary, we have identified an important physiological process, water balance, that exhibits age-related deterioration and genetic differentiation between aging-selected and control populations of *D. melanogaster*. Our understanding of the life histories and ecologies of these populations suggests that water stress is an important proximate cause of mortality in aging-selected populations. Several lines of evidence indicate that cuticular function is the primary site of selection on water balance. The cuticle therefore provides a concrete, tractable system for investigating mechanisms that influence age-related physiological performance, stress resistance and mortality.

Acknowledgements

Supported by NSF grants IBN-9317471 to A.G.G and IBN-9507435 to T.J.B.

References

Archer, M.A., 1998. Detection of a trade-off between high levels of desiccation resistance and longevity in *Drosophila melanogaster*. Master's thesis, University of California, Irvine, CA, USA.

Bradley, T.J., Williams, A.E., Rose, M.R., 1999. Physiological responses to selection for desiccation resistance in *Drosophila melanogaster*. Am. Zool. 39, 337–345.

Chippendale, A.K., Gibbs, A.G., Sheik, M., Yee, K.J., Djawdan, M., Bradley, T.J., Rose, M.R., 1998. Resource acquisition and the evolution of stress resistance in *Drosophila melanogaster*. Evolution 52, 1342–1352.

Djawdan, M., Sugiyama, T.T., Schlaeger, L.K., Bradley, T.J., Rose, M.R., 1996. Metabolic aspects of the trade-off between fecundity and longevity in *Drosophila melanogaster*. Physiol. Zool. 69 (5), 1176–1195.

Djawdan, M., Chippindale, A.K., Rose, M.R., Bradley, T.J., 1998. Metabolic reserves and evolved stress resistance in *Drosophila melanogaster*. Physiol. Zool. 71 (5), 584–594.

Gibbs, A.G., 1998. Water-proofing properties of cuticular lipids. Am. Zool. 38, 471–482.

Gibbs, A.G., 1999. Laboratory selection for the comparative physiologist. J. Exp. Biol. 202, 2709–2718.

Gibbs, A.G., Crowe, J.H., 1991. Intra-individual variation in cuticular lipids studied using Fourier transform infrared spectroscopy. J. Insect Physiol. 37, 743–748.

Gibbs, A.G., Chippindale, A.K., Rose, M.R., 1997. Physiological mechanisms of evolved desiccation resistance in *Drosophila melanogaster*. J. Exp. Biol. 200, 1821–1832.

Graves, J.L., Toolson, E.C., Jeong, C., Vu, L.N., Rose, M.R., 1992. Desiccation, flight, glycogen and postponed senescence in *Drosophila melanogaster*. Physiol. Zool. 65, 268–286.

Hoffmann, A.A., Parsons, P.A., 1989a. An integrated approach to environmental stress tolerance and life history variation. Desiccation tolerance in *Drosophila*. Biol. J. Linn. Soc. 37, 117–136.

Hoffmann, A.A., Parsons, P.A., 1989b. Selection for increased desiccation resistance in *Drosophila melanogaster*: additive genetic control and correlated responses for other stresses. Genetics 122, 837–845.

Hoffmann, A.A., Parsons, P.A., 1991. Evolutionary Genetics and Environmental Stress, Oxford University Press, New York.

Lighton, J.B., 1994. Discontinuous ventilation in terrestrial insects. Physiol. Zool. 67, 142–162.

Luckinbill, L.S., Arking, R., Clare, M.J., Cirocco, W.C., Buck, S.A., 1984. Selection for delayed senescence in *Drosophila melanogaster*. Evolution 38, 996–1003.

Rose, M., 1984. Laboratory evolution of postponed senescence in *Drosophila melanogaster*. Evolution 38, 1004–1010.

Rourke, B.C., Gibbs, A.G., 1999. Effects of lipid phase transitions on cuticular permeability: Model membrane and in situ studies. J. Exp. Biol. 202, 3255–3262.

Service, P.M., 1987. Physiological mechanisms of increased stress resistance in *Drosophila melanogaster* selected for postponed senescence. Physiol. Zool. 58, 321–326.

Service, P.M., Hutchison, E.W., MacInley, M.D., Rose, M.R., 1985. Resistance to environmental stress in *Drosophila melanogaster* selected for postponed senescence. Physiol. Zool. 58, 380–389.

Sohal, R.S., Orr, W.C., 1998. Role of oxidative stress in senescence. Aging-Clin. Exp. Res. 10, 149–151.

Toolson, E.C., 1982. Effects of rearing temperature on cuticle permeability and epicuticular lipid composition in *Drosophila pseudoobscura*. J. Exp. Zool. 222, 249–253.

Williams, A.E., Bradley, T.J., 1998. The effect of respiratory pattern on water loss in desiccation-resistant *Drosophila melanogaster*. J. Exp. Biol. 201, 2953–2959.

Williams, A.E., Rose, M.R., Bradley, T.J., 1997. CO_2 release patterns in *Drosophila melanogaster*: the effect of selection for desiccation resistance. J. Exp. Biol. 200, 615–624.

Williams, A.E., Rose, M.R., Bradley, T.J., 1998. Using laboratory selection for desiccation resistance to examine the relationship between respiratory pattern and water loss in insects. J. Exp. Biol. 201, 2945–2952.

III

REPRODUCTION, NUTRITION, AND AGING

Reproduction, Nutrition, and Aging

Michael R. Rose[1] and Adam K. Chippindale[2]

[1]*Department of Ecology and Evolutionary Biology, University of California, Irvine, CA 92697-2525, USA mrrose@uci.edu; also to whom correspondence should be sent*

[2]*Department of Biology, Queen's University at Kingston, ON K7L 3N6, Canada chippind@biology.queensu.ca.*

The effect of multiple generations of later reproduction on the evolution of longevity is an obvious corroboration of the evolutionary theory of aging developed by Hamilton (1966) and Charlesworth (e.g. 1980). However, it is clearly not the only evolutionary effect of late reproduction of fruit fly cultures. Early female fecundity is decreased and later fecundity is increased (e.g. Rose, 1984, reprinted here; Luckinbill *et al.*, 1984). Unsurprisingly, female reproduction evolves along with longevity in Methuselah flies. In addition, Service and his colleagues (e.g. Service, 1989, 1993; Service and Fales, 1993; Service and Vossbrink, 1996) showed that male fertility was also altered in Methuselah flies. When young, "O" flies (selected for late reproduction) had less ability or inclination to fertilize females, compared with the control flies. When older, the fertility of O flies was greater than that of their control, or "B," flies. These findings suggested the existence of a general "evolutionary trade-off" between early reproduction and later survival or reproduction, also called "antagonistic pleiotropy" between early and late life, in both females and males.

But the early demonstration of this trade-off was to raise more questions than answers. One of the complications has already been discussed in Section One: the relative reduction in early fecundity in O flies comes and goes as the environment changes (Leroi *et al.*, 1994; reprinted here). If there is an evolutionary trade-off between reproduction and longevity, it is one that depends on the environment, among other things.

A more interesting question is the biological mechanism(s) that underlies the evolutionary trade-off between reproduction and longevity, when it does occur. What are the physiological mechanisms that require a reduction in early fertility when aging is slowed or postponed?

The idea of an energetic trade-off between survival and reproduction was one clue. George C. Williams (e.g. 1966) had long argued for the inevitability of a trade-off between survival and reproduction because of the diversion of limiting resources, especially energy. If, for example, calories stored in the form of fat were used to produce gametes, then those same calories will not be available to the adult body for sustenance during periods of starvation.

The natural connection was to the calories stored in the fat used to survive acute starvation. We already knew that such stored calories were a reasonable predictor of longevity, so long as fat reserves did not rise to very high levels (cf. Service *et al.*, 1985, reprinted here; Djawdan *et al.*, 1998, reprinted here; Phelan *et al.*, 2003). But how could we show that these stored calories were also used in reproduction?

Chippindale and Leroi performed a simple manipulative experiment that revealed some of the physiology involved (Chippindale *et al.*, 1993; reprinted here). They supplied Methuselah flies and their controls with two nutritional regimes: high and low amounts of yeast paste spread over the surface of agar-rich food. They also performed experiments where the levels of yeast paste were changed abruptly, from high to low, and from low to high. Their results showed that well-fed female flies reproduce more, die sooner, and have *lower* levels of starvation resistance, relative to flies fed less. Apparently female flies shunt calories from fat reserves to eggs when they are given more food. They then reproduce more, but die sooner. This result implicated the transfer of resources between ovaries and fat bodies within the fly.

Studies of the ovaries of Methuselah flies have been somewhat ambiguous in their support for this interpretation. An early study (Rose *et al.*, 1984) found that the Methuselah flies had smaller ovaries when young, compared with normal flies, but later work (Chippindale *et al.* 1997, reprinted here) did not reproduce this finding. Carlson *et al.* (1998) found that oocytes matured slower in young

Methuselah flies, but their ovaries had marginally more ovarioles. This pattern suggests slower maturation of the Methuselah ovary, but a larger eventual structure. More ovarioles might facilitate later reproduction of these flies by numerically counteracting the loss of stem cells with age. This might account for some of the inconsistencies in the data. The overall picture of the Methuselah female reproductive system is one of somewhat slower maturation of a more durable structure. This conforms to the detailed pattern of Methuselah female fly reproduction, which is reduced when young, especially under stressful conditions (Rose, 1984; Leroi *et al.*, 1994; Djawdan *et al.*, 1996; all reprinted here).

One of the central parameters for any interpretation of bioenergetics is metabolic rate. We have already mentioned the controversy about metabolic rates that grips the nematode aging literature. The interpretation of fruit fly research on aging would be greatly complicated if metabolic rates fluctuate widely with longevity. Djawdan *et al.* (1996; reprinted here) made a careful study of the metabolic rate, caloric reserves, and energetic expenditures of Methuselah flies in the population cages within which are they are normally kept. While the caloric reserve levels of Methuselah flies were greater than those of normal flies, and their fecundity was much lower, the metabolic rate of the longer-lived flies was not significantly changed. The overall energy budget of the Methuselah flies has changed substantially, but metabolic rate is unusually steady in the cages the flies inhabit.

Another feature of reproduction and energetics that is quite different in Methuselah flies is their responsiveness to mating. The fecundity of Methuselah females is highly responsive to being mated for the first time, so long as there is abundant nutrition (Chippindale *et al.*, 1997; reprinted here). Methuselah females that receive abundant food increase their egg-laying more than ten-fold when they lose their virginity, compared with virgin females given abundant food. Normal females increase their egg-laying only two-fold when well-fed and mated for the first time. In a sense, the Methuselah females appear to hold back from laying eggs until they receive both nutritional and sexual cues. This reproductive restraint probably enables them to live a long time as adults, and then reproduce at late ages.

The effect of diet and mating status on longevity-related characters in male Methuselah flies differs in important ways from the pattern seen in females. For example, while diet has a strong influence on longevity in both sexes, and on starvation resistance in females, it has a slight or non-existent effect on starvation resistance or mating propensity in young males (Chippindale *et al.*, 1993, 1997; both reprinted here). Male B and O flies pursue females, losing weight and stored caloric reserves, early in life irrespective of the background medium provided. While inert with respect to nutrition early in life, males are sensitive to the presence of mates, with virgin males being more starvation tolerant than those that have cohabitated with females. This pattern is qualitatively the same for Methuselah males and their controls. Quantitatively, however, Methuselah males lose starvation resistance after adult eclosion at a reduced rate compared with B males and have lower overall mating propensity (Chippindale *et al.*, 1997; Service *et al.* papers cited above), implicating reduced early courtship or mating activity in the evolution of postponed senescence.

Methuselah flies appear to use food energy with restraint. They keep their caloric reserves at a higher level and they tend to reduce their early reproduction. Later on in adult life, after they are mated, they mobilize their resources for reproduction long after normal flies are already dead. It is a striking re-shaping of the fly physiology toward the goal of delayed aging.

References

Carlson, K.A., T.J. Nusbaum, M.R. Rose, and L.G. Harshman. 1998. Oocyte maturation and ovariole number in lines of *Drosophila melanogaster* selected for postponed senescence. *Funct Ecol* **12**: 514–520.

Charlesworth, B. 1980. *Evolution in Age-structured Populations*. Cambridge University Press, Cambridge.

Chippindale, A.K., A.M. Leroi, S.B. Kim, and M.R. Rose. 1993. Phenotypic plasticity and selection in *Drosophila* life-history evolution. I. Nutrition and the cost of reproduction. *J Evol Biol* **6**: 171–193.

Chippindale, A.K., A.M. Leroi, H. Saing, D.J. Borash, and M.R. Rose. 1997. Phenotypic plasticity and selection in *Drosophila* life-history evolution. 2. Diet, mates and the cost of reproduction. *J Evol Biol* **10**: 269–293.

Djawdan, M., A.K. Chippindale, M.R. Rose, and T.J. Bradley. 1998. Metabolic reserves and evolved stress resistance in *Drosophila melanogaster. Physiol Zool* **71**: 584–594.

Djawdan, M., T. Sugiyama, L. Schlaeger, T.J. Bradley, and M.R. Rose. 1996. Metabolic aspects of the trade-off between fecundity and longevity in *Drosophila melanogaster. Physiol Zool* **69**: 1175–1195.

Hamilton, W.D. 1966. The moulding of senescence by natural selection. *J Theor Biol* **12**: 12–45.

Leroi, A.M., A.K. Chippindale, and M.R. Rose. 1994. Long-term laboratory evolution of a genetic trade-off in *Drosophila melanogaster*. I. The role of genotype x environment interaction. *Evolution* **48**: 1244–1257.

Luckinbill, L.S., R. Arking, M.J. Clare, W.C. Cirocco, and S. A. Buck.1984. Selection for delayed senescence in *Drosophila melanogaster*. *Evolution* **38**: 996–1003.

Phelan J.P., M.A. Archer, K.A. Beckman, A.K. Chippindale, and M.R. Rose. 2003. Breakdown in correlations during laboratory evolution. I. Comparative analyses of *Drosophila* populations. *Evolution* **57**: 527–535.

Rose, M.R. 1984. Laboratory evolution of postponed senescence in *Drosophila melanogaster. Evolution* **38**: 1004–1010.

Rose, M.R., M.L. Dory, A.M. Coyle, and P.M. Service. 1984. The morphology of postponed senescence in *Drosophila melangaster. Can J Zool* **62**: 1576–1580.

Service, P.M. 1989. The effect of mating status on lifespan, egg laying, and starvation resistance in *Drosophila melanogaster* in relation to selection on longevity. *J Insect Physiol* **35**: 447–452.

Service, P.M. 1993. Laboratory evolution of longevity and reproductive fitness components in male fruit flies: Mating ability. *Evolution* **47**: 387–399.

Service, P.M. and A.J. Fales. 1993. Evolution of delayed reproductive senescence in male fruit flies: Sperm competition. *Genetica* **91**: 111–125.

Service, P.M., E.W. Hutchinson, M.D. MacKinley, and M.R. Rose. 1985. Resistance to environmental stress in *Drosophila melanogaster* selected for postponed senescence. *Physiol Zool* **58**: 380–389.

Service, P.M. and R. Vossbrink. 1996. Genetic variation in "first" male effects on egg-laying and remating by female *Drosophila melanogaster*. *Behav Genet* **26**: 39–47.

Williams, G.C. 1966. Natural selection, the costs of reproduction and a refinement of Lack's principle. *Am Nat* **100**: 689–690.

J. evol. Biol. 6: 171–193 (1993)
1010–061X/93/020171–23 $ 1.50 + 0.20/0

Phenotypic plasticity and selection in Drosophila life-history evolution. I. Nutrition and the cost of reproduction

Adam K. Chippindale, Armand M. Leroi, Sung B. Kim and Michael R. Rose*

Department of Ecology and Evolutionary Biology, University of California, Irvine, CA 92717, USA.

Key words: Life-history evolution; phenotypic plasticity; *Drosophila*; dietary restriction; trade-offs.

Abstract

Earlier experiments have shown that the evolution of postponed senescent populations can be achieved by selection on either demographic or stress resistance characters. Both types of selection have produced results in which survival characters (stress resistance and longevity) have apparently traded-off against early-life fecundity.

Here we present the results of a series of experiments in which an environmental variable – the level of live yeast inoculate applied to the substrate – produces a qualitatively similar phenotypic response: longevity and starvation resistance are enhanced by lower yeast levels, at the expense of fecundity. For the starvation resistance versus fecundity experiments we show a negative and linear relationship between the norms of reaction for each character across a gradient of yeast levels. This phenotypic trade-off is stable across the 20 populations and 4 selection treatments reported on here, and its general agreement with earlier selection results suggests that the evolutionary response and the phenotypically plastic response may share a common physiological basis. However, an important discrepancy in the lifetime fecundity data between the selection response and the dietary manipulations preclude strict analogy. The results broadly conform to a simple "Y-model" of allocation, in which a limited resource is divided between survival and reproduction; here the characters are starvation resistance and longevity versus fecundity.

---

* Author for correspondence.

One of the major issues in research on life-history evolution is the relationship between trade-offs inferred from responses to selection and trade-offs inferred from responses to phenotypic manipulation (Partridge and Harvey, 1988). Reznick (1985, 1992), among others (e.g. Rose et al., 1987), has argued that hypotheses about the role of trade-offs in evolution depend critically on the effects of alleles, which may be different from the physiological interactions that might give rise to a phenotypic trade-off. One factor that has limited the pertinence of this debate is the relative lack of experimental systems in which *both* genetic and non-genetic trade-offs have been delineated.

Laboratory evolution of life-history in *Drosophila melanogaster* has been extensively exploited in the study of evolutionary genetic trade-offs (e.g. Rose and Charlesworth, 1980; Rose, 1984; Luckinbill et al., 1984; Service and Rose, 1985; Luckinbill and Clare, 1985; Service et al., 1988; Hutchinson et al., 1991). In addition, there have been a number of *Drosophila* experiments that indicate considerable phenotypic plasticity for life-history characters under environmental manipulation (Partridge and Farquhar, 1981; Partridge, 1987; Partridge et al., 1986, 1987; Luckinbill et al., 1988; Service, 1989). An obvious strategy, given these results, is to attempt to identify common patterns in these two types of experiment, and then seek mechanisms uniting the genetic and manipulative findings.

A central theme of many theories concerning evolutionary and other trade-offs is that of energetic allocation (e.g. Gadgil and Bossert, 1970). In the context of life-history evolution, the phenomenon of energetic allocation has surfaced in research indicating the role of allocation of energetic reserves, specifically lipid and glycogen, between survival and reproductive functions in the evolution of postponed aging in *D. melanogaster* (Service, 1987, 1989; Graves et al., 1992). But possibly analogous phenomena arise in the dietary restriction literature of gerontology, in which reduction in the caloric intake of rodents appears to give rise to increased longevity as a result of a shift of resources from reproductive activity to adult survival (Holehan and Merry, 1985; Masoro, 1988; Rose, 1991). This parallelism suggests the possibiliy that genetically-mediated allocation of energetic reserves in *D. melanogaster* could proceed by similar mechanisms to those of nutritionally-mediated allocation, as in rodent dietary restriction. A problem facing this hypothesis is that dietary restriction in *Drosophila* has given inconsistent results, with some failures to discover any apparent enhancement of longevity with dietary restriction (e.g. Le Bourg and Medioni, 1991). Therefore, exploration of this possible line of research connecting plasticity and evolutionary response via energetic metabolism requires elucidation of the nature of dietary restriction responses in *D. melanogaster*.

In this article, we report dietary manipulation experiments with *D. melanogaster* that appear to mimic both evolutionary trade-offs between survival and reproduction as well as the phenotypic trade-offs observed in rodent caloric restriction, at least with respect to some characters. In addition, we present evidence that energetic metabolism determines at least part of the phenotypic trade-off between survival and reproduction in these populations.

Methods and materials

Populations employed

The *Drosophila* populations used in these experiments were all derived from a South Amherst, Massachusetts, population sampled by P. T. Ives in 1975, this population being designated "IV". This population has since been maintained on two-week discrete generations with abundant food at 25° C and census populations of at least one thousand adults per generation. In 1980, ten populations were derived from the basal IV stock: five replicate "B" populations, maintained in the same fashion as the IV stock, and five replicate "O" populations selected for later reproduction by extension of the total generation-time to 10 weeks, as outlined in Rose (1984). In 1988, ten additional populations were derived from the O populations. Five of these were selected for desiccation resistance in cages ("D") and five of these were controls handled in parallel except for the absence of desiccation induced motality ("C") (Rose et al., 1990). The D populations evolved greater desiccation resistance (Rose et al., 1990) and longevity (Rose et al., 1992), relative to the C populations. The total generation time of both D and C populations is about three weeks. All cultures are given banana-molasses medium for larval growth, as in Rose (1984).

Assay procedures

Since the imposition of radically different selection regimes can create parental effects, in all experiments the populations were sampled and placed on a common two-week cycle for two generations, after the fashion of the B populations.

Egg collection for the experimental flies was done from these synchronized population samples over 24 hours, and the larvae were then reared at 25° C at a density of 60–80 per vial, each vial containing approximately five mL of food. When the adults began to eclose, the earliest to emerge were discarded and collection of adults for assay proceeded over the next 24 hours, at approximately the peak of the emergence distribution. Reported age of the adults is from emergence ± 12 hours.

Age-specific mortality and fecundity were assayed as in Rose (1984), except as indicated below. Starvation resistance was assayed using the procedures of Service et al. (1985), except that a single mated pair of flies were starved together in each assay vial.

Statistical analysis

All statistical comparisons proceeded using the means of populations as basic observations, since our hypotheses concern those means, rather than the phenotypes of individuals. Given this procedure, the distribution of individual phenotypes within populations are not tested for normality. The distribution of population means cannot be usefully tested because of the small number of populations per treatment, typically five.

All statistical work was carried out using STATVIEW on MacIntosh computers.

Longevity and fecundity comparisons of B and O, D and C populations

In these experiments, dietary manipulation involved variation in the amount of live yeast supplied to a pair of adult flies, maintained in their own vial, with daily transfer to a new vial. Two levels of yeast were used: "high", the yeast level being 1.5 mg/vial, and "low", 0.15 mg/vial. The yeast paste was spread on the type of charcoal medium used in Rose (1984), in order to facilitate counting of eggs. In addition to the "high" and "low" food treatments, B_1 and O_1 populations were subjected to a third yeast level, consisting of undiluted yeast suspension applied with a pasteur pipette, as in Rose (1984). Vials treated in this manner have stickier surfaces; the concern was that longevity differences might arise between vials with different amounts of yeast because of stickiness alone. Comparison of this treatment with the "high" yeast treatment (t-test; analysis not shown) did not reveal any statistically significant difference for longevity or fecundity. This suggests that the mechanical features of the substrate are not the primary cause of differences in life-history when yeast levels vary.

A total of 1,600 flies were used in the comparison of B and O populations. Each mated pair was transferred each day at the same time, with any death or escape noted and replacements being made from widows or widowers within the experiment. Fecundity data were collected from all used vials for the first 16 days. After that, daily counts continued to be made from B_1 and O_1 vials, only. The other four populations in each treatment were counted once every four days on a rotating basis.

A total of 2,400 flies were used in the comparison of D and C populations. Only the fecundities from days three, four, and five from eclosion were counted. After day five, transfers occurred on Monday, Wednesday, and Friday, in the D/C comparison, although mortality checks were performed daily. Complete longevity records were kept for both B/O and D/C comparisons.

An experiment was performed to determine the time-course of the fecundity response to high and low nutrition. Forty mated pairs of flies from each of four populations (B_4, B_5, O_4 , O_5) were given either high or low nutrition for seven days after adult emergence. Half of the flies from each cohort were then switched to the opposite nutrition level for three additional days. Daily fecundity records were kept for all flies throughout the ten days of observation, for a total of 1,600 observations.

Response of starvation resistance and fecundity to varying yeast levels in B and O populations

Two experiments were performed in which starvation resistance and fecundity were measured over a range of yeast levels. Four mated pairs were kept per vial prior to assay, with 20 flies of each sex being employed for each population at each yeast level for each character assayed. The total number of flies employed was 2,400. The yeast conditioning period lasted four days; results from the nutritional switching experiment, discussed below, suggest that this is sufficient time to establish the fecundity levels characteristic of a particular nutritional regime. At the end of the yeast conditioning, the flies were split into starvation and fecundity groups. The experiment used six yeast levels: 0.0195, 0.391, 0.781, 1.562, 3.125, and 12.5 mg/vial. Fecundities were measured for 24 hours of egg-laying with one pair of flies per vial against a background of yeasted charcoal food. Starvation resistance was measured by placing a mated pair of flies in a sealed and humidified vial without food; mortality was scored every 6 hours. All ten B and O populations were compared.

Results

Effects of dietary restriction on longevity in B and O populations

Figure 1 gives the average survival curves for B and O flies, at both high and low yeast levels. The patterns of mean longevity among lines are summarized in Table 1. Analysis of variance shows that B and O populations differ significantly in longevity, as do high and low yeast treatments. There are no significant interaction effects between the B/O treatment effect and the dietary treatment effect, although the three-way interaction between sex, stock, and nutrition is near the threshold of significance ($P = 0.059$).

Effects of dietary restriction on fecundity in B and O populations

Figure 2 gives the daily fecundities of B and O females, at both high and low yeast levels. Analysis of variance of early fecundity, summarized in Table 2, shows that B and O populations differ significantly, as do high and low yeast treatments. There is no significant interaction effect between the B/O treament effect and the dietary treatment effect. This result generalizes over all age-classes. For example, if we compare daily fecundities around days 3, 15, and 30 (the imprecision about day arising from the staggered collection of later fecundity data), then the high-yeast group always has statistically greater fecundity. [Analysis not shown.] Unfortunately, the staggered data collection precludes the comparison of total lifetime fecundities, among treatments.

Temporal impact of dietary switching

Figure 3 shows the time-course of daily fecundity for B and O populations that undergo switched nutrition beginning at day 7 of adult life. It takes about three

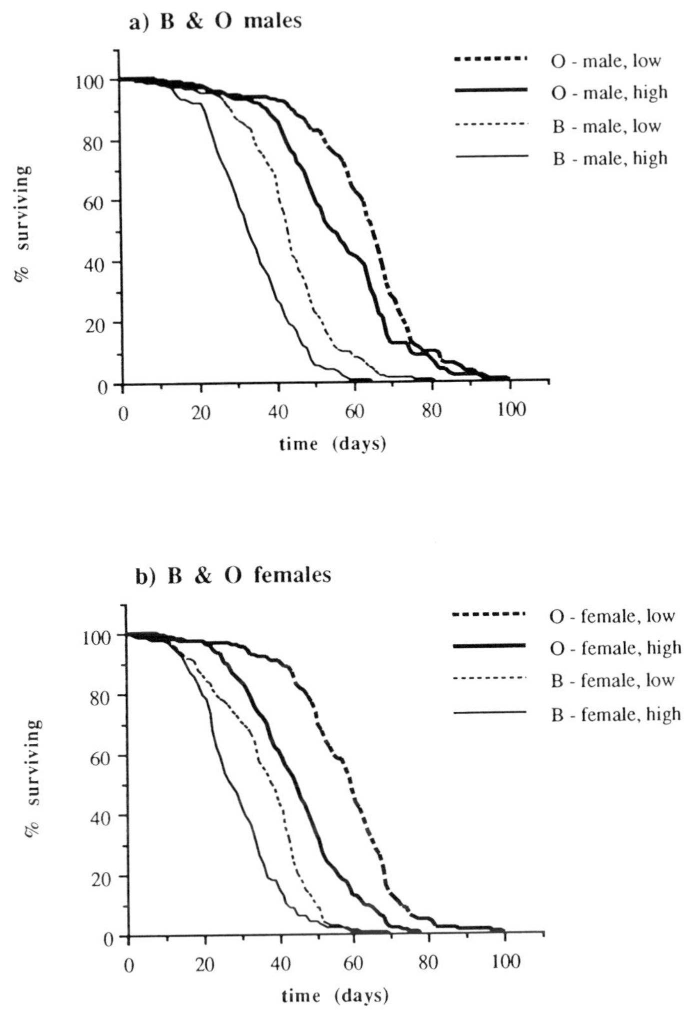

Fig. 1. Survivorship curves for B and O stocks under "low" and "high" yeast treatments. Each line represents the mean of the five replicate populations (B_{1-5} & O_{1-5}), thus 200 individuals at time zero. Mean longevities are significantly greater for the low treatments ($p < 0.01$; results & analysis are shown in Table 1).

days for fecundity to switch to the value characteristic of the unswitched control from the new yeast regime. The experiment was terminated at that point. This result suggests that the effect of yeast substrate upon fecundity is not simply a matter of immediate response to some environmental stimulus, such as the perceived value of the medium for the larval development of eggs that the female might lay upon it. It seems more plausible that the fecundity changes as a result of different levels of nutrition taken up from the substrate.

Table 1. B & O longevities under "low" and "high" yeast treatments

			Nutrition					
			Low			High		
Sex	Stock	Population	mean	s.d.	n	mean	s.d.	n.
		B1	40.1	7.7	30	34.3	10.6	34
		B2	44.3	14.4	27	31.4	10.1	31
	B	B3	42.4	14.5	31	33.0	11.9	32
		B4	46.9	10.8	32	32.0	9.7	33
		B5	43.5	13.3	35	39.1	11.6	32
M	Stock Summary:		43.4	2.5	155	34.0	3.1	162
		O1	62.1	13.0	26	55.3	10.9	30
		O2	58.5	15.9	35	52.7	17.1	29
	O	O3	68.5	14.9	30	57.7	14.4	26
		O4	60.9	16.3	28	52.5	24.2	28
		O5	68.9	18.7	28	61.2	11.3	25
	Stock Summary:		63.8	4.7	147	55.9	3.7	138
		B1	35.5	11.3	36	29.6	7.5	39
		B2	34.5	11.7	35	25.6	12.3	32
	B	B3	37.1	10.3	34	28.8	10.0	36
		B4	38.6	15.4	34	28.9	8.9	33
		B5	35.3	11.5	37	32.8	14.2	33
F	Stock Summary:		36.2	1.6	176	29.1	2.6	173
		O1	55.7	13.1	35	43.8	11.8	33
		O2	55.5	15.4	37	40.5	14.3	36
	O	O3	58.2	13.8	31	43.5	14.8	36
		O4	62.3	15.9	36	50.0	12.3	37
		O5	57.1	15.5	32	44.5	16.2	33
	Stock Summary:		57.8	2.8	171	44.5	3.5	175

Data from the longevity and fecundity experiment conducted on a total of 1600 flies at two different nutrition levels : ("low" = 0.15 mg/vial; "high" = 1.5 mg/vial). The analysis was a three-factor ANOVA with population means as replicates. All factors (stock, nutrition, & sex) were highly statistically significant ($p < 0.01$).

Effects of dietary restriction on longevity in D and C populations

Figure 4 gives the average survival curves for D and C flies of both sexes, at both high and low yeast levels. Three-way analysis of variance of the data summarized in Table 3 shows that D and C populations differ significantly in longevity, as do high and low yeast treatments. There is no significant interaction effect.

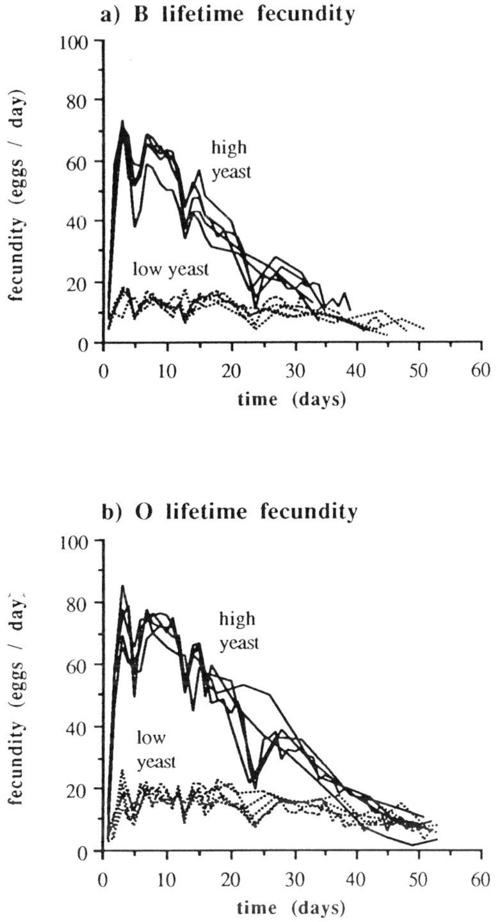

Fig. 2. Daily fecundities of B & O females. 24-hour fecundities are shown for each of the five replicate populations separately for all ages before the population fell below 8 individuals. The solid lines are the "high" treatment, and the dotted lines are the "low" treatment flies. The differences in early fecundity between yeast and selection (B vs O) treatments are both highly significant ($p < 0.01$; results and analysis are shown in Table 2).

Effects of dietary restriction on early fecundity in D and C populations

Table 4 gives the early fecundities for D and C females, at both high and low yeast levels. Analysis of variance shows that D and C populations differ significantly in fecundity, D flies laying more, as do high and low yeast treatments. There is no significant interaction effect between the D/C treatment effect and the dietary treatment effect.

Table 2. B & O early fecundities under "low" and "high" yeast treatments

		Nutrition					
		Low			High		
Stock	Population	mean	s.d.	n	mean	s.d.	n.
B	B1	13.5	6.6	40	63.6	12.7	40
	B2	14.5	3.6	39	55.9	10.9	36
	B3	13.4	2.5	38	62.6	8.7	39
	B4	12.6	2.4	40	59.5	15.0	40
	B5	11.8	4.7	40	64.5	8.5	40
			total:	197		total:	195
	Stock mean	13.15			61.20		
	s.d.	1.0			3.5		

		Low			High		
Stock	Population	mean	s.d.	n	mean	s.d.	n.
O	O1	15.2	3.5	40	70.8	8.8	37
	O2	18.7	4.8	39	70.2	16.4	38
	O3	16.3	4.9	39	70.7	9.7	38
	O4	12.9	5.6	40	62.4	13.7	40
	O5	9.9	4.8	37	61.9	10.1	36
			total:	195		total:	189
	Stock mean	14.59			67.19		
	s.d.	3.4			4.6		

ANOVA RESULTS

Source	F-test	P
Stock	5.95	0.030
Nutrition	1129.5	<0.01
Interaction	2.184	0.160

Data from the longevity and fecundity experiment conducted on a total of 1600 flies at two different nutrition levels: ("low" = 0.15 mg/vial; "high" = 1.5 mg/vial). Early fecundity values are computed as the average daily fecundities over days 3, 4, & 5. The analysis was a two-factor ANOVA with population means as replicates.

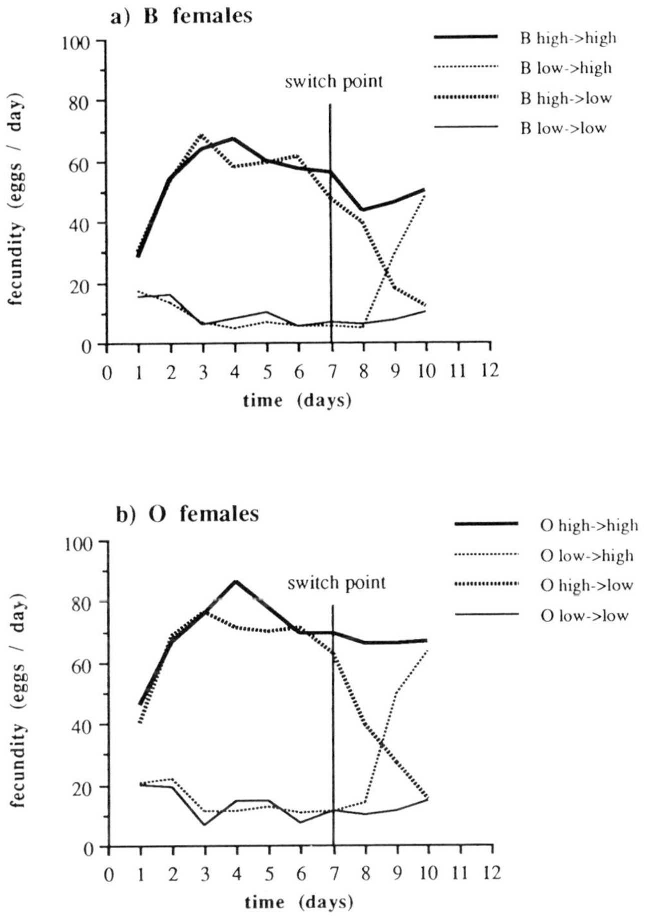

Fig. 3. Fecundity of females switched between yeast environments. Fecundities for four groups of flies, two constant-level controls (solid lines) and two switched lines (broken lines), are shown. The adjustment of fecundity in flies switched from one environment to the other takes about 3 days.

Effects of varying yeast levels on fecundity and starvation resistance in B and O populations

Figure 5 shows the effect of varying yeast levels upon fecundity and starvation resistance in the ten B and O populations, the latter character for both males and females. These results indicate that the effect of dietary variation "saturates" at about 3.0 mg of yeast per vial. Table 5 gives the analysis of these data, at each yeast level. Evidently, the B and O populations differ in fecundity significantly at higher yeast levels, but not lower yeast levels. Fecundity also falls significantly from higher to lower yeast levels, below five mg of yeast per vial. All two-way interactions are

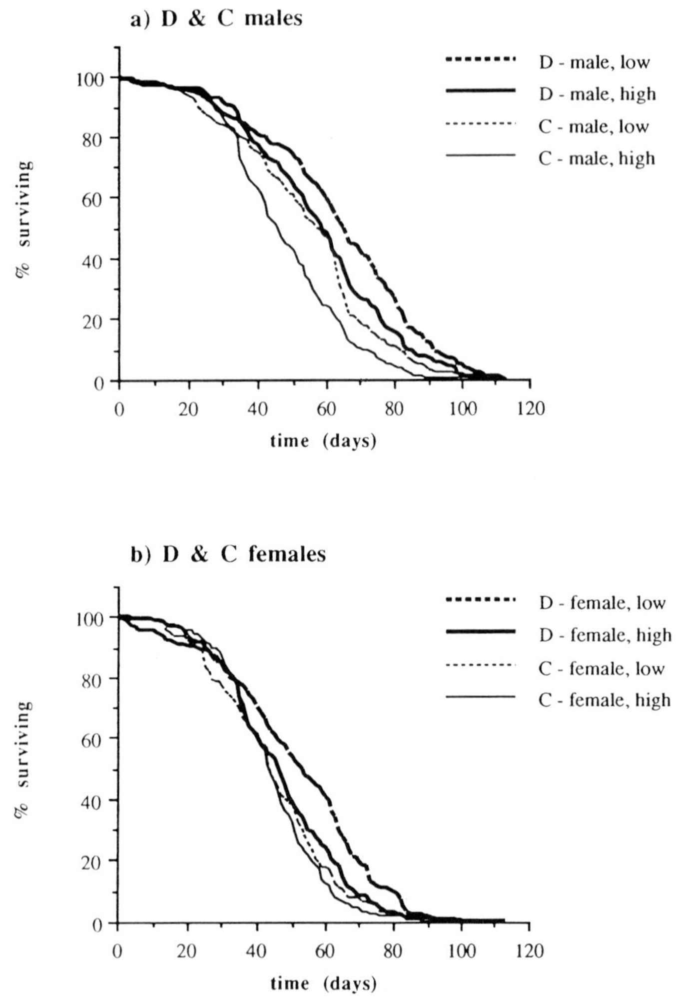

Fig. 4. Survivorship curves for D and C stocks under "low" and "high" yeast treatments. Each line represents the mean of the five replicate populations (D_{1-5} & C_{1-5}), thus 300 individuals at time zero. Mean longevities were greater under the low yeast conditions, significantly so in all cases except for C females (see Table 3, and discussion in text).

significant for starvation resistance, but not the three-way interaction. As a proportionate response, starvation resistance is extended more dramatically by reduced yeast in females compared with males. Males show increases of 26.4% and 37.9% in B and O stocks, while females show increases of 98.1% and 102.6% in B and O stocks, respectively. [The proportions are calculated as the difference between the highest and lowest starvation times, divided by the latter.] Thus the effect of dietary restriction varies with stock and gender, separately.

Table 3. D & C longevities under "low" and "high" yeast treatments

			Nutrition					
			Low			High		
Sex	Stock	Population	mean	s.d.	n	mean	s.d.	n.
		D1	53.1	21.9	56	53.3	15.8	56
		D2	54.6	22.2	52	56.7	18.1	52
	D	D3	56.3	21.3	53	51.6	21.7	53
		D4	76.1	25.2	56	60.9	23.2	58
		D5	68.1	16.2	54	67.7	24.5	54
M	Stock Summary:		61.6	10.0	271	58.0	6.5	273
		C1	51.2	14.9	54	42.9	17.1	54
		C2	46.5	18.2	51	51.2	18.4	51
	C	C3	54.9	23.7	52	49.6	24.2	52
		C4	58.6	23.8	57	49.2	14.8	57
		C5	53.2	20.5	56	46.2	14.9	56
	Stock Summary:		52.9	4.5	270	47.8	3.3	270
		D1	44.0	17.7	55	47.7	13.9	56
		D2	51.5	19.9	56	45.4	17.0	58
	D	D3	51.4	22.6	57	41.7	22.1	56
		D4	60.6	16.7	57	48.6	15.6	60
		D5	54.4	19.8	56	46.6	15.2	55
F	Stock Summary:		52.4	6.0	281	46.0	2.7	285
		C1	41.9	14.0	58	41.6	13.6	56
		C2	44.5	17.3	56	42.8	11.2	58
	C	C3	39.5	19.1	56	43.9	16.8	56
		C4	45.9	20.1	56	45.5	13.0	58
		C5	49.5	17.0	58	48.3	12.2	58
	Stock Summary:		44.3	3.8	284	44.4	2.6	286

Data from the longevity and fecundity experiment conducted on a total of 2400 flies at two different nutrition levels: ("low" = 0.15 mg/vial; "high" = 1.5 mg/vial). The analysis was a three-factor ANOVA with population means as replicates. All factors (stock, nutrition, & sex) were highly statistically significant ($p < 0.05$).

Figure 6 summarizes the pattern of covariation between fecundity and starvation resistance over varying yeast levels. Both B and O treatments exhibit the same qualitative trade-off pattern. Quantitatively, the slopes of these trade-offs are statistically indistinguishable (two-tailed t-test on the slopes of the regressions for each of the ten populations). The data from the O populations are shifted upward, substantially for starvation resistance and to a lesser extent for fecundity.

Table 4. D & C early fecundities under "low" and "high" yeast treatments

		Nutrition					
		Low			High		
Stock	Population	mean	s.d.	n	mean	s.d.	n.
D	D1	15.1	3.4	37	81.5	17.4	38
	D2	13.6	5.1	39	51.9	18.7	39
	D3	14.8	3.3	40	84.1	47.4	39
	D4	14.0	6.0	37	63.4	14.9	40
	D5	15.6	5.3	40	64.5	13.5	36
			total:	193		total:	192
	Stock mean	14.62			69.08		
	s.d.	0.8			13.5		

		Low			High		
Stock	Population	mean	s.d.	n	mean	s.d.	n.
C	C1	11.3	4.0	40	63.1	13.7	40
	C2	13.3	4.1	39	58.2	23.9	40
	C3	13.5	3.2	38	56.8	17.8	40
	C4	15.4	5.6	38	58.1	12.1	39
	C5	16.4	3.2	37	49.8	18.0	40
			total:	192		total:	199
	Stock mean	13.98			57.16		
	s.d.	2.0			4.8		

ANOVA RESULTS

Source	F-test	P
Stock	3.77	0.07
Nutrition	227.58	<0.01
Interaction	3.04	0.10

Data from the longevity and fecundity experiment conducted on a total of 2400 flies (D & C stocks) at two different nutrition levels: ("low" = 0.15 mg/vial; "high" = 1.5 mg/vial). The analysis was a two-factor ANOVA with population means as replicates. Early fecundity is computed as the average daily fecundity from days 3, 4, & 5 of the experiment.

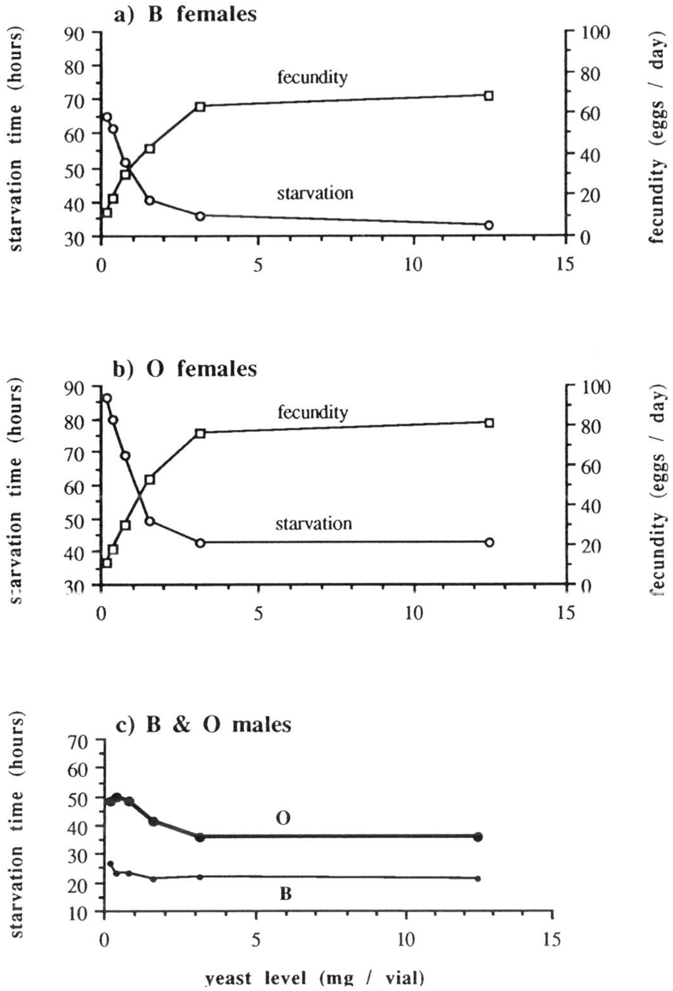

Fig. 5. Changes in starvation resistance and fecundity across a yeast gradient. Starvation resistance and fecundity covary in a striking mirror-image manner. Each data point is the mean of all five replicate populations (B or O), or approximately 100 individuals. Statistical comparison of the selection treatments, and errors, are given in Table 5; results of analyses of variance are given in the results section.

Discussion

Dietary restriction in Drosophila *can be analogous to caloric restriction in other genera*

Experiments with a diversity of invertebrates (e.g. waterstriders, Kaitala, 1991; carabid beetles, Ernsting & Isaaks, 1991; rotifers, Robertson & Salt, 1981) have

Table 5. B & O responses of starvation resistance and fecundity to a range of yeast concentrations
Table 5(a) : Male starvation resistance (hours)

#	Yeast Level (mg/vial)	B1	B2	B3	B4	B5	O1	O2	O3	O4	O5
1	0.195	28.55	23.50	27.70	28.40	25.89	45.65	53.50	41.80	50.50	50.84
2	0.391	17.75	24.10	23.26	27.10	24.60	54.65	45.65	43.90	52.35	50.84
3	0.781	22.75	22.80	21.75	25.67	24.05	49.75	47.70	44.52	49.99	49.86
4	1.562	19.59	20.40	25.25	20.45	21.59	35.05	45.63	40.41	42.85	43.31
5	3.125	23.75	23.60	19.55	22.75	20.40	37.35	32.32	38.75	38.12	33.05
6	12.500	21.15	23.79	19.85	23.35	17.95	39.45	38.00	35.15	34.25	32.65
population n:		117	119	119	118	120	120	115	118	117	116
treatment n:						593					586

#	B Mean	B s.d.	O Mean	O s.d.	O–B diff.	
1	26.81	2.13	48.46	4.68	21.65	**
2	23.36	3.45	49.48	4.54	26.12	***
3	23.40	1.51	48.36	2.35	24.96	***
4	21.46	2.24	41.45	4.03	19.99	***
5	22.01	1.92	35.92	3.00	13.91	**
6	21.22	2.43	35.90	2.78	14.68	***

grand N: 1179

Table 5(b) : Female starvation resistance (hours)

#	Yeast Level (mg/vial)	B1	B2	B3	B4	B5	O1	O2	O3	O4	O5
1	0.195	66.35	66.70	62.80	66.80	62.40	83.15	94.00	85.30	83.80	84.30
2	0.391	53.45	60.60	68.11	64.00	60.90	74.15	81.16	79.00	81.70	84.30
3	0.781	45.55	56.20	51.95	53.35	51.35	72.01	70.35	73.57	60.78	66.91
4	1.562	33.17	46.20	43.65	41.45	38.45	49.15	57.30	48.15	43.15	47.75
5	3.125	36.45	40.30	34.04	33.85	34.45	45.45	44.00	40.85	39.65	42.55
6	12.500	32.17	36.80	33.05	32.35	29.46	45.15	45.50	41.15	42.35	40.15
	population n:	118	120	118	119	119	119	119	119	119	119
	treatment n:					594					595

#	B Mean	B s.d.	O Mean	O s.d.	O–B diff.	
1	65.01	2.21	86.11	4.48	21.10	***
2	61.41	5.38	80.06	3.81	18.65	***
3	51.68	3.90	68.72	5.08	17.04	**
4	40.58	5.03	49.10	5.13	8.52	*
5	35.82	2.71	42.50	2.33	6.68	**
6	32.77	2.64	42.86	2.38	10.09	***

grand N: 1189

Table 5(c) : Female fecundity (eggs/day)

#	Yeast Level (mg/vial)	B1	B2	B3	B4	B5	O1	O2	O3	O4	O5
1	0.195	12.62	3.95	13.45	10.75	12.85	11.25	8.40	10.55	8.45	16.15
2	0.391	22.45	13.55	18.45	14.25	19.60	20.45	12.58	15.80	15.60	27.95
3	0.781	28.15	29.85	28.20	37.65	26.25	28.53	27.20	26.25	37.05	29.10
4	1.562	44.05	47.26	36.15	49.55	38.05	45.15	47.75	62.45	58.60	54.95
5	3.125	47.37	63.90	74.55	64.60	61.65	63.70	78.45	85.84	69.95	79.68
6	12.500	62.26	65.60	78.00	70.70	63.21	70.80	83.15	92.40	76.95	83.74
	population n:	118	119	120	120	118	119	119	119	120	117
	treatment n:					595					594

#	B Mean	B s.d.	O Mean	O s.d.	O–B diff.	
1	11.73	1.85	10.96	3.16	−0.77	
2	18.66	2.95	18.48	6.00	−0.18	
3	30.02	4.45	29.63	4.30	−0.39	
4	43.01	5.78	53.78	7.26	10.77	
5	63.41	10.20	75.52	8.71	12.11	*
6	67.95	6.50	81.41	8.09	13.45	*

grand N: 1189

Data from an experiment conducted upon the B & O stocks, in which 4800 flies were treated under six different dietary treatments. Each cell gives the mean value for a group of twenty flies from a single replicate population. Two-tailed hypothesis tests were performed for the difference between selection treatments at each yeast level ("*" $P < 0.05$; "**" $P < 0.01$; "***" $P < 0.001$). Results are summarized graphically in figs. 5 & 6.

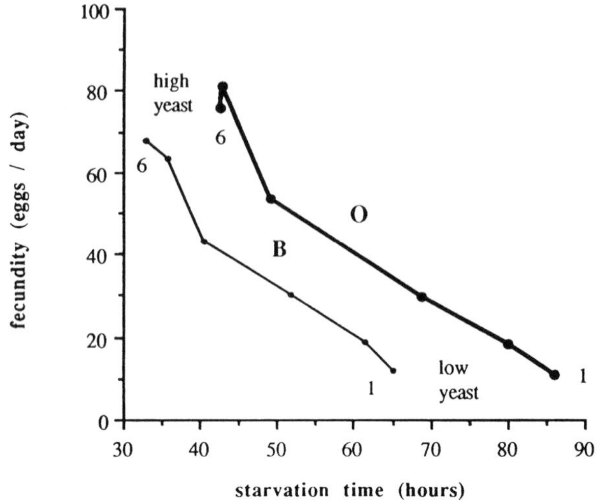

Fig. 6. Plot of fecundity against starvation resistance for B & O females, showing the negative and roughly linear relationship between the two traits, and the parallel responses of the two selection treatments (B & O), across a gradient of yeast levels (1 → 6, from Table 5). The two slopes (B = −0.57 & O = −0.64) are not statistically different, each giving an exchange rate of about 1 and 2/3 eggs per hour of starvation time under a simplistic allocative trade-off model.

provided evidence for trade-offs between survival and reproduction under dietary manipulation. These invertebrate results are consistent with the dietary restriction literature for rodents, where negative associations between fecundity and longevity under dietary manipulation have been found (Holehan and Merry, 1986). Unlike some earlier reports (e.g. David et al., 1971; LeBourg and Medioni, 1991), but like some other studies (e.g. Partridge et al., 1987), we have found that restricting the level of available nutrients can give rise to increased longevity, in association with reduced fecundity. This effect arises over a range of four different types of laboratory stock, for a total of 20 populations.

Some part of the mixed success of earlier studies on *Drosophila* may be owing to methodological differences, but a deeper point may lie in specific plastic responses to constant vs. temporally variable environments. LeBourg & Medioni (1991) manipulated food levels by alternation of feeding and starvation, while David et al. (1971) used a dilution of stock medium; both studies found a reduction in lifespan under the dietary restriction treatments, despite a reduction in fecundity. Interestingly, the early report of Alpatov (1930) found an absence of increased lifespan with reduced surface yeast, in both males and females, but an increased lifespan in males with provision of new medium every other day compared with provision of new medium every day. It is intuitively obvious that over-reduction of food levels will lead to a reduction in lifespan, for starvation time is merely longevity in the absence of food. In earlier experiments of our own (data not shown), using the same methods as those reported here, it was found that for extremely low-levels of yeast inoculate both longevity and fecundity were reduced. Similarly, McCauley et al. (1990) found a threshold level of feeding for *Daphnia* above which longevity increased over a range, and then declined again at higher levels. In the work of Kaitala (1991), waterstriders showed extended lifespan in a temporally fluctuating, low average food, environment, but not in a constant low food treatment. Such inconsistencies make broad taxonomic generalizations impossible. It is notable that the two *Drosophila* studies that do agree used similar methods, ours and that of Partridge (1987); a constant level of yeast added to the medium does affect the phenotypic trade-off with an extension of lifespan under lower-food conditions. Thus different experimental treatments may account for the considerable variation among previous publications on the effects of dietary restriction in *Drosophila*. On the other hand, variation between *Drosophila* stocks could be involved, particularly if stocks have been so inbred that normal forms of phenotypic plasticity have been destroyed. The present study does not provide any basis for discriminating among these, or other, hypotheses for dietary restriction findings from diverse laboratories.

Dietary restriction and the evolutionary trade-off between fecundity and longevity

Clearly, dietary restriction can greatly reduce fecundity through the early and middle part of the life span, as well as reducing the total egg output. Associated with this is an increase in lifespan. In previous work with *Drosophila*, both ours (e.g. Rose, 1984) and that of others (e.g. Luckinbill and Clare, 1985), significant increases in longevity are associated with depressed *early* fecundity, but no overall reduction in fecundity. There is thus some degree of analogy between the effects of selection and dietary restriction, at the level of overall life-history patterns, but there is no direct parallel.

A problem in much evolutionary research is that of genotype-by-environment interaction for functional characters (cf. Stearns et al., 1991). When genotype-by-environment interactions arise, they can vitiate comparative studies in which different species are compared in common laboratory environments and trade-offs are then inferred from the patterns of covariation among traits over species (e.g. Schnebel and Grossfield, 1988). In this context, a pertinent anomaly arose in the early fecundities of B and O stocks, as shown in Table 2. In these data, O fecundity is equal or significantly higher than B fecundity. The reasons for this result are explored in Leroi et al. (in prep.), in which it is shown that this fecundity difference involves genotype-by-environment interactions. B early fecundities are higher than O early fecundities under the environmental conditions normally used to culture B flies. In the present experiments, the high yeast environment corresponds to the "standard" fecundity assay used in Rose (1984) and subsequent publications. This standard environment more closely approximates the selection environment of the O's than the B's. It is striking that, even in laboratory selection in which many environmental variables have been controlled (light, temperature, larval density and so on), a genotype-by-environment interaction can severely complicate the comparison of populations with different selection histories.

Another discrepancy between the response to selection and phenotypic plasticity is that the overall fecundity of dietarily-restricted flies is reduced, relative to that of well-fed flies. In the original B/O comparison (Rose, 1984), it was only early fecundity that was reduced in the evolution of postponed aging.

A component of the response to dietary restriction involves a trade-off between starvation resistance and fecundity

Of greater interest, in terms of parallelism between dietary restriction and selection, is that starvation resistance responds strikingly to both. Moreover, it does so in a manner which suggests just the same kind of strict trade-off between starvation resistance and fecundity, given the pattern of Figure 6 and the large magnitude of the negative genetic correlation between starvation resistance and fecundity found by Service and Rose (1985). The self-evident hypothesis is that selection and dietary restriction might involve the same physiological mechanisms, at the organismal level. This then leads naturally to a discussion of more specific hypotheses for the relationship between life-history and physiological characters, to which we now turn.

Allocative and other hypotheses for the trade-off between survival and reproduction

The natural point of reference for any discussion of trade-offs is the Y-model of Noordwijk and de Jong (1985). In this model, some resource is allocated strictly between one of just two different life-history characters, where the two characters are typically rendered as either survival or reproduction. The present findings appear to correspond closely to the expectations of such Y-models, as illustrated by Figures 5 and 6. The increase in starvation resistance and fecundity characters in the O populations could be explained, in terms of the Y-model, as reflecting an increased net acquisition of energy for allocation between survival and reproductive functions. However, we have no direct evidence for any such increased resource acquisition in the O populations. It is also notable that B and O populations have similar trade-offs between starvation resistance and fecundity, despite some 350 generations of evolutionary divergence, which suggests the obduracy of this trade-off (cf. the "structured pleiotropy" of Stearns et al., 1991).

In the present case, the relevant resource would appear to be energy being allocated to either survival or reproductive functions. Earlier work with this system has shown that selection for late-life reproductive success causes an increase in total body lipid (Service, 1987). Several lines of evidence suggest that starvation resistance is linked to fecundity by a Y-shaped pathway where the common resource is lipid. Starvation resistance has often been shown to correlate positively with total body lipid when larval density or diet are manipulated (David et al., 1983; Zwaan et al., 1991). Surgical manipulations such as ovariectomy and allatectomy cause hypertrophy of the fat-body due to an increase in cellular lipid content (Butterworth and Bodenstein, 1968; Doane, 1960a, 1961; Bownes, 1988). Similarly, mutations such as *adipose* and *fes*, which prevent uptake of ovarian lipids, have fat-bodies engorged with excess cellular lipid (Butterworth and Bodenstein, 1968; Doane, 1960a, 1960c). Females homozygous for *adp* also have a far greater starvation resistance than wild-type flies (Doane, 1960b). There is also evidence for the involvement of the neuroendocrine system in the control of this pathway (Kelly et al., 1987; Doane, 1960a, 1960b, 1960c). The formation of yolk proteins and their uptake by ovaries can be curtailed by depriving flies of nutritional protein, a response that is also mediated by neuroendocrine hormones (Bownes and Blair, 1986; Bownes et al., 1988; Bownes and Reid, 1990). Thus the response of females to varying nutrition found here mimics that of surgery and mutations in *Drosophila* in its effect on starvation resistance and fecundity. The relatively smaller response of male starvation resistance might then be attributed to differences in the gonadal lipid requirements between the sexes, complicated by differences in mating behavior (cf. Service, 1989).

A somewhat different line of reasoning would be that variation in nutrition could induce variation in overall metabolic rate, with lower metabolic rates at lower levels of nutrition. Starvation resistance would then be increased due to a slower rate of depletion of energetic reserves. Fecundity would be reduced, due to a slower processing of metabolites into eggs. This type of effect could also arise in conjunction with an allocative shift between fat body and reproductive functions. In the context of the earlier selection experiments, this interpretation of the differentiation of starvation resistance has been refuted by the absence of metabolic rate differences between B and O flies at ages at which there is a pronounced difference in starvation resistance (Service et al., 1985; Service, 1987). However, this finding may not generalize to the dietary manipulation experiments. We do not, at present, prefer one hypothesis over the other; the physiological mechanisms underlying the effects of diet and selection may or may not correspond.

Does phenotypic plasticity mimic the response to selection?

One of the more intense disputes in life-history research is that between those who advocate phenotypic manipulation as a means of revealing evolutionary constraints, such as Partridge (Partridge and Harvey, 1988), and those who reject its use for that purpose (e.g. Reznick, 1992). The present results can be viewed as supporting either side of this debate. In support of the manipulation approach, there are a number of illuminating parallels between dietary restriction and life-history evolution in our *D. melanogaster* stocks. Against this approach, there are important disanalogies between the two, particularly where age-specific fecundity is concerned. However, it has been possible to uncover interesting features of the evolutionary physiology of life-history from these experiments, and therefore experiments of this kind are worth performing, though their interpretation may be demanding.

Acknowledgements

The authors thank D. Reznick, L. Partridge, L. D. Mueller and T. Nusbaum for their comments on the manuscript. We are grateful to J. E. Nelson, L. N. Vu, A. Azad, J. Shiotsugu, and A. Olvido for their help in the laboratory. This research was supported by US-PHS Grants AG06346 and AG09970, from the National Institute on Aging.

References

Alpatov, W. W. 1930. Experimental studies on the duration of life: influence of different feeding during the larval and imaginal stages on the duration of life of the imago of *Drosophila melanogaster*. Am. Nat. 64: 37–55.

Bownes, M., and M. Blair. 1986. The effects of a sugar diet and hormones the expression of *Drosophila* yolk-protein genes. J. Insect Physiol. 32: 493–501.

Bownes, M., A. Scott, and A. Shirras. 1988. Dietary components modulate yolk protein gene transcription in *Drosophila melanogaster*. Development 103: 109–128.

Bownes, M., and G. Reid. 1990. The role of the ovary and nutritional signals in the regulation of fat-body yolk-protein gene expression in *Drosophila melanogaster*. J. Insect Physiol. 36: 471–479.

Butterworth, F. M., and D. Bodenstein. 1968. Adipose tissue of *Drosophila*. III. The effect of the ovary on cell growth and the storage of lipid and glycogen in the adult tissue. J. Exp. Zool. 167: 207–218.

David, J., J. Van Herrewege, and P. Fouillet. 1971. Quantitative under-feeding of *Drosophila*: effects on adult longevity and fecundity. Exp. Geront. 6: 249–257.

David, J. R., R. Allemand, J Van Herrewege, and Y. Cohet. 1983. Ecophysiology: abiotic factors. *In* Ashburner, M., H. L. Carson and J. N. Thompson (eds.), The Genetics and Biology of Drosophila. Academic Press, London, pp. 105–170.

Doane, W. W. 1960a. Developmental physiology of the mutant female *sterile (2) adipose* of *Drosophila melanogaster* I. Adult morphology, longevity, egg production, and egg lethality. J. Exp. Zool. 145: 1–22.

Doane, W. W. 1960b. Developmental physiology of the mutant female *sterile (2) adipose* of *Drosophila melanogaster* II. Effects of altered environment and residual genome on its expression. J. Exp. Zool. 145: 23–42.

Doane, W. W. 1960c. Developmental physiology of the mutant female *sterile (2) adipose* of *Drosophila melanogaster* III. Corpus allatum – complex and ovarian transplantation. J. Exp. Zool. 145: 275–298.

Doane, W. W. 1961. Starvation of alletectomized female *Drosophila melanogaster*. J. Exp. Zool. 146: 275–298.

Ernsting, G., and J. A. Isaaks. 1991. Accelerated aging: a cost of reproduction in the carabid beetle *Notiophilus biguttatus* F. Funct. Ecol. 5: 299–303.

Gadgil, M., and W. H. Bossert. 1970. Life historical consequences of natural selection. Am. Nat. 102: 52–64.

Graves, J. L., E. C. Toolson, C. Jeong, L. N. Vu, and M. R. Rose. 1992. Desiccation, flight, glycogen, and postponed senescence in *Drosophila melanogaster*. Physiol. Zool. 65: in press.

Holehan, A. M., and B. J. Merry. 1985. Lifetime breeding studies in fully fed and dietarily restricted CFY Sprague-Dawley rats. I. Effects of age, housing conditions and diet on fecndity. Mech. Ageing Dev. 33: 19–28.

Hutchinson, E. W., A. J. Shaw, and M. R. Rose. 1991. Quantitative genetics of postponed aging in *Drosophila melanogaster* II. Analysis of selected lines. Genetics 127: 728–736.

Kaitala, A. 1991. Phenotypic plasticity in reproductive behaviour of waterstriders: trade-offs between reproduction and longevity during food stress. Funct. Ecol. 5: 12–18.

Kelly, T. J., T. S. Adams, M. B. Schwartz, M. J. Birnbaum, E. C. Rubenstein, and R. B. Imberski. 1987. Juvenile hormone and ovarian maturation in the Diptera: a review of recent results. Insect Biochem. 17: 1089–1093.

Le Bourg, E., and J. Medioni. 1991. Food restriction and longevity in *Drosophila melanogaster*. Age & Nutrition 2: 90–94.

Luckinbill, L. S., R. Arking, M. J. Clare, W. C. Cirocco, and S. A. Buck. 1984. Selection for delayed senescence in *Drosophila melanogaster*. Evolution 38: 996–1003.

Luckinbill, L. S., and M. J. Clare. 1985. Selection for lifespan in *Drosophila melanogaster*. Heredity 55: 9–18.

Luckinbill, L. S., J. L. Graves, A. H. Reed, and S. Koetsawang. 1988. Localizing genes that defer senescence in *Drosophila*. Evol. Ecol. 1: 37–46.

Masoro, E. J. 1988. Food restriction in rodents: An evaluation of its role in the study of aging. J. Gerontol. 43: 59–64.

McCauley, E., W. W. Murdoch, and R. M. Nisbet. 1990. Growth, reproduction, and mortality in *Daphnia pulex* Leydig: life at low food. Funct. Ecol. 4: 505 –514.

van Noordwijk, A. J., and G. de Jong. 1986. Acquisition and allocation of resources: their influence on variation in life-history tactics. Am. Nat. 128: 137–142.

Partridge, L. 1987. Is accelerated senescence a cost of reproduction? Funct. Ecol. 1: 317–320.

Partridge, L., and M. Farquhar. 1981. Sexual activity reduces lifespan of male fruitflies. Nature 294: 580–582.

Partridge, L., K. Fowler, S. Trevitt, and W. Sharp. 1986. An examination of the effects of males on the survival and egg-production rates of female *Drosophila melanogaster*. J. Insect Physiol. 32: 925–929.

Partridge, L., A. Green, and K. Fowler. 1987. Effects of egg-production and of exposure to males on female survival in *Drosophila melanogaster*. J. Insect Physiol. 33: 745–749.

Partridge, L., and P. H. Harvey. 1988. The ecological context of life history evolution. Science 241: 1449–1455.

Reznick, D. 1985. Costs of reproduction: an evaluation of the empirical evidence. Oikos 44: 257–267.

Reznick, D. 1992. Measuring the costs of reproduction. Trends in Ecology and Evolution 7: in press.

Robertson, J. R., and G. W. Salt. 1981. Responses in growth, mortality and reproduction to variable food levels by the rotifer, *Asplanchna girodi*. Ecology 62: 1585–1596.

Rose, M. R. 1984. Laboratory evolution of postponed senescence in *Drosophila melanogaster*. Evolution 38: 1004 –1110.

Rose, M. R. 1991. *Evolutionary Biology of Aging*. Oxford University Press, New York.

Rose, M. R., and B. Charlesworth. 1980. A test of evolutionary theories of senescence. Nature 287: 141–142.

Rose, M. R., P. M. Service, and E. W. Hutchinson. 1987. Three approaches to trade-offs in life-history evolution. *In* V. Loeschcke (ed.), Genetic Constraints on Adaptive Evolution, V. Springer-Verlag, Berlin.

Rose, M. R., J. L. Graves, and E. M. Hutchinson. 1990. The use of selection to probe patterns of pleiotropy in fitness characters. *In* F. Gilbert (ed.), Insect Lifecycles: Genetics, Evolution, and Co-ordination, Springer-Verlag, London.

Rose, M. R., L. N. Vu, S. U. Park, and J. L. Graves. 1992. Selection on stress resistance increases longevity in *Drosophila melanogaster*. Exp. Gerontol. 27: in press.

Schnebel, E. M., and J. Grossfield. 1988. Antagonistic pleiotropy: an interspecific *Drosophila* comparison. Evolution. 42: 306–311.

Service, P. M. 1987. Physiological mechanisms of increased stress resistance in *Drosophila melanogaster* selected for postponed senescence. Physiol. Zool. 60: 321–326.

Service, P. M. 1989. The effect of mating status on lifespan, egg laying, and starvation resistance in *Drosophila melanogaster* in relation to selection on longevity. J. Insect Physiol. 35: 447–452.

Service, P. M., and M. R. Rose. 1985. Genetic covariation among life-history components: the effect of novel environments. Evolution 39: 943–945.

Service, P. M., E. W. Hutchinson, M. D. Mackinley, and M. R. Rose. 1985. Resistance to environmental stress in *Drosophila melanogaster* selected for postponed senescence. Physiol. Zool. 58: 380–389.
Service, P. M., E. W. Hutchinson, and M. R. Rose. 1988. Multiple genetic mechanisms for the evolution of senescence in *Drosophila melanogaster*. Evolution 42: 708–716.
Stearns, S., G. de Jong, and R. Newman. 1991. The effects of phenotypic plasticity on genetic correlations. Trends in Ecology and Evolution. 6: 122–126.
Zwaan, B. J., R. Bijlsma, and R. F. Hoekstra. 1991. On the developmental theory of ageing. I. Starvation resistance and longevity in *Drosophila melanogaster* in relation to pre-adult breeding conditions. Heredity 66: 29–39.

Received 20 February 1992;
accepted 17 August 1992.
Corresponding Editor: W. Gabriel

Metabolic Aspects of the Trade-Off between Fecundity and Longevity in *Drosophila melanogaster*

Minou Djawdan
Tina T. Sugiyama
Lisa K. Schlaeger
Timothy J. Bradley*
Michael R. Rose
Department of Ecology and Evolutionary Biology, University of California, Irvine, California 92717

Accepted 12/15/95

Abstract

An antagonistic pleiotropy between early fecundity and longevity has been well demonstrated in Drosophila melanogaster. *It has been suggested that this pleiotropy may in part be due to a trade-off in energy allocation between reproduction and survival. We have tested this hypothesis by examining five populations of* Drosophila melanogaster *that show increased longevity following laboratory selection for postponed reproduction. These were compared to five control populations. We measured body mass, lipid and carbohydrate content, metabolic rate, and fecundity over the adult life span of the flies. Energy content brought forward into the adult stage from the pupae and adult metabolic rate were identical for short- and long-lived flies. Long-lived flies showed decreased fecundity and increased storage of lipid and carbohydrate compared to short-lived flies. Our study is the first to measure simultaneously allocations of metabolic energy resources between various forms of metabolic storage and reproduction in lines of* Drosophila melanogaster *and to measure their equivalence using a common energy unit (J). Energy equivalence analysis demonstrated that the trade-off in energy allocation between fecundity and metabolic storage was not quantitatively perfect with regard to energy values. Long lived flies accumulated additional lipid and carbohydrate by forgoing early reproduction, but these accumulated metabolic resources were far lower in energy content than were the additional eggs produced by short-lived flies.*

* To whom correspondence should be addressed; E-mail: tbradley@uci.edu.

Physiological Zoology 69(5):1176–1195. 1996. 0031-935X/96/6905-95103$02.00

Introduction

The evolution of life history characteristics has been studied extensively in a variety of organisms (reviewed by Roff [1992]). Life history strategies can influence reproductive strategy and timing, longevity, energy expenditures, and a variety of other biological traits related to fitness. Antagonistic pleiotropies are traits that can have a positive effect on fitness at a given time or under one environmental condition and a negative effect at another time or place (Koufopanou and Bell 1984; Tinbergen et al. 1985; Reznick 1992). An example is the much-studied antagonistic pleiotropy between longevity and early fecundity. This antagonistic pleiotropy has been identified in many organisms (Clutton-Brock et al. 1982, 1983; Pinero et al. 1982; Smith and Young 1982; Kaitala 1991) and has been shown to have a genetic basis in *Drosophila melanogaster* (Rose 1984).

The evolution of postponed senescence in the fruit fly *Drosophila melanogaster* has been studied by a number of workers (Rose and Charlesworth 1980; Partridge and Farquhar 1981; Luckinbill et al. 1984, 1988, 1990; Rose 1984; Rose et al. 1984, 1992). Rose (1984) found that flies selected for postponed senescence had a lower fecundity rate early in life than control populations. Further studies by Service et al. (1985) demonstrated that flies selected for postponed senescence had a significantly higher tolerance for desiccating conditions, starvation, and high ethanol concentration than their controls.

Service (1987) explored the physiological mechanisms of the increases in stress resistance of flies selected for postponed senescence. He compared the oxygen consumption, rate of activity, and lipid content of long-lived flies and their controls. Oxygen consumption was found to be significantly higher in young control flies; however, at later ages no significant difference was seen in the oxygen consumption of control and long-lived flies. Rates of activity were also found to be higher in young control flies. The long-lived flies had a higher activity rate at later ages than controls. The lipid content of long-lived flies was found to be significantly higher than in control populations over the range of ages tested. On the basis of these results, Service (1987) concluded that none of the physiological traits he investigated could account for all of the stress resistance patterns reported by Service et al. (1985). However, he did conclude that the greater starvation resistance of long-lived flies could be accounted for by the higher lipid content at all ages of the long-lived flies (Service 1987). A subsequent study by Graves et al. (1992) found that resistance to desiccation was positively correlated with flight duration and glycogen content. Long-lived flies were found to have higher glycogen content than their controls. Graves et al. (1992) suggested a trade-off between different physiological functions that make use of limited glycogen reserves. For example, flies flown to exhaustion prior to desiccation resistance testing showed reduced glycogen levels and reduced desiccation resistance times.

These previous studies indicated that flies selected for postponed senescence show an overall pattern of low fecundity and oxygen consumption early in life and increased lipid and glycogen content. These results suggest a trade-off of energetic reserves between survival and reproduction, as has been proposed by Service (1987), Graves et al. (1992), Chippindale et al. (1993), and Chippindale (1994). Chippindale et al. (1993) suggest that energy allocations may be divided between starvation resistance and fecundity, conforming to a simple Y model of allocation, where a limited resource is divided between survival and reproduction (van Noordwijck and de Jong 1986).

No single study has simultaneously examined metabolic energy expenditure, reproduction, and the storage of metabolic substrates (lipid and glycogen) as a means of testing the role of energy allocation in the pleiotropy between reproduction and longevity. In the present study we report our investigations of this energetic trade-off in *Drosophila melanogaster*. The experiments described here seek to examine simultaneously the allocation of metabolic energy resources between various forms of metabolic storage and reproduction in lines of *Drosophila melanogaster* selected for postponed senescence and their controls. These studies therefore provide the opportunity to quantify metabolic allocations to various activities and to test the hypothesis that the trade-off involved has a mechanistic basis in differential energy allocation. If the Y model of resource allocation presented by van Noordwijk and de Jong (1986) applies in this system, we expect the pleiotropy between early fecundity and longevity to lead to a quantitative trade-off in energy used for reproduction and energy stored for subsequent use. In other words, we expect that the energy placed into eggs by control populations will be conserved in populations selected for postponed senescence and will appear as carbohydrate and lipid stores.

Material and Methods

Fly Culture Regime

The flies used for these studies are derived from lines selected for postponed senescence and their controls (Rose 1984). In these studies, five populations of *Drosophila melanogaster* were selected for postponed senescence (O populations) and another five populations were derived as controls (B populations) from a founding population of outbred flies. Selection began in 1980. Currently, B populations are kept at a discrete generation of 2 wk and O populations at a discrete generation of 10 wk, all being maintained at 25°C and 24 h light.

Maintenance of the Selected and Control Lines

Flies in the B populations are propagated in 8-dram vials filled with approximately 5 mL of banana molasses food at a density of 60–100 eggs and allowed to develop and reach adulthood for 14 d following oviposition. On day 15, adults are removed from rearing vials and allowed to lay eggs on banana molasses medium for 1 h in fresh 8-dram vials. Egg density is controlled at 60–100 per vial. Adults are discarded after egg laying.

O populations are also propagated as larvae in 8-dram vials on banana molasses medium at 60–100 eggs per vial. Adults are moved to cages on day 15 after egg collection, after which they are given a petri dish of banana molasses food changed three times a week. On day 68, flies are given banana molasses food enriched with yeast for 2 d prior to egg collection. This addition of yeast seems critical to the reproductive effort of the O populations. Without this yeast, O flies will lay extremely low numbers of eggs throughout their lives.

Production of the Experimental Generation

In order to examine genetically as opposed to environmentally induced differences between the lines, flies to be used for these specific studies were raised in a common environment for two generations prior to experimentation, as described below, to avoid parental and grandparental effects. A uniform rearing schedule was chosen that was distinct from either the B or O rearing protocol. Eggs from each population were obtained to rear experimental generation 1. Egg density was controlled by placing approximately 60–80 eggs into each of 25 8-dram banana molasses food vials per treatment. Adult flies were moved to cages 14 d after egg collection, where fresh banana molasses food was provided and changed three times a week. On days 15–18 after egg collection, flies were given banana molasses food in a petri dish plus a layer of yeast for 3 d to induce increased egg laying. After 3 d of extra yeast, flies were allowed to lay eggs for 4 h. These eggs were used to raise experimental generation 2. In this generation, flies were cultured as described above except that density was precisely controlled by counting 60 eggs into each of 100 8-dram vials of banana molasses food per population type. The eggs of these flies were used to produce experimental generation 3, the generation on which measurements were actually made. For this generation eggs were collected by matching an O population with a B population on each collection day. Replicate O1 was matched with B1, O2 with B2, and so on. Egg collection was started with the O1, B1 pair. Egg collection for the following pairs was staggered by 1 d per pair of populations such that O5 and B5 eggs were collected 5 d after the O1, B1 pair. This staggering was necessary because of equipment limitations. For each pair of populations, 150 vials of 60 eggs each were collected per population type. On formation of pupae, vials were checked every 12 h for darkened pupae and emerging flies. For each treatment, within 12 h of their emergence, some flies were removed and frozen for physiological measurements (day 0 samples). At the same time, other flies were placed in clean vials with food at approximately 60 flies per vial at 25°C, for 24 h, after which they were either frozen and stored or used for live physiological measurements (day 1 samples). All other flies were transferred to cages within 12 h of eclosion. Fresh food petri plates containing banana molasses food with no additional yeast were added three times a week for the duration of the life span of flies. All populations were kept at 25°C and 24 h light. No additional yeast was given to any of the experimental populations.

Flies previously selected for postponed senescence (O flies) were measured for physiological characteristics at 0–12 h after eclosion, as well as 1, 4, 7, 20, 35, and 60 d after eclosion. Control populations (B flies) were measured at 0–12 h after eclosion, as well as 1, 4, 7, and 20 d after eclosion.

For the purpose of constructing an energy budget, the following physiological measurements were done on all five O and five B populations on the days mentioned above: lipid content, carbohydrate content, fecundity, and metabolic rate. At 0–12 h after eclosion (day 0), only lipid and glycogen content were measured. Flies required for live measurements (metabolic rate, fecundity assays) were removed from the cages on the day of measurement.

Lipid Measurements

Flies from paired treatments of O and B were removed from the colony on the appropriate day and frozen. The genders were separated and dried at 60°C for 45 min. For each gender, six groups of 10 flies were dried and weighed on a Cahn electrobalance to within 0.001 mg. Each group of 10 flies was placed in a 10 × 50 mm Whatman thimble. Thimbles were then stapled shut and placed in the extractor of a Soxhlet apparatus with petroleum ether as the solvent. Petroleum ether was used as the extracting solvent because it extracts fewer nonlipids than other solvents (Dobush et al. 1985). Flies underwent extraction for at least 20 h, at which point they were removed from the extractor and thimble and placed in an oven at 60°C to dry for 30 min. Flies were then weighed a second time on the Cahn electrobalance to within 0.001 mg. The difference between the initial dry weight and the postextraction dry weight was considered the total lipid content of flies. The initial dry mass and lean dry mass were also recorded. In each run, paired B and O flies of the same gender and age were analyzed simultaneously.

Carbohydrate Measurements

Flies were removed from the colonies and frozen on the appropriate day. The colorimetric method of carbohydrate extraction of Van Handel (1965) was modified as described below. Flies were separated by gender and dried in an oven for 45 min at 60°C. For each gender, five groups of five flies were weighed and placed in 1.7-mL microcentrifuge tubes. Then 500 μL of water was added to each tube, and flies were ground with a handheld, battery-operated grinder. Tubes were placed in boiling water for 5 min. Next, 100 μL from each tube was transferred to a 13 × 100 mm test tube. Three milliliters of anthrone reagent (150 mg anthrone per 100 mL of 72% sulfuric acid) was then added to each tube, and the tubes were incubated in a water bath at 90°C for 20 min. In each run, paired B and O flies of the same gender and age were assayed simultaneously. For each assay, two duplicate standard curves were prepared along with three samples of a known amount of glycogen assayed with the flies. Although the standard curves did not differ significantly, the standard curve that predicted the known concentration with an error of 10% or less was used for calculating the carbohydrate content of the flies.

Estimates of Metabolic Rate

We were interested in obtaining estimates of metabolic rate of flies in their population cages under normal fly culture conditions. Therefore, CO_2 production was measured from males and females in a 1-L population cage with food present. This environment allowed the flies to fly and carry out normal colony activities while measurements were being made with the

Sable System Li-Cor CO_2 analyzer and control system. Flies were anesthetized with CO_2. Fifty males and 50 females were counted and placed in respirometry cages approximately 1 L in volume with banana molasses food present. Flies were allowed to recover from anesthesia for 1 h prior to measurement. Preliminary studies indicated that the flies recovered from anesthesia within an hour. The air in an empty cage with food was measured in each run as a baseline. Measurements were made at 25°C and 24 h light. Dry, CO_2-free air was drawn into each chamber at 100 mL min^{-1}. For each replicate per treatment, CO_2 production of the flies in each cage was measured once for 4 h. After the chamber was flushed and a steady state had been reached, the longest time period within the last hour of CO_2 production was used to calculate mean CO_2 production for each population. The last hour of CO_2 production was used because of the long washout time. The negligible CO_2 value of the empty cage with food was subtracted from all results of cages with flies. At the end of the experiment, dead flies, if any, were removed from the cage. The live flies were anesthetized, dried for 1 h at 60°C, and weighed. Mortality from anesthesia was very low but did increase with age to a maximum value of 10% mortality. In all cases, estimates of metabolic rate are expressed per milligram for live flies only.

Energy Content

The energy content of adults 4 d after eclosion, of food, and of eggs was measured with a Phillipson microbomb calorimeter as described by Paine (1971) (Gentry Instruments, Aiken, S.C.). The calorimeter was calibrated with known weights of benzoic acid (benzoic acid = 26.502 J mg^{-1}). Flies, food, and eggs were dried in an oven at 60°C for 45 min. Genders were separated and weighed in groups of 10 flies. For each gender, 10 groups of 10 flies each were analyzed by mixing the weighed flies with a known amount of benzoic acid and forming a pellet. Since the energy content of benzoic acid pellets was known, these were subtracted from the pellet's total energy content. The remainder was assumed to be the fly, food, or egg energy content.

Fecundity

The fecundity of B and O flies was measured by placing known numbers of males and females (750–1,500 flies total) in a cage within 12 h of eclosion. Flies were maintained on petri dishes of banana molasses food, which were changed every other day for the duration of the measurement.

On the day of measurement, the dead flies in each cage were removed, sexed, and counted. Five petri dishes of food were placed at 25°C for 1 h to bring them up to temperature. A fresh petri dish was then placed in the cage for 1 h. Measurements were repeated for 5 consecutive hours. The number of eggs on each dish was counted and recorded. To calculate fecundity, the number of eggs per hour was divided by the number of females alive in the cage.

Statistical Analysis

Statistical analyses were conducted with the SAS statistical package. B and O populations were compared at ages 0, 1, 4, 7, and 20 d. Data are presented for O populations at ages 35 and 60 d, but these data are not used in the statistical analysis. For each B and O population, replicate values of each measurement were used for analysis. B and O populations were raised in pairs (a block) and then tested over a range of ages. Statistical analyses were done by ANOVA with population, age, and block as factors in the analysis of all measurements except metabolic rate measurements. Metabolic rates were measured once per replicate B or O population, and ANOVA was done with only population and age as factors in the analysis. With the exception of metabolic rate measurements, male and females were analyzed separately.

Results

Body Mass

B and O females eclosed at similar average dry body masses. The body mass of O females increased up to day 7 after eclosion and remained at that level during the rest of their life span (Fig. 1). B females maintained their body mass at the time of eclosion during their life span (Fig. 1). ANOVA indicates a significant difference in dry body mass of B and O females (df = 1, 4; $F = 38.26$; $P > 0.005$). Body mass also differed significantly with age (df = 4, 16; $F = 9.61$; $P > 0.001$). The interaction between age and treatment was also significant (df = 4, 16; $F = 8.94$; $P > 0.001$), indicating a difference in the pattern of change in mass with age in B and O females.

B and O males eclosed at similar average dry masses and lost body mass by day 7. This low body mass was maintained for the duration of their life span for both treatments, with B males losing more mass than O males (Fig. 1). ANOVA indicated that there was no significant difference in the mass of

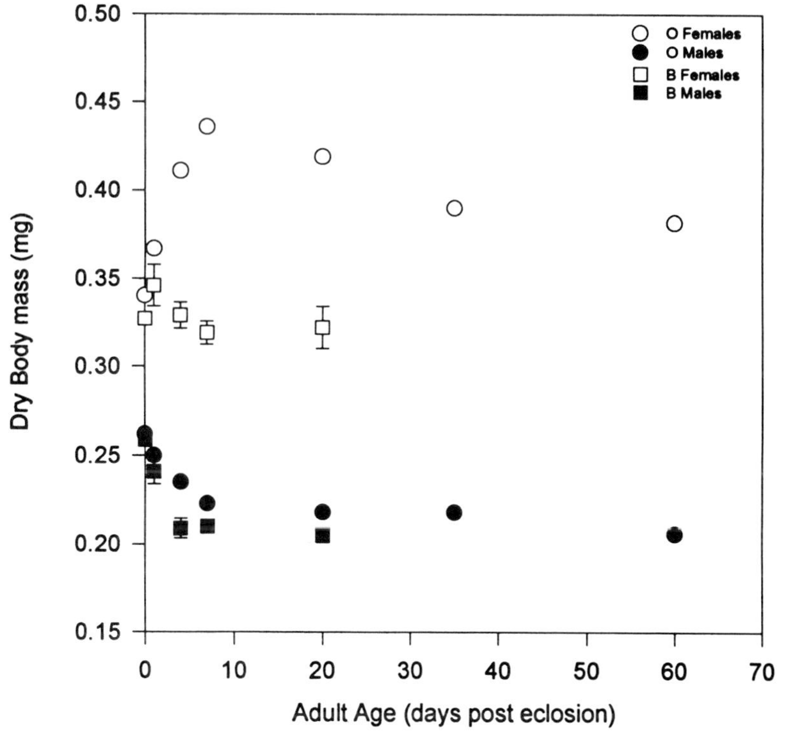

Fig. 1. Mean dry body mass (mg) of B and O flies as a function of age, with standard error bars. O flies are represented by circles, B flies by boxes. Male values are given in closed symbols. Female symbols are open. Female O flies gained significantly more mass with age than B females. There was a significant decrease in body mass with age in B and O males, but they did not differ from each other in body mass at any point during their life spans.

B and O males; however, there was a significant effect of age (df = 4, 16; $F = 41.96$; $P > 0.001$). The interaction between treatment and age was not significant in males.

Lean dry body mass of flies was obtained after extraction of lipids. The lean dry body mass of both B and O females increased initially and then dropped over time (data not shown). B and O females differed significantly in lean dry mass (df = 1, 4; $F = 19.74$; $P > 0.025$), with O females having a higher lean dry mass than B females. Lean dry mass also changed significantly over time (df = 4, 16; $F = 6.47$, $P > 0.005$). The interaction between population and age was also significant (df = 4, 16; $F = 13.49$; $P > 0.001$). These changes in lean dry mass with age and the differences between B and O females were entirely due to the effect of carbohydrate accumulation. Subtraction of the known weights of carbohydrates (see below) from the lean dry body mass eliminated the observed differences in lean dry body mass between B and O females. B and O males did not differ significantly from each other in lean dry body mass.

Lipid Content

On emergence from pupae, B and O populations regardless of gender showed no significant differences in total lipid content (mg of lipid per fly). Total lipid content in both B and O females increased up to day 20 after eclosion, after which lipid content was maintained at a high level up to day 60 for O females (Fig. 2). ANOVA indicated no significant difference in lipid content between B and O females; however, the increase in lipid with age was significant (df = 4, 16; $F = 20.07$; $P > 0.001$). The population by age interaction was not significant, which indicates that B and O females had the same pattern of lipid increase with age.

The lipid content of both B and O males dropped by day 7 after eclosion. Males from both lines maintained their lipid content at this low level until the end of their life span (Fig. 2). ANOVA indicated no significant difference in lipid content with age or between populations. B and O males had similar patterns of lipid content decrease with age, as indicated by the nonsignificance of the population by age interaction.

Carbohydrate Content

Female O flies increased in carbohydrate content rapidly up to day 7 after eclosion. By day 20 after eclosion, the carbohydrate content of these females had dropped, and it continued to decreased to day 60 (Fig. 3). B females also increased in carbohydrate content up to day 7, with no further change to day 20 (Fig. 3). There was a significant difference in carbohydrate concentration between B and O females (df = 1, 4; $F = 20.53$, $P > 0.001$) and with adult age (df = 4, 16; $F = 12.23$; $P > 0.001$). The interaction between treatment and age was also significant (df = 4, 16; $F = 4.32$, $P > 0.025$).

Male O flies also increased in carbohydrate content to day 4 after eclosion and subsequently lost carbohydrate by day 60 (Fig. 3). A pattern of carbohydrate increase similar to that of B females was seen in male B flies (Fig. 3). B and O males differed significantly in carbohydrate concentration (df = 1, 4; $F = 10.33$; $P > 0.05$) and with adult age (df = 4, 16; $F = 3.29$; P

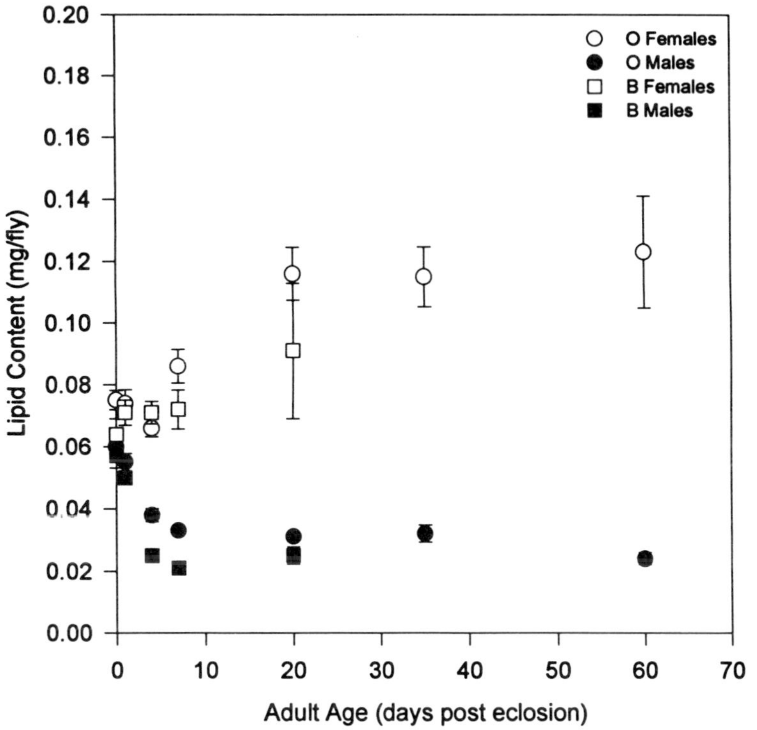

Fig. 2. Mean total lipid content (mg per fly) of B and O flies as a function of adult age, with standard error bars. O flies are represented by circles, B flies by boxes. Male values are given in closed symbols. Female symbols are open. Female O flies had a higher lipid content than female B flies. When B and O flies were compared, regardless of gender, total lipid content did not differ significantly. However, a significant effect of age was seen both in male and female flies.

> 0.05). The interaction between treatment and age was also significant (df $= 4, 16$; $F = 6.55$; $P > 0.005$).

Estimates of Metabolic Rate

Carbon dioxide production in both B and O populations started at 17.5 μL CO_2 mg^{-1} h^{-1} at 24 h after emergence, peaked at 27 μL CO_2 mg^{-1} h^{-1} at 7 d of age, and subsequently dropped for both populations (Fig. 4). An ANOVA indicated significant differences in CO_2 production with age (df = 3; $F = 7.68$; $P > 0.001$) but no significant difference between B and O populations in estimates of metabolic rate.

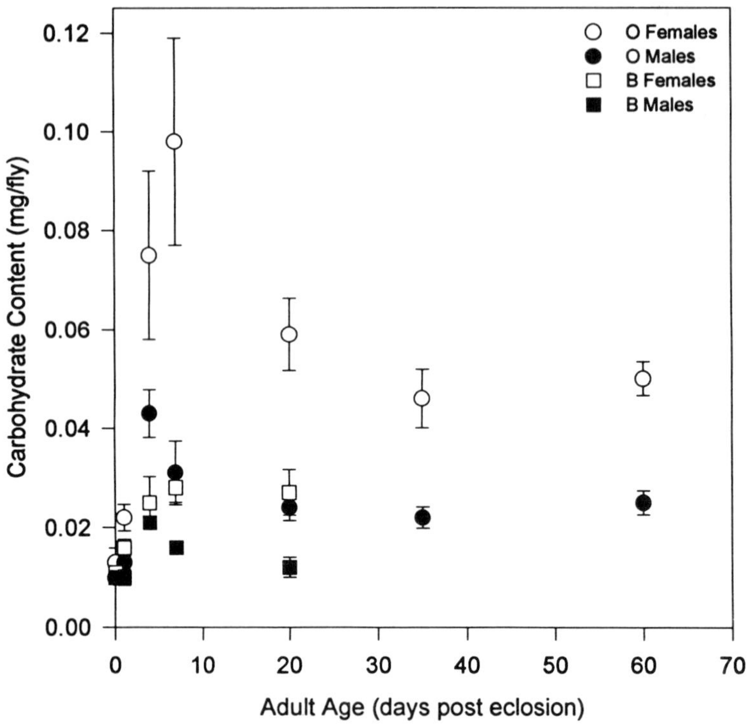

Fig. 3. Mean total carbohydrate (mg per fly) content of B and O flies as a function of adult age, with standard error bars. O flies are represented by circles, B flies by boxes. Male values are given in closed symbols. Female symbols are open. Both O males and females had significantly higher carbohydrate contents than their B counterparts. Carbohydrate content of both genders of both treatments was significantly affected by age.

Energy Content

Eggs had a mean energy content of 29.3 ± 1.14 J mg^{-1} dry weight ($n = 10$). Food energy content was 25.2 ± 0.70 J mg^{-1} dry weight ($n = 8$). B and O flies had a energy content of 27.8 ± 0.45 J mg^{-1} dry weight ($n = 18$). ANOVA indicated no significant difference in the energy content of eggs, food, or B and O flies.

Fecundity

The fecundity of B females averaged 1.8 eggs laid per female per hour at 24 h of adult age and decreased steadily with age until it reached 0.32 at 20

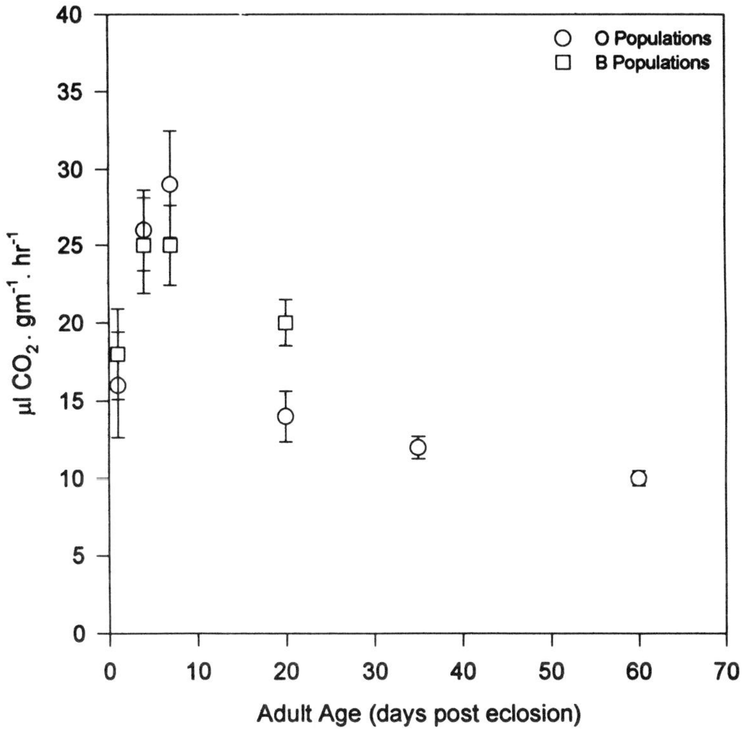

Fig. 4. Estimates of metabolic rate (µL CO_2 mg^{-1} h^{-1}) of B and O flies as a function of age, with standard error bars. Males and females were measured together. O flies are represented by circles, B flies by boxes. Rates of CO_2 release by B and O populations did not differ significantly from each other over the first 20 d of adult life.

d of age (Fig. 5). O females, on the other hand, started at about 0.24 eggs laid per females per hour at 24 h and decreased to near zero by 7 d of age (Fig. 5). ANOVA indicated a significant difference in fecundity with age (df – 1, 3; F – 59.42; $P > 0.005$) and between populations (df = 3, 12; $F = 4.37$; $P < 0.025$). The interaction between population and age was not significant.

Discussion

The antagonistic pleiotropy between fecundity and longevity was clearly manifested under the conditions used in this experiment. The B lines, which exhibited rates of fecundity 10 times those of the O lines at early ages,

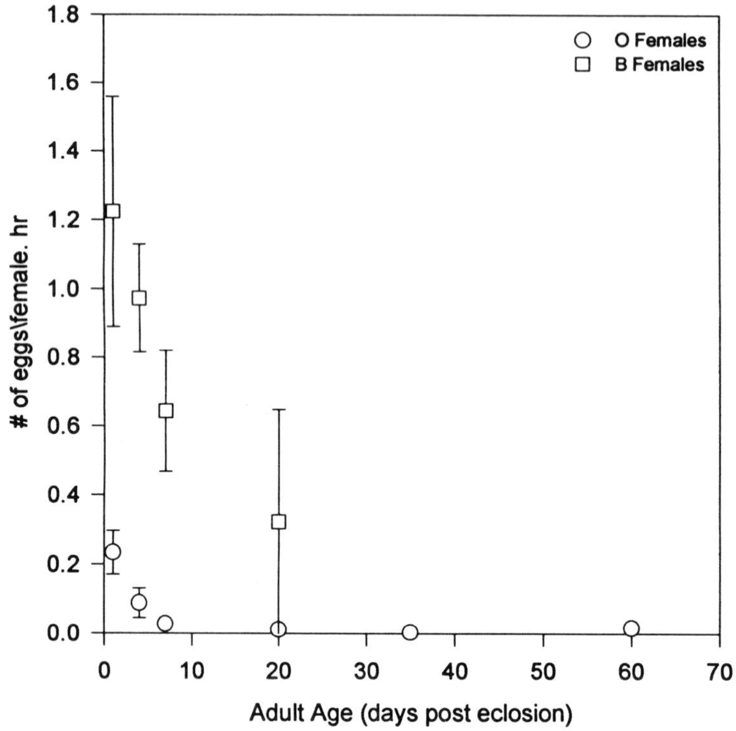

Fig. 5. Mean fecundity (number of eggs per female alive per hour) of B and O females as a function of age, with standard error bars. O flies are represented by circles, B flies by boxes. B females had a significantly higher rate of fecundity than their O counterparts.

showed such high mortality after day 20 that inadequate numbers for experimentation were available on day 35, while the O lines survived in large numbers through day 60. This is a very important point for all of our subsequent measurements, since the expression of life history pleiotropies can be very sensitive to environmental conditions (Robertson and Salt 1981; Kaitala 1991; Chippindale et al. 1993; Leroi et al. 1994). This study is the first to compare the distribution of resources in long-lived and short-lived flies under conditions in which the antagonistic pleiotropy is well expressed and energy distribution can be quantitatively compared.

Energy accumulated during larval feeding may be carried forward through the pupal stage to the adult. If long-lived O flies accumulated more metabolic resources during larval feeding than short-lived B flies, one would expect newly eclosed B and O adults to differ in this respect. Our results indicate that B and O flies at the time of eclosion do not differ in mass, carbohydrate, or lipid content. Therefore, B and O flies bring similar resources forward into the adult stage from the larval feeding period. Any subsequent differences between B and O flies must, therefore, be the result of differences in behavior and physiology expressed in the adult life stage. Since the flies were subjected to two generations of identical rearing procedures, it is also clear that the observed differences are genetically determined.

We measured egg production, metabolic rate, and carbohydrate and lipid accumulation in B and O flies under conditions in which the antagonistic pleiotropy between early fecundity and longevity was intensely expressed. We wished to be able to correlate and balance the energy used for metabolic rate, for reproduction, and for metabolic storage. This is most easily done in females, where the metabolic investment in eggs is relatively easily quantified. We have quantified these values for female flies in our study and expressed them in a universal energy unit (J) for the purpose of comparing differential energy distribution in B and O flies.

As shown in Figure 3, no differences in metabolic rate were observed between B and O flies. Significant differences were observed over time, with higher rates observed at 4 and 7 d of adult life, but this occurred in both treatments. It is clear, therefore, that differences in life history and metabolic patterns between B and O flies do not include differences in metabolic rate. Assuming that the primary fuel for metabolism in the flies is carbohydrate (a reasonable assumption given respiratory quotient measurements in other studies), total metabolic rate in the B flies is equivalent to 77.1 J per female and in the O flies is equivalent to 82.3 J per female over the first 20 d of adult life (Fig. 6).

Fecundity levels, unlike metabolic rate, were highly different in B and O flies. B females were found to produce an average of 312 eggs per female over the first 20 d of adult life, while the O flies produce 23 eggs per female. These low levels of egg production in the O flies become understandable when the distinct rearing conditions of the B and O flies are considered. In the normal selection environment, B flies are reared to adulthood in vials for 14 d, at which time they are transferred to clean vials containing fresh banana molasses medium without supplemental yeast. The B flies are thus selected for a capacity to lay eggs early in adulthood on fresh medium in the absence of yeast. Their high fecundity in the present study in the absence of yeast is therefore not surprising. The O flies in the selection environment are reared to adulthood in vials and removed to cages on day 14 of age. In the cages, they are presented with plates of banana molasses food three times a week for 8 additional weeks. Three days before egg collection, flies are given yeast in order to promote and elicit egg production. The O flies

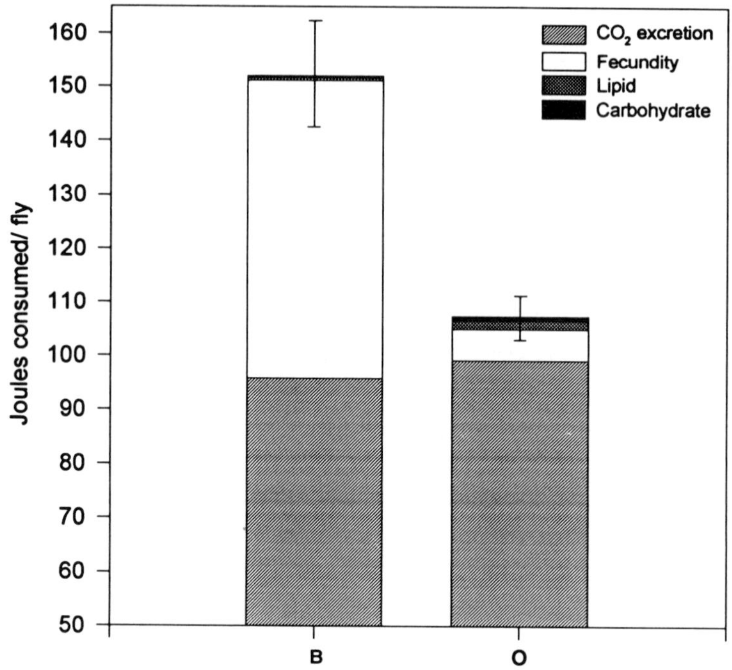

Fig. 6. Calculated values for the average energy output over the first 20 d of adult life for a single B or O female. B females expended considerably more energy in the form of eggs than O females. O females stored much more metabolic reserves in the form of carbohydrate and lipid compared to B flies. Note that the Y-axis values begin at 50 J consumed per female in order to make clear the difference between B and O females with regard to the storage of lipid and carbohydrate. The error bar indicates the standard error.

are thus selected for high fecundity late in life, particularly in the presence of yeast supplementation of the food. The O flies in the present study therefore showed low fecundity when presented with fresh food in the absence of yeast.

The eggs were found to have an energy content of 29.3 J mg^{-1} dry weight, yielding a total energetic output in the form of eggs over the first 20 d of adult life of 55.2 J for a B female and 5.8 J for an O female.

B females do not increase in mass or carbohydrate content during their life span. They do show a small but statistically significant increase in lipid with age. O females gain weight rapidly in the first few days of life and tend to maintain an elevated body mass for the duration of their life span. This initial weight gain in the O females seems related primarily to the accumulation of carbohydrates early in life and is maintained by the subsequent increase in the amount of lipid accumulated at a slower rate. We calculated the net energetic accumulation of B and O females over the first 20 d of life assuming a value of 17.6 J mg^{-1} for carbohydrate and 39.4 J mg^{-1} for lipid. B flies accumulate 0.27 J as carbohydrate and 1 J as lipid, while O flies accumulate 0.77 J as carbohydrate and 1.55 J as lipid over the first 20 d of adult life. Figure 6 illustrates the energetic values of B and O carbohydrate and lipid accumulation over the first 20 d of adult life. It is noteworthy that O females accumulate more than twice as much energy in metabolic reserves as do the B females. Neither B nor O flies show any changes in lean dry body mass during their adult lifetimes.

Our results differ from those of Service et al. (1985) and Service (1987) in two ways. First, we found no difference in the metabolic rates of B and O flies. Service (1987) found significantly higher oxygen consumption in young B flies than in young O flies and no difference in oxygen consumption between B and O flies at later ages. Second, although B and O females both increased in lipid content with adult age, we did not find a significant difference in lipid content or rates of lipid accumulation between B and O flies. This is true even if we analyze our data as percent lipid as did Service. These differences presumably derive from the distinct conditions under which flies were cultured and fly metabolic rate was measured. Service raised and maintained flies in 8-dram vials with banana food and measured metabolic rate in a Gilson respirometer, where fly mobility is somewhat restricted, and with food absent. We raised our flies to adulthood in vials and then transferred them to cages for the duration of their adult life. Metabolic rate was measured by flow-through respirometry in flight cages with food present. These differences in protocol and fly culture between our study and Service (1987) are sufficient to account for the differences in metabolic rate observed between the two studies, as also seen in Leroi et al. (1994), where changes in the culture environment of B and O flies resulted in very different physiological responses. For the purposes of this article, the important point is that metabolic rate, fecundity, and metabolic storage were measured on the same populations under conditions where the antagonistic pleiotropy between fecundity and longevity was intensely expressed.

If the Y model of resource allocation described by van Noordwijk and de Jong (1986) is in force in this system, one might expect the strong antagonistic pleiotropy between early fecundity and longevity expressed under the conditions of our experiment to lead to a similar trade-off between energy used for reproduction and energy stored for subsequent use. Our data, as presented in Figure 6, demonstrate such a qualitative trade-off. O flies do not show high early fecundity, and they store more energy. By quantifying the energy placed in the eggs and stored in reserve, we are able to demonstrate, however, that O flies are not storing an amount of energy equivalent to that placed in the eggs by the B flies. This observation leads to two important conclusions.

The first conclusion derives from the observation that the amount of energy stored by the O flies as lipid and carbohydrate is less than that used by the B flies in their eggs. As a result, the energy stored by the O flies is insufficient to produce the number of eggs that was produced by the B flies. Under normal conditions of culture, egg production by O flies is stimulated late in life by the addition of yeast to the food. Therefore, if O flies are going to display late fecundity equivalent in magnitude to the early fecundity of the B flies, their metabolic stores are calorically inadequate, and they will have to cover the metabolic cost of these eggs through additional feeding. This demonstrates that the antagonistic pleiotropy between early and late fecundity demonstrated by selection in *Drosophila melanogaster* (Rose and Charlesworth 1981; Luckinbill et al. 1984) is not a quantitatively perfect trade-off, with regard to energetic reserves, under the conditions of this experiment. The failure of the O female to accumulate still more lipid is not due to a lack of genetic variance for additional lipid storage. Lines selected from the O flies for starvation resistance (Rose et al. 1992) show additional lipid storage in adult females compared to normal O females (A. K. Chippindale, unpublished data).

The second conclusion that can be drawn from Figure 6 derives from the observation that the B flies have a greater energy output than do the O flies. The total of aerobic metabolism, egg energetic content, and storage is higher in the B than in the O flies. Our results are therefore consistent with the hypothesis that selection for postponed senescence is associated with a change in energy allocation patterns.

Acknowledgments

We would like to thank Larry Mueller, Allen Gibbs, Adam Chippindale, and Ted Nusbaum for helping us improve earlier drafts of the manuscript. The authors also wish to thank Cora Cash-Clark for her help with the carbohydrate assay, Matthew Nelson, Kellen Murphy, and Cristina Zuazua for assistance with fly maintenance. Ted Nusbaum was a great help with statistical analysis and Allen Gibbs was infinitely generous with space and other resources. We would also like to acknowledge Hardip Brar and all the people in the

Rose lab who supplied us with flies. Funding for this research was provided by United States Public Health Service grant AGO-9970.

Literature Cited

CHIPPINDALE A.K. 1994. The Evolution of Life History Trade-Offs in *Drosophila melanogaster*. PhD diss., University of California, Irvine.

CHIPPINDALE A.K., A.M. LEROI, S.B. KIM, and M.R. ROSE. 1993. Phenotypic plasticity and selection in *Drosophila* life-history evolution. I. Nutrition and the cost of reproduction. J. Evol. Biol. 6:171–193.

CLUTTON-BROCK T.H., E.F. GUINNES, and S.D. ALBON. 1982. Red Deer Behavior and Ecology of the Two Sexes. University of Chicago Press, Chicago.

———. 1983. The costs of reproduction in red deer hinds. J. Anim. Ecol. 52:367–383.

DOBUSH G.R., C.D. ANKNEY, and D.G. KREMENTZ. 1985. The effect of apparatus, extraction time, and solvent type on lipid extractions of snow geese. Can. J. Zool. 63:1917–1920.

GRAVES J.L., E.C. TOOLSON, C. JEONG, L.N. VU, and M.R. ROSE. 1992. Desiccation, flight, glycogen, and postponed senescence in *Drosophila melanogaster*. Physiol. Zool. 65:268–286.

KAITALA A. 1991. Phenotypic plasticity in reproductive behaviour of waterstriders: trade-offs between reproduction and longevity during food stress. Funct. Ecol. 5: 12–18.

KOUFOPANU V. and G. BELL. 1984. Measuring the cost of reproduction. IV. Predation experiments with *Daphnia pulex*. Oecologia 64:81–86.

LEROI A.M., A.K. CHIPPINDALE, and M.R. ROSE. 1994. Long term evolution of a life history trade-off in *Drosophila melanogaster*. I. The role of genotype by environment interaction. Evolution 48:1230–1243.

LUCKINBILL L.S., R. ARKING, M.J. CLARE, W.C. CIROCCO, and S.A. BUCK. 1984. Selection for delayed senescence in *Drosophila melanogaster*. Evolution 38:996–1003.

LUCKINBILL L.S., J.L. GRAVES, A.H. REED, and S. KOETSAWANG. 1988. Localizing genes that defer senescence in *Drosophila*. Evol. Ecol. 1:37–46.

LUCKINBILL L.S., V. RIHA, S. RHINE, and T.A. GRUDZIEN. 1990. The role of glucose-6-phosphate dehydrogenase in the evolution of longevity in *Drosophila melanogaster*. Heredity 65:29–38.

PAINE R.T. 1971. The measurement and application of the 4021 calorie to ecological problems. Annu. Rev. Ecol. Syst. 2:145–164.

PARTRIDGE L. and M. FARQUHAR. 1981. Sexual activity reduces lifespan of male fruit flies. Nature 294:580–582.

PINERO D., J. SARUKHAN, and P. ALBERDI. 1982. The costs of reproduction in a tropical palm *Astrocaryum mexicanum*. J. Ecol. 70:473–481.

REZNICK D. 1992. Measuring the costs of reproduction. Trends Ecol. Evol. 7:42–45.

ROBERTSON J.R. and G.W. SALT. 1981. Responses in growth, mortality, and reproduction to variable food levels by the rotifer *Asplanchna girodi*. Ecology 62:1585–1596.

ROFF D.A. 1992. The Evolution of Life Histories: Theory and Analysis. Chapman & Hall, London.

ROSE M.R. 1984. Laboratory evolution of postponed senescence in *Drosophila melanogaster.* Evolution 38:1004–1110.

ROSE M.R. and B. CHARLESWORTH. 1980. A test of evolutionary theories of senescence. Nature 287:141–142.

———. 1981. Genetics of life history in *Drosophila melanogaster.* II. Exploratory selection experiments. Genetics 97:187–196.

ROSE M.R., M.L. DOREY, A.M. COYLE, and P.M. SERVICE. 1984. The morphology of postponed senescence in *Drosophila melanogaster.* Can. J. Zool. 62:1576–1580.

ROSE M.R., L.N. VU, S.U. PARK, and J.L. GRAVES, JR. 1992. Selection on stress resistance increases longevity in *Drosophila melanogaster.* Exp. Gerontol. 27:241–250.

SERVICE P.M. 1987. Physiological mechanisms of increased stress resistance in *Drosophila melanogaster* selected for postponed senescence. Physiol. Zool. 60:321–326.

SERVICE P.M., E.W. HUTCHINSON, M.D. MACKINLEY, and M.R. ROSE. 1985. Resistance to environmental stress in *Drosophila melanogaster* selected for postponed senescence. Physiol. Zool. 58:380–389.

SMITH A.P. and T.P. YOUNG. 1982. The cost of reproduction in *Senecio keniodendron,* a giant rosette species of Mt. Kenya. Oecologia 55:243–247.

TINBERGEN J.M., J.H. VAN BALEN, and H.M. VAN ECK. 1985. Density-dependent survival in an isolated great tit population: Kluyver's data reanalyzed. Ardea 73:38–48.

VAN HANDEL E. 1965. Microseparation of glycogen, simple sugars, and lipids. Ann. Biochem. 11:266–271.

VAN NOORWIJK A.J. and G. DE JONG. 1986. Acquisition and allocation of resources: their influence on life-history tactics. Am. Nat. 128:137–142.

J. evol. biol. 10 (1997) 269–293
1010–061X/97/030269–25 $ 1.50 + 0.20/0

Journal of Evolutionary Biology

Phenotypic plasticity and selection in *Drosophila* life history evolution. 2. Diet, mates and the cost of reproduction

A. K. Chippindale,[1,*] A. M. Leroi[2], H. Saing,[3] D. J. Borash and M. R. Rose

Department of Ecology and Evolutionary Biology, University of California, Irvine, CA 92717, USA
[1]*Present address: Department of Biology @N.S. IV, University of California, Santa Cruz, CA 95064, USA, e-mail:adam@darwin.ucsc.edu*
[2]*Present address: Department of Biology, Imperial College at Silwood Park, Ascot, Berks SL5 7PY, UK, e-mail: a.leroi@ic.ac.edu*
[3]*Present address: School of Medicine, The University of Hong Kong, Sassoon Road, Hong Kong*

Key words: Costs of reproduction; phenotypic plasticity; starvation resistance; mating; dietary manipulation; longevity; *Drosophila melanogaster*; life-history.

Abstract

While it is commonplace for biologists to use the response to environmental manipulation as a guide to evolutionary responses to selection, the relationship between phenotypic plasticity and genetic change is not generally well-established. The life-histories of laboratory *Drososphila* populations are among the few experimental systems which simultaneously afford information on phenotypic plasticity and evolutionary trajectories. We employed a combination of two replicated selectively differentiated stocks (postponed aging stocks and their controls; 10 populations in total) and two different environmental manipulations (nutrition and mating) to explore the empirical relationship between phenotypic plasticity and evolutionary trajectories. While there are a number of parallels between the results obtained using these two approaches, there are important differences. In particular, as the detail of the biological characterization of either type of response increases, so their disparities multiply. Nonetheless, the combination of environmental

* Author for correspondence

manipulation with evolutionary divergence provides valuable information about the biological connections between life-history, caloric reserves, and reproductive physiology in *Drososphila*.

Introduction

The existence and nature of trade-offs among life-history traits has been the subject of considerable theoretical and empirical study since Williams (1957, 1966a, b) clearly articulated the idea that reproduction may be costly to organisms. These costs, whether affecting longevity or future reproduction, are now widely thought to constrain the evolution of life-history traits (Reznick, 1985; Partridge and Harvey, 1985; Bell and Koufopanou, 1986; Lessels, 1991; Parker and Maynard Smith, 1990; Smith, 1991; Stearns, 1992; Roff, 1992).

But the apparent simplicity of this idea has not precluded a long-standing debate as to the kinds of trade-offs that should be measured if we are to understand life-history evolution. There are two main camps in this dispute. On the one hand are those who advocate the use of "phenotypic" or environmental manipulations to reveal trade-offs (e.g., Bell and Koufopanou, 1986; Partridge and Harvey, 1988; Lessels, 1991; Partridge, 1992). These workers point out that genetic correlations among life-history traits are sensitive to the effects of environment and population structure, and are difficult to determine for most organisms. According to this viewpoint, the best guide to the evolutionary options available to a population is the phenotypic covariance among traits. On the other hand are those who suggest that only those trade-offs which manifest themselves as negative genetic correlations among life-history traits should be considered of evolutionary importance (e.g., Charlesworth, 1990; Reznick, 1992; Rose, 1991). These authors favor quantitative genetic experiments and the observation of correlated selection responses. The second group has argued that the phenotypic plasticity revealed by single-generation manipulations need not indicate the effects of alleles within populations; hence, patterns of plasticity may, or may not, represent the genetic variance and covariance shaping the response to selection.

Analogously, considerable debate has focused on the question of whether or not phenotypic variances and covariances are adequate substitutes for genetic variances and covariances in evolutionary models. Using a large data set, Cheverud (1988) suggested that phenotypic correlations may be substituted "in many situations" for genetic correlations when the latter are not available, providing the variances are appropriately scaled. Willis et al. (1991) protested against Cheverud's proposition, pointing out that Cheverud's data were strongly biased towards laboratory studies, where V_e is likely to be smaller than in the field, and towards morphological characters, which may behave differently than fitness characters. These, among other factors, may make phenotypic correlations unrepresentative of natural populations. But if the contribution of the additive genetic variance to the total phenotypic variance is small due to environmental variation, then the estimation of genetic parameters will be complicated. In particular, extreme environments (experimentally imposed or not) will damp the expression of genetic variation in the phenotypic correlation. The kernel of this debate, however, truly hinges on the fact that phenotypic parameters are easier to quantify and can be more accurately measured, particularly for those working on species for which quantitative genetic measures are difficult to obtain. The question remains: can one predict the evolution of quantitative characters without genetics?

One way to directly address the efficacy of using phenotypically measured costs to estimate genetic costs is to use populations for which both phenotypic and genetic results are obtainable. Replicated, laboratory-evolved stocks of *Drososphila* are ideal experimental material for this purpose, having both known selection histories and defined environmental conditions which can be quantitatively varied within a physiologically relevant range. This was the strategy of Luckinbill et al. (1988) and Service (1989) in response to a challenge from Partridge and Andrews (1985). Based on survivorship data from mate-restriction experiments, Partridge and Andrews suggested that the decline of early-life fecundity in populations selected for postponed senescence might be due to the removal of the direct physiological costs of exposure to mates (i.e., behaviourally mediated) rather than antagonistic pleiotropy of the sort suggested by the original experiments (Rose and Charlesworth, 1981a, b; Rose, 1984; Luckinbill et al., 1984). Luckinbill et al. (1988) and Service (1989) confirmed that mate restriction increases *Drososphila* survivorship, but both failed to find statistically significant interactions between selection and mating treatments. Partridge and Fowler (1992) also failed to find significant interaction between mating-status and longevity in virgin experiments. In other words, the increases in longevity of individuals kept as virgins (relative to mated controls) were parallel in the long- and short-lived lines, indicating that the deleterious effects of exposure to mates alone could not explain the response to selection of longevity.

We have also investigated the relationship between physiological and genetic manipulations. In a previous paper (Chippindale et al., 1993), we compared the results of laboratory selection for postponed senescence in *D. melanogaster* with environmental manipulations of diet that modify lifespan, female fecundity, and environmental stress resistance. We found that altering the availability of dietary yeast against a constant carbohydrate-rich medium caused phenotypic changes in life-history which mimicked, to a considerable extent, the evolutionary changes revealed by selection. In particular, adult flies maintained on low levels of dietary yeast had much reduced fecundities, but also increased longevities, relative to flies kept at high yeast levels, a result found earlier by several workers (Alpatov, 1932; Partridge et al., 1987; Trevitt et al., 1988). We also found that, just as longevity increased with decreased yeast ration, so too did starvation resistance (Chippindale et al., 1993). Since longevity and starvation resistance are known to be positively genetically correlated in these populations (Service et al., 1985; Rose et al., 1992), it seemed plausible that this environmental response was representative of the physiological pathways governing the correlated response of these characters under selection.

However, though the qualitative pattern of change was found to be consistent between the two methodologies, the quantitative results were not. For one thing, where the long-lived stocks initially showed a decline in early, but not late, fecundity, maintenance of females at low yeast levels causes an immediate decline in fecundity at all ages (Chippindale et al., 1993). Like the studies of mate-restriction (e.g., Service, 1989), the differences between the long- and short-lived selection treatments were not diminished by food restriction. We also found that the responses of these characters to environment were highly asymmetrical, in that the decline in fecundity under low food conditions was much more severe than the increases in the survival characters. We have since shown that phenotypic correlations among some of these same characters may be substantially modified by selection (Leroi et al. 1994b, c). The problem is that, if the genetic relationships among traits subject to environmental manipulation can change in the course of evolution, even in the short term, how can such manipulative experiments predict the response to selection?

Though the evidence points strongly towards the need to study evolution by doing selection experiments, the prevalence of physiological manipulation experiments in research on life-history evolution leads us to the conclusion that further comparison is useful. We may find that some kinds of phenotypic manipulations are good predictors of evolutionary responses, or that, once again, the claims of proponents of manipulative experiments are refuted. In this paper, we extend our investigation of the relationship between evolutionary responses and physiological manipulations. We examined the consequences for life-history and morphological traits of mating status, and we compared these to dietary manipulation responses. Since virgin *D. melanogaster* females have decreased fecundity, as well as an increased longevity and starvation resistance relative to mated ones (Partridge and Farquar, 1981; Luckinbill et al., 1988; Service, 1989), these results suggest that females deprived of mates and females deprived of dietary yeast respond in much the same way. There may be a generalized response to egg-laying substrate and to mate availability. Such generalized responses might then reveal central physiological controls linking reproduction and survival . . . controls which might be of general evolutionary importance.

Furthermore, we investigated the effects of diet upon male mating success. We previously showed that male longevity and starvation resistance are greater at low yeast levels than high (Chippindale et al., 1993). But these effects are far less dramatic in males than in females, and may thus have different features under selection and manipulation.

Materials and methods

Populations employed

The ten populations employed in this study were derived from a large, laboratory adapted population, called IV, in 1980 (Rose, 1984). Five of these populations, designated *B*, have been maintained on a generation time of two weeks; the other five, called *O*, have been maintained on a generation time of ten weeks since 1981. As a consequence of selection for late-life fertility, the *O*'s evolved postponed senescence relative to the *B*'s (Rose, 1984). These experiments reported herein were performed between 1990 and 1993. The experiments described below were not necessarily executed in the order of their numbering.

General procedures

We synchronized selection treatments and eliminated non-genetic (parental and grandparental) effects by experimentally assaying all populations after two generations of parallel rearing in a common environment. Because size may influence a variety of fitness components, experimental flies were raised at 25 °C with carefully controlled densities of 60–80 larvae per vial. Adult age was standardized by discarding very early eclosing flies and then collecting adults at 6 hour intervals over the course of the subsequent 24 hours. Virginity (in appropriate experimental treatments) was ensured by isolating sexes from each other upon collection of the young adults, and further confirmed by searching for evidence of fertile eggs (i.e., larvae) in conditioning vials after two days. Adults were dietarily conditioned for four days in one of three ways: (1) Plain banana/molasses medium (Rose, 1984), (2) banana food plus "low" yeast, a supplement of highly dilute yeast suspension equaling 0.195 mg of yeast/vial, or (3) banana food plus the "high" yeast, equaling 3.125 mg of yeast/vial (Chippindale et al., 1993). All conditioning treatments initially featured 8 flies; 4 pairs in the case of mated treatments. Yeasted vials were dried overnight after inoculation, so that yeast stickiness would not differentially affect the performance of the flies. The flies were transferred to fresh medium on day 2 of conditioning. Both the plain and low food treatments are known to result in low female fecundities and high starvation resistance values, relative to the high treatment.

The specific assay procedures were as follows: For the fecundity assay, females were placed individually into vials containing sucrose-agar medium darkened with charcoal (Rose, 1984), which facilitates egg-counting. The female was permitted to lay eggs for 24 hours (± 10 minutes), and the vial containing the eggs was then refrigerated for counting.

For the starvation resistance assay, flies were handled in the standard fashion (Service et al., 1985), and all procedures noted above were employed before the females were separated and placed into the stress assay vials. The adults were sealed into the closed end of the vial with clipped foam plug. Moistened cotton balls were added to humidify the vial. Checks on mortality were made every six hours. 40 females/population/conditioning treatment were assayed for starvation resistance. Flies were considered dead when they failed to respond to a jarring tap on two successive checks.

For measurement of ovary weight, the females were frozen at −4 °C for later dissection. The ovaries were then separated from the bodies by careful dissection under water in a well slide. Both ovaries and bodies were dried at 40 °C for 48 hours prior to weighing on a Cahn electrobalance.

For any one of the assayed characters we compared the effects of mating- and dietary-manipulations in several ways, first, and most simply, we asked whether the direct affects of these environmental manipulations upon that character are similar. Second, we asked whether there is any evidence for interaction among the environmental manipulations. Finally, we asked whether there is any evidence that the phenotypic response to these manipulations evolved independently. Possibly, the two selection treatments differed in their response to dietary manipulation, but not to mating status. This would be evidence that the phenotypic responses are not tightly genetically correlated, and hence probably differ in their physiological basis. We consider each of these levels of analysis for each character, in turn.

Experiment 1: Effect of dietary and mating manipulations upon starvation resistance

Like longevity and fecundity, resistance to starvation is easily modified by dietary manipulation in *D. melanogaster* (Zwaan et al., 1991; Chippindale et al., 1993; Leroi et al., 1994a, b, c). In the first starvation experiment, we investigated: (a) the responses of males and females to dietary manipulation, (b) the responses of males and females to mates, and (c) the change in starvation resistance from adult emergence to four days of age. Newly emerged adults 3 ($\pm$3) hours old were sorted into an immediately starved and a conditioned group, the latter consisting of virgin males, virgin females, and mixed sex vials. After four days, flies were removed from the conditioning vials and placed in groups of four, same sex, flies, into clean, humidified vials where they were starved (e.g., Service et al., 1985). Mortality checks were made every 6 hours, at 1:00, 7:00, 13:00, and 19:00 until all flies were dead.

Experiment 2: Fecundity, starvation resistance, and ovary weight under dietary and mating manipulations

This experiment consisted of three fully crossed factors: selection treatment (*B* vs. *O*); dietary yeast level (high yeast vs. plain food, as defined above), and mating status (mated vs. virgin). Flies from each of the 10 assayed populations were removed from their conditioning vials and assigned to one of the following conditioning treatments: yeast-mated; yeast-virgin; no yeast-mated; no yeast-virgin. At the end of the conditioning period, the flies were CO_2 anaesthetized, and assigned randomly to one of the following three assays: fecundity, starvation resistance or ovary weight. The fecundity of approximately 40 females/population/conditioning treatment were measured, 793 females in total. The ovaries of a total of 2 920 females were weighed in groups of ten in this component of the study.

Experiment 3: Starvation resistance of virgin and mated females

This experiment was conducted to test the replicability of the earlier results for mature females, as these differed from experiment 1 to experiment 2, and both experiments were qualitatively different from those of Service (1989). The assay procedures were identical to those reported above for starvation resistance, except that only females were tested after conditioning.

Experiment 4: Mating propensity and diet

This experiment compared the propensity to mate of the two selection treatments after different dietary conditioning treatments. Virgin males were placed in groups of eight in the conditioning vials containing either plain food or food with a high yeast supplement (see above). Virgin females were all placed in groups of eight in conditioning vials containing the high level of dietary yeast. Conditioning took place over four days, during which time the flies were transferred to fresh medium on alternate days. The males were then separated and each was added to a vial containing 10 virgin females randomly selected from within the same population. Each male was permitted 6 hours ($\pm$ 10 minutes) to mate with these females in an undisturbed environment. The male was then discarded and the females were separated and placed individually into yeasted vials for a period of several days. The mating success of each male was scored by checking these vials for the presence of larvae after four days. Observation of larvae was taken to signify mating, but no attempt was made to quantify number of offspring sired by any given male (i.e., fertility).

Statistical analysis

All statistical analysis was carried out using Abacus' StatView 4 or SuperAnova software on Macintosh computers. Since *B* and *O* populations were all derived from a common ancestral population (IV), all analyses of selection treatment differences were unpaired with respect to replicate number. The hypotheses tested in this paper all concern the evolution of populations. Therefore, in general, we have not addressed levels of variation subordinate to the means of populations within selection treatments. Our application of statistics tends to the conservative, as even the largest experiments are reduced to relatively few degrees of freedom in analysis. Analysis of variance was the exclusive form of data analysis. In all cases, the design of the ANOVA was factorial and fully crossed, using type III sums of squares.

Results

Starvation resistance

Given the well-established superiority of the *O* selection treatment's starvation resistance (Service et al., 1985; Service, 1989; Chippindale et al., 1993; Leroi et al., 1994a), it was surprising to find that the *B* selection treatment's starvation resistance was slightly higher than the *O*'s upon eclosion. Nonetheless, the starvation resistance of the newly-emerged imago was remarkably uniform across selection treatments and sexes: We found mean starvation survival times ($\pm$st. error) of 56.5 ($\pm$1.19) and 60.0 ($\pm$0.99) hours for *B*'s, and 53.9 ($\pm$1.35) and 58.4 (+0.60) hours for *O*'s, males and females, respectively. Factorial analysis of variance with selection treatment and sex as fixed effect factors revealed that both factors were statistically significant ($P < 0.01$; no significant interaction between sex and selection). In previously reported data for these same selected lines, there were no significant differences for either the selection treatment or sex factors upon eclosion (Chippindale et al., 1994, 1996). Here, though significant, the differences at emergence from pupa were small, indicating that most of the differentiation in starvation resistance between selection treatments and sexes results from differences in adult

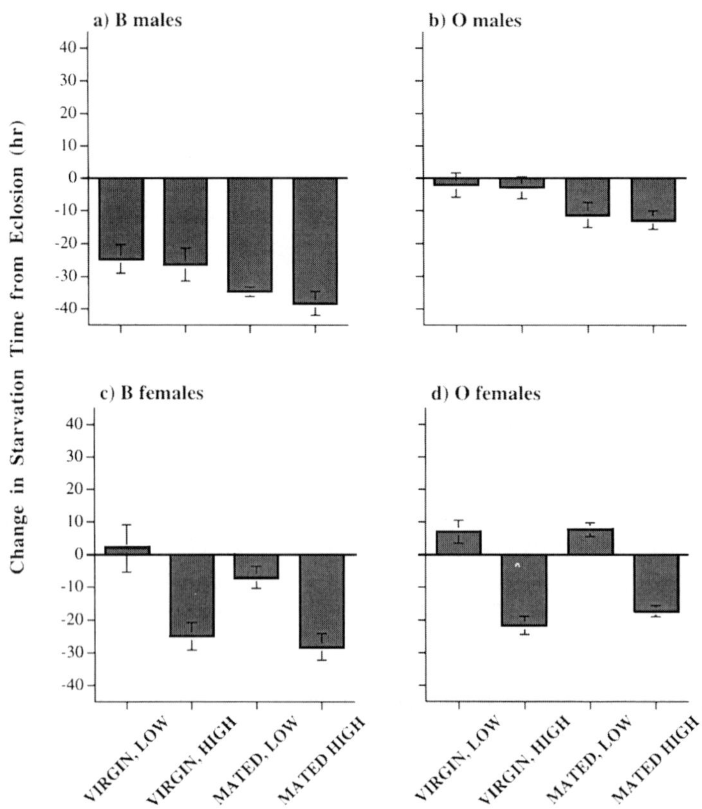

Fig. 1. Starvation resistances of *B* and *O* selection treatments were similar at eclosion, but differentiated over the first 4 days of adult life, giving rise to the characteristic difference between the selection treatments ($B < O$). The starvation resistance is shown as a difference value to the starvation time at pupal eclosion. Apparently, most differences resulted from a loss of starvation resistance over the early-adult period, relative to the eclosion values, and were dependent upon both diet and mating status. For males (parts a and b) it is evident that the presence of mates is more stressful than nutritional factors. For females (parts c and d) the reverse is true: high yeast levels had more deleterious effects than did the presence of mates. Bars are selection treatment means ($\pm$ standard errors).

Table 1. Results of analysis of variance for females in experiment 2. *F*-ratios are given, with statistical significance (p) represented by asterisks. All traits were measured on *B* and *O* adults which were then treated for four days with one of two diets (low or high) and one of two mating treatments (virgin or mated), in all possible combinations. Starvation refers to mean survival time (± 3 hr) "propensity" is a measure based upon the proportion of females laying eggs in a population, and all sizes were dry weight measures.

Source of variation	df	Starvation	24 hr Fecundity		Size (dry weight)		
			mean	propensity	ovary	body	% ovary
Selection	1	29.43***	48.99***	94.57***	0.65	3.784†	0.06
Mating	1	15.51**	737.84***	429.99***	8.29**	0.150	13.78***
Diet	1	194.01***	1657.98***	214.40***	2.23	2.570	6.04*
Selection × mating	1	0.80	70.17***	63.34***	0.01	5.902*	1.07
Selection × diet	1	4.43*	33.18***	30.16***	0.49	0.210	0.27
Mating × diet	1	4.68*	518.35***	9.64**	38.38***	0.010	51.54***
Selection × mating × diet	1	0.54	79.28***	45.36***	0.33	0.001	1.75

† $0.05 < p < 0.1$; * $p < 0.05$; ** $p < 0.01$; *** $p < 0.001$

behavior, resource acquisition, or resource allocation; these differences do not manifest themselves during the pre-adult period to any substantial degree. In Figure 1, we portray the starvation resistance of the *B* and *O* selection treatments in terms of their deviation from the baseline (zero) provided by starvation time at emergence from pupa. The plot shows the change from eclosion to 4 days of adult age, in response to all 4 possible combinations of the environmental treatments.

In females, the differentiation of selection treatments for starvation resistance was critically dependent on nutritional environment and the presence of males over the first 4 days of adult life (Fig. 1; Tab. 1). The factorial ANOVA for this first experiment (Selection × Mates × Diets) revealed that the *O* females attain their characteristically greater survival times over this four-day period [$O > B$; $F_{(1,32)} = 56.73$; $P < 0.0001$]. Similarly, while the treatment means of males and females for starvation are very similar upon eclosion, female starvation times are considerably greater after 4 days of feeding. It is apparent that most differences between the selection treatments and between the sexes are generated by differential loss of resistance by young adults. Thus, in general, *B*'s show a greater decline in resistance with age than *O*'s, and male flies lose more than female flies do, at least in the first 4 days of adult life. The high food treatment decreased resistance to starvation in females of both selection treatments.

As in our previous work (Chippindale et al., 1993), survival-related traits of males from both selection treatments were less affected by diet than were females. Indeed, in the present study, males were virtually inert with respect to dietary environment (Fig. 1; Tab. 2). For starvation resistance, the treatment mean for males conditioned on low food was only 1.4 hours higher than the mean for males conditioned on high food [$F_{(1,32)} = 0.26$; $P = 0.61$]. Mating status, however, had a highly significant effect on the average starvation time of males [$F_{(1,32)} = 17.12$; $P < 0.001$]. Averaging across all ten populations, virgin males survived for an average of 40.75 hours, while mated males survived for 29.97 hours; virgins were 36% more resistant than their mated counterparts. More significant still was the difference between the selection treatments [$O > B$; $F_{(1,32)} = 56.73$; $P < 0.0001$], as *O* males were 76.8% (19.6 hr) more starvation resistant than *B* males at four days of age.

The overall effect of virginity was to increase starvation resistance relative to the mated treatment, however the results were mixed. For males, the data obtained in experiment 1 indicated a strongly negative average effect of mating, comparable to the findings of Service (1989). For females, Service reported a negative but non-significant effect of mating for the *B*'s, but a significantly positive effect of mating on the *O*'s starvation resistance. The pattern from our experiment 1 was similar to Service's result: mated *B* females were lower in average starvation times than were virgins, and mated *O* females were longer starving than their virgin counterparts. However, neither of these differences was statistically significant; nor did we find any overall effect of mating on *B* and *O* females [$F_{(1,32)} = 0.394$; $P = 0.53$]. As Table 1 shows, in experiment 2, the effect of mating was significantly negative, manifest as lower mean starvation time for females of both *B* and *O* selection treatments [$F_{(1,32)} = 15.51$; $P < 0.001$]. For *O* females considered alone, these data were opposite in direction to those of Service and to our own first result [mated < virgin; $F_{1,16} = 7.01$; $P < 0.05$]. Thus, there was some inconsistency among the first two data sets presented herein, and between these data and those of Service, particularly for the *O* females. Therefore, we conducted a third assay of female starvation resistance.

In experiment 3, there was no significant effect of mating on *O* average starvation resistance [$F_{(1,16)} = 27$; $P < 0.6$]. The results for females across the three experiments are summarized in Table 3. A significant interaction between the selection treatment and mating factors reflected the inconsistency of the *O*'s response to the mating

Table 2. Results of analysis of variance for male traits. Mean squares and *F*-ratios are given, with statistical significance (*p*) represented by asterisks. *B* and *O* adults were treated for four days following eclosion with one of two diets (low or high) or mating treatments (virgin or mated) in the starvation resistance assay. Starvation resistance (part a) refers to mean survival time (± 3 hr), and mating propensity (part b) is the mean number of females mated by males of their respective *B*- or *O*-selected population.

Source of variation	a) Starvation resistance			b) Mating success		
	df	MS	*F*	df	MS	*F*
Selection	1	3850.03	56.73***	1	5.08	11.90***
Diet	1	17.89	0.26	1	0.88	1.80
Mating	1	1161.76	17.12***	1	–	–
Selection × diet	1	16.29	0.24	1	0.11	0.64
Selection × mating	1	0.40	0.01	1		
Mating × diet	1	14.73	0.22	1		
Selection × mating × diet	1	0.56	0.01	1		
Residual	32	67.86		16	0.49	

† $0.05 < p < 0.1$; * $p < 0.5$; ** $p < 0.01$; *** $p < 0.001$

Table 3. Effects of selection, diet and mates upon female early-life starvation resistance. Selection for postponed ageing significantly increased resistance to starvation stress. Across all environmental treatments, the *O*'s (long-lived) survived 11.7 hours (25%) longer than the *B*'s (short-lived baseline). Addition of yeast, which stimulates egg-laying, caused starvation resistance to drop dramatically in all ten populations. Thus, the "low" food treatment increased starvation time by 23.7 hours (67.7%) in *B*'s, and by 32 hours (76.6%) in *O*'s, compared to the "high" (yeasted) treatment. Mating had an asymmetrical effect upon the two selection treatments: Virgins were significantly longer starving than were mated females in *B*'s (+7.0 hr; +16.4%) but were equal in *O*'s (+1.1 hr; 2.0%). Overall the effect of mates was significant ($p < 0.01$), but there was a strong and significant interaction between selection and mating treatments. See Results section for analysis.

Experiment		Virgin				Mated			
		low		high		low		high	
		B	*O*	*B*	*O*	*B*	*O*	*B*	*O*
1	mean	71.60	89.23	39.86	43.92	56.60	75.02	31.72	43.56
	s.e.	5.21	2.83	3.04	1.84	4.83	3.17	1.28	2.98
2	mean	61.84	64.48	35.07	35.84	53.05	65.10	31.87	40.18
	s.e.	7.24	3.52	4.20	2.76	3.41	2.09	4.07	1.72
3	mean	53.69	73.69	37.32	44.75	51.59	77.51	32.28	43.74
	s.e.	3.69	1.98	2.02	3.24	1.75	2.08	1.14	3.18
	grand mean	62.38	75.80	37.42	41.50	53.75	72.54	31.96	42.49
	s.e.	5.18	7.22	1.39	2.84	5.18	7.22	1.39	2.84

treatment. Over all three experiments, when *O* females were analyzed separately, there was no significant effect of mating [$F_{(1,48)} = 0.536$; $P = 0.48$] on starvation. In summary, although mating had significant deleterious effects upon male starvation for both selection treatments, and upon females from the B selection treatment, there was no reproducible response of females from the *O* selection treatment to the mating manipulation.

Female fecundity

We examined the fecundity data in two ways: (1) Overall fecundity was computed as the population's mean number of eggs per female per 24 hours. (2) The propensity of females to oviposit was computed as the proportion of females that laid one or more eggs. As expected, both mate and yeast deprivation cause a significant decline in overall fecundity in both selection treatments (Tab. 1; Figs. 2 and 3). In general, the effect of diet was much more severe than the effect of mates. Thus, over both selection treatments, the addition of yeast to the medium increased average female fecundity 13.6 times, while the addition of males increased it just 3.8 times.

A three-factor ANOVA (fully crossed factors: Selection Treatment, Diet, and Mating) was performed on each of these measures of early-life fecundity (Tab. 1). For the unconditional fecundity data, all of the main effects were highly significant

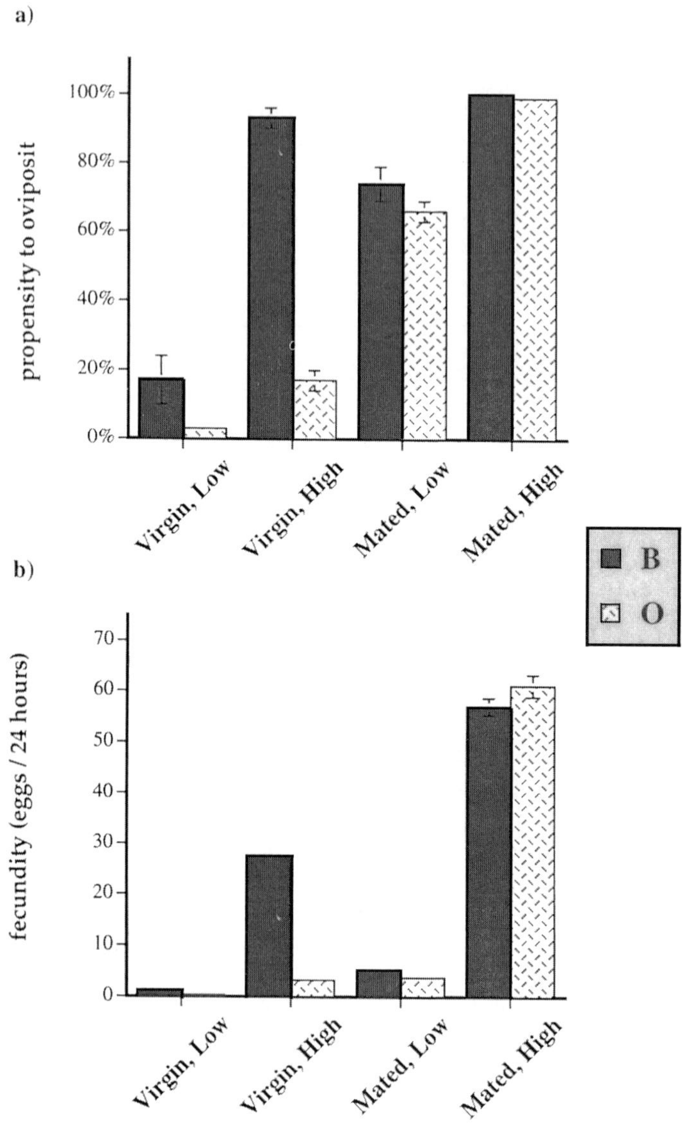

Fig. 2. The propensity of females to oviposit (a) and the average number of eggs layed (b) in *B* and *O* females under 4 different treatments. Propensity to oviposit was measured as the proportion of females that layed at least one egg. Daily fecundity was the 24 hr average across all females in the population, including zeroes. Females from the *O* selection treatment displayed a strong dependence on the presence of both rich substrate and mates for oviposition. Bars are selection treatment means (±standard errors).

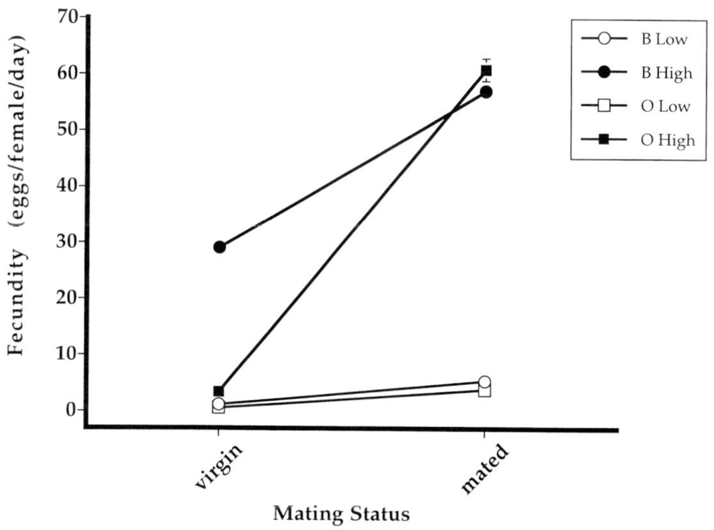

Fig. 3. The interaction between diet and mating status is clear in this plot: *O* females require *both* mates and high food levels to produce substantial numbers of eggs, while *B* females layed eggs at more than 50% of the maximum rate as virgins when provided with a rich substrate. Error bars are standard errors.

($B > O$; High Food > Low Food; Mated > Virgin; all $P < 0.0001$), as were all of the interactions ($P < 0.0001$). For propensity to oviposit, again all of the main effects were highly significant (all $P < 0.0001$). In the ANOVA for laying propensity, all of the interactions were significant at $P < 0.0001$, except for Mating × Diet ($P < 0.001$). Owing to the complexity of these interactions, we show the results for averaging fecundity as both a bar chart (Fig. 2) and an interaction plot (Fig. 3).

To summarize the results, the selection treatments showed strong differentiation in their reactions to the two types of environmental treatment. In the high yeast environment, *B* females produced 50.0% as many eggs when virgin compared to when mated, while *O* females produced 5.6% as many eggs. Under unyeasted (low) food conditions these numbers were 21.9% (*B*) and 14.4% (*O*). For propensity to oviposit, a similar pattern is observed, as 93% of *B* females laid eggs in the high food/virgin treatment despite the absence of mates; only 17% of *O* females produced eggs under the same conditions. Thus, the *O* females displayed considerably greater sensitivity to mating status than did the *B* females. The effects of food level were relatively symmetrical for the two selection treatments when mated, as yeast elevated *B* fecundity by 52.6 and *O* fecundity by 57.0 eggs/female/day. As virgins, however, yeast affected B fecundity by plus 27.3 eggs, while *O*'s only increased by 2.8 eggs/female/day. The *O*'s therefore required *both* mates and a protein rich diet to reach high fecundity, while the *B's* were stimulated by food to lay substantially, even in the absence of males.

Ovary weight

The results of the ovary weight assay (experiment 2), shown in Figure 4, were analyzed using a fully crossed, three factor ANOVA (selection treatment, food level, and mating; Tab. 1). The effect of selection treatment was nonsignificant [$F_{(1,32)} = 0.646$; $P = 0.43$], contrary to the result of Rose et al. (1984), nor was food

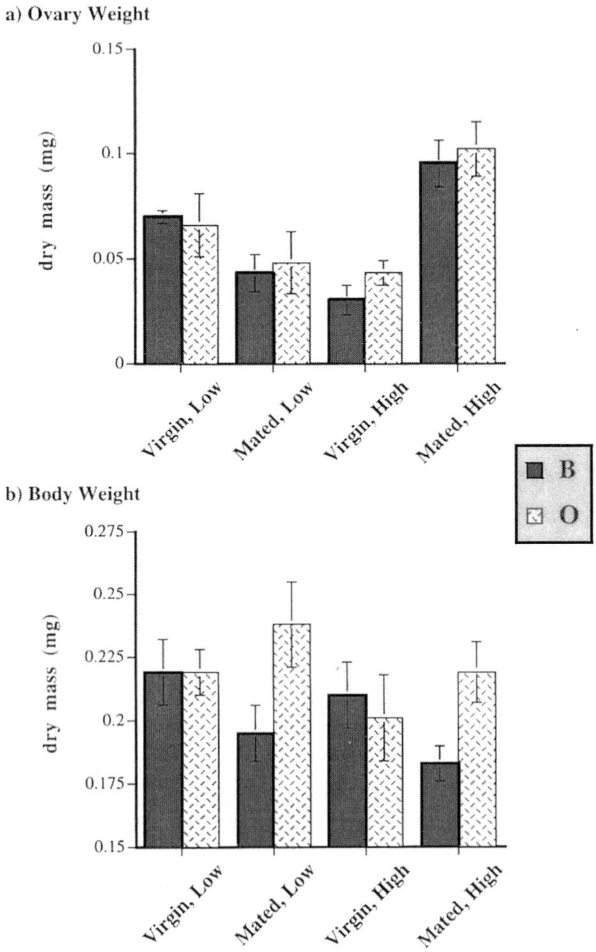

Fig. 4. Ovary (a) and body (b) dry weights of *B* and *O* females under 4 different treatments. The ovaries of mated females were significantly larger than those of virgins, over both diet treatments. It is notable, however, that the ovary weight of virgins was greater than mated flies under the low dietary treatment. This difference in weight relative to mating treatment across dietary treatments gave rise to a significant interaction between the two environmental treatments. Bars are selection treatment means ($\pm$standard errors).

level significant [$F_{(1,32)} = 2.27$; $P = 0.14$]. Mating, however, produced a significant effect on ovary weight [$F_{(1,32)} = 8.29$; $P < 0.01$]. Mated females had ovaries 1.7 times larger than did virgins under high food conditions, but virgins under low food conditions had ovaries 50% larger than mated flies. The interaction between food and mating factors for this trait was highly significant ($P < 0.0001$).

Body weight

The analysis of female body weight (minus the ovaries) was an ANOVA of identical form to that described above for the ovaries.

These data are presented in Figure 4. We found no significant differences among the main factors for this trait, although "selection" was nearly significant [$B < O$; $F_{(1,32)} = 3.78$; $P = 0.06$]. A mating by selection interaction was evident [$F_{(1,32)} = 5.90$; $P < 0.05$] as B's were slightly heavier than O's when virgin, but lighter when mated (Tab. 1).

Relative ovary weight

The relative size of the ovary may be a more sensitive indicator of reproductive vs. somatic investment than the unscaled ovary weight. The ratio of ovary mass to total body mass was computed for each group [ovary/(ovary + body)]. Ovary weights ranged between 12% and 34% of overall female weight. The 3-way ANOVA (factors as above) revealed significant effects of mating and food, as expected given the differences in ovary weights and the absence of differences in body weight (Tab. 1). The interaction between these two factors reflects the peculiar fact that relative ovary weight was 200–300% greater in mated flies over virgins at high yeast, but approximately 33% lower in mated vs. virgin flies at the low yeast level.

Effects of diet on male mating success

Figure 5 shows the results of the mating propensity assay. In general, the number of matings resulting in fertilization was very high, with some males mating all 10 females in the 6 hours of the experiment. The female conditioning environment (high yeast, virgin) was, presumably, highly conductive to mating (e.g., Trevitt et al., 1988), particularly given its effect on the fecundity of mated females (see above). An analysis of variance (Factors: Selection, Food) was performed. On average (standard error in brackets), males from the B selection treatment fertilized 6.30 (± 0.19) out of the possible 10 females, while O males fertilized 5.23 (± 0.25) females. The effect of selection treatment was significant [$F_{(1,16)} = 11.9$; $P < 0.01$]. Unfortunately, the design of our experiment did not permit us to separate the mating success of males from the receptivity of females from that population.

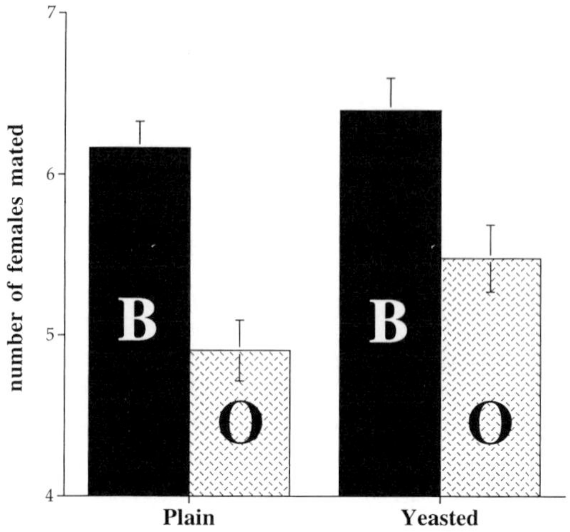

Fig. 5. Mating propensity early in adult life. Males which had been fed low (no live yeast) or high (live yeast) food were given 6 hours to mate with 10 virgin females which had been maintained on *ad libitum* yeast for 4 days. The mating success of *B* males was greater than that of *O* males in both food environments, befitting their selection treatments. No effect of dietary treatment was detected. Bars are selection treatment means ($\pm$standard errors).

Because we conditioned all females in a common environment, yet varied male diet, we can legitimately ask whether the effect of diet on male mating success is the same in the *B*'s as in the *O*'s. Diet did not affect mating propensity significantly [$F_{(1,16)} = 1.80$; *n.s.*], nor was there a significant stock $\times$ yeast interaction. The average number of fertilizations under the "low" treatment was 5.56 ($\pm$0.31). The average number of fertilizations under the "high" food treatment was 5.98 ($\pm$0.25). Since the females from each population were not subjected to dietary manipulation, while males were, we conclude that the *B* and *O* males' mating ability was not significantly affected by diet.

Summary of key findings

The survival-related trait measured in these experiments was starvation resistance. Starvation resistance time was found to be nearly identical in the two selection treatments upon pupal eclosion, but became highly differentiated in the imago after 4 days of conditioning. Thus, the previously-documented difference in starvation resistance ($B < O$) was manifest only after maturation. While males declined in all selection and environmental treatments, the negative effect of mates was much stronger than that of high dietary yeast levels. For females of both selection treatments, the relative impacts of the two environmental treatments were reversed: females were strongly negatively affected by the high yeast treatment, but only the *B* selection treatment displayed a repeatable reduction in survivorship after being exposed to males.

The early-reproduced *B* selection treatment displayed significantly higher female fecundity and propensity to oviposit across experimental treatments (Fig. 2 and 3). However, under the most permissive circumstances (high food with mates) the fecundity of the *O* females was slightly higher than that of the *B*'s. The *B* selection treatment therefore displayed lower sensitivity to fertility-limiting environmental circumstances (low diet and mate-restriction). These patterns of fecundity were not closely mirrored in the weight of the ovary (Fig. 4). Specifically females from both selection treatments had large ovaries in the absence of yeast or mates, and the large difference in fecundity between *B* and *O* females under high food conditions, but without mates, was not matched by an ovary weight difference.

Males from the *B* selection treatment were significantly more successful than *O*-selected males in mating females, whether or not the males had received a dietary yeast supplement (Fig. 5). There was no significant effect of the dietary manipulation upon the propensity of these selection treatments to mate, despite the large female fecundity differences noted above.

Interactions in the analyses were common. For example, analysis of the female 24 hour fecundity data revealed that all two- and three-way interaction terms were statistically significant (Tab. 1, see also Fig. 3). Indeed, the Mate × Diet interaction was significant for 5 of 6 traits given in Table 1. The pervasiveness of these interactions suggests that the environmental manipulations differed substantially in their physiological effects upon the fly.

Discussion

The evolution of the *O* (long-lived) populations from a short-lived ancestor was initially characterized in terms of the antagonistic pleiotropy between early fertility, on one hand, and survival and later reproduction, on the other (Rose, 1984). While these initial results have proven robust in most respects, the *B* and *O* lines have also become a study in the multifarious nature of adaptation. Central to this work has been the investigation of costs of reproduction, particularly as manifest in selection responses that reveal trade-offs between fitness traits. The present results shed light on three key areas in the study of the evolutionary costs of reproduction. The first area is the extension of our record of responses to selection in long- and short-lived populations, specifically the further dissection of the female response to selection and the evolution of male mating ability. The second area is the response of these selected populations to two types of environmental manipulation: diet and mates. The third area of investigation is the somewhat refractory question of the correspondence between the genetic response to selection and the immediate phenotypic response to environment. We show that the genetic and environmental responses may interact significantly, as may the responses to the two different manipulations themselves. We now discuss each of these three themes in greater detail.

Evidence for adaptation to demographic selection regime

The greatest difference between *B* and *O* selection regimes is the timing of reproduction. In this study, most experimental assays were conducted on flies at approximately 14 days from oviposition, the same age at which *B* populations are selected to reproduce. Several of our results appear to reflect adaptation to early reproduction in the *B* populations. For example, *B* populations showed significantly higher propensity to mate early in adulthood than did *O* populations, under both dietary conditioning regimes (Fig. 5). This difference between selection treatments may have resulted from differences in male mating ability, female receptivity, copulation duration, or latency between copulations. We cannot identify which of these potential mechanisms was involved because each type of male was mated to females of the same genotype (population), rather than to a common female genotype (as in Service, 1993). In fact, there is evidence that all of these mechanisms may be involved in the difference observed between *B*- and *O*-selected populations in early-life propensity to mate. In Service's (1993) experiments, *B*-selected males displayed higher competitive and non-competitive mating ability, although only the former result was statistically significant. And, there is considerable evidence that the selection treatments differ significantly in remating speed, female receptivity and copulation duration, each in the direction of increased early mating in the B's (Service and Vossbrink, 1996; P. M. Service pers. comm. 1996). Irrespective of the particular mechanism involved, it is apparent that the propensity to mate early in adult life is lower in postponed aging lines (*O*) than their early-reproduced (*B*) counterparts. Although we did not measure later-life mating propensity, this result would be predicted from an adaptationist viewpoint, given a stressful effect of early reproduction on future reproduction and survival.

Another point of interest in comparing the selection treatments was their differentiation for starvation resistance. Imagos of *B* and *O* lines emerged from pupa with similar starvation resistance times (Fig. 1). It was during the first four days after eclosion that the selection treatments diverged. *B* populations displayed a greater loss of starvation resistance across all environmental treatments. The physiological basis of starvation endurance in *Drosophila* is well known, having been connected to whole-body and fat-body lipid content (Wigglesworth, 1949; Doane, 1960; Service, 1987; Zwaan et al., 1991; Chippindale et al., 1996) and to overall caloric content (Djawdan et al., in prep.). In other words, starvation time is a good index of the caloric stores on board the fly.

The pattern of differentiation we observed suggests either that *B*'s expend more energy in activities that prevent them from recharging their caloric reserves early in life, such as courtship and mating, or that they are simply not selected for storage beyond the larval stage. This interpretation is strengthened by the observation of greater sensitivity to the presence of mates in the *B*'s. *O*-selected females were completely unaffected by the presence of males, and even displayed significantly higher starvation times when mated in one study (Service, 1989), although this result was ephemeral in our study. Since starvation resistance and longevity are positively genetically correlated (e.g., Rose et al., 1992), these results suggest that changes in mating behavior may be partially involved in the increased longevity of *O* females. To the contrary, however, Service (1989) demonstrated that virginity affected the longevity of *B* and *O* treatments to an equal degree. Luckinbill et al. (1988) found that the difference between long- and short-lived selected lines was reduced, but not eliminated, when the lines were maintained as virgins. There is, therefore, some evidence for the partial involvement of mate avoidance in the selection response of long-lived lines, as suggested by Partridge and Andrews (1985).

As previously reported (Service, 1989), *B* females in the present study were much more likely to lay eggs than *O* females when both groups were kept as virgins. The difference we observed was particularly striking under the "high" food condition, where more than 90% of virgin *B* females laid eggs, while fewer than 20% of virgin *O* females would. Apparently, either the presence of rich medium or mates will stimulate most *B* females to lay eggs. The *O* females showed a strong dependence upon mating status alone in their propensity to oviposit. These findings are somewhat surprising in view of the fact that part of the *O* selection response has been an increased responsiveness of female fecundity to live dietary yeast. When provided with mates and abundant dietary yeast, *O* females produced more eggs than *B* females (present results; Chippindale et al., 1993; Leroi et al., 1994a) despite the observation that mating with *B* males induces higher female fecundity early in life (Service and Vossbrink, 1996). The reason that *O*'s have higher fecundity than *B*'s when given abundant yeast has to do with differences in conditioning prior to egg-laying in routine selection (see Leroi et al., 1994a for extended discussion). Although the fecundity response of *O*'s to dietary yeast is inhibited when females are not mated, the ovary weight data clearly indicate that *O*'s build equal- or larger-sized ovaries than *B*'s do under higher food conditions. The sum of these observations suggests that *O* females are more sensitive than *B* females are to specific environmental conditions in their oviposition behaviour early in life, but fecundity observations do not necessarily reflect reproductive investment.

Evidence for a cost of mating for males

In females, costs of reproduction have been inferred from a variety of different types of data: comparitive studies, phenotypic manipulations, phenotypic correlations, selection experiments, and genetic correlations (reviewed by Bell and Koufopanou, 1986). Although there are many presumed instances of costs of reproduction for males stemming from anecdote, particularly in the pursuit of reproduction (rutting, elaborate displays, and so on), there is a paucity of empirical demonstrations of these costs. One reason for this asymmetry is that it is typically considered more difficult to evaluate male reproductive effort physiologically and more difficult to keep track of progeny production. These difficulties translate into a very limited picture of the direct costs of mating and gamete production for males.

Van Voorhies (1992) appeared to have found a cost to mating in males of the nematode *C. elegans*, manifest as a decrement in longevity when males were mated to hermaphrodites. Van Voorhies argued that the observed diminution in lifespan of males was a direct consequence of a cost to sperm production. Curiously, there was no apparent cost to mating for the hermaphrodites in Van Voorhies' experiment, contrary to conventional wisdom that eggs cost more to produce than sperm do. Gems and Riddle (1996), using the same worm strain and a more extensive protocol, found exactly the reverse: a cost of mating for the hermaphrodites but not for the males. The result of Van Voorhies was ephemeral, and evidence for a direct cost to sperm production in nematode males remains elusive. In *Drososphila*, Partridge and Farquar (1981), Partridge et al. (1986, 1987), Luckinbill et al. (1988) and Service (1989), have independently shown that male flies have a significantly shorter lifespan when housed with females than when kept as virgins. Similarly, we have shown here that males kept in mixed-sex cohorts have significantly lower starvation resistance than those kept as virgins. In all of these examples from *Drososphila*, it is impossible to differentiate between the costs associated with mating and the costs associated with courtship and associated activities.

New complexity is added to the story for fruit flies by the present results: While males and females of these selection treatments were previously found to be sensitive to the presence of mates and quality of diet in terms of longevity (Chippindale et al., 1993), and here both sexes were highly sensitive to the presence of mates in terms of starvation resistance, only female starvation resistance was significantly affected by diet. Although previously we have found a significant negative effect of richer diet on this trait in males (Chippindale et al., 1993; Leroi et al., 1994b, c), the magnitude of the effect was still very slight relative to its magnitude in females. While the high fertility environment (mates and yeast) produced a roughly symmetrical decline in survival probability for both sexes in longevity (Chippindale et al., 1993), the response of the sexes in early-life starvation resistance was asymmetrical. In other words, it would appear that there is little or no additional cost to males resulting from being housed with more fertile females over the short-term, but a cost is manifest over the long-term. Conceivably this result is mediated by cumulative remating damage to males. However, more work will be required to resolve the specific sources of differences in male mate tolerance among the selection and environmental treatments discussed herein.

The strongest evidence for a direct cost to mating in males comes from the work of Service and colleagues. These workers have shown reductions in a variety of male early-life reproductive performance traits, and increases in late-life performance, in the *O* selection treatment. For example, recently Service and Vossbrink (1996) demonstrated that, early in life, males of the late-reproduced *O* selection treatment are less effective at inducing egg-laying, and produce a higher sperm-dependent remating threshold in the females they inseminate. Service and Vossbrink's results suggest that there are evolved differences in seminal fluid composition and its interaction with sperm *in utero*. Coupled with their earlier work (Service, 1989; Service 1993; Service and Fales, 1993), which demonstrated lower competitive mating ability of *O* males early in life, among other things, there is substantial evidence for a cost of reproduction in *Drosophila* males. In Service's studies this cost has been demonstrated through the observation of age-dependent trade-offs in replicated and divergently selected populations, and corresponds to what is known from females of these selected populations.

Phenotypic manipulations and selection responses: Limited parallels

Here, as in our other studies (Chippindale et al. 1993; Leroi et al. 1994a, b), lower nutrition dietary treatments had the general effect of suppressing reproduction and improving survival traits. In females of both selection treatments, we observed both dramatically lower fecundity and a reduced propensity to oviposit under low food conditions, accompanied by a reduction of ovary weight, and increases in starvation resistance. Thus, in broad view, the present dietary plasticity results mimic the antagonistic evolutionary relationship between survival and reproduction previously demonstrated for the *B* and *O* lines (Rose, 1984; Rose et al., 1984).

However, there are important differences between the genetic responses to selection and the responses produced by dietary manipulation. First among these is the incongruence of results for the two sexes. While diet affected females strongly, it had no measurable influence upon male mating ability or starvation resistance. The absence of a difference in starvation resistance across food levels was particularly surprising in view of the strongly positive evolutionary correlation between longevity and starvation resistance (Service et al., 1985; Rose et al., 1992). In our earlier work (Chippindale et al., 1993), we found that the longevities of both sexes negatively covaried with yeast level.

Despite the superficial similarity of dietary and mating effects on survival and fertility, the two treatments had opposite effects on ovary mass. For example, mate-restricted females given the high yeast treatment have larger ovaries than mated females, suggesting that well-fed virgin females stored eggs rather than not making them. This result is consistent with results from other *Drosophila* stocks and indicates that the effects of mating and dietary manipulations differed, at least in part, physiologically. There are precedents in the literature for this conclusion. For example, juvenile hormone, produced by the corpora allata, is known to play a significant role in mediating the response of vitellogenesis to nutrition (Bownes and Blair, 1986; Bownes et al., 1988; Bownes and Reid, 1990). However the egg-retention behavior of virgin females is thought to be under the control of a myotropic hormone produced in the pars intercerebrallis (e.g., Boulétreau-Merle, 1986). Thus we find no master control unifying responses to selection, mating status manipulation, and dietary manipulation. The hope for environmental manipulation revealing the "controls" or "directions" for evolution, in this case, has proven chimerical.

Are phenotypic manipulations useful in the study of costs of reproduction?

The effect of single-generation environmental manipulation experiments is to modify the distribution of phenotypes within populations. Such changes may or may not be indicative of evolutionarily relevant variation, as it is seldom clear whether or not the actual pattern of allelic effects has been revealed. There are no strong theoretical guidelines to determine which manipulations or patterns of variation at the phenotypic level reflect the effects of alleles. When experimental manipulations are performed, the environmental component of the total phenotypic variation is increased, and this may further aggravate the extraction of genetic information. However, it seems possible that, if rudimentary physiological pathways are stimulated by the environmental change introduced by the experimenter, a set of options available to selection could be revealed. But which conditions are relevant? For example, in the present experiments, changes in food level led to a variety of changes in life-history and morphological traits. We varied the level of yeast suspension against a constant carbohydrate-rich background medium. If instead we had diluted the base medium across a range of values for which the flies experienced starvation, perhaps we would have observed a positive association between fecundity and starvation resistance across low food levels (as in David et al., 1971; Le Bourg and Medioni, 1991). Another way that survival-related characters have been manipulated is through the experimental variation of larval density. Recently, Zwaan et al. (1991) were able to repeat and extend the experimental results of Lints and Lints (1969) by showing that, within a certain range of densities, crowding increases relative fat fraction, longevity and starvation resistance at the same time. The three types of manipulation – larval crowding, mate restriction and dietary restriction – generate similar increases in survival traits, especially longevity of the imago, but have completely different effects on other traits, such as ovary weight, known to be genetically correlated with longevity from selection experiments. Thus, the phenotypic covariance under environmental manipulations need not bear any resemblance to the pattern observed under selection, even when a major trait is affected in the same way, once the full range of responding characters is studied.

The process of deciding which experimental manipulation of the environment reveals relevant patterns of plasticity may often be as much dictated by *a priori* expectations about what one is supposed to see as it is by any good set of mechanistic criteria. The only way to evade these difficulties may be to show an exact correspondence between the physiological mechanisms underlying the selection response and the manipulation (Partridge and Sibly, 1991). This is of course the task that we have not been able to accomplish here, for we have shown the absence of such an exact correspondence. But even if it were possible to accomplish this goal, one might ask why one would do the manipulation at all, given that one will have the results of selection anyway. One reason one might perform phenotypic manipulations is that they may reveal valuable information about the short-term physiological and ecological trade-offs operating in a population. The costs revealed by such manipulations may indicate important mechanistic connections among traits. Another reason to perform phenotypic manipulations is to test the robustness of results from selection. For example, if a trade-off is inferred using quantitative genetic methods under one set of environmental circumstances, it is reinforced if it is reproducible in others.

Interpreting the biology underlying multiple-character, multiple-manipulation plasticity is a formidable task. We are not so confident as some authors, such as those who invent optimization models, who readily decide that total caloric expenditure or some other single variable is minimized or maximized over all manipulations. This is not to criticize attempts to probe the physiological workings of the organism by the use of multiple assays under multiple conditions. Our view tends to be that environmental manipulations will be useful to the extent to which they can be related to known selection histories of well-defined populations, as illustrated by our use of *B* and *O* populations in this work and analogous experiments in some of our other papers (Leroi et al., 1994a, b, c).

Acknowledgements

This work was supported by NSF and NIH grants AG06346 and AG09970 to MRR. We alphabetically thank J. Alipaz, P. Campos, W. Chen, Q. Nguyen, J. Shiotsugu, and L. Vu for help in conducting the experiments. T. J. Bradley kindly provided advice on weighing ovaries, as well as the use of laboratory space and equipment to do so. B. Holland, D. Templin and two anonymous reviewers provided many useful comments that improved the manuscript.

References

Alpatov, W. W. 1932. Egg production in *Drosophilamelanogaster* and some factors which influence it. J. Exp. Zool. 63: 85–111.

Bell, G. and V. Koufopanou. 1986. The cost of reproduction. *In* R. Dawkins and M. Ridley (Eds.), Oxford Surveys in Evolutionary Biology V.3, Oxford Univ. Press, Oxford, UK.

Boulétreau-Merle, J. 1986. Seasonal genetic variations in fecundity and egg-laying controls in temperate *D. melanogaster* populations. Advances in Invertebrate Reproduction 4: 461–470.

Bownes, M. and M. Blair. 1986. The effects of a sugar diet and hormones the expression of *Drosophila* yolk-protein genes. J. Insect Physiol. 32: 493–501.

Bownes, M., A. Scott and A. Shirras. 1988. Dietary components modulate yolk protein transcription in *Drososphila melanogaster*. Development 103: 109–128.

Bownes, M. and G. Reid. 1990. The role of ovary and nutritional signals in the regulation of fat-body yolk-protein gene expression in *Drosophila melanogaster*. J. Insect Physiol. 36: 471–479.

Charlesworth, B. 1990. Optimization models, quantitative genetics, and mutation. Evolution 44: 520–538.

Cheverud, J. M. 1988. A comparison of genetic and phenotypic covariances. Evolution 42: 958–968.

Chippindale, A. K., A. M. Leroi, S. B. Kim and M. R. Rose. 1993. Phenotypic plasticity and selection in *Drosophila* life-history evolution. I. Nutrition and the cost of reproduction. J. Evol. Biol. 6: 171–193.

Chippindale, A. K., D. T. Hoang, P. M. Service and M. R. Rose. 1994. The evolution of development in *Drosophila melanogaster* selected for postponed senescence. Evolution 48(6): 1880–1899.

Chippindale, A. K., T. J. F. Chu and M. R. Rose, 1996. Complex trade-offs and the evolution of starvation resistance in *Drososphila*. Evolution 50 (2): 753–766.

David, J., J. Van Herrewege and P. Fouillet. 1971. Quantitative under-feeding of *Drosophila*: Effects on adult longevity and fecundity. Exp. Geront. 6: 249–257.

Doane, W. W. 1960. Developmental physiology of the mutant female *sterile (2) adipose* of *Drosophila melanogaster*. II. Effects of altered environment and residual genome on its expression. J. Exp. Zool. 145: 23–42.

Gems, D. and D. L. Riddle. 1996. Longevity in *Caenorhabditis elegans* reduced by mating but not gamete production. Nature 379: 723–725.

Le Bourg, E. and J. Medioni. 1991. Food restriction and longevity in *Drosophila melanogaster*. Age & Nutrition 2: 90–94.

Leroi, A. M., A. K. Chippindale and M. R. Rose. 1994a. Long-term laboratory evolution of a life-history trade-off in *Drosophila melanogaster*. 1. The role of genotype by environment interaction. Evolution 48 (4): 1244–1257.

Leroi, A. M., W. R. Chen and M. R. Rose. 1994b. Long-term laboratory evolution of a life-history trade-off in *Drosophila melanogaster*. 2. Stability of genetic correlations. Evolution 48 (4): 1257–1268.

Leroi, A. M., S. B. Kim and M. R. Rose. 1994c. Do phenotypic manipulations reveal evolutionary trade-offs? An experimental test using *Drosophila melanogaster*. American Naturalist 144: 661–676.

Lessels, C.M. 1991. The evolution of life histories. *In* J. R. Krebs and N. B. Davies (Eds.), *Behavioural Ecology* (3rd Edition). Blackwell Scientific, Oxford.

Lints, F. A. and C. V. Lints. 1969. Influence of preimaginal environment on fecundity and ageing in *Drosophila melanogaster* hybrids. I. Preimaginal population density. Experimental Gerontology 4: 231–238.

Luckinbill, L. S., M. J. Arking, M. J. Clare, W. C. Cirocco and S. Buck. 1984. Selection for delayed senescence in *Drosophila melanogaster*. Evolution 38: 996–1003.

Luckinbill, L. S., J. L. Graves Jr., A. Tomkiw and O. Sowirka. 1988. A qualitative analysis of some life-history correlates of longevity in *Drosophila melanogaster*. Evolutionary Ecology 2: 85–94.

Parker, G. A. and J. Maynard Smith. 1990. Optimality theory in evolutionary biology. Nature 348: 27–33.

Partridge, L. 1987. Is accelerated senescence a cost of reproduction: Funct. Ecol. 1: 317–320.

Partridge, L. 1992. Measuring reproductive costs. Trends in Ecology and Evolution 7: 99–100.

Partridge, L. and M. Farquhar. 1981. Sexual activity reduces lifespan of male fruitflies. Nature 294: 580–582.

Partridge L. and R. Andrews. 1985. J. Insect Physiology 31: 393–395.

Partridge, L. and P. H. Harvey. 1985. Costs of reproduction. Nature 316: 20–21.

Partridge, L. and P. H. Harvey. 1988. The ecological context of life history evolution. Science 241: 1449–1455.

Partridge, L., K. Fowler, S. Trevitt and W. Sharp. 1986. An examination of the effects of males on the survival and egg-production rates of female *Drosophila melanogaster*. J. Insect Physiol. 32: 925–929.

Partridge, L., A. Green and K. Fowler. 1987. Effects of egg-production and of exposure to males on female survival in *Drosophila melanogaster*. J. Insect Physiol. 33. 745–749.

Partridge, L. and R. Sibly. 1991. Constraints in the evolution of life-histories. Phil. Trans. R. Soc. Lond. B 332: 3–13.

Partridge, L. and K. Fowler. 1992. Direct and correlated responses to selection on age at reproduction in *Drosophila melanogaster*. Evolution 46 (1): 76–91.

Reznick, D. 1985. Costs of reproduction: an evaluation of the empirical evidence. Oikos 44: 257–267.

Reznick, D. 1992. Measuring the costs of reproduction. Trends in Ecology and Evolution 7: 42–45.

Roff, D. 1992. The Evolution of Life Histories. Chapman and Hall, NY, USA.

Rose, M. R. 1984. Laboratory evolution of postponed senescence in *Drosophila melanogaster*. Evolution 38: 1004–1010.

Rose, M. R. 1991. Evolutionary Biology of Aging. Oxford University Press, New York.

Rose, M. R. and B. Charlesworth. 1981a. Genetics of life-history in *Drosophila melanogaster*. I. Exploratory selection experiments. Genetics 97: 173–186.

Rose, M. R. and B. Charlesworth. 1981b. Genetics of life-history in *Drosophila melanogaster*. II. Sib analysis of adult females. Genetics 97: 187–196.

Rose, M. R., M. L. Corey, A. M. Coyle and P. M. Service, 1984. the morphology of postponed senescence in *Drosophila melanogaster*. Can. J. Zool. 62: 1576–1580.

Rose, M. R., L. N. Vu, S. U. Park and J. L. Graves. 1992. Selection on stress resistance increases longevity in *Drosophila melanogaster*. Expt. Gerontol. 27: 241–250.

Service, P. M. 1987. Physiological mechanisms of increased stress resistance in *Drosophila melanogaster* selected for postponed senescence. Physiol. Zool. 60: 321–326.
Service, P. M. 1989. The effect of mating status on lifespan, egg laying, and starvation resistance in *drosophila melanogaster* in relation to selection on longevity. J. Insect Physiology 35(5): 447–452.
Service, P. M. 1993. Laboratory evolution of longevity and reproductive fitness components in male fruit flies: mating ability. Evolution 47(2): 387–399.
Service, P. M., E. W. Hutchinson, M. D. MacKinley and M. R. Rose. 1985. Resistance to environmental stress in *Drosophila melanogaster* selected for postponed senescence. Physiol. Zool. 58: 380–389.
Service, P. M. and A. J. Fales. 1993. Evolution of delayed reproductive senescence in male fruit flies: Sperm competition. Genetica 91: 111–125.
Service, P. M. and R. Vossbrink. 1996. Genetic variation in "first" male effects on egg-laying and remating by female *Drosophila melanogaster*. Behavior Genetics 26 (1): 39–47.
Smith, R. H. 1991. Genetic and phenotypic aspects of life-history evolution in animals. Advances in Ecological Research 21: 63–115.
Stearns, S. C. 1992. The Evolution of Life Histories. Oxford University Press. Oxford, U.K.
Trevitt, S., K. Fowler and L. Partridge. 1988. An effect of egg-deposition on the subsequent fertility and remating frequency of female *Drosophila melanogaster*. J. Insect Physiol. 34(8): 821–828.
Van Voorhies, W. A. 1992. Production of sperm reduces nematode lifespan. Nature 360: 456–458.
Wigglesworth, V. B. 1949. The utilization of reserve substances in *Drososphila* during flight. J. Exp. Biol. 26: 150–163.
Williams, G. C. 1957. Pleiotropy, natural selection, and the evolution of senescence. Evolution 11: 398–411.
Williams, G. C. 1966a. Adaptation and Natural Selection. Princeton University Press. Princeton.
Williams, G. C. 1966b. Natural selection, the costs of reproduction and a refinement of Lack's principle. American Naturalist 100: 689–690.
Willis, J. H., J. A. Coyne and M. Kirkpatrick. 1991. Can one predict the evolution of quantitative characters without genetics? Evolution 45(2): 441–444.
Zwaan, B. J., R. Bijlsma and R. F. Hoekstra. 1991. On the developmental theory of ageing. I. Starvation resistance and longevity in *Drosophila melanogaster* in relation to pre-adult breeding conditions. Heredity 66: 29–39.

Received 28 March 1996;
accepted 25 July 1996.

IV

GENETICS AND MOLECULAR BIOLOGY OF METHUSELAH FLIES

Genetics and Molecular Biology

Hardip B. Passananti[1] and Michael R. Rose[2]

[1]*Orrick, Herrington & Sutcliffe LLP,*
4 Park Plaza, Suite 1600, Irvine, CA 92614
hpassananti@orrick.com

[2]*Dept. of Ecology and Evol. Biology,*
University of California, Irvine, Irvine, CA 92697-2525
mrrose@uci.edu; also to whom correspondence should be sent

The genetics and molecular biology of *Drosophila* aging have made rapid progress only in the last decade. Before 1990, the known large-effect mutants almost always reduced life span. The *Drosophila* mutants that enhanced lifespan mostly eliminated the reproductive organs of females (e.g. Maynard Smith, 1958), making such mutants of doubtful relevance to normal aging.

Methuselah flies offered a way out of this impasse, because they evidently possessed allelic variants that enhance longevity and postpone aging. Three main approaches have been used to reveal the genetics of these flies. First, individual genetic loci have been chosen as candidates for involvement in the increased lifespan of Methuselah flies. Among these, superoxide dismutase (Tyler *et al.*, 1993; reprinted here) and phosphoglucomutase (Deckert-Cruz *et al.*, 1997, reprinted here; Deckert-Cruz *et al.*, this volume) have given promising results, although there have been mixed findings for phosphoglucomutase (Allikian *et al.*, 2002). It has been shown that specific alleles at these loci attain higher frequencies in replicate lines of Methuselah flies. This does not demonstrate that such alleles are causally involved in aging, because these allele frequency differences could be due to selection at nearby loci. However, Deckert-Cruz *et al.* (this volume) find that "Slow" phosphoglucomutase alleles are associated with postponed aging in several *Drosophila* species. This result makes wholly accidental linkage a less plausible explanation for the association between such alleles and aging, because it is unlikely that accidental linkages would be sustained over millions of years of evolution within the *Drosophila* genus.

The second approach to the genetics of aging in Methuselah flies has been the study of a range of loci, with a view to characterizing the number and effects of such loci in general. Some of the earlier work on this question used quantitative genetic methods (e.g. Hutchinson and Rose, 1990). Later work used two-dimensional gel electrophoresis (Fleming *et al.*, 1993; reprinted here). All this work suggested that the differentiation of Methuselah flies was not the result of genetic changes at a single Mendelian locus. Instead, aging appears to have been increased through changes occurring at many loci, possibly a few hundred.

The third approach that has been used is to study changes in gene expression at specific loci. Among those studied so far, heat shock protein 22 appears to have increased expression at later ages (Kurapati *et al.*, 2001; reprinted here).

The work most related to that with the Methuselah fly system is that of Luckinbill and his colleagues (e.g. Luckinbill *et al.*, 1990) on similarly selected fruit flies. They find that loci involved in energetic metabolism, such as glucose-6-phosphate dehydrogenase (G6PD), are associated with genetic increases in lifespan. This is a start on the genetic dissection of Methuselah-type flies, but there is clearly a great deal more to be done.

Other work on the genetics and molecular biology of aging in *Drosophila* is naturally relevant to that with the Methuselah flies. This other work falls under three main headings.

One of the first molecular approaches to the genetics of aging was the use of *transgenics*. An early study using EF-1α found a significant extension to lifespan with the insertion of an additional copy of the gene (Shepherd *et al.*, 1989), though subsequent work found that this result depended critically on the position of insertion, mating status, and genetic background (Stearns and Kaiser, 1993). The effect of transgenic manipulation is thus fairly unpredictable. Work on superoxide dismutase transgenics has unearthed many of the same issues (e.g. Sun and Tower, 1999; Tower, 2000; Sun *et al.*, 2002; Orr and Sohal, 2003; Orr *et al.*, 2003; Spencer *et al.*, 2003). Among the many possible problems that afflict this kind of work, epistasis may interact with the position of insertion, producing an extremely complex problem in molecular genetics. There are also undoubtedly other, more obscure, difficulties with these genetic constructs.

A second alternative for the genetic study of aging in *Drosophila* is the identification of *mutants* that increase lifespan without disrupting fitness. The problem with this approach is *not* the identification of mutants that increase lifespan; that has been done already (e.g. Lin *et al.*, 1998; Rogina *et al.*, 2000). The problem is to establish that these mutants do not have lives of protracted debilitation or impaired fitness, like earlier *Drosophila* mutants (Maynard Smith, 1958).

A third alternative is the use of *molecular survey methods* that test a range of loci for their involvement in aging. One method is the use of high-density gene expression chips to survey large numbers of loci for any association between intensity of gene expression and age. Pletcher and colleagues (2002) have used this technology to identify gene pathways that show an activity pattern that correlates with adult age. They identify aging-correlated gene pathways involved in protease inhibition, antibacterial peptides, the mitochondrion, cell growth, and cell maintenance. There are problems, of course. Mere age is correlated with components of adult physiology other than aging, including reproduction. While reproduction exhibits aging as well, it substantially complicates the analytical problem to have age-dependent effects on survival confounded with effects on reproduction. Tower and colleagues (Bieschke *et al.*, 1998; Landis *et al.*, 2001; Landis *et al.*, 2003) use random insertion of transposable P-elements to test thousands of genomic sites for possible effects on aging. Such screens identify additional genes that affect aging, where these genes have involvement in membrane transport, phospholipid metabolism, and actin organization, among other things. Both of these studies indicate that a considerable diversity of molecular mechanisms affects aging in fruit flies, a conclusion fully compatible with estimates of gene number derived from Methuselah flies. Whatever else it is, aging is not a biological problem that is defined by a few simple genetic mechanisms.

References

Allikian, M.J., D. Deckert-Cruz, M.R. Rose, G.N. Landis, and J. Tower. 2002. Doxycycline-induced expression of sense and inverted-repeat constructs modulates *phosphogluconate mutase (Pgm)* gene expression in adult *Drosophila melanogaster*. *Genome Biol* **3**(5): 0021.1–0021.10.

Bieschke, E.T., J.C. Wheeler, and J. Tower. 1998. Doxycycline-induced transgene expression during *Drosophila* development and aging. *Mol Gen Gene* **258**: 571–579.

Deckert-Cruz, D.J., L.M. Matzkin, J.L. Graves, and M.R. Rose. 2004. Electrophoretic analysis of Methuselah flies from multiple species. This volume.

Deckert-Cruz, D.J., R.H. Tyler, J.E. Landmesser, and M.R. Rose. 1997. Allozymic differentiation in response to laboratory demographic selection of *Drosophila melanogaster*. *Evolution* **51**: 865–872.

Fleming, J.E., G.S. Spicer, R.C. Garrison, and M.R. Rose. 1993. Two dimensional protein electrophoretic analysis of postponed aging in *Drosophila*. *Genetica* **91**: 183–198.

Hutchinson, E.W. and M.R. Rose. 1990. Quantitative genetic analysis of *Drosophila* stocks with postponed aging. In: *Genetic Effects on Aging II*, D.E. Harrison (ed.). Telford Press, Caldwell, N.J., pp. 66–87.

Kurapati, R., H.K. Brar Passananti, M.R. Rose, and J. Tower. 2000. Increased *hsp22* RNA levels in *Drosophila* lines genetically selected for increased longevity. *J Gerontol: Biol Sci* **55**: B552–B559.

Landis, G., D. Bhole, L. Lu, and J. Tower. 2001. High-frequency generation of conditional mutations affecting *Drosophila melanogaster* development and life span. *Genetics* **158**: 1167–1176.

Landis, G., D. Bhole, and J. Tower. 2003. A search for doxycycline-dependent mutations that increase *Drosophila melanogaster* life span identifies the *VhaSFD, Sugar baby, filamin, fwd* and *Cct1* genes. *Genome Biol* **4**: R8.

Lin, Y.J., L. Seroude, and S. Benzer. 1998. Extended life-span and stress resistance in the *Drosophila* mutant methuselah. *Science* **282**: 943–946.

Luckinbill, L.S. V. Riha, S. Rhine, and T.A. Grudzien. 1990. The role of glucose-6-phosphate dehydrogenase in the evolution of longevity in *Drosophila melanogaster*. *Heredity* **65**: 29–38.

Maynard Smith, J. 1958. The effects of temperature and of egg-laying on the longevity of *Drosophila subobscura*. *J Exper Biol* **35**: 832–842.

Orr, W.C., R.J. Mockett, J.J. Benes, and R.S. Sohal. 2003. Effects of overexpression of copper-zinc and manganese superoxide dismutases, catalase, and thioredoxin reductase genes on longevity in *Drosophila melanogaster*. *J Biol Chem* **278**: 26418–26422.

Orr. W.C. and R.S. Sohal. 2003. Does overexpression of Cu,Zn-SOD extend life span in *Drosophila melanogaster*? *Exper Gerontol* **38**: 227–230.

Pletcher, S.B., S.J. Macdonald, R. Marguerie, U. Certa, S.C. Starns, D.B. Golstein, and L. Partridge. 2002. Genome-wide transcript profiles in ageing and calorically restricted *Drosophila melanogaster*. *Curr Biol* **12**: 712–723.

Rogina B., R. A. Reenan, S.P. Nilsen, and S.L. Helfand. 2000. Extended life-span conferred by cotransporter gene mutations in *Drosophila*. *Science* **290**: 2137–2140.

Shepherd, J.C.W., U. Walldorf, P. Hug, and W.J. Gehring. 1989. Fruit flies with additional expression of the elongation factor EF-1α live longer. *Proceedings of the National Academy of Sciences USA* **86**: 7520–7521.

Spencer, C.C., C.E. Howell, A.R. Wright, and D.E.L. Promislow. 2003. Testing an "aging gene" in long-lived *Drosophila* strains: Increased longevity depends on sex and genetic background. *Aging Cell* **2**: 123–130.

Stearns, S.C. and M. Kaiser. 1993. The effects of enhanced expression of elongation factor EF-1α on lifespan in *Drosophila melanogaster*. IV. A summary of three experiments. *Genetica* **91**: 183–198.

Sun, J., and J. Tower. 1999. FLP recombinase-mediated induction of Cu/Zn-superoxide dismutase transgene expression can extend the life span of adult *Drosophila melanogaster* flies. *Mol Cell Biol* **19**(1): 216–228.

Sun, J., D. Folk, T.J. Bradley, and J. Tower. 2002. Induced overexpression of mitochondrial Mn-superoxide dismutase extends the life span of adult *Drosophila melanogaster*. *Genetics* **161**: 661–672.

Tower, J. 2000. Transgenic methods for increasing *Drosophila* lifespan. *Mech Ageing Dev* **118**: 1–14.

Tyler, R.H., H. Brar, M. Singh, A. Latorre, J.L. Graves, L.D. Mueller, M.R. Rose, and F.J. Ayala. 1993. The effect of superoxide dismutase alleles on aging in *Drosophila*. *Genetica* **91**: 143–149.

Genetica **91**: 143–149, 1993.

The effect of superoxide dismutase alleles on aging in *Drosophila*

Robert H. Tyler, †, Hardip Brar, Meena Singh, Amparo Latorre[1], Joseph L. Graves[2], Laurence D. Mueller, Michael R. Rose & Francisco J. Ayala
Department of Ecology and Evolutionary Biology, University of California, Irvine, CA 92717, USA
[1] *Present address: Departamento de Genetica, Facultad de Biologia, Universidad de Valencia, Spain*
[2] *Author for correspondence*

Received and accepted 22 June 1993

Key words: superoxide dismutase, aging, *Drosophila*, evolutionary genetics, senescence

Abstract

The effects of superoxide dismutase on aging were tested using two differt experimental approaches. In the first, replicated populations with postponed aging were compared with their controls for frequencies of electrophoretic alleles at the SOD locus. Populations with postponed aging had consistently greater frequencies of the allele coding for more active SOD protein. This allele was not part of a segregating inversion polymorphism. The second experimental approach was the extraction of *SOD* alleles from different natural populations followed by the construction of different *SOD* genotypes on hybrid genetic backgrounds. This procedure did not uncover any statistical effect of *SOD* genotype on longevity or fecundity. There were large effects on longevity and fecundity due to the family from which a particular *SOD* genotype was derived. To detect the effects of *SOD* genotypes on longevity with high probability would require a ten-fold increase in the number of families used.

Introduction

Studies of the temporal correlates of the aging process do not necessarily indicate the causal mechanisms controlling it, while treatments that shorten life span may kill as a result of novel pathologies (Maynard Smith, 1966; Rose, 1991). For this reason, other experimental strategies are required to unravel the mechanisms that normally control aging. Postponed aging arising from selection for survival to, and reproduction at, later ages requires genetic mitigation of normal aging mechanisms, and therefore constitutes a model system of choice for the analysis of the genetics of aging (Hutchinson & Rose, 1987).

Choice of a good model system for the genetics of aging has not, however, led to immediate breakthroughs in causal understanding. Some electrophoretic studies of the genetic basis of selectively postponed aging in *Drosophila melanogaster* have already been performed (Luckinbill *et al.*, 1989). Some significant correlations between postponed aging and the loci affecting energetic metabolism were found by Luckinbill *et al.*, though that work suffered from a lack of replication. Only one or two selected populations were compared to their control populations. In addition, the founding population sizes were fairly small. Fortunately, these problems are easily remedied using the larger set of stocks created by Rose (e.g. 1984), in which there are five selected lines and five controls, the entire set of stocks coming from a large wild-caught sample of *D. melanogaster* from the endemic Ives population of South Amherst, Massachusetts.

A more profound problem with electrophoretic comparisons of postponed-aging stocks with their controls is that of linkage disequilibrium. When postponed-aging stocks are created by selection from a common founding population, the particular linkage disequilibria of that population will be in common among all derivatives undergoing selection. Allelic state correlations between physically proximal loci will be maintained for some part of the selection process, causing nearby loci to undergo parallel changes of allele frequency in all selected stocks. Therefore, stocks with selectively postponed aging are not necessarily, by themselves, reliable systems for inferring the causal involvement of particualar loci in aging.

One potential solution to this difficulty is to use stocks with postponed aging to identify initial candidate genes, genes that might be involved in postponed aging, or that might change in allelic composition because of linkage disequilibrium alone. Then genotypes at this locus can be arbitrarily assembled from different natural populations, and tested for their effects upon aging (cf. Serradilla & Ayala, 1983). In this way, the problem of the original linkage relations in the founding population before selection can be overcome. In the present paper, we apply this approach to one locus of great interest to gerontologists, that coding for Cu, Zn superoxide dismutase. This free radical scavenging molecule is of interest because of its role in catalyzing the conversion of superoxide radicals to hydrogen peroxide, the latter then undergoing conversion to water due to the action of catalase (McCord & Fridovich, 1969). Since superoxide radicals are highly damaging for macromolecules such as protein, lipid, and nucleic acid, the cumulative effects of such free radicals have long been proposed as a major factor in aging (e.g. Harman, 1956). For this reason, we were particularly interested in applying the combination of stocks with postponed aging and sampled alleles from wild populations to test for the effects of superoxide dismutase alleles upon aging. In natural populations of *D. melanogaster*, Cu, Zn superoxide dismutase has *F* ('fast') and *S* ('slow') allelomorphs, with *S* being rarer but of greater *in vitro* activity (Lee, Misra & Ayala, 1981). On the free radical theory of aging, this allelomorph should have elevated frequencies in populations that are longer-lived. In addition, the *S* allele should give rise to increased longevity, on average, in flies of *SS* or *FS* genotype, relative to *FF* genotypes. The present article tests these predictions. We find that some of these predictions are borne out, but others are not.

Materials and methods

Postponed aging stocks

The postponed aging populations used in the present study were derived from a long-standing wild population that had been sampled in 1975 (Rose, 1984). All fly culture proceeded at 25 °C with abundant food, little crowding, and moderate humidity. In Feb. 1980, one generation of the base population was used to found ten outbred populations, five of which (called B_1-B_5) were kept under the same conditions as the base population, with two-week discrete generations. The other five were cultured using females of increasingly greater ages as the generations progressed, such that by late 1981 all reproducing females had to attain 70 days of age (Rose, 1984). Such populations are designated O_1-O_5, with no correspondence between B and O subscripts. Since 1981, cohorts sampled from the O populations have had significantly greater mean and maximum longevities, relative to B populations (Rose, 1984; Hutchinson & Rose, 1991).

In addition to this longevity difference, B and O populations have a number of other differences which indicate that O flies live longer because of postponed aging. These differences include increases in: tethered flight duration (Graves & Rose, 1990), spontaneous locomotion at later ages (Service, 1987), 24-h egg-laying at later ages (Rose, 1984), and resistance to stress (Service *et al.*, 1985). These findings strongly suggest that the longer-lived O populations indeed possess postponed aging, through both general and late-age enhancements in physiological functions. Derivatives of the O populations have also been subjected to relaxed selection (R populations), in which cultivation returned to 2 week generations (Service, Hutchinson & Rose, 1988). The phenotypes of these R populations have largely returned to those of the control B populations (Graves *et al.*, 1992; and unpub. data), including mean longevity. If superoxide dismutase is a causal factor postponing aging, then the electrophoretic profile of the R populations should be identical to that of the B populations.

Extraction of SOD genotypes from natural populations

D. melanogaster females were collected from Culver City, Los Angeles County, in southern California and El Rio Vineyard, San Joaquin Country, in northern California. From each population, 20 isofemale lines were made homozygous for each of both the *F* and *S* superoxide dismutase alleles. This was done by inbreeding and electrophoretic screening over multiple generations.

These inbred lines were then crossed with each other in pairs to create lines of known SOD genotype but hybrid background. This procedure mitigates problems of inbreeding depression. Such crosses were performed by crossing in each case 20 males from one line with 20 females from a different line. Four different genotypes were constructed: *FF, SS, FS,* and *SF*, where the latter pair of genotypes differ according to the maternal genotype.

Assay procedures
All electrophoretic procedures used the methods described in Ayala *et al.* (1972). Polytene chromosomes were extracted from larvae of B and O populations and stained. Surveys were made of 29-55 different chromosomes for the presence of inversion heterozygotes on each arm. All assays of life-history phenotypes used control-density larval rearing, with 25-50 larvae per vial. Adult females were kept together with two males, in vials of banana-molasses food for the Culver City samples and corn meal-flour food for the El Rio samples. Fecundity assays were made periodically, using a charcoal-colored high-agar medium without yeast. Vials were examined each day for survival of the adults.

In these experiments there were block effects which confounded two sources of experimental variation. The first of these is the population of origin and the second is laboratory handling. The assays for each population were conducted in different laboratories. Thus, different personnel, incubators and type of food were used for collecting data from each population. Consequently, all statistical analyses treated results from El Rio and Culver City as two blocks.

Table 1. Frequency of the S('slow') allele in B and O populations for Cu,Zn superoxide dismutase (SOD). Only two electromorphs were detected. Electrophoresis was performed according to Ayala *et al.* (1972); *n* is the number of individuals sampled.

Line	n	*FF*	*FS*	*SS*	*S* Frequency
B_1	50	50	0	0	0.00
B_2	48	50	0	0	0.00
B_3	50	50	0	0	0.00
B_4	24	24	0	0	0.00
B_5	24	24	0	0	0.00
Average allele frequency					0.00

Line	n	*FF*	*FS*	*SS*	*S* Frequency
O1	48	26	17	5	0.28
O2	48	39	8	1	0.10
O3	48	22	15	11	0.39
O4	49	20	26	3	0.33
O5	48	36	9	3	0.16
Average allele frequency standard error					0.25 ± 0.05

Results

Allele frequencies in postponed-aging stocks
As shown in Table 1, the *S* allele is significantly more frequent in O populations, averaging about 25%, with no *S* alleles detected in the B population samples. Indeed, this average frequency of *S* in the O populations is greater than has been observed before for this allele in natural populations of *Drosophila* (Smit-McBride, Moya & Ayala, 1988; Singh, Hickey & David, 1982 and unpub. data). While these samples suggest that the B populations entirely lack the *S* allele, further sampling of these populations has produced a few instances where flies from B populations carry the *S* allele heterozygously. The R populations have the same profile as the B sample shown here, the *S* allele being absent. All of these findings are in accord with the free-radical theory of aging, in that the more active allele is significantly more frequent in populations with postponed aging.

But the conclusion that SOD is causally related to aging could be erroneous if the *S* allele is tightly linked to an allele which itself postpones aging. The total number of generations separating B from O populations is over 300, the two population-types have been replicated five-fold, and the census population sizes have been in the thousands. These experimental design features facilitate attainment of linkage equilibrium at unselected loci. Nonetheless, if the *S* allele were associated with an inversion polymorphism, recombination could be too weak to prevent linkage disequilibrium between the *S* allele and a selected allele at another locus within the same inversion. Table 2 shows the results from

Table 2. Number of polytene chromosome arms sampled and inversions observed for B and O populations.

	Chromosome arm					
Line	X	2R	2L	3R	3L	Inversions
B_1	38	48	48	48	48	None
B_2	30	41	41	41	41	None
B_3	30	41	41	41	41	None
B_4	36	50	50	50	50	None
B_5	36	48	48	48	48	None
O_1	35	49	49	49	49	None
O_2	28	44	44	44	44	None
O_3	29	40	40	40	40	None
O_4	36	52	52	52	52	One on 2R
O_5	40	55	55	55	55	None

a study of the cytology of 2205 polytene chromosomes. Only one inversion variant was detected (In(2R)NS), on the right arm of chromosome 2 in the O_4 population. Moreover, the superoxide dismutase locus is located on chromosome 3, in any case. Therefore, the only remaining countervailing hypothesis would be linkage disequilibrium with a favored allele that is physically very close to SOD on chromosome 3. Any such allele is unlikely to have the same linkage relationship with SOD in the unrelated California populations of *D. melanogaster* discussed below.

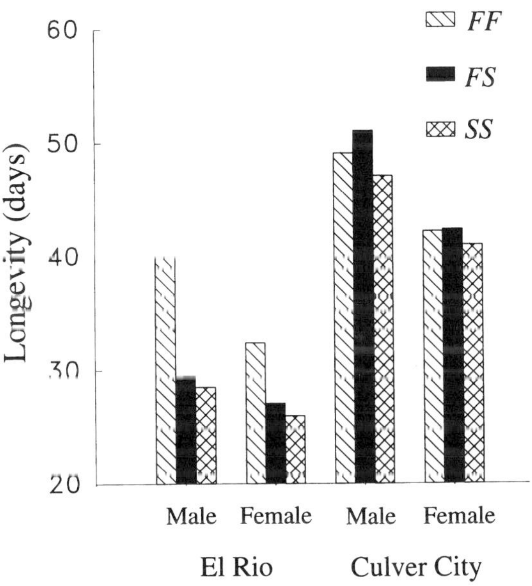

Fig. 1. The mean longevity for each SOD genotype as a function of sex and population.

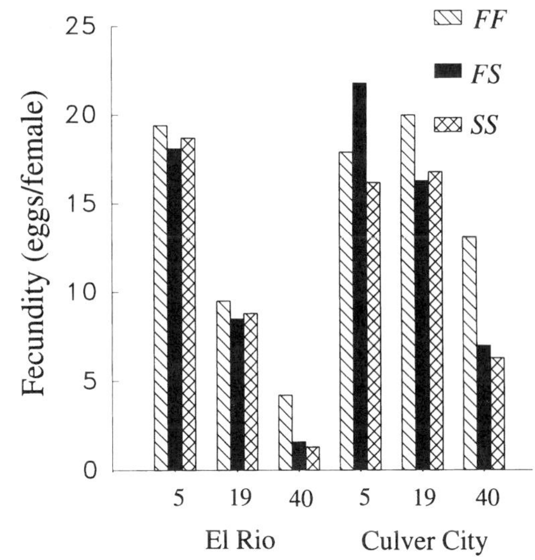

Fig. 2. Female fecundity for each SOD genotype at three different ages 5, 19 and 40 days in the El Rio and Culver City populations.

Effects of SOD genotype on aging phenotypes

The results of analyses of variance for longevity and age-specific fecundity are shown in Tables 3

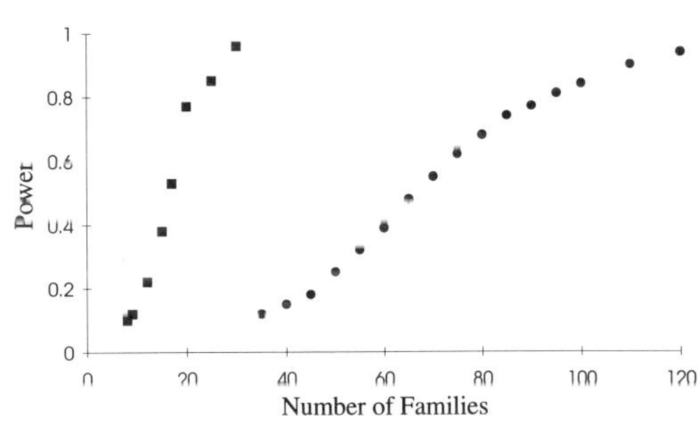

Fig. 3. The power of detecting significant genotype effects on longevity (circles) and fecundity (squares) as a function of the number of families sampled within each genotype. These calculations utilized the longevity data from the El Rio population to estimate the differences in longevity between genotypes and the variance between families and the fecundity data at day 40 from Culver City.

Table 3. Results of the analysis of variance (ANOVA) on longevity. Population was treated as a block effect.

Source of variation	Sum of squares	Degrees of freedom	Mean square	F	Significance
Population	139,030	1	139,030		
Genotype	1,697	2	849	0.255	NS
Sex	25,418	1	25,418	80	<0.001
Family (Genotype)	56,475	17	3,322	10.5	<0.001
Sex by Genotype	219	2	109	0.03	NS
Error	907,114	2,867	316		

NS = not significant

and 4 respectively. Throughout these analyses, it is apparent that there is little significant effect of genotype upon aging patterns (Figs. 1-2). To the extent to which there is a directional effect, the *F* allele is associated with increased longevity, although not significantly so.

In these experiments within each genotypic class were nested several families (5-10). In every case (Tables 3-4) there was a statistically significant effect of families on either longevity or fecundity. Genotypic effects were assessed by placing the family mean square in the denominator of the F ratio. Consequently, it appears that the number of families sampled will have a great impact on the statistical power of these tests. We should add that ANOVAs on the separate experimental results from El Rio or Culver City do not reveal significant effects of genotype due to the large between-family variation.

In Figure 3 the statistical power of the longevity assay on the El Rio population and the day 40 fecundity assay on the Culver City population are shown. These particular examples were chosen since they show the greatest differences between genotypes. For instance, the power of the ANOVAs on fecundity at day 5 and 19 is still below 20%, even when 100 families are sampled within each genotype.

These results suggest that the genetic background that is unique to each family is causing variation that obscures any differences that the SOD locus might produce. Except for the possibility of late fecundity, samples of about 100 families per genotype would be required to insure a high

Table 4. Results of the analysis of variance (ANOVA) on fecundity. The ANOVAs were carried out on fecundity at day 5, 19 and 40 separately. Population was treated as a block effect.

Source of variation	Sum of squares	Degrees of freedom	Mean square	F	Significance
		Day 5 Fecundity			
Population	347	1	347		
Genotype	12,325	2	6,163	2.76	NS
Family (Genotype)	37,910	17	2230	10.67	<0.001
Error	357,348	1,709	209		
		Day 19 Fecundity			
Population	26,413	1	26,413		
Genotype	749	2	374	0.428	NS
Family (Genotype)	14,884	17	876	5.76	<0.001
Error	228,493	1,503	152		
		Day 40 Fecundity			
Population	6,360	1	6,360		
Genotype	1,536	2	768	3.40	0.06
Family (Genotype)	3,842	17	226	2.89	<0.001
Error	53,281	681	78		

NS = not significant

probability of detecting differences between genotypes of the magnitude observed in this study.

Discussion

The present study finds that more active superoxide dismutase alleles are associated with postponed aging in laboratory stocks. We have failed to demonstrate that such alleles directly increase life span or later fecundity, but this failure arose from a lack of experimental power. That is to say, this negative result is not a pertinent refutation of the free radical mechanism for aging. The lack of inversion polymorphisms involving *SOD* indicates that alleles at the locus are not in linkage disequilibrium with distal loci. However, linkage disequilibrium with physically proximal loci could nevertheless be the factor responsible for the association between superoxide dismutase and postponed aging. Luckinbill *et al.* (1989) found no difference in *SOD* allele frequencies between their long- and short-lived populations of *D. melanogaster*. Luckinbill's study employed two replicates of each treatment, and thus has less statistical power then the present study. Moreover, the frequency of the *S* allele is low in natural populations ($\approx$ 1%), so the *S* allele may have been absent in the original founding sample, or lost by chance during the early generations of selection.

Of great relevance to the present study is the finding of Reveillaud, Niedzwiecki and Fleming (1991), who transformed *D. melanogaster* stocks with highly active bovine SOD DNA to which a strong actin promoter had been linked. In addition to greatly increasing SOD activity levels in transformed flies, adult life span was significantly increased among replicate transformants. Although this increase was of no greater magnitude than 10%, it constitutes a powerful demonstration that more active SOD can give rise to increased lifespan.

Comparing the results with (i) allele frequencies of postponed aging stocks, (ii) alleles extracted from natural populations, and (iii) transformation experiments with extremely active SOD genes (Reveillaud, Niedzwiecki & Fleming, 1991), some conclusions suggest themselves. These disparate results can be reconciled if it is supposed that the *FS SOD* electromorphs have differential effects on aging that are smaller in magnitude than the differential effects between bovine and dipteran *SOD*. If bovine *SOD* transformation can increase life span by at most 10%, relative to dipteran *SOD*, then the effect of the normal dipteran allelic difference might be about 2-4%. The problem for experiments in which alleles are extracted from natural populations is that the hybrid lines constructed from the isofemale inbreds have substantial differences in life span arising from different accidents of fixation in their ancestral inbred lines. Such differences among lines would thus swamp the effect of longevity differences between genotypes at the *SOD* locus, because the latter differences are so small.

From this line of argument, two final inferences can be drawn. Firstly, this paper provides additional evidence supporting the involvement of *SOD* in postponing, and thus controlling, aging in *Drosophila*, particularly in light of the bovine DNA transformation results. Secondly, the extraction of *SOD* alleles from natural populations has been unable to demonstrate a quantitative effect of this locus on aging phenotypes due to variation in genetic backgrounds between families. Transformation experiments might be preferable in any such tests.

Acknowledgements

Lisa M. Chiu and Sarah S. Kwak provided considerable technical help in the performance of these experiments. This research was supported in part by grants PHS AG06346 and AG09970 from the National Institute of Aging (USA) to MRR and the Department of Energy (USA) to FJA, and in part by the estate of Robert H. Tyler. We are grateful to the Tyler family and attornies for their support, cooperation, and encouragement.

References

Ayala, F. J., J. R. Powell, M. L. Tracey, C. A. Mourao & S. Perez-Salas, 1972. Enzyme variability in the *Drosophila willistoni* group. IV. Genic variation in natural populations of *Drosophila willistoni*. Genetics 70. 113-139.

Graves, J. L. & M. R. Rose, 1990. Flight duration in *Drosophila melanogaster* selected for postponed senescence, pp. 59-65 in Genetic effects on aging. II, edited by D. E. Harrison. Telford Press, Caldwell, N.J.

Graves, J. L., E. C. Toolson, C. Jeong, L. N. Vu & M. R. Rose 1992. Desiccation, flight, glycogen, and postponed senescence in *Drosophila melanogaster*. Phys. Zool. 65: 268-286.

Harman, D., 1956. Aging - a theory based on free radical and radiation chemistry. J. Gerontol. 11: 298-300.

Hutchinson, E. W. & M. R. Rose, 1987. Genetics of aging in insects. Rev. Biol. Res. Aging 3: 63-70.

Hutchinson, E. W. & M. R. Rose, 1991. Quantitative genetics of postponed aging in *Drosophila melanogaster*. I. Analysis of outbred populations. Genetics 127: 719-727.

Lee, Y. M., H. P. Misra & F. J. Ayala, 1981. Superoxide dismutase in *Drosophila melanogaster*. Biochemical and structural characterization of allozyme variants. Proc. Natl. Acad. Sci. USA 78: 7052-7055.

Luckinbill, L. S., T. A. Grudzien, S. Rhine & G. Weisman, 1989. The genetic basis of adaptation to selection for longevity in *Drosophila melanogaster*. Evol. Ecology 3: 31-39.

Maynard Smith, J., 1966. Theories of aging, pp. 1-35 in Topics in the Biology of Aging, edited by P. L. Krohn. Interscience, NY.

McCord, J. M. & I. Fridovich, 1969. Superoxide dismutase: An enzymatic function for erythrocuprin. J. Biol. Chem. 224: 6049-6055.

Reveillaud, I., A. Niedzwiecki & J. E. Fleming, 1991. Expression of bovine superoxide dismutase in *Drosophila melanogaster* augments resistance to oxidative stress. Mol. Cell. Biol. 11: 632-40.

Rose, M. R., 1984. Laboratory evolution of postponed senescence in *Drosophila melanogaster*. Evolution 38: 1004-1010.

Rose, M. R., 1991. Evolutionary Biology of Aging. Oxford University Press, NY.

Serradilla, J. M. & F. J. Ayala, 1983. Ecological and evolutionary divergence in five species of *Drosophila*. Z. zool. Syst. Evolut.-forsch. 21: 194-200.

Service, P. M., 1987. Physiological mechanisms of increased stress resistance in *Drosophila melanogaster* selected for postponed senescence. Physiol. Zool. 60: 321-326.

Service, P. M., E. W. Hutchinson, M. D. MacKinley & M. R. Rose, 1985. Resistance to environmental stress in *Drosophila melanogaster* selected for postponed senescence. Physiol. Zool. 58: 380-389.

Service, P. M., E. W. Hutchinson & M. R. Rose, 1988. Multiple genetic mechanisms for the evolution of senescence in *Drosophila melanogaster*. Evolution 42: 708-716.

Singh, R. M., D. A. Hickey & J. David, 1982. Genetic differentiation between geographically distant populations of *Drosophila melanogaster*. Genetics 101: 235-256.

Smit-McBride, Z., A. Moya & F. J. Ayala, 1988. Linkage disequilibrium in natural and experimental populations of *Drosophila melanogaster*. Genetics 120: 1043-1051.

Genetica **91**: 183–198, 1993.

Two-dimensional protein electrophoretic analysis of postponed aging in *Drosophila*

James E. Fleming[1]*, Greg S. Spicer[1], Roger C. Garrison[2] & Michael R. Rose[2]
[1] *Linus Pauling Institute of Science and Medicine, 440 Page Mill Road, Palo Alto, CA 94306, USA*
[2] *Institute of Molecular Medical Sciences, 460 Page Mill Road, Palo Alto, CA 94306, USA*
[3] *Department of Ecology and Evolutionary Biology, University of California, Irvine, CA 92717, USA*
* *Address for correspondence: Department of Biology, Eastern Washington University, Cheney, WA 99004, USA*

Received and accepted 22 June 1993

Key words: aging, *Drosophila*, electrophoresis, evolution, longevity

Abstract

Five populations of *Drosophila melanogaster* that had been selected for postponed aging were compared with five control populations using two-dimensional protein gel electrophoresis. The goals of the study were to identify specific proteins associated with postponed aging and to survey the population genetics of the response to selection. A total of 321 proteins were resolvable per population; these proteins were scored according to their intensity. The resulting data were analyzed using resampling, combinatoric, and maximum parsimony methods. The analysis indicated that the populations with postponed aging were different from their controls with respect to specific proteins and with respect to the variation between populations. The populations selected for postponed aging were more heterogeneous between populations than were the control populations. Maximum parsimony trees separate the selected populations, as a group, from their controls, thereby exhibiting a homoplastic pattern.

Introduction

The identification of the genes involved in the control of aging has been one of the major goals of gerontology. One method for the pursuit of this goal is the analysis of the genetic basis of accelerated aging or shortened lifespan. This approach has led to the identification of metabolic deficiencies that can hasten death, such as those of some *Drosophila* mutants (Bozcuk, 1981). However, deterioration and death in such stocks may depend on the introduction of novel pathologies (Maynard Smith, 1966; Hutchinson & Rose, 1987) rather than an acceleration in the mechanisms of normal aging. While this certainly need not always occur, the problem of ambiguity as to whether or not it has occurred in any particular case undermines the reliability of inferences concerning the mechanistic basis of aging from genetic analyses of shortened lifespan.

An alternative approach is to create stocks with genetically postponed aging. This has been done by means of mutagenesis and selection on recombinant inbreds in *Caenorhabditis elegans*, a nematode (Johnson & Hutchinson, 1990). It has also been achieved by selection on outbred *Drosophila* populations (Rose, 1989). A particularly important feature of the *Drosophila* research has been the replication of these selection experiments between laboratories (Rose, 1984; Luckinbill *et al.*, 1984), in repeated experiments (Rose & Charlesworth, 1981; Rose, 1984), and within experiments (Rose, 1984, Luckinbil *et al.*, 1984; Hutchinson & Rose, 1991). The *Drosophila* postponed-aging stocks as a group thus will not be overly dependent on the specific genetic accidents arising from mutagenesis, linkage disequilibrium, and the like in individual lines.

There are two additional advantages to the use

of these stocks. The first is that they have been generated by the use of natural selection in the laboratory. Normal *Drosophila* cultures are strongly selected for high early fertility, with negligible selection for later fertility. The *Drosophila* stocks with postponed aging have been created by selection for later reproduction, thereby producing an increase in the force of natural selection at later ages (Rose, 1991, Ch. 1, 3). These stocks are thus directly interpretable in terms of the evolutionary theory of aging, and may be used to test this theory. This makes findings obtained from them of relevance both to mechanistic gerontology and to evolutionary biology.

The second advantage to working with these stocks is that they have been characterized in terms of a number of biological variables: life-history (e.g. Rose, 1984; Luckinbill *et al.*, 1984), morphology (Rose *et al.*, 1984; Luckinbill *et al.*, 1988), physiology (Service *et al.*, 1985; Service, 1987), and quantitative genetic transmission (Clare & Luckinbill, 1985; Hutchinson & Rose, 1991). One-dimensional protein electrophoresis has been used to test for the involvement of specific loci in postponed aging (Luckinbill *et al.*, 1989; Tyler *et al.*, this volume), and significant results have been obtained for a few loci.

One-dimensional electrophoretic surveys of proteins allow the detection of genetic associations between biochemically characterized proteins and postponed aging, but not other types of protein. To overcome this limitation, the present study used two-dimensional protein electrophoresis to survey a significant fraction of the polypeptides found in *Drosophila*.

Material and methods

Stocks

All populations employed derive ultimately from an endemic *Drosophila melanogaster* population from Massachusetts, studied by Ives (e.g. 1970) for more than 40 years. In 1975, a sample from this population was established in the laboratory and kept free of systematic inbreeding for five years. In 1980, ten populations were derived from this base population, five then selected for postponed aging (01-05) and five controls (B1-B5). The O stocks underwent selection for late fertility, while the B stocks were selected for early fertility, as was the ancestral population. This type of selection is expected to generate increased O stock life spans and did (Rose, 1984). Recent data (unpub.) indicate a longevity increase in the O stocks of 60-80%, along with numerous other biological changes. A cytological point of note for the present study is that these populations largely lack inversion polymorphism, within or between populations (Tyler *et al.*, this volume). This makes it unlikely that allele frequencies change in parallel because of pervasive linkage disequilibrium ranging over many loci.

Two-dimensional gels and radiolabelling

Flies were labelled with [^{35}S]-methionine (specific activity 1151 Ci/mmole; 1 Ci = 37 GBq; New England Nuclear) for 18 hours at 25 °C as described in Fleming *et al.* (1988). This labelling procedure is limited in the number of proteins detected compared with silver staining, which could label one thousand or more proteins per gel. However, for this initial survey, it was decided to handle a smaller number of spots on the gels, to reduce the difficulty of scoring. The electrophoretic technique was essentially that of O'Farrell (1975), making use of the Anderson Iso-Dalt system (Anderson & Anderson, 1978). After the labelling period, 50 flies of each population were homogenized in 200 microliters of O'Farrell's lysis buffer A in a dounce homogenizer. The isoelectric focusing gel contained LKB ampholines (pH range 3.5-10, lot 99). Ten microliter samples (approximately 10^6 dpm) were applied under the electrolyte at the negative ends of the gels and then run at 500 V for 20 h. Each gel was loaded with the same amount of protein so as to minimize variation in spot migration due to protein loading, which is common for two-dimensional gel electrophoresis. The O populations took up slightly less label per milligram of protein than the B population, but relative spot intensity is adjusted by different exposure times of the film, as in Fleming *et al.* (1986). All first dimension samples were run simultaneously in the same apparatus under identical conditions. The voltage was then increased to 1000 V for an additional hour. Immediately after this, the gels were removed from the glass tubes and frozen in Anderson's equilibration buffer at -76 °C.

The second-dimension electrophoretic procedure was based on O'Farrell's original technique. All second-dimension gels were run simultaneously in the Anderson ten-gel Dalt tank using the Anderson Iso-Dalt system (Anderson & Anderson, 1978). These gels consisted of 10% acrylamide and 1% $NaDodSO_4$. After electrophoresis, gels were dried with a Hoeffer model SE1150 slab-gel dryer. The dried gels were exposed to a Kodak XAR-2 x-ray film. Three different exposures of these gels were carried out in order to score lower intensity spots. These exposure times were 3 days, 9 days, and 13 days. The resulting autoradiographs were developed with a Konica automatic film processor. The gel shown in Figure 1 is a nine-day exposure.

Gels were scored as previously described by Fleming *et al.* (1986). The autoradiograph from population B_1 was arbitrarily chosen as the master

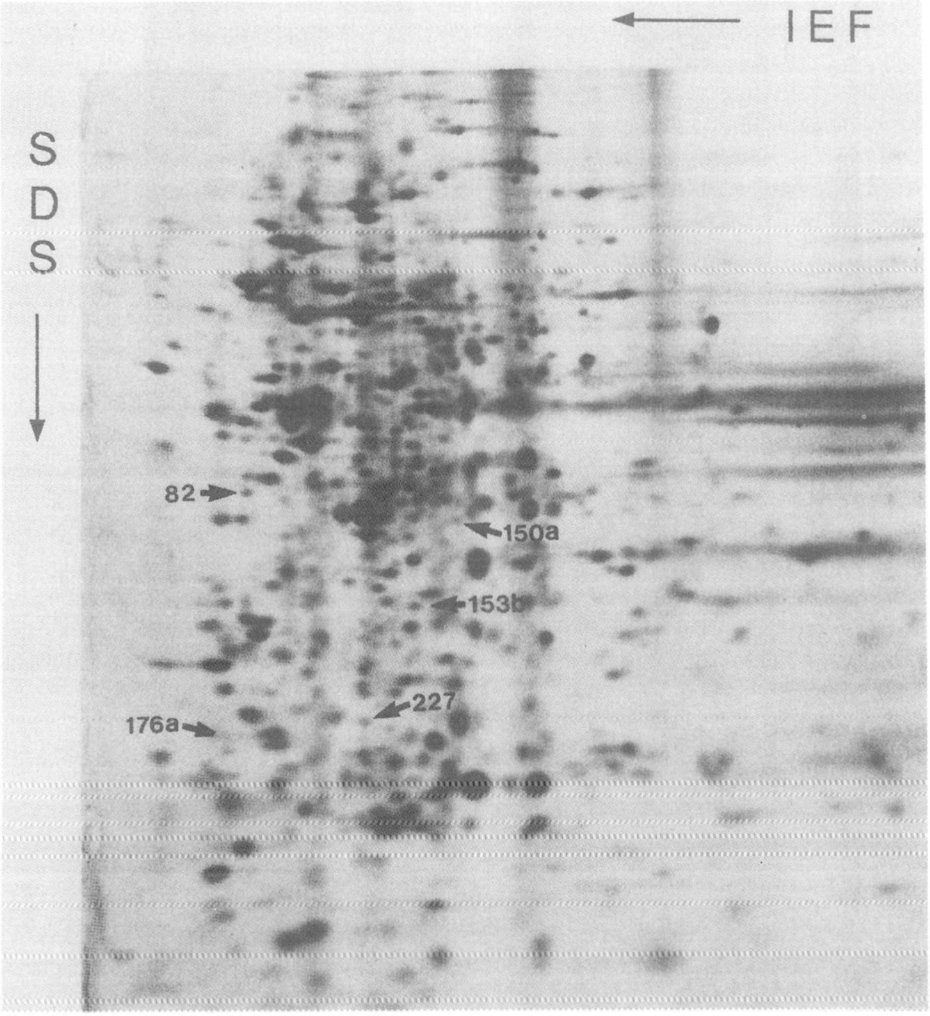

Fig. 1. Autoradiograph of a two-dimensional gel electrophoretic pattern of soluble [^{35}S]-methionine-labelled polypeptides from adult *Drosophila melanogaster*, strain B1. Several of the proteins that are significantly differentiated between the B and O stocks are indicated by their corresponding numbers from Table 1. IEF, isoelectric focusing dimension: SDS, sodium dodecyl sulfate, molecular weight dimension.

186

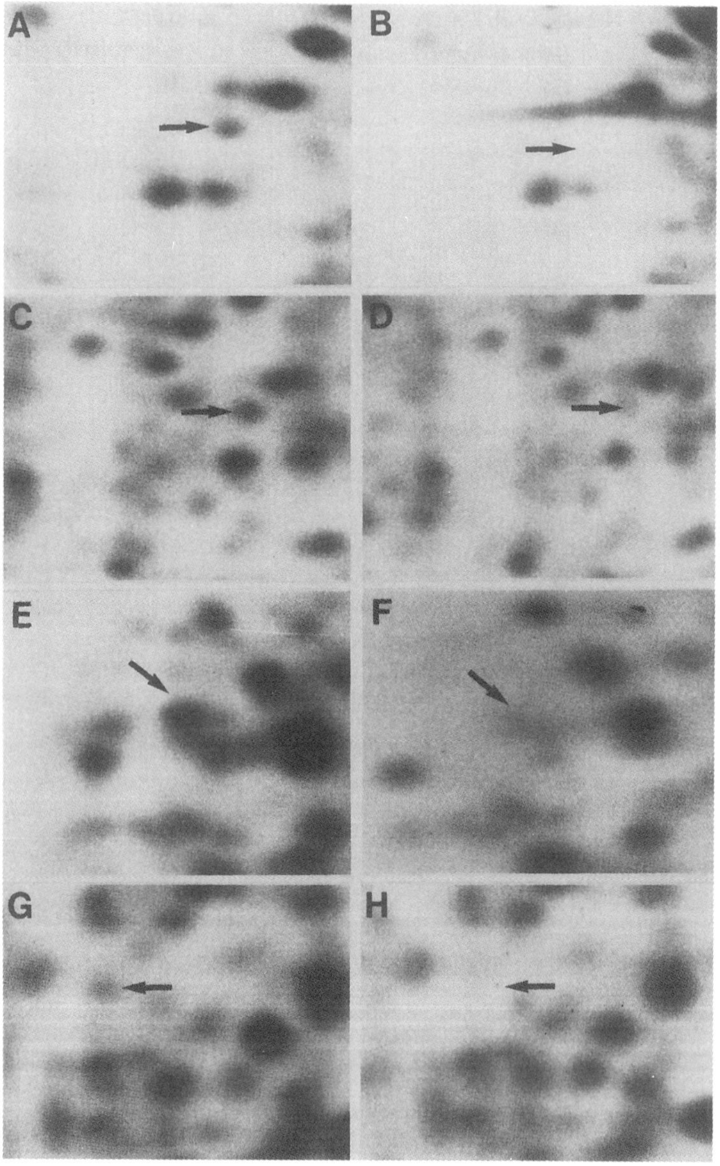

Fig. 2. Regions of the autoradiographs selected from the B and O stocks that show several of the proteins that are significantly differentiated. It is evident from these exposures that some of the differences may be quantitative. A, spot ('sp') 82, stock ('st') B1; B, sp 82, st 02; C sp 153b, st B1; D, sp 153b, st 05; E sp 176a, st B3; F, sp 176a, st 02; G, sp 227, st B1; H, sp 227, st 02.

gel. Each autoradiogram from the ten gels was then divided into nine equal sections and the protein spots were scored in each section for each gel beginning with the high molecular weight acidic proteins and ending with the low molecular weight basic proteins. Spots near the edge of the gel at the low molecular weight end were not scored. The protein scoring was carried out independently by three individuals in order to minimize variation due to any one individual's scoring.

Initially, all gels were compared for the presence or absence of spots relative to this gel. Relative quantitative differences were scored after the spots were numbered and the initial comparison of presence/absence status was made. For semi-quantitative comparisons, spots were given scores ranging from 0 to 5. A value of 3 represents the 'normal' abundance for a particular spot as determined from the master gel. Values of 4 and 5 were assigned to spots with apparently increased quantity, while values of 1 and 2 were assigned to spots with apparently decrease quantity, values 5 and 1 representing more extreme differences. A value of 0 means that a particular spot was not detected on that gel. The majority of spots were scored using the three-day exposure autoradiograph. Weaker spots were scored from the nine-day and thirteen-day exposures. This permitted accurate scoring for the weakest spots, where their presence was concerned. For the most abundant spots, the three-day exposure allowed the best scoring, because none of these spots were obscured by overexposure at that time, as they were in the nine-day or thirteen-day exposure autoradiograms. The nonlinearity of the X-ray film was not considered an important issue for the semi-quantitative comparisons, because only relative values were assigned to spots based on comparison with spot size on the master gel. These relative values were also assigned only when they were easily scored. Since the O populations incorporated less radiolabel than the B populations, it is possible that some of the proteins that received scores of zero in the O population might show up on longer exposures of the gel. However, as noted by Garrels (1979), quantitative differences of this magnitude can be considered significant qualitative differences in protein expression nonetheless. Illustrative spots are shown in Figure 2.

Results

A total of 321 abundant proteins were resolved and coded by number. The 109 that exhibited heterogeneity over populations were analyzed for patterns of differentiation among populations. Please note that, in population genetics generally, the term 'differentiation' refers to statistical separation of populations, not biological development or its effects.

Resampling analysis of quantitative scores

The problem facing any inference from these scores is the large amount of between-population differentiation within each of the B and O groups, because some differences between Bs and Os could thus be the result of chance alone. This problem can be alleviated by comparing the observed data with all possible resamplings of the data from a statistical universe consisting of the original gel scores rearranged in every possible combination of 'pseudo-Os' and 'pseudo-Bs', each a group of five populations. For example, one such arrangement would be B1, B3, B5, O1, and O2 in the first group of five populations, with B2, B4, O3, O4, and O5 in the second. There are 3,628,800 unique permutations of this kind. However, only membership in the two distinct groups of five matters, so the permutations reduce to 252 resampling classes, each consisting of 14,400 equivalent permutations.

The data for these 252 resamplings were analyzed for comparison with the particular pattern observed. Three different statistics were analyzed in this manner: (i) mean protein intensity score for each group of five populations, original or resampled; (ii) variance in protein intensity scores over the group of five populations, original or resampled; and (iii) the difference between the total protein intensity scores of the two groups of five populations, original or resampled. Each will be discussed in turn.

The resampling pattern for the mean protein intensity scores placed the B and O means at opposite extremes of the distribution, with the B mean (2.57) the highest case of all and the O mean (2.08) part of a doubleton tied for the lowest value. The original B populations and O populations have mean scores that seem unlikely to have been so extremely differentiated by chance alone, given that they are located at the two extremes of the resampling distribution. The B population is located as a singleton

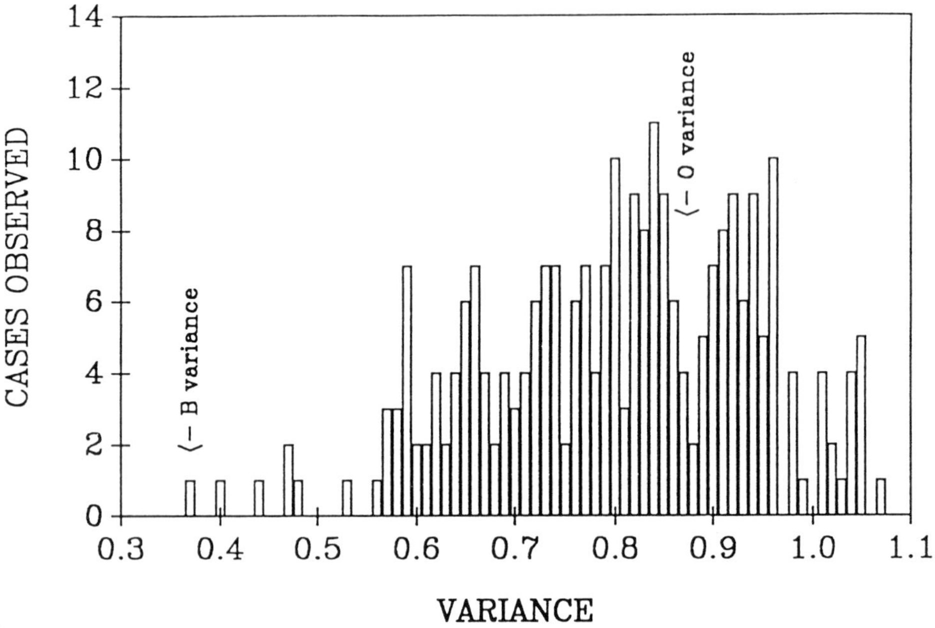

Fig. 3. The between-population variances of all 252 samplings of five populations from the original set of ten populations. The B and O sets of populations are included in this group of 252, and their between-population variances are indicated.

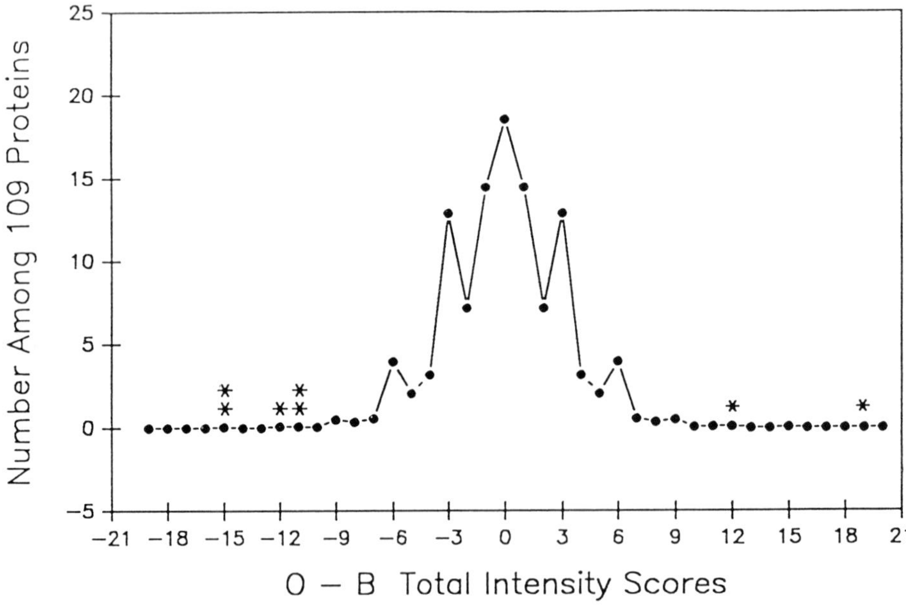

Fig. 4. The distribution of differences between total protein intensity scores of two groups of five populations each sampled from the original ten populations. All 252 possible resamplings were performed. These give the expectations for the number of extreme differences expected to arise by chance. The asterisks indicate seven proteins for which the observed difference between O and B total intensity scores is greater than 10. A difference this great or greater is expected to occur in less than one protein out of 109 polymorphic proteins in a random sample constructed from these same data. The occurrence of seven such proteins out of 109 polymorphic proteins is therefore not likely to be purely accidental.

at its extreme, while the O population is part of a doubleton. If there is just one population at either extreme, then the likelihood of those being the B and O populations is 2(1/252)(1/251), assuming uniformity of probability of occurrence over the distribution (this is a conservative assumption). If one of the extremes is a doubleton and the other a singleton, then the probability of B and O populations being located at the extremes is 2[1/252 + 1/251] (1/250), which is less than 0.0001. Therefore, the B populations have average protein intensity that is significantly greater than that of the O populations.

Figure 3 shows the distribution of average variances over the 109 proteins in the resampled data. The B populations have the lowest average variance, which has a probability of 1/252 of occurring by chance alone, assuming uniform likelihood over the distribution. However, the O populations are located well within the distribution of resampled variances. This seems to indicate that the B populations are particularly lacking between-populations, within-treatment heterogeneity, while the O populations are not.

Figure 4 shows the distribution of the total intensity scores over five pseudo-O populations minus the total pseudo-B protein score, which we will define as 'DT'. Extreme values of DT indicate proteins that may be associated with postponed aging. In the resampled data, the frequency of DT values greater than nine in magnitude is 0.0045, so nine constitutes a threshold for probabilities less than 0.01. Out of 109 proteins, 0.49 proteins would be expected to have this degree of differentiation, by chance alone. Seven proteins had DT values this extreme: 22a, 73a, 82, 150a, 153b, 227, and 230.

An alternative method of data analysis is to consider the resampling of these proteins individually, and then assign a probability to the observed protein differentiation pattern *per protein*. When this is done for the proteins that were found to be significant in the overall resampling analysis, the DT values of proteins 22a, 82, 153b, 227, and 230 have probabilities of less than 0.01 of occurring by chance, while the DT values of proteins 73a and 150a have probabilities of less than 0.05 of occurring by chance, in two-tailed tests. This qualitatively corroborates the results of the overall analysis, though the specific significance values are reduced in two cases.

One concern with these data is that these proteins may be differentiating as a result of overall differentiation in mean protein intensity scores. That is, there might be a global effect on protein intensities, rather than a protein-specific one. Of course, the global effect could alternatively be a side-effect of individual substitutions proceeding in the same direction. In any case, if the significance thresholds for the protein differentiation scores (DT) are moved by the 2.5 units required to allow for the difference in five individual B and O protein intensity differences, proteins 22a and 150a are no longer significantly differentiated at the 0.01 level. However, they are still significantly differentiated at the 0.05 level. In any case, the parsimony analysis given below also indicates that proteins 22a and 150a are statistically associated with the differentiation of B and O populations.

Combinatoric analysis

A different method of analysis can be used if the data are reduced to presence/absence scores of 1/0, respectively. Thus the present combinatoric analyses of protein scores ignore quantitative variation in spot size, given only that the protein is present. Note also that this analysis ignores the preceding resampling analysis, and therefore is not limited to considering the proteins already found to be differentiated.

There are exactly five proteins scored with five 1s and five 0s: 82, 111a, 200d, 225, and 227. Two out of five of these cases involve five Bs of state 1 and five Os of state 0. The probability of either of these cases arising by chance, out of the 252 equally likely combinations of five 1s and five 0s, is 1/126, since there are only two possible combinations of five 1s and five 0s with this differentiation pattern. The probability of getting two or more cases like proteins 82 and 227 out of five is 0.00062. Therefore, proteins 82 and 227 are not likely to have differentiated by chance alone.

A similar calculation for proteins 73a, 150a, and 176a, which are only one spot away from a similar five-present/five-absent spot differentiation, indicates that the probability that they are differentiated by chance alone is 0.000416. But the parallel calculation for protein 200d, which is *two* spots away from five-present/five-absent differentiation, reveals that such a differentiation pattern has a probability of 0.669 of being observed by chance alone.

Table 1. Resampling analysis of the pattern of concordance between proteins in patterns of within-treatment, between-populations differentiation. The 'steps' indicate how many proteins along the ordered list of proteins, observed or shuffled, concordance was checked. Note that monomorphic proteins are included in the observed and resampled data, giving rise to most cases of concordance. In particular, this is responsible for the greater concordance among the B populations, since more of these populations exhibit an absence of change in protein score. The hypothesis test is for differences between observed and resampled mean number of concordants (* , P < 0.05; **, P < 0.01), so that *either* too much or too little concordance is detected.

Steps	Observed concordants	B Resampled Mean	Variance	Significance of difference
1	238	227.5	6.95	**
2	231	226.88	6.667	
3	228	226.11	6.78	
4	227	225.28	7.45	
5	224	224.66	7.54	
6	223	223.84	7.39	
7	222	223.18	7.39	
8	224	222.58	7.62	
9	226	221.85	7.29	
10	222	221.20	7.91	
11	226	220.32	8.29	*
12	223	219.66	8.21	
13	227	219.18	9.20	*
14	225	218.33	8.03	*
15	223	217.48	8.57	
145	123	125.41	6.25	
146	123	124.73	5.93	
147	123	124.22	6.17	
148	124	123.27	5.33	
149	121	122.59	5.42	
150	122	121.76	5.18	
151	120	121.22	5.20	
152	119	120.47	4.89	
153	120	119.87	4.47	
154	119	119.10	4.80	
155	120	118.35	4.26	
245	51	54.78	9.01	
246	48	53.86	9.37	*
247	49	53.20	9.49	
248	46	52.44	9.16	*
249	49	51.69	8.86	
250	47	51.08	8.42	
251	46	50.25	8.49	
252	45	49.40	8.19	
253	43	48.91	9.39	*
254	43	48.12	9.33	
255	42	47.46	8.95	*
285	21	26.24	5.80	*
286	20	25.49	5.80	*
287	20	24.75	5.50	*
288	18	24.19	5.40	**
289	17	23.48	5.80	**
290	16	22.61	5.04	**
291	16	21.90	5.68	*
292	16	21.33	5.32	*

Steps	Observed concordants	B Resampled Mean	Variance	Significance of difference
293	14	20.59	5.10	**
294	14	19.76	4.56	**
295	13	19.16	4.74	**
296	13	18.47	4.66	*
297	14	17.66	4.40	
298	13	17.03	4.09	*
299	11	16.32	4.30	**
300	10	15.55	4.03	**
1	176	158.58	15.89	**
2	167	158.20	15.41	*
3	172	157.67	15.78	**
4	170	157.16	16.52	**
5	165	156.64	17.69	*
6	167	156.12	16.57	**
7	162	155.84	16.51	
8	169	155.07	16.61	**
9	161	154.53	17.18	
10	158	154.25	17.36	
11	161	153.67	15.64	*
12	159	153.13	17.38	
13	160	152.68	16.83	*
14	166	152.06	18.54	**
15	154	151.86	17.35	
145	89	87.71	11.49	
146	88	86.91	12.24	
147	85	86.57	10.47	
148	88	86.01	11.40	
149	85	85.53	9.81	
150	83	85.01	9.86	
151	83	84.35	10.00	
152	83	84.03	10.01	
153	87	83.55	9.45	
154	88	82.94	9.90	
155	79	82.79	9.66	
245	31	38.20	11.58	*
246	30	37.59	11.81	*
247	28	36.87	11.17	**
248	29	36.50	11.64	*
249	30	36.06	11.93	
250	25	35.36	11.55	**
251	26	35.14	11.15	**
252	27	34.40	11.74	*
253	25	34.23	10.97	**
254	25	33.51	11.87	*
255	22	32.99	10.73	**

Table 1. Continued.

Steps	Observed concordants	B Resampled Mean	Variance	Significance of difference
285	13	18.34	8.30	
286	10	17.70	6.73	**
287	11	17.32	7.60	*
288	9	16.89	7.29	**
289	8	16.40	7.48	**
290	9	15.75	6.57	**
291	7	15.25	6.32	**
292	6	14.93	6.88	**
293	5	14.27	6.27	**
294	6	13.80	6.10	**
295	4	13.20	5.77	**
296	4	12.93	5.50	**
297	2	12.34	5.60	**
298	3	11.94	4.90	**
299	2	11.38	4.79	**
300	1	10.43	5.09	**

Thus there are only five proteins that have significant differentiation *when scoring is by presence/absence only*, and the data can be analyzed combinatorically: 73a, 82, 150a, 176a, and 227.

Valid inference of B/O differentiation depends on comparisons of multiple populations of each type, as opposed to comparisons of single populations of each type. Five wholly independent comparisons of this kind can be made with the presence/absence scoring, each comparison using one of the five B_i/O_i pairs. When this is done, comparisons of single pairs over all 321 proteins yield an average of 21 proteins for which differentiation in single pairs is not sustained over the ten populations, 4.4 cases for which the single pair is an adequate guide to the general pattern, and 0.6 proteins where there is overall differentiation, but not in the pair examined. The reason these disparities arise is the differentiation of populations within the selected and control groups, that is, the within-treatment variance.

The resampling and combinatoric methods of analysis give two different groups of proteins that could be involved in postponed aging: 22a, 73a, 82, 150a, 153b, 227, 230 and 73a, 82, 150a, 176a, 227, respectively. Of the second group, only 176a is absent from the first. However, the DT score for 176a is nine, for which the resampling probability is only about 0.05. The first-group proteins missing from the second group, 22a, 153b, and 230, consist of cases with only two or three populations lacking the protein, cases that have high probability of occurring by chance when the data are reduced to presence/absence scores.

Effects of proximity on differentiation

One of the natural concerns facing any analysis of data from two-dimensional protein electrophoresis is the degree to which the protein spots represent genetically interdependent phenotypes. That is, perhaps differentiation detected in one spot is genetically associated with that of another spot, so that both are only markers of one underlying genetic change.

We tested for this possibility using protein intensity scores from the full set of 321 protein spots. Again, the data were extensively resampled, but in this case resampling shuffled the data by rows, so that different proteins were placed in juxtaposition in the table of intensity scores. We then tested how often adjacent protein scores matched over populations of a given type, B or O, in both the original data and the resampled. One thousand resamplings, obtained by 321 shufflings each, were calculated; the total number of possible combinations was 321!, since order mattered, an astronomical number. The proteins were originally labelled in rough sequence, so that protein spots 1 and 2 are near each other, while 1 and 200 are far apart. The most obvious test is to check whether adjacent proteins from the original data are more likely to resemble each other than randomly shuffled data. In both B and O data, immediately adjacent protein scores resemble each other more than they do in any of the shuffled cases. (See Table 1) This seems to indicate that there may be some genetic associations between adjacent proteins in the original data.

But Figure 5 reveals that the actual pattern is more complex than that. The graph displays the pattern of correspondence between proteins a variable number of steps away from each other in the data table. Proteins that are within 100 steps of each other generally resemble each other more than expected by chance, while proteins 200 to 300 steps apart resemble each other less than expected by chance. Table 1 indicates the outcome of statistical testing from the resampling.

The obvious interpretation of these results is that the areal resemblance effect is a result of variation

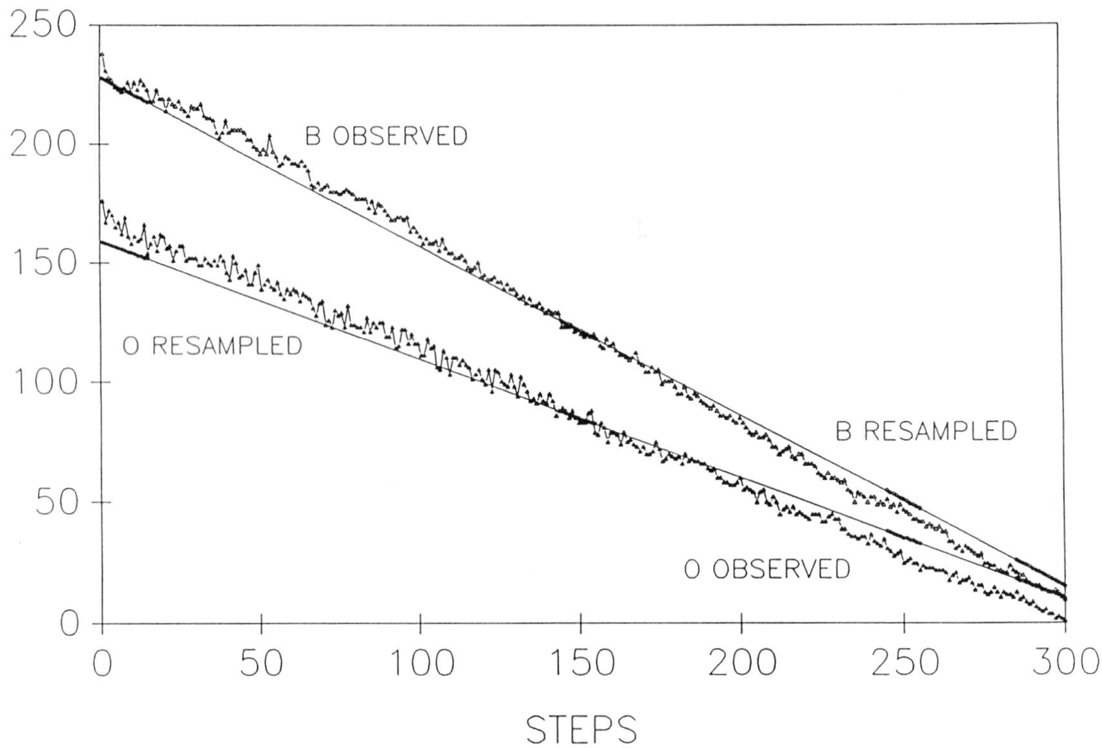

Fig. 5. The number of concordant patterns of between-population differentiation among B and O stocks. The thick solid line gives the observed number in the B stocks. The dashed line gives the observed number in the O stocks. The boxed and starred segments give the areas of resampling, as outlined in the text. The thin solid lines give the interpolated resampled pattern between regions actually resampled. Evidently, the observed patterns of concordance indicate that nearby proteins tend to resemble each other in between-population differentiation more than expected by chance, with distant proteins tending to resemble each other less. See Table 2 for numerical detail.

in autoradiographic intensities during autoradiogram preparation. There are several factors that could be at work: different gel thicknesses, variation in proximity of the photographic plate to the gel, and so on. In combination, they appear to make nearby protein intensities resemble each other. In terms of scientific inference, effects like these would constitute sources of noise that would obscure the detection of differences between populations, since they would vary within treatments, and thus obscure between-treatment differentiation. But such effects would not lead to spurious inferences of consistent differentiation.

Tree analysis

The data were analyzed using both phenetic and parsimony approaches. The phenetic analyses were conducted using the NTSYS-pc programs of Rohlf (1990), and the parsimony analyses were performed using the PAUP program of Swofford (1989).

The presence/absence data were analyzed phenetically by calculating a genetic distance. It is difficult to homologize loci among populations using two-dimensional electrophoresis, so the generally employed genetic distance methods could not be employed. Instead, we used the association coefficient $F = 2n_{xy} / (n_x + n_y)$, where n_x and n_y are the number of protein spots scored for populations x and y, respectively, and n_{xy} is the total number of shared protein spots between populations x and y (Aquadro & Avise, 1981; cf. Dice, 1941; Sokal &

Table 2. Genetic similarity and distance values calculated from the 321 protein spots scored in the two-dimensional electrophoretic study. The values below the diagonal are the genetic similarity values (F), which are derived from the Dice association coefficient, and the values above the diagonal are the genetic distance values (-ln F). See text for more detail concerning the calculation of these genetic measures.

	B1	B2	B3	B4	B5	O1	O2	O3	O4	O5
B1	-	.0195	.0212	.0261	.0279	.0434	.0330	.0544	.0712	.0528
B2	.9807	-	.0081	.0197	.0214	.0369	.0231	.0443	.0573	.0426
B3	.9791	.9919	-	.0147	.0198	.0421	.0182	.0391	.0592	.0410
B4	.9742	.9805	.9854	-	.0148	.0508	.0266	.0479	.0646	.0428
B5	.9724	.9788	.9804	.9853	-	.0423	.0250	.0464	.0560	.0377
O1	.9575	.9638	.9588	.9505	.9585	-	.0304	.0451	.0546	.0541
O2	.9675	.9771	.9820	.9738	.9753	.9701	-	.0238	.0543	.0326
O3	.9470	.9567	.9616	.9532	.9546	.9559	.9764	-	.0702	.0404
O4	.9313	.9444	.9426	.9374	.9456	.9468	.9472	.9322	-	.0611
O5	.9486	.9583	.9599	.9581	.9630	.9474	.9680	.9604	.9408	-

Sneath, 1963; Sneath & Sokal, 1973). To convert this association coefficient into a genetic distance, we have used a logarithmic transformation to linearize the distance measure, to give -ln F. This genetic distance is analogous to Nei's (1972) genetic distance, and has the same interval. The present genetic distance measure was then clustered using UPGMA (Sokal & Sneath, 1963), with the 'Find' option to detect ties.

The F and -ln F values are given in Table 2. The mean genetic similarity among the ten populations is 0.962, with a range of 0.931 to 0.992. However, there is a substantial difference between the B and O groups in the amount of genetic differentiation. The B lines show a within-group mean genetic similarity of 0.981, but the O lines, with a mean of 0.954, show much greater within-group genetic distances. Figure 6 gives the phenogram resulting from the UPGMA clustering of values given in Table 2. The cophenetic correlation for this tree is 0.923, which is in the very good fit range (Rohlf, 1990).

One problem with the phenetic analysis just presented is that it requires a uniform rate of change over populations, an unlikely assumption given the known generation numbers of B and O populations, as well as the disparity in genetic distances within the B and O groups just presented. Accordingly, a parsimony analysis of these data was also performed. The analysis was performed using the branch-and-bound algorithm, which guarantees that the most parsimonious tree(s) will be found (Hendy & Penny, 1982). Both the presence/absence data and the multistate data were analyzed, the latter with the character states ordered. To assess confidence limits for the branching patterns, a bootstrap analysis was also performed (Felsenstein, 1985; Sanderson, 1989). A total of 300 replications were performed using the branch-and-bound algorithm. The result is presented as a majority-rule consensus tree (Margush & McMorris, 1981), which shows the most frequently occurring branching orders.

The presence/absence parsimony analysis produced three equally most parsimonious trees, which together define the strict consensus tree given in Figure 7a. These three trees have length 105, with

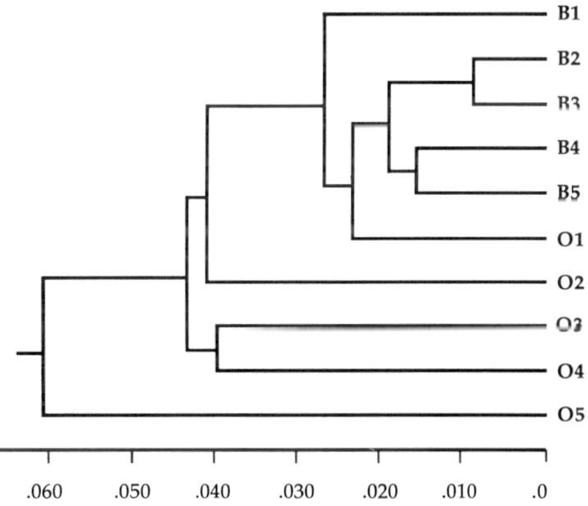

Fig. 6. Phenetic analysis. Phenogram produced when the binary data are analyzed by clustering the genetic distance values of Table 2 with UPGMA. The cophenetic correlation is 0.923.

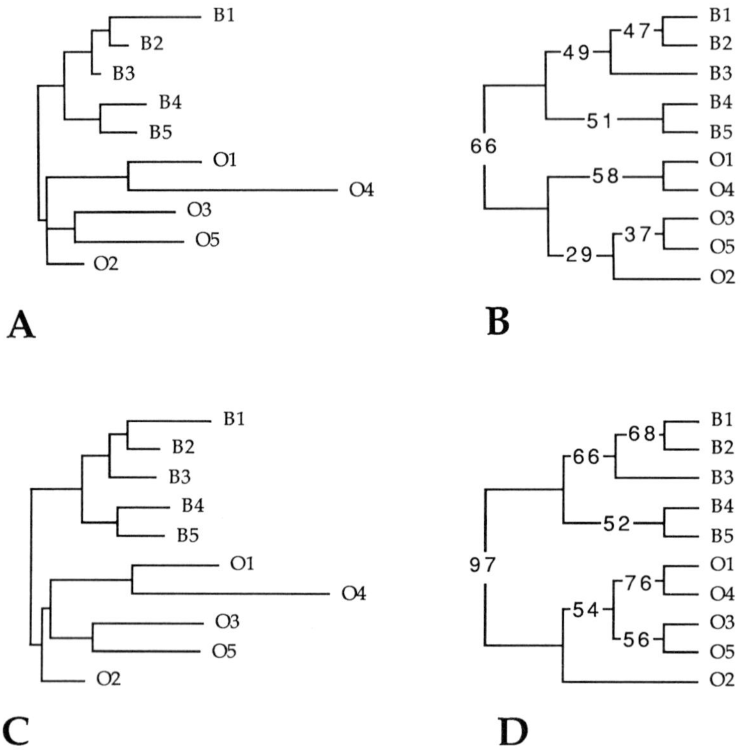

Fig. 7. Parsimony trees. A. Strict consensus tree based on the three most parismonious trees for the binary data set. Branches are drawn proportional to their length. B. Majority rule consensus tree from the bootstrap of the binary data set. The numbers on the nodes represent the frequency of occurrence. C. Most parsimonious tree from the ordered multistate data. Branches are drawn proportional to their length. D. Majority rule consensus tree from the bootstrap of the ordered multistate data. The numbers on the nodes represent the frequency of occurrence.

an overall consistency index of 0.667 and a consistency index excluding uninformative characters of 0.527. The bootstrap majority-rule consensus tree is topologically identical to one of the most parsimonious trees, and is shown in Figure 7b. Note that only one of the bifurcations, that between B and O lines, has a confidence level greater than 65%, though the 66% confidence level of that branching does not inspire particularly great certainty. In any case, the other branchings are of such low occurrence that little confidence can be placed in them. Table 3 gives the proteins that are important in determining the node separating the B group from the O group. Three different optimization methods were used to calculate this node: minimum F-value, accelerated transformation, and delayed transformation (Swofford & Maddison, 1987). With three different calculation methods and three equally parsimonious trees, the number of trees in which a given protein can contribute to the resolution of B and O groups ranges between one and nine. The ordered multistate data produce a unique most parsimonious tree, given in Figure 7c. The tree has a length of 379, with an overall consistency index of 0.675 and a consistency index excluding uninformative characters of 0.595. The bootstrap majority-rule consensus tree is topologically identical to the most parsimonious tree, which is frequently not the

Table 3. Protein spots that define the separation of the B and O groups of flies as determined by the parsimony analysis of the binary data set. Spots are listed in order of their importance. See text for more details.

Protein spot number	Number of trees present	Optimization[1]	Consistency index
82	9	F, A, D	1.000
227	9	F, A, D	1.000
176A	9	F, A, D	.500
73A	7	F, A, D	.500
150A	3	D	.500
22A	3	A	.333
168	3	A	.333
200D	2	D	.333

[1] F = minimum f-value transformation optimization; A = Accelerated transformation optimization; D = Delayed transformation optimization.

case (Felsenstein, 1985; Sanderson, 1989), and is shown in Figure 7d. The split between the B and O populations arises 97% of the time. However, once again the levels of significance for the rest of the tree are much lower, only three of the six branches exceeding the 65% confidence level. Table 4 gives the proteins involved in the B and O group split from the multistate tree analysis. Recall that there is only one most parsimonious tree per calculation method, so the three different calculation methods give a maximum of three trees on which a protein can be placed at the B-O node.

Table 4. Protein spots that define the separation of the B and O groups of flies as determined by the parsimony analysis of the multistate data set. Spots are listed in order of their importance, as defined by their consistency index. See text for more details.

Protein spot number	Number of trees present	Optimization (see Table 3)	Consistency index
54	3	F, A, D	1.000
82	3	F, A, D	1.000
227	3	F, A, D	1.000
230	3	F, A, D	1.000
20c	1	D	0.750
45c	3	F, A, D	0.750
153b	3	F, A, D	0.667
150a	1	D	0.600
176a	3	F, A, D	0.600
73a	3	F, A, D	0.500
155a	3	F, A, D	0.500
181a	3	F, A, D	0.500
197b	3	F, A, D	0.500
209a	3	F, A, D	0.500
18	3	F, A, D	0.429
22a	3	F, A, D	0.429
198a	1	A	0.375
200d	1	D	0.375
168	1	A	0.333

Discussion

Resolving proteins that postpone aging

One of the elementary questions that can be asked of the present study is whether or not it indicates those loci or proteins that are causally involved in postponing aging. The answer to this can only be somewhat equivocal, partly because of the different results obtained with different methods and partly because of fundamental population-genetic limitations.

The resampling study delimited a set of seven proteins that are statistically associated with the differentiation of stocks having postponed aging, based on the total score for B and O groups: 22a, 73a, 82, 150a, 153b, 227, and 230. Combinatoric analysis of presence/absence data found the following proteins: 73a, 82, 150a, 176a, and 227. Protein 176a would also have been regarded as significantly different in the resampling analysis at a significance level of about 0.05. Thus the combinatoric analysis essentially reveals a subset of the proteins detected in the resampling analysis of the multistate data. Tables 4 and 5 give the proteins found to differentiate B and O groups in the tree analyses. Ordered by position number, Table 3 gives proteins 22a, 73a, 82, 150a, 168, 176a, 200d, and 227. The only two new proteins in this list are 168 and 200d, though it is notable that these are among the lowest ranked proteins in Table 3, with respect to both incidence over maximum parsimony trees and consistency index. The DT scores are -9 and -7 for proteins 168 and 200d, respectively, the former being marginal for statistical significance in the resampling analysis. Table 4 gives a total of 19 proteins, the ones not already listed being 18, 20c, 45c, 54, 155a, 181a, 197b, 198a, and 209a. It is noteworthy that Table 4 includes all ten proteins identified by the other methods.

Thus there are four different lists of proteins associated with B and O stock differentiation. All four of these lists indicate the involvement of proteins 73a, 82, 150a, 176a, and 227. If one relaxes stringency to presence in three out of the four lists, then protein 22a should be added. If two out of four lists is regarded as sufficient, then proteins 153b, 168, 200d, and 230 are to be added to those involved in differentiation. For now it seems most prudent to confine further consideration to proteins delineated by at least three out of four methods of data analysis: 22a, 73a, 82, 150a, 176a and 227.

Statistical association between protein differences and B-O differentiation does not necessarily imply causal involvement. One point is that there is a non-trivial possibility that one or two statistically associated proteins are only associated by chance, this being inherent in any statistical testing procedure. In addition, other detected proteins might not be causally involved in the physiological changes postponing aging, instead being products of alleles in tight linkage disequilibrium with alleles at the loci causally involved in the response to selection. In this case, the alleles detected here would at least be a guide to the location of causally important alleles. Another possibility is that differentiated proteins could be undergoing changes as a result of the effects of genetic changes at other loci that then give rise to modification of the detected protein at some point after transcription of the protein's gene sequence. Such modification could be a pleiotropic effect of the original genetic change which is unrelated to the effect of that change upon aging.

However, we would argue that while there may be reasonable doubt concerning the causal involvement of any of the proteins adduced here, what has been accomplished is a considerable reduction in the number of candidate proteins worthy of further investigation. That is to say, further tests of six proteins is a considerably more feasible project than further tests of 321 proteins. This is not to say that only these six proteins are of interest. Undoubtedly our methods have failed to identify numerous other loci that associated with postponed aging in our longer-lived stocks.

Pattern of evolution

One of the interesting features of the data is that there is an abundance of variance between populations within selection regimes. Part of this variation probably arises from 'regional' autoradiogram intensity variation, as revealed by the correlation between protein intensities over regions (Fig. 5). However, the variation between B populations is strikingly less than that between O populations, as shown in Figure 3, suggesting that genuine biological differences play a role in determining this type of variation. Evidently, the O populations have differentiated from each other over the course of the experiment more than B populations have. Yet the O populations have undergone many fewer generations, about 50 O generations versus more than 200 B generations at the time the samples were obtained. Thus a simple drift hypothesis is not an obvious explanation for these results, unless the O populations have had sufficiently lower effective population sizes than the B populations, which is not known at present. An alternative hypothesis might be one of heterogeneous responses to selection in the different O populations. If this has in fact arisen, then some of the inconsistent differentiation of the O populations may be due to the heterogeneous responses of particular loci in individual populations to selection for postponed aging, responses that are not general over all O populations. However, the present data can do little to test the validity of this conjecture, which must await the analysis of specific proteins in greater detail than the present study.

Number of loci

One of the more interesting questions in research on the genetics of aging is the number of loci that control aging. Previous work in *Drosophila* has used the segregation indices for F_1, F_2, and backcross progeny (e.g. Luckinbill *et al.*, 1987; Hutchinson & Rose, 1990). Such studies appear to have had little statistical power, and cannot distinguish the number of loci involved from infinitely many loci.

The present data also provide an extremely crude perspective on this question. Suppose that each of the six robustly differentiated proteins is so differentiated because of genetic differentiation at a distinct locus. This may be a linked locus or it may be a locus that modifies the transcription or post-synthetic alteration of the labelled protein. Thus, this locus does not have to be the locus coding for the primary sequence of the protein. Granting this assumption of genetic distinctness, then our findings suggest that 6 out of about 300 loci were involved in postponing aging, or 2%. If the genome size is

10,000 to 20,000 transcribed loci in *Drosophila*, and our loci are a representative sample, then the total number of loci that can postpone aging in these flies is about 200 to 400. The assumptions used to obtain these numbers are of course speculative, and we do not wish to defend them. However, these results at least suggest the possibility that on the order of 10^2 loci may be involved in the control of aging in *D. melanogaster*. This further suggests the necessity of methods that can identify loci in the face of considerable polygenic variation, where the genetics of aging are concerned.

Acknowledgements

We are grateful to W. M. Fitch, R. R. Hudson, R. E. Lenski, and L. D. Mueller for suggestions concerning the analysis of the data. We thank S. Dinan and M. Hartsuck for technical assistance. This research was supported in part by a US Public Health Service grant (AGO6346) to MRR and donations provided to the Linus Pauling Institute of Science and Medicine.

References

Anderson, N. G. & N. L. Anderson, 1978. Analytical techniques for cell fractions. XXI. Two-dimensional analysis of serum and tissue proteins: multiple isoelectric focusing. Anal. Biochem. 85: 331-340.

Aquadro, C. F. & J. C. Avise, 1981. Genetic divergence between rodent species assessed by using two-dimensional electrophoresis. Proc. Natl. Acad. Sci. 78: 3784-3788.

Bozcuk, A. N., 1981. Genetic longevity in *Drosophila*. V. The specific and hybridized effect of *rolled*, *sepia*, *ebony* and *eyeless* autosomal mutants. Exp. Gerontol. 11: 103-112.

Clare, M. J. & L. S. Luckinbill, 1985. The effects of gene-environment interaction on the expression of longevity. Heredity 55: 19-29.

Dice, L. R., 1941. Measures of the amount of ecologic association between species. Ecology 26: 297-302.

Felsenstein, J., 1985. Confidence limits on phylogenies: an approach using the bootstrap. Evolution 39: 783-791.

Fleming, J. E., P. S. Melnikoff & K. G. Bensch, 1984. Identification of mitochondrial proteins on two-dimensional electrophoresis gels of extracts of adult *Drosophila melanogaster*. Biochim. Biophys. Acta 802: 340-345.

Fleming, J. E., E. Quaottrocki, G. Latter, J. Miquel, R. Marcuson, E. Zuckerkandl & K. G. Bensch, 1986. Age-dependent changes in proteins of *Drosophila melanogaster*. Science 231: 1157-1159.

Fleming, J. E., J. K. Walton, R. Dubitsky & K. G. Bensch, 1988. Aging results in an unusual expression of *Drosophila* heat shock proteins. Proc. Nat. Acad. Sci. USA 85: 4099-4103.

Garrels, J., 1979. Two-dimensional gel electrophoresis and computer analysis of proteins synthesized by clonal cell lines. J. Biol. Chem. 254: 7961-7977.

Hendy, M. D. & D. Penny, 1982. Branch and bound algorithms to determine minimal evolutionary trees. Math. Biosci. 59: 277-290.

Hutchinson, E. W. & M. R. Rose, 1987. Genetics of aging in insects. Rev. Biol. Res. Aging 3: 62-70.

Hutchinson, E. W. & M. R. Rose, 1990. Quantitative genetic analysis of postponed aging in *Drosophila melanogaster*, p. 66-87 in Genetic effects on aging II, edited by D. A. Harrison. Telford, Caldwell, N.J.

Hutchinson, E. W. & M. R. Rose, 1991. Quantitative genetics of postponed aging in *Drosophila melanogaster*. I. Analysis of outbred populations. Genetics 127: 719-727.

Ives, P. T., 1970. Further studies of the South Amherst population of *Drosophila melanogaster*. Evolution 24: 507-518.

Johnson, T. E., 1987. Aging can be genetically dissected into conponent processes using long-lived lines of *Caenorhabditis elegans*. Proc. Natl. Acad. Sci. USA 84: 3777-3781.

Johnson, T. E. & E. W. Hutchinson, 1990. Aging in *Caenorhabditis elegans*: Update 1988. Rev. Biol. Res. Aging 4: 15-27.

Luckinbill, L. S., M. J. Clare, W. L. Krell, W. C. Cirocco & S. A. Buck, 1987. Estimating the number of genetic elements that defer senescence in *Drosophila melanogaster*. Evolution 38: 996-1003.

Luckinbill, L. S., T. A. Grudzien, S. Rhine & G. Weisman, 1989. The genetic basis of adaptation to selection for longevity in *Drosophila melanogaster*. Evol. Ecology 3: 31-39.

Luckinbill, L. S., J. L. Graves, A. H. Reed & S. Koetsawang, 1988a. Localizing the genes that defer senescence in *Drosophila*. Heredity 60: 367-374.

Luckinbill, L. S., J. L. Graves, A. Tomkiw & O. Sowirka, 1988b. A qualitative analysis of some life-history correlates of longevity in *Drosophila melanogaster*. Evol. Ecol. 2: 85-94.

Maynard Smith, J., 1966. Theories of aging, p. 1-35 in Topics in the biology of aging, edited by P. L. Krohn. Interscience, New York.

Margush, T. & F. R. McMorris, 1981. Consensus n-trees. Bull. Math. Biol. 43: 239-244.

Nei, M., 1972. Genetic distances between populations. Am. Nat. 106: 283-292.

O'Farrell, P. H., 1975. High resolution two-dimensional electrophoresis of proteins. J. Biol. Chem. 250: 4007-4021.

Ohnishi, S., M. Kawanishi & T. I. Watanabe, 1983. Biochemic phylogenies of *Drosophila*: protein differences detected by two-dimensional electrohoresis. Genetica 61: 55-63.

Parker, J., J. Flanagan, J. Murphy & J. Gallant, 1981. On the accuracy of protein synthesis in *Drosophila melanogaster*. Mech. Age. Dev. 16: 127-139.

Rohlf, F. J., 1990. NTSYS-pc. Numerical taxonomy and multivariate analysis system (ver. 1.60). Exeter Pub., Setauket, New York.

Rose, M. R., 1984. Laboratory evolution of postponed senescence in *Drosophila melanogaster*. Evolution 38: 1004-1010.

Rose, M. R., 1989. Genetics of increased lifespan in *Drosophila*. Bioessays 11: 132-135.

Rose, M. R., 1991. Evolutionary biology of aging. Oxford University Press, New York.

198

Rose, M. R. & B. Charlesworth, 1981. Genetics of life-history in *Drosophila melanogaster*. II. Exploratory selection experiments. Genetics 97: 187-196.

Rose, M. R., M. L. Dorey, A. M. Coyle & P. M. Service, 1984. The morphology of postponed senescence in *Drosophila melanogaster*. Can. J. Zool. 62: 1576-1580.

Sanderson, M. J., 1989. Confidence limits on phylogenies: the bootstrap revisited. Cladistics 5: 113-129.

Service, P. M., E. W. Hutchinson, M. D. MacKinley & M. R. Rose, 1985. Resistance to environmental stress in *Drosophila melanogaster* selected for postponed senescence. Physiol. Zool. 58: 380-389.

Service, P. M., 1987. Physiological mechanisms of increased stress resistance in *Drosophila melanogaster* selected for postponed senescence. Physiol. Zool. 60: 321-326.

Sneath, P. H. A. & R. R. Sokal, 1973. Numerical taxonomy. W. H. Freeman and Co., San Francisco.

Sokal R. R. & P. H. A. Sneath, 1963. Principles of numerical taxonomy. W. H. Freeman and Co., San Francisco.

Swofford, D. L., 1990. PAUP. Phylogenetic analysis using parsimony (ver. 3. On). Ill. Nat. Hist. Surv., Champaign, Illinois.

Swofford, D. L. & W. P. Maddison, 1987. Reconstructing ancestral character states under wagner parsimony. Math. Biosci. 87: 199-229.

Evolution, 51(3), 1997, pp. 865–872

ALLOZYMIC DIFFERENTIATION IN RESPONSE TO LABORATORY DEMOGRAPHIC SELECTION OF *DROSOPHILA MELANOGASTER*

DENISE J. DECKERT-CRUZ,[1] ROBERT H. TYLER,[2] JACOB E. LANDMESSER, AND MICHAEL R. ROSE
Department of Ecology and Evolutionary Biology, University of California, Irvine, California 92697-2525
[1]*E-mail: ddeckert@ea.oac.uci.edu*

Abstract.—*Drosophila melanogaster* populations that exhibit constrasting life histories as a result of laboratory selection were compared at several potentially relevant enzyme loci. Selection regimes included postponed reproduction, accelerated development, and intermediate generation time. Each selection regime was represented by fivefold replicated populations maintained for between 50 and 500 generations. For each population, allele frequencies were calculated from frequencies of electrophoretically distinguishable allozymes of alcohol dehydrogenase, α-glycerol-3-phosphate dehydrogenase, phosphoglucomutase, and CuZn-superoxide dismutase. Based on allozyme frequency changes consistent across replicate populations, two of the studied loci responded to both selection for postponed reproduction and selection for accelerated development. The responses to contrasting selection regimes were in opposing directions, suggesting antagonistic pleiotropy.

Key words.—Aging, alcohol dehydrogenase, *Drosophila*, glycerol-3-phosphate dehydrogenase, life history, phosphoglucomutase, superoxide dismutase.

Received June 12, 1996. Accepted January 28, 1997.

Evolutionary theory has explained aging as a result of the decrease in the force of natural selection with age (Haldane 1941; Medawar 1946, 1952; Hamilton 1966; Charlesworth 1980; Rose 1991). One or both of two mechanisms may be operating: mutation accumulation and antagonistic pleiotropy. Either mechanism could maintain genetic variation, such that aging is subject to selection. Mutation accumulation is a result of recurrent deleterious mutation and random genetic drift maintaining deleterious variants that have effects only at late ages. Such alleles are nearly neutral insofar as their deleterious expression occurs after most reproduction, when the strength of natural selection is greatly diminished (Charlesworth 1980). According to the theory of antagonistic pleiotropy, an allele with deleterious effects late in life can be maintained in a population through its favorable effects early in life, when the strength of natural selection is greater. Antagonistic pleiotropy can contribute to polymorphism through heterozygote superiority, particularly when deleterious effects are recessive (Rose 1985).

Just as there are alleles likely to cause aging through their deleterious effects late in life, so there may be alleles that enhance longevity through beneficial effects late in life. Such alleles should be kept at a low frequency by being neutral or slightly deleterious early in life. Of course, such alleles should be much rarer than alleles causing aging, since most mutations are deleterious and even a slight disadvantage early in life could eliminate an allele from the population.

In *Drosophila melanogaster*, postponed aging can be accomplished by allowing only the flies that can reproduce late in life to contribute to the next generation. Selecting flies for late-life reproduction has resulted in postponed aging in sets of *Drosophila* spp. stocks with independent origins (Wattiaux 1968a,b; Rose and Charlesworth 1981; Luckinbill et al.1984; Rose 1984). For *D. melanogaster* there is no statistical evidence for the involvement of fewer than 100 loci in postponed aging (Hutchinson and Rose 1990; Fleming et al. 1993).

[2] Deceased.

Because flies selected for postponed aging have enhanced resistance to stresses such as starvation, desiccation, and forced flight (Service et al. 1985; Graves and Rose 1990), gene loci that encode metabolic enzymes are expected to be among the loci at which allele frequencies change in response to selection. Electrophoretic surveys of a dozen metabolic enzymes have been performed on the long-lived and short-lived *D. melanogaster* populations of Luckinbill et al. (1989). They found allele frequencies to differ only at the glucose-6-phosphate dehydrogenase locus, though their low effective population sizes may have limited the diversity of genotypes available to selection.

Enzyme electrophoresis has been used previously to correlate laboratory selection with allele frequency changes at allozyme loci. Foundational work in this area included the demonstration that allele frequencies at the amylase locus could be changed by maintaining *D. melanogaster* populations on food varying in starch content (Hickey 1977). Cavener and Clegg (1981) reported responses to ethanol selection at the alcohol dehydrogenase, alpha-glycerophosphate dehydrogenase, malate dehydrogenase, and 6-phosphogluconate dehydrogenase loci. The alcohol dehydrogenase locus has also been shown to respond to temperature selection (Oakeshott 1979) and accelerated development (Knibb 1987). Carfagna et al. (1980) used the phosphoglucomutase locus to demonstrate frequency-dependent and heterotic selection. Bijlsma and Kerver (1983) showed differential survival among genotypes in the presence of DDT for the pentose phosphate shunt enzymes glucose-6-phosphate dehydrogenase and 6-phosphogluconate dehydrogenase.

While all of the above studies deal with laboratory evolution, a much greater number of studies have utilized electrophoretic techniques to support hypotheses of past evolution in natural environments. Watt et al. have provided the best example of the evolutionary importance of an allozyme locus to natural insect populations. They showed that phosphoglucose isomerase (PGI) allozymes in *Colias* butterflies differ consequentially in kinetics and thermal stability. These

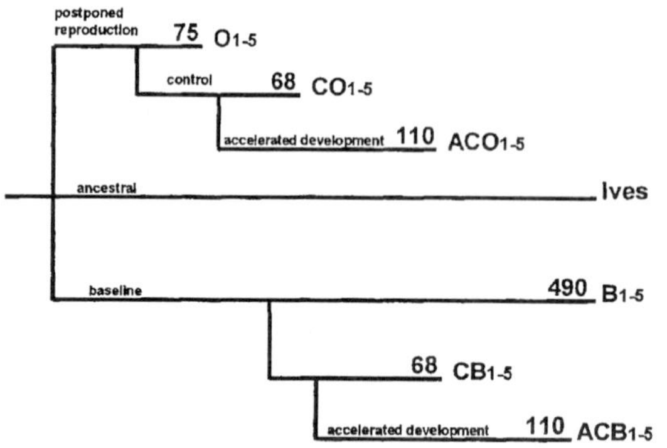

FIG. 1. Generations of selection completed at time of genotype sampling.

biochemical differences affect daily flight capacity such that they can be used to predict differences among PGI genotypes in daily flight capacity. The flight differences result in genotypic differences in survival, male mating success, and female fecundity (Watt 1992, 1994). Using models assuming fitness is proportional to flux through a biochemical pathway, Hastings (1992) has suggested that selection pressures acting on enzyme variants may be much higher than previously thought.

This study evaluates responses to selection at several loci that encode enzymes of possible relevance to life history. Allele frequencies at these loci are calculated for populations of *D. melanogaster* kept on discrete generations of characteristic lengths. All the populations have in common the genetic background of a single founding population. Their current selection regimes represent an entire spectrum of different generation times. Each generation time is represented by five replicate populations.

MATERIALS AND METHODS

This study utilized six fivefold-replicated *D. melanogaster* stocks: B, O, CB, CO, ACB, and ACO. The B and O populations were formed in February 1980 (Rose 1984) from a single ancestral generation of the Ives stock obtained in 1975 from Massachusetts (Ives 1970) (Fig. 1). The long-lived Os have been maintained on 10-week generations since 1981, while their controls, the Bs, have been maintained on two-week generations like the Ives stock. The mean female longevity is 32.8 days for the Bs and 55.6 days for the Os (Chippindale et al., pers. comm.).

CB and CO are stocks of intermediate generation length derived from the Bs and Os, respectively (Fig. 1), in 1988 (Rose et al. 1992). They serve as controls for stocks derived from them, including both the accelerated stocks mentioned below and stress selected stocks which determine the CB/CO generation time but otherwise are not relevant to this study. The CB/CO generation time is on the order of five weeks, which favors relatively late fecundity, compared to the two-week generation time of the Bs. The mean female longevity is 38.8 days for the CBs and 46.5 days for the COs.

The ACB and ACO stocks were derived from the CB and CO lines, respectively, such that the replicate number of each line corresponds to the CB or CO line from which it was derived. The ACB and ACO lines had completed approximately 100 generations of selection for accelerated development at the time of this study, each generation lasting less than two weeks. Mean female longevity equals 33.0 days for the ACBs and 31.1 days for the ACOs.

Selection for accelerated development was accomplished by allowing only approximately the first 13% of flies to emerge from pupa to contribute eggs to the next generation. Development time had decreased significantly after ten generations of selection. Viability is lower for these stocks than for stocks with longer development times (Chippindale 1994; Chippindale et al. 1994).

Individual female flies were homogenized and subjected to starch gel electrophoresis. The sample size was approximately one hundred flies for every population except O1 and O5; for these two populations five hundred flies were analyzed. For the five B populations and the five O populations, gels were stained for alcohol dehydrogenase, α-glycerol-3-phosphate dehydrogenase, phosphoglucomutase, and CuZn-superoxide dismutase, among other enzymes (Pasteur et al. 1988; Manchenko 1994). Flies from the CO, CB, ACO, and ACB populations were stained for only the first four enzymes. Allele frequencies were calculated from the allozymes resolved on the gels, and tested for deviation from Hardy-Weinberg equilibrium using the χ^2 statistic.

The two main hypotheses to be tested are: (1) do the O populations, selected for postponed reproduction, have different allele frequencies from those of the baseline B populations; and (2) do the populations selected for accelerated development have different allele frequencies from those of the control populations? By using a factorial ANOVA of arcsine-transformed allele frequencies, we were able to condense our analyses to a single statisitical test on the data for each locus, thus reducing the error associated with multiple comparisons. The three population types (parental, control, and accelerated development) were regarded as one factor, and the two original lineages (B and O) as another. The differentiation associated with population type was subjected to a Tukey-HSD post hoc test in order to identify differentiation associated with selection for accelerated development. The statistical tests were performed using the multivariate general linear hypotheses section of Systat software (Wilkinson 1990).

For the O1 and O5 lines, maximum likelihood estimation of gamete frequencies permitted evaluation of linkage between the phosphoglucomutase and CuZn-superoxide dismutase loci, which are both on the third chromosome. The overall hypothesis that none of the disequilibrium coefficients differ from zero was tested using the method of Weir (1990). In this equation, n is the gamete count, D is the linkage disequilibrium constant of loci u and v, and k and l are the number of alleles at locus u and locus v, respectively.

$$X_T^2 = \sum_{u=1}^{k} \sum_{v=1}^{l} \frac{n\hat{D}_{uv}^2}{\tilde{p}_u \tilde{p}_v}$$

TABLE 1. Allele frequencies.

	B series stocks							O series stocks						
Population type	Fly line	*Adh F*	*Gpdh F*	*Sod S*	*Pgm F*	*Pgm M*	*Pgm S*	Fly line	*Adh F*	*Gpdh F*	*Sod S*	*Pgm F*	*Pgm M*	*Pgm S*
Parental lines	B1	0.81	0.59	0.00	0.19	0.56	0.25	O1	0.91	0.93	0.17	0.06	0.38	0.56
	B2	0.87	0.38	0.00	0.27	0.73	0.003	O2	0.89	0.9	0.14	0.053	0.553	0.394
	B3	0.95	0.37	0.00	0.17	0.82	0.02	O3	0.865	0.78	0.22	0.04	0.31	0.65
	B4	0.91	0.75	0.07	0.43	0.56	0.002	O4	0.76	0.735	0.215	0.09	0.45	0.46
	B5	0.88	0.996	0.00	0.33	0.66	0.01	O5	0.92	0.94	0.08	0.29	0.21	0.50
	Mean	0.885	0.62	0.01	0.28	0.67	0.05	Mean	0.87	0.86	0.165	0.11	0.38	0.51
Control lines	CB1	0.90	0.9	0.04	0.082	0.663	0.255	CO1	0.94	0.72	0.21	0.1	0.56	0.34
	CB2	0.98	0.83	0.05	0.05	0.95	0	CO2	1.00	0.87	0.06	0.08	0.6	0.32
	CB3	0.80	0.64	0.00	0.25	0.72	0.02	CO3	0.945	0.93	0.13	0	0.87	0.14
	CB4	0.76	0.79	0.00	0.11	0.59	0.3	CO4	0.87	0.78	0.125	0.118	0.566	0.316
	CB5	0.93	0.8	0.07	0.17	0.81	0.02	CO5	0.88	0.9	0.13	0.00	0.82	0.18
	Mean	0.88	0.79	0.03	0.13	0.75	0.12	Mean	0.93	0.84	0.13	0.06	0.68	0.26
Accelerated development lines	ACB1	0.925	0.87	0.00	0.00	1.00	0.00	ACO1	0.99	0.74	0.09	0.00	1.00	0.00
	ACB2	0.88	0.84	0.00	0.00	1.00	0.00	ACO2	1.00	0.59	0.19	0.00	1.00	0.00
	ACB3	1.00	0.97	0.00	0.00	1.00	0.00	ACO3	0.96	1.00	0.00	0.00	1.00	0.00
	ACB4	1.00	1.00	0.00	0.00	1.00	0.00	ACO4	0.98	1.00	0.00	0.00	1.00	0.00
	ACB5	1.00	1.00	0.00	0.00	1.00	0.00	ACO5	0.99	1.00	0.00	0.00	1.00	0.00
	Mean	0.96	0.94	0.00	0.00	1.00	0.00	Mean	0.98	0.90	0.06	0.00	1.00	0.00
ANOVA p-values		*Adh F*	*Gpdh F*	*Sod S*			*Pgm S*							
B vs. O stock lineage		0.503	0.350	0.000			0.000							
Population type		0.001	0.031	0.024			0.000							
Interaction		0.553	0.415	0.130			0.000							

RESULTS

Throughout the stocks, there are two allozymes each for alcohol dehydrogenase (ADH), α-glycerol-3-phosphate dehydrogenase (GPDH) and CuZn-superoxide dismutase (SOD). They are designated F and S for faster and slower migration in the gel. Three allozymes for phosphoglucomutase (PGM) are present in the B and O stocks; they are designated F, M, and S.

Genotype frequencies in all the populations are at Hardy-Weinberg equilibrium. No deviations from Hardy-Weinberg equilibrium were supported, after Bonferroni correction to the χ^2 tests. Linkage disequilibrium between the *Pgm* and *Sod* loci is small. The linkage disequilibrium coefficients are 0.017 for the O1 population and 0.012 for the O5 population. The χ^2 values are 0.62 for the O1 population and 0.36 for the O5 population; at the $P = 0.05$ level, the critical χ^2 value is 3.84. Hence, the linkage disequilibrium coefficients do not differ significantly from zero.

Selection for Postponed Reproduction

Comparison of the O lines to the B lines shows allele frequencies diverging at two of the four polymorphic enzyme loci. While change was insignificant at the *Adh* and *Gpdh* loci, selection for postponed reproduction changed allele frequencies significantly at the *Sod* and *Pgm* loci (Table 1). At the *Adh* locus, allele frequencies are on average the same in the O lines as in the B lines. While the mean *Gpdh* allele frequencies of the Bs and Os appear divergent, allele frequencies at this locus are highly variable, and there is no statistically significant differentiation associated with postponed reproduction (Table 1). In contrast, at the *Sod* locus, an increase in the frequencies of the previously rare *S* allele is highly statistically significant ($P < 0.001$). Significant change also occurred in the frequencies of the *Pgm S* allele ($P < 0.001$). In all five B populations, *M* is the most abundant allele, *F* is relatively common, and *S* is generally rare. In all five O populations, the *S* allele is much more common, often exceeding the *M* allele in abundance.

Flies with Intermediate Generation Lengths

At the *Adh*, *Gpdh*, and *Sod* loci, the CB and CO lines do not differ significantly from the B and O lines (Appendix). However, the *Pgm* locus is exceptional (Table 1). In the CB lines, relative to the B lines, the abundance of the *F* allele has been halved. In the CO lines, relative to the O lines, the *S* allele has become substantially less common, while the *M* allele has increased in frequency. In both the CB and the CO lines, the frequency of the S allele has changed in the same direction longevity has changed (Fig. 2).

Flies Selected for Accelerated Development

In the accelerated development lines, the more abundant allele generally increased in frequency at all four loci, and fixations were common (Table 1). At the *Adh* and *Pgm* loci, the increases represent significant differentiation between accelerated development and control stocks (Appendix). At the *Gpdh* locus, there is a trend toward fixation from the parental to the most derived stocks, but the accelerated development stocks are not significantly different from their controls, the CBs and COs (Appendix). At the *Sod* locus, the *S* allele is undetectable in every ACB line as well as three ACO lines (Table 1). These apparent fixations constitute significant divergence of the accelerated development lines, at least compared to the parental lines (Appendix), but for the ACB lines they represent small increases in frequency of the *F* allele, since the *S* allele was already rare in the B parental stock.

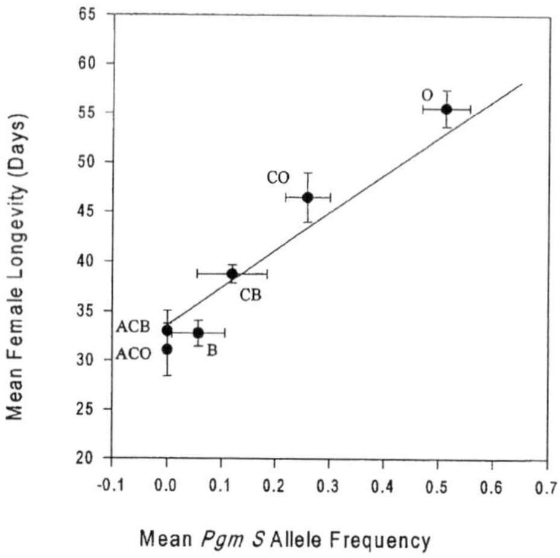

FIG. 2. *Pgm S* allele frequency versus longevity.

Consistent fixation across accelerated development lines is more apparent at the *Pgm* locus (Table 1). At this locus, not only do the fixations represent highly significant changes in allele frequency, but some constitute large increases in the proportion of the *M* allele. An overview of the demographically selected stocks shows a progressive increase in the frequency of the *Pgm S* allele with mean longevity (Fig. 2).

DISCUSSION

Both selection for postponed reproduction and selection for accelerated development can be correlated with significant differentiation in allele frequencies. Responses to selectively postponed aging were seen at the *Sod* and *Pgm* loci, while responses to selectively accelerated development were seen at the *Adh* and *Pgm* loci, and possibly the *Sod* locus. The allele frequency changes associated with the postponed aging lines represent robust responses to selection, as they are both substantial and consistent in direction across fivefold replicated populations. Since allele frequency changes in the accelerated development lines are relatively small in some cases, it may be desirable to consider genetic drift as a cause for allozymic differentiation of the accelerated development lines from their controls. Subsequently the results will be discussed by locus.

The lines selected for accelerated development were derived from populations in which the more common allozyme was already encoded by a strong majority of alleles. However, the initial allele frequencies of the accelerated development lines cannot be assumed to equal the allele frequencies of the CB and CO lines, even if the ACBs and ACOs were representative samples, because the ACB and ACO lines are separated from the CB and CO lines studied here by over 25 generations of selection in the CB and CO lines. Nevertheless, a combination of founder effect and finite population size could make genetic drift a viable explanation of the fixation of common alleles.

Effective population size is a key issue for the *Adh* and *Sod* loci, since the *S* allele at each of these loci is relatively rare throughout all the stocks. If we assume that the initial allele frequencies in the accelerated development lines are equal to the allele frequencies of their controls, changes resulting in fixations range from less than 1% to greater than 20%. These changes should require up to 2N generations for a neutral allele (Hartl and Clark 1989, p. 72). While the effective sizes of the populations are not precisely known, the population size is approximately 1000 flies. If we conservatively assume that the effective population size is 500, up to 1000 generations are still required for the fixations observed, given neutral alleles. Moreover, for both *Adh* and *Sod*, the direction of allele frequency change is consistent for nine of 10 accelerated development populations. Thus the significant increase in frequency of the *Adh F* allele is likely to be a real effect of selection.

Alcohol Dehydrogenase and α-Glycerophosphate Dehydrogenase

Heinstra et al. (1987) suggested that natural selection may act on the *Adh* polymorphism in larvae, and indeed the selection regime imposed on the ACB and ACO populations acted upon the larval stage. Our association of *Adh F* allele frequency increases with accelerated development concurs with the findings of Knibb et al. (1987) who selected *D. melanogaster* for fast, intermediate, and slow development times. Their fast development selection regime, like our own, reduced the time to the first emergence of adults by using only the earliest flies to emerge from the pupa as parents for the next generation. Over the course of selection the frequency of the *Adh F* allele rose in three of four fast development lines. Cavener's (1983) study of developmental rate selection provides a contrasting result. He selected for fast and slow development in fivefold replicated populations from each of two different geographical localities. The *Adh F* allele frequency did not change significantly in the populations Cavener selected for fast development. McKechnie and Geer (1988) have demonstrated epistasis between *Adh* and *Gpdh* in the determination of larval ethanol tolerance. Larvae of the Adh^{FF} $Gpdh^{FF}$ and Adh^{SS} $Gpdh^{SS}$ allelic combinations are more ethanol tolerant than larvae with other combinations. So the tendency for the *Gpdh F* allele to increase in the ACB and ACO populations may have been affected by the increase in the *Adh F* allele. Izquierdo and Rubio (1989) found the allele frequencies at α*Gpdh* to be determined to a considerable extent by genotypic combinations with the *Adh F* allele. On the other hand, they detected a difference in development rate between α*Gpdh* alleles, with the α*Gpdh S* allele associated with a faster rate of development. Barnes et al. (1989) also found that flies with the α*Gpdh S/S* genotype developed slightly faster. Hence, selection for accelerated development may place conflicting pressures on the α*Gpdh* locus, which could explain its lack of response to selection in our study. Regardless, the *Adh* results found here support earlier findings indicating that the *F* allele at the locus fosters rapid development.

CuZn-Superoxide Dismutase

One of the most consistent and significant results of the present study is the association between increased *S* allele frequencies of superoxide dismutase and increased lifespan. Given this finding, what interpretation can be placed on it? Superoxide dismutase, as a free radical scavenger, has long been regarded as a probable factor in aging according to the free radical theory of aging (Harman 1956). Superoxide dismutase (SOD) catalyzes the conversion of superoxide anion radicals to hydrogen peroxide, which is converted to water by catalase. Overexpression of SOD and catalase retards accumulation of oxidative damage of age. Flies with an extra dose of the *Sod* and *Cat* genes exhibit a more gradual increase in mitochondrial hydrogen peroxide generation and slower accumulation of 8-hydroxydeoxyguanosine, a product of DNA oxidation (Sohal et al. 1995). They also had less protein oxidative damage according to two different measures: protein carbonyl levels (Orr et al. 1994) and rate of decrease in glucose-6-phosphate dehydrogenase activity oxidation (Sohal et al. 1995). Bovine as well as *Drosophila* SOD can make *D. melanogaster* more resistant to oxidative stress (Fleming et al. 1992). The flies treated with a bovine *Sod* gene averaged SOD expression 32.5% greater than normal. These transgenic flies had a longer lifespan than the controls; the difference was slight but significant (Reveillaud et al. 1991). For an increase in longevity it may be necessary to enhance the activity of both SOD and catalase; it was hypothesized that the toxicity of hydrogen peroxide limited SOD activity. Seto et al. (1990) reported that there was no change in oxygen metabolism or longevity when they added an additional copy of *Sod*, despite increases in *Sod* transcription and SOD activity. On the other hand, Orr et al. (1994) reported that the lifespan of transgenic flies with three copies of *Drosophila Sod* genes was extended up to one third.

Either *Sod* or a locus linked to it indeed responds to selection for postponed reproduction. If the favored locus is not *Sod* itself, it must be closely linked; Tyler et al. (1993) eliminated the possibility of *Sod* falling within an inversion polymorphism by showing a general absence of inversion polymorphisms in the Ives *D. melanogaster*. The S allozyme, which corresponds to the allele found at elevated frequencies in the fly lines with postponed aging, is known to be more active than the average F allozyme (Lee et al. 1981). In any case, the transgenic studies strongly indicate the causal involvement of more active *Sod* alleles in increasing lifespan. The results from selection for this locus, then, corroborate the transgenic findings.

The *Sod F* allele actually consists of a series of haplotypes indistinguishable by starch gel electrophoresis, while the *S* allele consists of an invariant sequence that differs from one of the Fast haplotypes at only the nucleotide site that encodes the amino acid difference between the two electrophoretically distinguishable allozymes (Hudson et al. 1994). While the invariance of the *S* allele supports selection, both the Slow haplotype and the Fast haplotype most like it are abundant. Therefore, the variant favored by selection is probably not the amino acid difference that changes electrophoretic mobility, such that the differentiation between B and O lines presented here is an underestimate. The lack of identity between the electrophoretic polymorphism and the selected variant may explain the failure of Tyler et al. (1993) to find a significant effect of *Sod* genotype on longevity.

Tests for linkage disequilibrium could also be confounded by the existence of a selectively favored *Sod* allele that includes the *S* allele but is not delimited by it. While no linkage disequilibrium between the *Sod* and *Pgm* loci was detected, it could have been obscured by our inability to distinguish between subtypes of the *Sod F* allele. The number of gametes carrying *Pgm S* and *Sod S*, the two alleles hypothesized to be selectively favored, is an underestimate of the number of gametes carrying the *Pgm S* allele and the selectively favored *Sod* allele. Smit-McBride et al. (1988) showed that linkage disequilibrium between *Pgm* and *Sod* is possible. They detected it in two of six experimental populations studied but in neither of two natural populations studied.

Phosphoglucomutase

While the possibility of linkage to an unidentified selected locus exists, phosphoglucomutase is likely to be important for life-history parameters because of its key role in metabolism. Gu (1991) associated the polymorphism at *Pgm* with variation in flight capacity of the moth *Epiphyas postvittana*. Goulson (1993) found that the genotype at *Pgm* affected the length of time for which individual *Maniola jurtina* butterflies could fly continuously. Watt et al. (1985) discovered a *Pgm* heterozygote advantage in male mating success. Carfagna et al. (1980) provided evidence that the *Pgm* polymorphism has selective significance in *D. melanogaster*. They started two populations at each of three different frequencies of the "Pgm^B" allele, 0.02, 0.50, and 0.98. All six populations converged to 10% to 15% "Pgm^B."

In our study, the phosphoglucomutase locus responded to demographic selection with electrophoretically detectable divergence of substantial magnitude. The S allozyme appears strongly favored by selection for postponed reproduction, and the longevity of a fly stock can be predicted from the frequency of the *Pgm S* allele (Fig. 2). Furthermore, the frequency of the *S* allele averages less than 0.01 in four of five of the B lines, making a substantial increase without direct selection dependent on strong linkage disequilibrium. Selection intensity must exceed recombination fraction in order for linkage disequilibrium to increase, and an increase in linkage disequilibrium would be expected primarily when the favored allele is rare (Asmussen and Clegg 1981). In order for the *Pgm S* allele to increase from an average frequency of 0.01 to an average frequency of 0.51, a high proportion of the favored alleles would have to be linked to the *Pgm S* allele. If every copy of the favored allele can be traced back to a single copy, 100% of the copies of the *Pgm S* allele should be linked to the favored allele. This could happen only if the mutant later to be favored by selection for postponed aging were linked to the very rare *Pgm S* allele in every newly formed replicate O line. This is particularly unlikely given the large sizes and long establishment of these lines, and thus opportunities for recombination.

The *Pgm M* allele has become fixed in every line selected for accelerated development. Not only can it be argued that selection for accelerated development favors the *M* allele, relative to the *S* allele, but two pieces of evidence support recessive expression of the M allele. Firstly, its presence in the O lines may signify that *M/S* heterozygotes have fitness under selection for postponed reproduction comparable to that of *S/S* homozygotes. Secondly, its consistency of fixation over all the accelerated development lines might require more generations for a dominant allele. To corroborate this hypothesis, four sets of fifty simulations were performed using the genetic drift and selection option of the Populus software developed by Don Alstad of the University of Minnesota. All four sets of simulations assumed a selection coefficient of 0.05 and a starting frequency of the *M* allele of 0.75. These values were chosen based on the assumption that the real values, which are unknown, should be less than these values, making the estimates conservative. I set the population size at either 500 or 300 and assumed the *M* allele to be either completely dominant or completely recessive. For a population size of 500, fixation required 311 ± 158 generations when the allele was assumed to be dominant. When the allele was assumed to be recessive, fixation required 85 ± 30 generations. For a population size of 300, a dominant allele required 286 ± 200 generations for fixation, while a recessive allele required 77 ± 26 generations. Since the accelerated development lines were formed about one hundred generations ago, dominance of the *M* allele is highly unlikely, in the absence of a selection coefficient substantially greater than 5%. Codominance is possible, as the average fixation time with codominance is on the order of 135 ± 50 generations. However, the consistency of fixation across five populations argues against the likelihood of codominance explaining fixation within one-hundred generations.

Since genetic drift appears to be unable to account for the fixation of the *Pgm M* allele in all accelerated development populations, this allele must be favorable rather than neutral in a short generation-time environment. This *Pgm* allele or an allele at a linked locus appears to be subject to antagonistic pleiotropy, since it is favorable for reproduction at early ages but relatively deleterious for reproduction at late ages. The *Pgm S* allele, or a linked allele, is favorable at late ages but kept at a low frequency in the absence of selection for postponed reproduction. It appears to be relatively deleterious in flies reproduced before late ages, as shown by the CO populations. The COs were derived from the Os, and are maintained on a shorter generation time. The frequency of the *Pgm S* allele has fallen substantially in these populations (Table 1).

The dominance of the *Pgm S* allele relative to the *Pgm M* allele may slow its progress toward fixation in populations selected for postponed aging. Moreover, heterozygotes may have a greater probability than homozygotes of reproducing late in life, such that the *Pgm S* allele is prevented from reaching fixation despite its correlation with longevity. With regard to the *Sod* locus, Peng et al. (1991) found that *F/S* heterozygotes were more productive than homozygotes. While these flies were reproduced early in life only, the result is consistent with our findings of relatively modest increases in the *Sod S* allele.

Dominance of the alleles that responded to selection (*Sod S* and *Pgm S*) over the alleles that would affect lifespan in most flies (*Sod F* and *Pgm M*) is consistent with the findings of Buck et al. (1993) who studied the localization and regulation of longevity determinant genes in a selected strain of *D. melanogaster*. They showed that mean longevity is determined in large part by recessive genes on the third chromosome. Luckinbill et al. (1988) had used chromosome substitution lines to reveal that the third chromosome accounts for 66% to 72% of the observed variation in female longevity. *Sod* and *Pgm* are indeed located on the third chromosome. Buck et al. (1993) further showed that the second chromosome can inhibit the third chromosome in its longevity determinant capacity, and that the first chromosome can negate this action of the second chromosome. Graf and Ayala (1986) have identified a second chromosome modifer of *Sod*, which further emphasizes the consistency of the *Sod* locus with the expression pattern described by Buck et al.

We have identified two loci which contribute to longevity in *D. melanogaster*: the locus encoding phosphoglucomutase and the locus encoding CuZn-superoxide dismutase. Both of the loci may be subject to antagonistic pleiotropy, as selection for the longevity-opposing life history of accelerated development appears to render the longevity-associated alleles deleterious. Currently we are expanding our study to other species of *Drosophila* to determine whether the same or different loci are associated with postponed aging in other species.

Acknowledgments

Many hours of technical assistance were provided by M. Arias, D. T. Frank, S. J. Mee, G. M. Vazquez, and C. P. You. This research was supported by the estate of Robert H. Tyler and by US-PHS grant AG09970 to M.R.R. Helpful comments on previous versions of the manuscript were provided by R. R. Hudson, L. D. Mueller, D. N. Reznick, and three anonymous reviewers.

Literature Cited

Asmussen, M. A. and M. T. Clegg. 1981. Dynamics of the linkage disequilibrium function under model of gene frequency hitchhiking. Genetics 99:337–356.

Barnes, P. T., B. Holland, and V. Courreges. 1989. Genotype-by-environment and epistatic interactions in *Drosophila melanogaster*: the effects of glycerol phosphate dehydrogenase allozymes, genetic background, and rearing temperature on larval development time and viability. Genetics 122:859–868.

Bijlsma, R., and J. W. Kerver. 1983. The effect of DDT on the polymorphism at the *Glpd* and *Pgd* loci in *Drosophila melanogaster*. Genetics 103:447–464.

Buck, S., R. A. Wells, S. P. Dudas, G. T. Baker III, and R. Arking. 1993. Chromosomal localization and regulation of the longevity determinant genes in a selected strain of *Drosophila melanogaster*. Heredity 71:11–22.

Carfagna, M., L. Fucci, L. Gaudio, G. Pontecorvo, and R. Rubino. 1980. Adaptive value of PGM polymorphism in laboratory populations of *Drosophila melanogaster*. Genet. Res. 36:256–276.

Cavener, D. R. 1983. The response of enzyme polymorphisms to developmental rate selection in *Drosophila melanogaster*. Genetics 105:105–113.

Cavener, D. R., and M. T. Clegg. 1981. Multigenic response to ehanol in *Drosophila melanogaster*. Evolution 35:1–10

Charlesworth, B. 1980. Evolution in age-structured populations. Cambridge Univ. Press, Cambridge.

Chippindale, A. K. 1994. The evolution of trade-offs in vomplex life cycles: the *Drosophila* model. Ph.D. diss. Univ. of California, Irvine.

Chippindale, A. K., D. T. Hoang, P. M. Service, and M. R. Rose. 1994. The evolution of development in *Drosophila melanogaster* selected for postponed senescence. Evolution 48:1880–1899.

Fleming, J. E., I. Reveillaud, and A. Niedzwiecki. 1992. Role of oxidative stress in *Drosophila* aging. Mut. Res. 275:267–279.

FLEMING, J. E., G. S. SPICER, R. C. GARRISON, AND M. R. ROSE. 1993. Two-dimensional protein electrophoretic analysis of postponed aging in *Drosophila melanogaster*. Genetica 91:183–198.

GOULSON, D. 1993. Allozyme variation in the butterfly, *Maniola jurtina* (Lepidoptera: Satyrinae) (L.): evidence for selection. Heredity 71:386–393.

GRAF, J., AND F. AYALA. 1986. Genetic variation for superoxide dismutase level in *Drosophila melanogaster*. Biochem. Genet. 24:153–168.

GRAVES, J. L. AND M. R. ROSE. 1990. Flight duration in *Drosophila melanogaster* selected for postponed senescence. Pp. 59–65 *in* D. E. Harrison, ed. Genetic effects on aging II. Telford Press, Inc., Caldwell, NJ.

GU, H. 1991. Electrophoretic variation at flight related enzyme loci and its possible association with flight capacity in *Epiphyas postvittana*. Biochem. Genet. 29:345–354.

HALDANE, J. B. S. 1941. New paths in genetics. Allen and Unwin, London.

HAMILTON, W. D. 1966. The moulding of senescence by natural selection. J. Theor. Biol. 12:12–45.

HARMAN, D. 1956. Aging: a theory based on free radical and radiation chemistry. J. Gerontol. 11:298–300.

HARTL, D. L., AND A. F. CLARK. 1989. Princilpes of population genetics. Sinauer, Sunderland, MA.

HASTINGS, I. M. 1992. The population genetics of alleles afecting enzyme activity. J. Theor. Biol. 157:305–316.

HEINSTRA, P. W. H., W. SCHARLOO, AND G. E. W. THORIG. 1987. Physiological significance of the alcohol dehydrogenase polymorphism in larvae of *Drosophila*. Genetics 117:75–84.

HICKEY, D. A. 1977. Selection for amylase allozymes in *Drosophila melanogaster*. Evolution 31:800–804.

HUDSON R. R., K. BAILEY, D. SKARECKY, J. KWIATOWSKI, AND F. J. AYALA. 1994. Evidence for positive selection in the superoxide dismutase (*Sod*) region of *Drosophila melanogaster*. Genetics 136:1329–1340.

HUTCHINSON, E. W., AND M. R. ROSE. 1990. Quantitative genetics of postponed senescence in *Drosophila melanogaster*. I. Analysis of outbred populations. Genetics 127:719–727.

IVES, P. T. 1970. Further genetic studies of the South Amherst population of *Drosophila melanogaster*. Evolution 24:507–518.

IZQUIERDO, J. I., AND J. RUBIO. 1989. Allozyme polymorphism at the α*Gpdh* and *Adh* loci and fitness in *Drosophila melanogaster*. Heredity 63:343–352.

KNIBB, W. R., J. G. OAKESHOTT, AND S. R. WILSON. 1987. Chromosome inversion polymorphisms in *Drosophila melanogaster*. IV. Inversion and *Adh* allele frequency changes under selection for different development times. Heredity 59:95–104.

LEE, Y. M., H. P. MISRA, AND F. J. AYALA. 1981. Superoxide dismutase in *Drosophila melanogaster*: biochemical and structural characteristics of allozyme variants. Proc. Nat. Acad. Sci. USA 78:7052–7055.

LUCKINBILL, L. S., R. ARKING, M. J. CLARE, W. C. CIROCCO, AND S. A. BUCK. 1984. Selection for delayed senescence in *Drosophila melanogaster*. Evolution 38:996–1003.

LUCKINBILL, L. S., J. L. GRAVES, A. H. REED, AND S. KOETSAWANG. 1988. Localizing genes that defer senescence in *Drosophila melanogaster*. Heredity 60:367–374.

LUCKINBILL, L. S., T. A. GRUDZIEN, S. RHINE, AND G. WEISMAN. 1989. The genetic basis of adaptation to selection for longevity in *Drosophila melanogaster*. Evol. Ecol. 2:85–94.

MANCHENKO, G. P. 1994. Handbook of detection of enzymes on electrophoretic gels. CRC Press, Inc., Boca Ratón, FL.

MCKECHNIE, S. W., AND B. W. GEER. 1988. The epistasis of *Adh* and *Gpdh* allozymes and variation in the ethanol tolerance of *Drosophila melanogaster* larvae. Genet. Res. 52:179–184.

MEDAWAR, P. B. 1946. Old age and natural death. Mod. Quart. 1: 30–56.

———. 1952. An unsolved problem of biology. H. K. Lewis, London.

OAKESHOTT, J. G. 1979. Selection affecting enzyme pomorphisms in laboratory populations of *Drosophila melanogaster*. Oecologia 143:341–354.

ORR, W. C., AND R. S. SOHAL. 1994. Extension of life-span by overexpression of superoxide dismutase and catalase in *Drosophila melanogaster*. Science 263:1128–1130.

PASTEUR, N., G. PASTEUR, F. BONHOMME, J. CATALAN, AND J. BRITTON-DAVIDIAN. 1988. Practical isozyme genetics. Ellis Horwood Ltd., Chichester, UK.

PENG, T. X., A. MOYA, AND F. J. AYALA. 1991. Two modes of balancing selection in *Drosophila melanogaster*: overcompensation and overdominance. Genetics 128:381–391.

REVEILLAUD, I., A. NIEDZWIECKI, AND J. E. FLEMING. 1991. Expression of bovine superoxide dismutase in *Drosophila melanogaster* augments resistance to oxidative stress. Mol. Cell. Biol. 11:632–640.

REVEILLAUD, I., J. PHILLIPS, B. DUYF, A. HILLIKER, A. KONGPACHITH, AND J. E. FLEMING. 1994. Phenotypic rescue by a bovine transgene in CuZn-superoxide dismutase-null mutant of *Drosophila melanogaster*. Mol. Cell. Biol. 14:1302–1307.

ROSE, M. R. 1984. Laboratory evolution of postponed senescence in *Drosophila melanogaster*. Evolution 38:1004–1010.

———. 1985. Life-history evolution with antagonistic pleiotropy and overlapping generations. Theor. Popul. Biol. 28:342–358.

———. 1991. Evolutionary biology of aging. Oxford Univ. Press, Oxford.

ROSE, M. R., AND B. CHARLESWORTH. 1981. Genetics of life history in *Drosophila melanogaster*. II. Exploratory selection experiments. Genetics 97:187–196.

ROSE, M. R., L. N. VU, S. U. PARK, AND J. L. GRAVES JR. 1992. Selection on stress resistance increases longevity in *Drosophila melanogaster*. Exp. Gerontol. 27:241–250.

SERVICE, P. M., E. W. HUTCHINSON, M. D. MACKINLEY AND M. R. ROSE. 1985. Resistance to environmental stress in *Drosophila melanogaster* selected for postponed senescence. Physiol. Zool. 58:380–389.

SETO, N. O. L., S. HAYASHI, AND G. M. TENER. 1990. Overexpression of Cu, Zn SOD in *Drosophila* does not affect life-span. Proc. Nat. Acad. Sci. USA 87:4270–4274.

SMIT-MCBRIDE, Z., A. MOYA, AND F. J. AYALA. 1988. Linkage disequilibrium in natural and experimental populations of *Drosophila melanogaster*. Genetics 120:1043–1051.

SOHAL, R. S., A. AGARWAL, S. AGARWAL, AND W. C. ORR. 1995. Simultaneous overexpression of Cu and Zn containing superoxide dismutase and catalase retards age-related oxidative damage and increases metabolic potential in *Drosophila melanogaster*. J. Biol. Chem. 270:15671–15674.

TYLER, R. H., H. BRAR, M. SINGH, A. LATORRE, J. L. GRAVES, L. D. MULLER, M. R. ROSE, AND F. AYALA. 1993. The effect of superoxide dismutase alleles on aging in *Drosophila*. Genetica 91:141–149.

WATT, W. B. 1992. Eggs, enzymes, and evolution-natural genetc variants change insect fecundity. Proc. Nat. Acad. Sci. USA 89: 10608–10612.

———. 1994. Allozymes in evolutionary genetics: self-imposed burden or extraordinary tool? Genetics 136:11–16.

WATT, W. B., P. A. CARTER, AND S. M. BLOWER. 1985. Adaptation at specific loci. IV. Differential mating success among glycolytic allozyme genotypes of *Colias* butterflies. Genetics 109:157–175.

WATTIAUX, J. M. 1968a. Parental age effects in *Drosophila pseudoobscura*. Am. Nat. 117:1035–1039.

———. 1968. Cumulative parental age effects in *Drosophila subobscura*. Evolution 22:406–421.

WEIR, B. S. 1990. Genetic data analysis. Sinauer, Sunderland, MA.

WILKINSON, L. 1990. SYSTAT: the system for statistics. SYSTAT, Evanston, IL.

Corresponding Editor: R. DeSalle

APPENDIX

Tukey-HSD tests on population type.

Alcohol dehydrogenase F allele frequencies			
	Accelerated	Control	Parental
Accelerated	p = 1.000		
Control	p = 0.013	p = 1.000	
Parental	p = 0.001	p = 0.532	p = 1.000
α-glycerol-3-phosphate dehydrogenase F allele frequencies			
	Accelerated	Control	Parental
Accelerated	p = 1.000		
Control	p = 0.100	p = 1.000	
Parental	p = 0.035	p = 0.866	p = 1.000
Cu, Zn-Superoxide dismutase S allele frequencies			
	Accelerated	Control	Parental
Accelerated	p = 1.000		
Control	p = 0.066	p = 1.000	
Parental	p = 0.031	p = 0.931	p = 1.000
Phosphoglucomutase S allele frequencies			
	Accelerated	Control	Parental
Accelerated	p = 1.000		
Control	p = 0.001	p = 1.000	
Parental	p = 0.000	p = 0.051	p = 1.000

Journal of Gerontology: BIOLOGICAL SCIENCES
2000, Vol. 55A, No. 11, B552–B559

Increased *hsp22* RNA Levels in *Drosophila* Lines Genetically Selected for Increased Longevity

Raj Kurapati,[1] Hardip Brar Passananti,[2] Michael R. Rose,[2] and John Tower[1]

[1]Department of Biological Sciences, University of Southern California, Los Angeles.
[2]Department of Ecology and Evolutionary Biology, University of California, Irvine.

RNAs for the small heat shock protein (hsp) genes *hsp22* and *hsp23* are induced during *Drosophila* aging, suggesting that these genes might have specific functions at late ages. To determine if *hsp22* and *hsp23* gene expression might correlate with life span, RNA levels for these and additional genes were analyzed throughout the adult life span in a set of five outbred "O" lines, which have been genetically selected for increased longevity, and in five matched control "B" lines. Control ribosomal protein genes *rp49* and *AP3/RpPO* RNA levels were similar in O and B lines. In contrast, *hsp22* RNA levels were twofold–tenfold higher in all five O lines relative to all five B lines, while *hsp23* exhibited a smaller but significant increase. Thus increased *hsp22* and *hsp23* RNA levels correlate with the increased life span and increased stress resistance of the genetically selected O lines.

AMONG the proteins induced in response to heat stress in eukaryotes are the "small" heat shock proteins (shsps), with molecular masses ranging from 15–30 kD (1). The shsps in eukaryotes are related by a conserved region of the protein called the α-crystallin domain. Mammalian shsps include αA-crystallin, αB-crystallin, and hsp27. There are four predominant shsps in *Drosophila* induced in response to heat stress, called hsp22, hsp23, hsp26, and hsp27. These shsps also have characteristic developmental expression patterns and are regulated by the hormone ecdysone. The function of the *Drosophila* hsp22 and hsp23 proteins has not been directly examined. However, *Drosophila* hsp27 and human hsp27 and αB-crystallin have each been found to confer increased resistance to heat and oxidative stress when expressed in cultured cells (1–5). Human hsp27 and bovine αA- and αB-crystallins have been shown to act in vitro as molecular chaperones, in that they can suppress thermally induced aggregation of other proteins and can facilitate refolding of denatured proteins (6–8). Taken together, the data suggest that one role of the shsp family may be the prevention of protein aggregation and denaturation during thermal and oxidative stress. The hsp70 heat shock protein is also induced in response to heat stress. The well-characterized hsp70 family proteins can prevent protein aggregation, facilitate protein refolding, and facilitate entry of damaged proteins into proteolytic pathways (9).

Drosophila exhibit a characteristic pattern of heat shock gene expression during aging, involving *hsp70*, *hsp22*, and *hsp23*. hsp70 protein is induced approximately 5-fold, specifically in thoracic muscle tissue, and may be a response to oxidative damage (10). Induction of *hsp70* during aging appears to involve both transcriptional and post-transcriptional regulation and requires functional heat shock response elements (HSEs) in the promoter region (11). *hsp22* RNA and, to a lesser extent, *hsp23* RNA were found to be induced during normal aging throughout the fly (10). *hsp22* expression during aging has been analyzed in detail (12). *hsp22* is induced in all tissues with regulation at both the transcriptional and post-transcriptional levels. *hsp22* RNA is induced up to 60-fold and hsp22 protein more than 150-fold. Like *hsp70*, transcriptional upregulation of *hsp22* during aging requires functional HSEs in the promoter. However, the greater magnitude of *hsp22* induction suggested the possible involvement of additional element(s). Aging also affects the expression of hsps in response to a heat stress, with old flies exhibiting a prolonged response and greater levels of protein turnover (13,14).

Aging is associated with the accumulation of abnormal and/or malfolded proteins and oxidatively damaged proteins in many organisms, including nematodes, flies, and mammals (15–17). This suggested the hypothesis that induction of hsp genes during *Drosophila* aging might have beneficial effects and help prolong the functional life span of the fly (10,12). At least two lines of evidence support this hypothesis and suggest a function for hsps in regulating *Drosophila* aging and life span. First, mild heat stress induces multiple hsps and increases resistance to subsequent stresses, an effect known as induced thermotolerance (18). Such mild heat stresses can also cause small increases in life span (19–21). Second, mild heat stress of *Drosphila* transgenic for extra copies of the *hsp70* gene exhibited increased hsp70 expression and an increase in life span several percent greater than controls (22,23).

The selection of genetically heterogeneous *Drosophila* lines over many generations for late-life fecundity results in increased longevity (24–27). The long-lived lines also exhibit a number of correlated phenotypes, including increased resistance to stresses such as starvation, desiccation, and oxidative stress (28–33). RNA levels for several antioxidant genes have been analyzed in a single long-lived line which exhibits increased oxidative stress resistance relative to its matched control line. Expression of Cu/Zn-superoxide dismutase, catalase, and xanthine dehydrogenase were found to be increased (34).

The studies presented here utilize a specific set of five genetically selected long-lived (O) lines, which have been extensively characterized and exhibit up to twice the life span of their five paired control (B) lines (35). The analysis of five independent sets of lines allows the identification of characteristics that are consistently altered by selection. To determine if selection of the long-lived O lines affected the aging-related expression of the *hsp22* and *hsp23* genes, these genes' RNA levels were quantitated throughout the adult life span of the O and B lines. Although there was considerable variability in expression of these genes between lines and time points, *hsp22* expression was consistently found to be 2–10-fold higher in each of the five long-lived O lines relative to the B control lines. Smaller, but significant, increases in expression were also observed for the *hsp23* gene.

Methods

Drosophila Stocks and Culture

O and B lines were cultured as previously described (35). O lines were taken off of selection for three generations prior to analysis. To obtain flies of defined ages, newly eclosed males were maintained at 25°C at a density of 40 per vial and were transferred to fresh vials every 4 days to prevent growth of bacteria or mold.

Northern Blot Analysis

Northern blot analysis was performed using standard methods (36). Briefly, for each age time point, 25 males were dissected into head, thorax, and abdomen. Body segments were homogenized in guanidine hydrochloride extraction buffer and extracted with phenol/chloroform. Nucleic acids were precipitated with ethanol, and the pellet was resuspended directly into deionized formamide/formaldehyde gel loading buffer. The total RNA was fractionated using 1% agarose/formaldehyde gels and transferred to GeneScreen membranes (Dupont, Boston, MA/NEN). Membranes were hybridized successively with radiolabeled gene-specific probes: *hsp22*, *hsp23*, *hsp26*, and *hsp27* (10,37); ribosomal protein 49 gene *rp49* (38); ribosomal protein/apurinic endonuclease gene *AP3*, also called *RpPO* (39–40); and the myosin heavy chain gene *Mhc* (41). Northern blot results were quantitated by phosphoimager analysis and/or autoradiography and densitometry. Densitometry was performed on autoradiographs exposed within the linear range of the film. Darker exposures are presented in the figures to allow visualization of signal from B line samples. All data are presented in relative band intensity units normalized to the signal with the *rp49* probe, as a control for loading.

Statistical Analyses

Five long-lived (O) and 5 control (B) populations were used in this study. Each time point for each population represented the average of 50 flies, as RNA was isolated from 50 flies for each time point and approximately one fifth of this RNA was used for each Northern blot gel lane. Statistical analyses were then based on the averages of the five replicate O populations versus the averages of the five replicate B populations.

In the statistical analyses the dependent variable was always the character under study (e.g., *hsp22*, *hsp23*, *Ap3*), and the factors were selection treatment (B vs O), age, and region (head, thorax, abdomen). Results of the experiment shown in Table 1 and Figure 1, involving data from day 5 only, were analyzed using a two-way factorial analysis of variance (ANOVA), with selection treatment and region as factors. These same data were also analyzed by region using a one-way ANOVA with selection treatment as the only factor.

Results of the experiments shown in Figures 2 and 3, analysis of RNA levels throughout the adult life span, were analyzed using a three-way factorial ANOVA with selection treatment, age, and region as factors. All ANOVAs were performed using Statview statistical analysis software (Abacus Concepts, Berkeley, CA). Results are presented in the legends to Figures 2 and 3 and Table 1.

To address possible differences in variance between groups of data, all statistical analyses were repeated using log-transformed data, and the results are presented in brackets in the legends to Table 1 and Figure 3. The significance of the results was unchanged, with two minor exceptions that tend to further support the general conclusions of the study. In Table 1, the difference in Myosin expression between O and B lines in thorax becomes insignificant, and in Figure 3, the difference in *hsp23* expression between O and B lines becomes significant.

Results

Northern blot analysis was used to compare RNA expression in long-lived O lines and short-lived B control lines. Each of the five O lines and five B lines was taken off of selection for three generations to avoid any possible direct effects of the culture conditions on gene expression. Adult male flies of each line were maintained at 25°C until 5 days, and then 25 flies were dissected into head, thorax, and abdomen. RNA was isolated from the dissected body segments and analyzed by Northern blots. The Northern blot of thorax RNA is presented (Figure 1). Each Northern blot was hybridized with a probe specific for the constitutively expressed control gene ribosomal protein 49 (*rp49*) as a loading control. *rp49* RNA levels are constant with age relative to several other housekeeping-type gene RNAs (10) and rRNA (34). The blots were hybridized with a probe specific for *hsp22*, and as a control for the specificity of *hsp22* expression, expression of the non-hsp genes ribosomal protein/apurinic endonuclease *AP3* and the myosin heavy chain *Mhc* were also analyzed. Results were quantitated by densitometry, normalized to *rp49* expression, and are presented in Table 1. The largest differences in expression between the five O populations and the five B populations were for *hsp22*. *hsp22* expression was found to be increased by an average 3.7-fold in thorax of O lines relative to B line controls (Figure 1 and Table 1). *hsp22* was also increased in head in O lines by an average 4.14-fold (Table 1). *hsp22* RNA levels were similar in O and B line abdomen in this

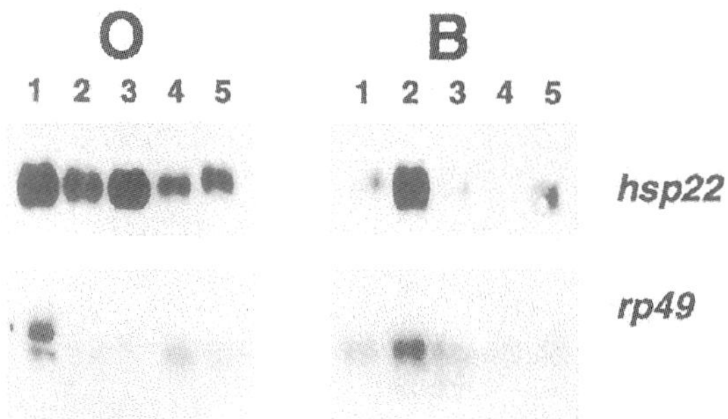

Figure 1. Northern blot analysis of 5-day-old O and B line fly thoraces. Total RNA was isolated from 25 thoraces at 5 days of age for each of the O1-5 and B1-5 lines, as indicated. RNA was transferred to Northern blot and hybridized successively with probes specific for *hsp22* and the *rp49* loading control, as indicated. Quantitation of data is presented in Table 1.

experiment. Smaller differences were observed for expression of the control genes. On average, *AP3* expression was increased by 85% in O line heads and decreased by 8% in O line abdomens. *Mhc* was increased by 57% in O line thorax. However, as shown below, the small differences in control gene expression were not reproduced in a larger experiment, whereas dramatically increased expression of *hsp22* in O line flies was consistently observed.

To confirm the upregulation of *hsp22* RNA expression in the O line flies, RNA expression was analyzed throughout the life span of independently cultured cohorts of the O and B lines. Each of the five O lines and five B lines was again taken off of selection for three generations, and adult male flies of each line were maintained at 25°C until defined ages. Then at various time points, 25 flies were dissected into head, thorax, and abdomen, and RNA was isolated from the dissected body segments and analyzed by Northern blot.

Blots for each O line and its matched B control line (O1 and B1, O2 and B2, etc.) were always hybridized together in the same container with a given probe so that the intensity of signal could be directly compared between the O and B lines of each set. Representative Northern blots are shown in Figure 2. Thorax RNA was isolated from flies of O and B lines 1–5. Each blot was hybridized successively with probes specific for *hsp22* and *rp49*. For each set of O and B lines, *rp49* expression was found to be approximately equal between the O and B lines (Figure 2). In O line flies, *hsp22* expression was 2–10-fold higher than in the B line flies throughout the period from about 2–40 days. The life span of the B line flies corresponds roughly to this period. *hsp22* RNA levels in the O line flies subsequently decreased between days 50–70.

Northern blot analysis was also performed for the head and abdomen body segments. The results for the head, thorax, and abdomen were quantitated by densitometry, normalized relative to *rp49* RNA levels, and are plotted in Figure 3. The greatest differences in expression were observed in the thorax, where *hsp22* RNA levels were 2–10-fold higher in 27 out of 30 (90%) of the time points for which both strains had detectable expression. A similar, but somewhat smaller difference was observed in the head. While at a subset of time points *hsp22* RNA levels were similar between O and B lines in the head, at the majority of time points *hsp22* RNA levels were 2–10-fold higher in O lines. Finally, the smallest difference was observed in the abdomen. Increased *hsp22* RNA levels in the O line flies' abdomens were observed at many time points for each set of lines except O2/B2. However, the increases were generally smaller and observed at fewer time points than for the thorax or head. A three-way factorial ANOVA demonstrates that *hsp22* expression is significantly increased in O lines

Table 1. Quantitation of RNA Levels in O and B Line Heads, Thoraces and Abdomens, at 5 Days of Age

	Population													
	O1	O2	O3	O4	O5	Oavg (*SD*)	B1	B2	B3	B4	B5	Bavg (*SD*)	O/B	*p*
Head														
AP3	236	227	231	215	189	219.6 (18.78)	145	90	130	126	102	118.6 (22.22)	1.85	<.0001 [.0002]
Mhc	45	42	39	72	45	48.6 (13.32)	60	17	34	73	71	51.4 (25.0)	0.94	.8306 [.8829]
hsp22	259	212	64	188	155	175.6 (73.0)	10	49	87	30	36	42.4 (28.6)	4.14	.0052 [.0072]
Thorax														
AP3	139	130	111	124	143	129.4 (12.7)	125	111	123	139	122	124.0 (10.0)	1.04	.4765 [.5237]
Mhc	170	155	176	162	126	157.8 (19.4)	126	102	25	102	148	100.6 (46.4)	1.57	.0346 [.1031]
hsp22	165	286	225	58	68	160.4 (98.7)	30	122	24	9	34	43.8 (44.8)	3.7	.0428 [.0241]
Abdomen														
AP3	103	92	93	98	98	96.8 (4.44)	100	99	111	106	112	105.6 (6.02)	0.92	.0302 [.0183]
Mhc	32	30	ND	60	38	40.0 (13.8)	41	37	36	39	34	37.4 (2.7)	1.07	.6873 [.6408]
hsp22	95	108	103	115	83	100.8 (12.3)	79	107	102	88	66	88.4 (16.7)	1.14	.2192 [.2186]

Notes: Northern blot data were quantitated for *hsp22*, *Mhc*, and *AP3* gene RNA levels for head, thorax and abdomen (data of Figure 1 and additional data), normalized to *rp49* levels as loading controls, and are presented as a table of band intensity values in relative units. Two-way factorial ANOVA was used to analyze the significance of the differences in RNA levels between O and B lines for each gene combining all data: *hsp22*, $p = .0002$ [.0001]; *Mhc*, $p = .055$ [.1499]; *AP3*, $p < .0001$ [<.0001]. One-way ANOVA was also used to analyze the significance of the differences in RNA levels between O and B lines for each gene in each body segment, and *p* values are presented above. When the significance of the differences in RNA levels between O and B lines for each gene in each body segment were analyzed using one-tailed paired *t* tests, the results were the same with the exception of *hsp22* in abdomen, which was then found to be significant ($p = .0348$). Bracketed numbers are results of statistical analyses using log-transformed data. Oavg = average value for O lines; Bavg = average value for B lines; O/B = Oavg divided by Bavg; ND = not done.

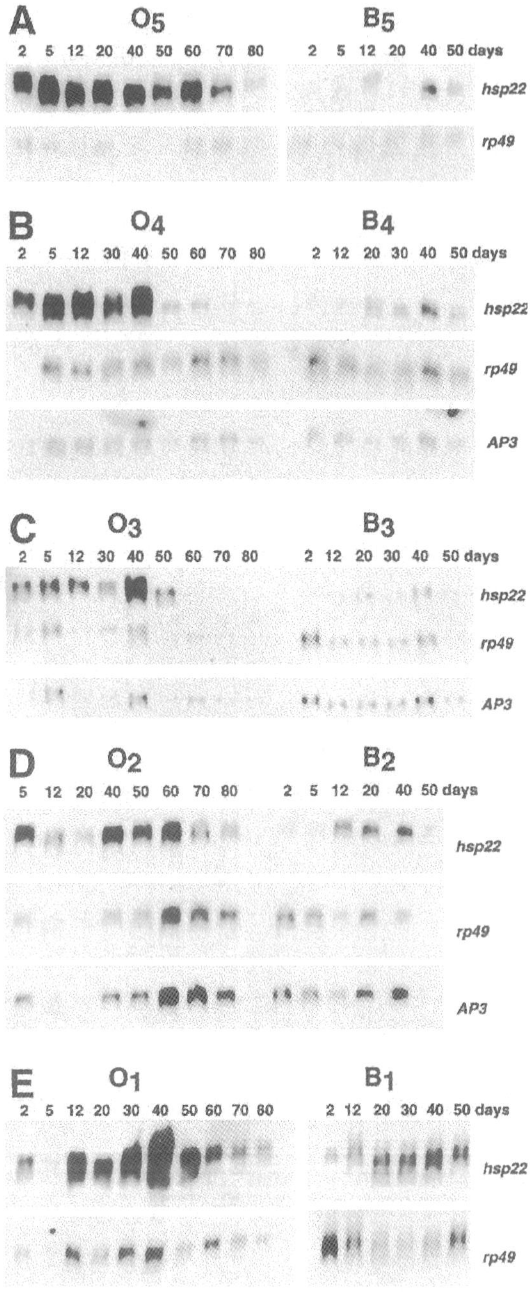

Figure 2. Northern blot analysis of O and B line thoraces throughout the adult life span. Total RNA was isolated from 25 thoraces at each of the indicated adult age time points for O and B line flies. Matched O and B samples were always hybridized together with the indicated probes so that band intensities could be directly compared between O and B. **A**, O5 and B5 thorax. **B**, O4 and B4 thorax. **C**, O3 and B3 thorax. **D**, O2 and B2 thorax. **E**, O1 and B1 thorax. The multiple *hsp22* bands in **E** were not reproducible, and appear to be an artifact of RNA isolation and/or southern conditions. Signal from the multiple bands was summed during quantitation. Quantitation of data is presented in Figure 3.

when the body segments are considered together or separately (Figure 3). Control *AP3* RNA levels were found to be equal between each of the five sets of O and B line flies, in the head, thorax and abdomen, at virtually every time point assayed. In the rare instances where *AP3* RNA levels differed between O and B line flies (c.f., Figure 3, A-04 and H-01) the differences were neither consistent across time points nor reproducible across the five lines. ANOVA demonstrates that *AP3* expression is not significantly different between the O and B lines when body segments are considered either together or separately (Figure 3).

Each Northern blot was also hybridized to a probe specific for *hsp23*. *hsp23* expression is somewhat more difficult to quantitate because its expression level in the adult fly is lower than that for *hsp22*. *hsp23* RNA expression was also found to be significantly greater in the O lines than in the B lines (Figure 3); however, the difference between O and B lines was generally smaller for *hsp23* than for *hsp22*. ANOVA demonstrates that *hsp23* expression is significantly increased in O lines when the body segments are considered together, and is significantly increased in the head and thorax when the body segments are analyzed separately (Figure 3). Finally, the Northern blots were also hybridized to probes specific for *hsp26* and *hsp27*. In general, the *hsp26* and *hsp27* RNAs were too rare to be quantitated in the O and B lines. In the cases where a Northern signal was observed, there was no reproducible difference in expression between the O and B lines (data not shown).

Discussion

The data of two independent Northern blot experiments demonstrates a consistent upregulation of *hsp22* RNA levels by 2–10-fold in five O lines relative to their B line controls. This increase was observed in the head and thorax, and in one of two experiments (the larger experiment) in the abdomen. A smaller, but significant, increase in expression was also observed for *hsp23* in head and thorax. In the O line samples, *hsp22* RNA levels are high throughout the first part of the life span, from 0 to ~50 days, which corresponds to the entire length of adult life span of the shorter-lived B lines. At around 50 days, *hsp22* expression in the O lines generally declines and remains lower for the rest of the O line life span. This dynamic pattern of expression may be occurring in each fly, with a high level of expression followed by a decrease. Alternatively, expression may be different in different subpopulations of the O line flies, and the decrease in RNA levels at later ages in the Northern blot analysis may represent a longer-lived subpopulation characterized by lower levels of *hsp22* RNA expression. Distinguishing between these two possibilities will require the development of a means to quantitate *hsp22* expression at the level of individual flies, such as by use of transgenic GFP reporter constructs (42,43).

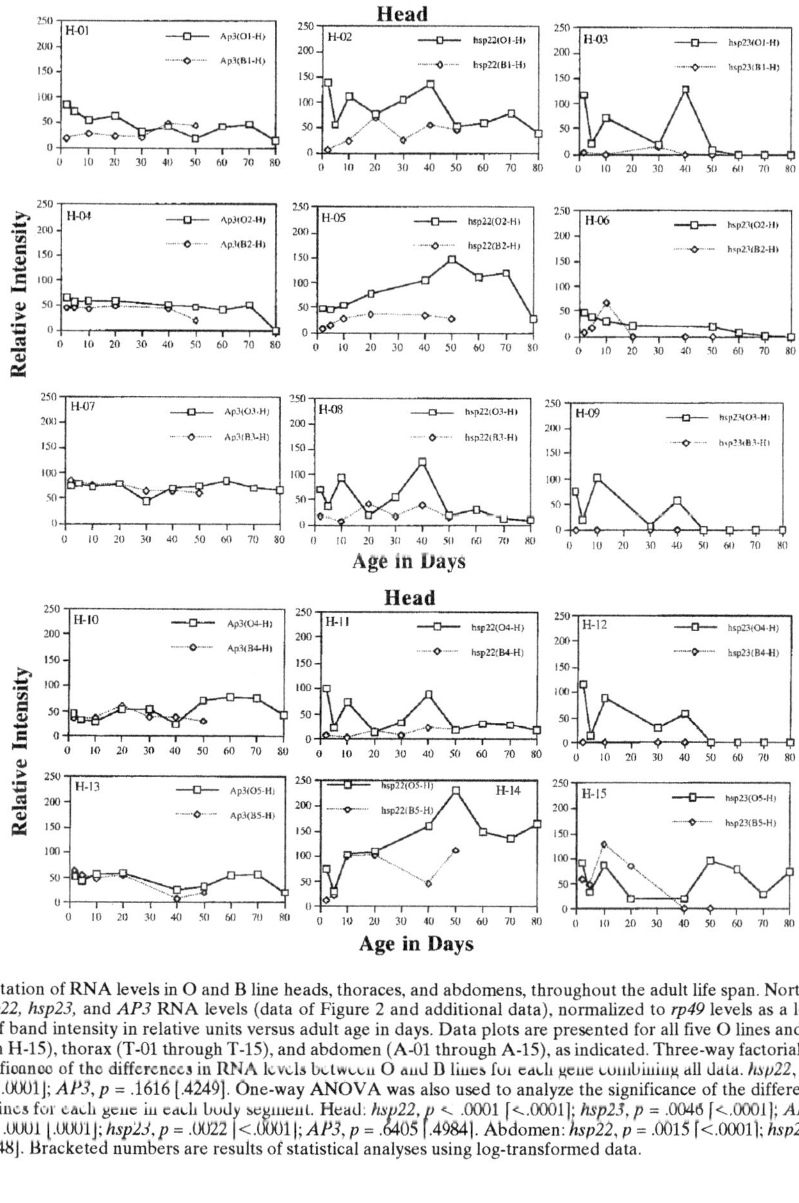

Figure 3. Quantitation of RNA levels in O and B line heads, thoraces, and abdomens, throughout the adult life span. Northern blot data were quantitated for *hsp22, hsp23,* and *AP3* RNA levels (data of Figure 2 and additional data), normalized to *rp49* levels as a loading control, and plotted. Plots are of band intensity in relative units versus adult age in days. Data plots are presented for all five O lines and all five B lines, for head (H-01 through H-15), thorax (T-01 through T-15), and abdomen (A-01 through A-15), as indicated. Three-way factorial ANOVA was used to analyze the significance of the differences in RNA levels between O and B lines for each gene combining all data. *hsp22*, $p < .0001$ [<.0001]; *hsp23*, $p = .0011$ [<.0001]; *AP3*, $p = .1616$ [.4249]. One-way ANOVA was also used to analyze the significance of the differences in RNA levels between O and B lines for each gene in each body segment. Head: *hsp22*, $p < .0001$ [<.0001]; *hsp23*, $p = .0046$ [<.0001]; *AP3*, $p = .0941$ [.055]. Thorax: *hsp22*, $p < .0001$ [.0001]; *hsp23*, $p = .0022$ [<.0001]; *AP3*, $p = .6405$ [.4984]. Abdomen: *hsp22*, $p = .0015$ [<.0001]; *hsp23*, $p = .4143$ [.0174]; *AP3*, $p = .6505$ [.7648]. Bracketed numbers are results of statistical analyses using log-transformed data.

In previous studies, two general patterns of *hsp22* RNA expression during aging have been observed (10,12). With Oregon-R wild type and *white*^{1118} strain flies cultured at low density in culture bottles, *hsp22* RNA levels were low or undetectable in young flies. Under these optimal conditions the increases in RNA levels observed during aging were large, ranging from 2.5–60-fold, depending on the body region assayed. In contrast, with the same strain of Oregon-R flies cultured at higher density using population cage protocols, the starting levels of *hsp22* RNA in young flies were higher, particularly in the head and abdomen, and the fold inductions observed during aging were small or undetectable (10). In the experiments presented in this article, the O and B line flies were cultured using higher density, population cage type protocols. Consistent with previous observations, *hsp22* RNA levels were relatively high in young animals and no significant induction with age was detectable. We conclude that these presumably more stressful culture conditions result in abundant *hsp22* RNA expression in young animals, which in turn masks the induction due to ag-

hsp22 RNA IS INCREASED IN LONG-LIVED LINES B557

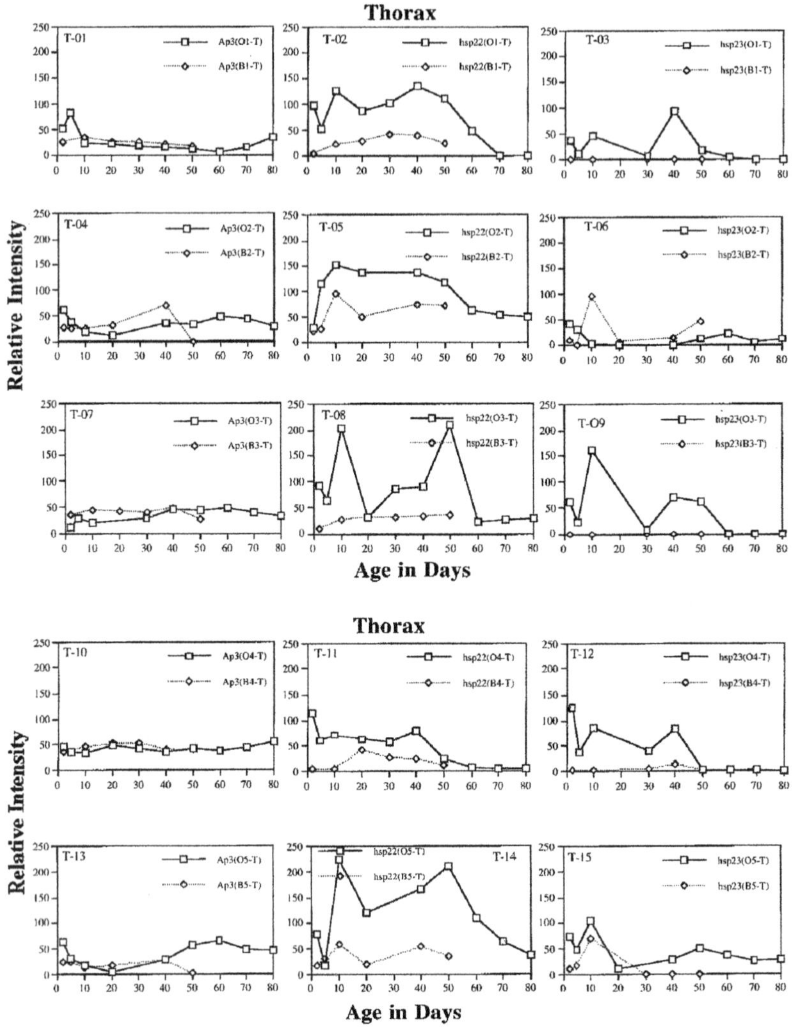

Figure 3. continued

ing. Further experiments will be required to test this idea conclusively. This situation does not alter the interpretation of the present result of increased *hsp22* RNA levels in O lines relative to B lines, because their culture conditions were identical. However, it is important to note that this difference is being observed under conditions that may be more stressful than the optimum.

hsp22 and, to a lesser extent, *hsp23* are induced at the RNA level during *Drosophila* aging under the appropriate conditions (10,12). Other hsps, such as *hsp26*, *hsp27*, and *hsp70* do not exhibit this induction at the RNA level or exhibit markedly smaller inductions. These results suggest a specialized mechanism or pathway for preferential induction of *hsp22* and *hsp23* RNAs during aging. The fact that it is *hsp22* and *hsp23* that are upregulated in the O line flies suggests that it may be this same aging-associated mechanism or pathway that is upregulated in the O line flies relative to the B lines.

Selection for increased life span in the O line flies is associated with a correlated phenotype of increased stress resistance. In particular, O line flies are more resistant to stresses such as desiccation, starvation and forced flight (28,32). Increased stress resistance is also associated with increased life span in other organisms. For example, specific mutations in genes of the dauer formation pathway in *C. elegans* increase both life span and resistance to stresses such as heat stress, oxidative stress, and ultraviolet irradiation (44–48). A problem afflicting all of these experiments, including

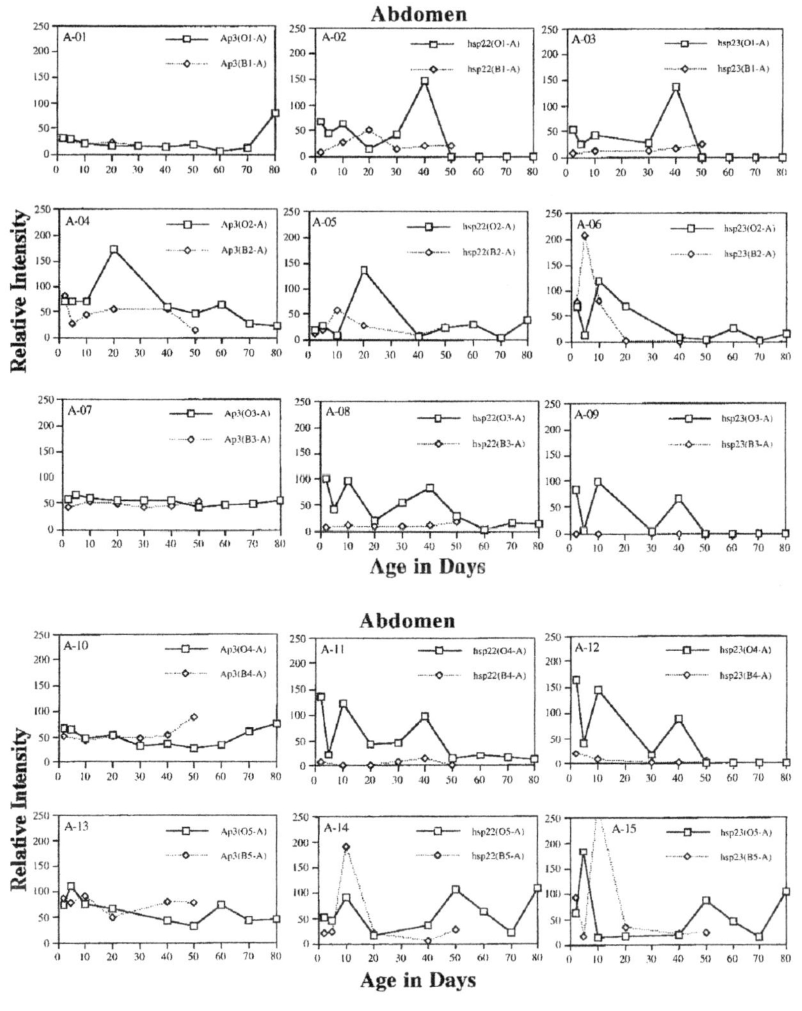

Figure 3. continued

the work presented here, is genotype by environment interaction. The effects of genetic differences can depend critically on assay environments. This has been shown to confuse inferences of character correlations in certain earlier studies of aging (24). The resolution of this difficulty requires additional experimentation, with a variety of genotypes and environments. However, initial significant correlations between specific characters such as stress resistance, hsp expression, and aging are not without interest because of this problem.

The correlation between increased stress resistance and increased life span suggests a testable model for the function of increased *hsp22* and *hsp23* expression in the O line flies. Other members of the shsp family have been demonstrated to be molecular chaperones, capable of conferring increased resistance to stresses such as heat and oxidative damage when overexpressed in cultured cells. We hypothesize that the upregulated expression of *hsp22* and *hsp23* in the O line flies contributes directly to their increased stress resistance and increased life span. A critical test of this hypothesis will require experimentally upregulating and/or downregulating the expression of *hsp22* and *hsp23* in transgenic flies, and assaying the effects of this altered expression on stress resistance and life span.

ACKNOWLEDGMENTS

This work was supported by grants from the Department of Health and Human Services (AG11644 and AG11833). R. K. was supported by a postdoctoral fellowship, through a National Institute on Aging training grant (AG00093).

Address correspondence to John Tower, Department of Biological Sciences, SHS172, University of Southern California, University Park, Los Angeles, CA 90089-1340. E-mail: jtower@USC.edu

REFERENCES

1. Arrigo AP, Landry J. Expression and function of the low-molecular-weight heat shock proteins. In: Morimoto RI, Tissieres A, Georgopoulos C, eds. *The Biology of Heat Shock Proteins and Molecular Chaperones*. Cold Spring Harbor: Cold Spring Harbor Laboratory Press; 1994:335–373.
2. Landry J, Chretien P, Lambert H, Hickey E, Weber LA. Heat shock resistance conferred by expression of the human HSP27 gene in rodent cells. *J Cell Biol*. 1989;109:7–15.
3. Huot J, Roy G, Lambert H, Chretien P, Landry J. Increased survival after treatments with anticancer agents of Chinese hamster cells expressing the human Mr 27,000 heat shock protein. *Cancer Res*. 1991; 51:5245–5252.
4. Mehlen P, Briolay J, Smith L, Diatz-Latoud C, Fabre N, Pauli D, Arrigo A-P. Analysis of the resistance to heat and hydrogen peroxide stresses in COS cells transiently expressing wild type of deletions mutants of the *Drosophila* 27-kDa heat shock protein. *Eur J Biochem*. 1993;215:277–284.
5. Mehlen P, Preville X, Chareyron P, Briolay J, Klemenz R, Arrigo A-P. Constitutive expression of human hsp27, Drosophila hsp27, or human αB-crystallin confers resistance to TNF- and oxidative stress-induced cytotoxicity in stably transfected murine *L929* fibroblasts. *J Immunol*. 1995;215:363–374.
6. Horwitz J. α-Crystallin can function as a molecular chaperone. *Proc Natl Acad Sci USA*. 1992;89:10449–10453.
7. Jakob U, Gaestel M, Engel K, Buchner J. Small heat shock proteins are molecular chaperones. *J Biol Chem*. 1993;268:1517–1520.
8. Merck KB, Groenen PJTA, Voorter CEM, de Haard-Hoekman WA, Horwitz J, Bloemendal H, et al. Structural and functional similarities of bovine α-crystallin and mouse small heat-shock protein. *J Biol Chem*. 1993;268:1046–1052.
9. Parsell DA, Lindquist S. The function of heat-shock proteins in stress tolerance: degradation and reactivation of damaged proteins. *Annu Rev Genet*. 1993;27:437–456.
10. Wheeler JC, Bieschke ET, Tower J. Muscle-specific expression of Drosophila hsp70 in response to aging and oxidative stress. *Proc Natl Acad Sci USA*. 1995;92:10408–10412.
11. Wheeler JC, King V, Tower J. Sequence requirements for upregulated expression of Drosophila hsp70 transgenes during aging. *Neurobiol Aging*. 1999;20:545–553.
12. King V, Tower J. Aging-specific expression of Drosophila hsp22. *Devel Biol*. 1999;207:107–118.
13. Fleming JE, Walton JK, Dubitsky R, Bensch KG. Aging results in an unusual expression of Drosophila heat shock proteins. *Proc Natl Acad Sci USA*. 1988;85:4099–4103.
14. Niedzwiecki A, Fleming JE. Changes in protein turnover after heat shock are related to accumulation of abnormal proteins in aging Drosophila melanogaster. *Mech Ageing Dev*. 1990;52:295–304.
15. Gershon H, Gershon D. Detection of inactive enzyme molecules in ageing organisms. *Nature*. 1970;227:1214–1217.
16. Finch CE. *Longevity, Senescence and the Genome*. Chicago: University of Chicago Press; 1990.
17. Stadtman ER. Protein oxidation and aging. *Science*. 1992;257:1220–1224.
18. Lindquist S. The heat-shock response. *Ann Rev Biochem*. 1986;55: 1151–1191.
19. Maynard Smith J. The effects of temperature and of egg-laying on the longevity of Drosophila subobscura. *J Exp Biol*. 1958;35:832–843.
20. Maynard Smith J. Prolongation of the life of Drosophila subobscura by brief exposure of adults to a high temperature. *Nature*. 1958;181:496–497.
21. Khazaeli AA, Tatar M, Pletcher SD, Curtsinger JW. Heat-induced longevity extension in Drosophila. I. Heat treatment, mortality, and thermotolerance. *J Gerontol Biol Sci*. 1997;52A:B48–B52.
22. Tatar M, Khazaeli AA, Curtsinger JW. Chaperoning extended life. *Nature*. 1997;390:30.
23. Tatar M. Transgenes in the analysis of life span and fitness. *The American Naturalist*. 1999;154(suppl):S67–S81.
24. Rose MR. Laboratory evolution of postponed senescence in Drosophila melanogaster. *Evolution*. 1984;38:1004–1010.
25. Luckinbill LS, Arking R, Clare M, Cirocco W, Buck S. Selection for delayed senescence in Drosophila melanogaster. *Evolution*. 1984;38: 996–1003.
26. Arking R. Successful selection for increased longevity in Drosophila: analysis of the survival data and presentation of a hypothesis on the genetic regulation of longevity. *Exp Geron*. 1987;22:199–220.
27. Partridge L, Fowler K. Direct and correlated responses to selection on age at reproduction in Drosophila melanogaster. *Evolution*. 1992;46: 76–91.
28. Service PM, Hutchinson MD, MacKinley MD, Rose MR. Resistance to environmental stress in Drosophila melanogaster selected for postponed senescence. *Physiol Zool*. 1985;58:380–389.
29. Service PM. Physiological mechanism of increased stress resistance in Drosophila melanogaster selected for postponed senescence. *Physiol Zool*. 1987;670:321–326.
30. Arking R, Buck S, Berrios A, Dwyer S, Baker GT. Elevated paraquat resistance can be used as a bioassay for longevity in a genetically based long-lived strain of Drosophila. *Dev Genet*. 1991;12:362–370.
31. Graves JL, Toolson EC, Jeong C, Vu LN, Rose MR. Dessication, flight, glycogen and postponed senescence in Drosophila melanogaster. *Physiol Zool*. 1992;65:268–286.
32. Rose MR, Vu LN, Park SU, Graves JL. Selection on stress resistance increases longevity in Drosophila melanogaster. *Exp Geron*. 1992;27: 241–250.
33. Force AG, Staples T, Soliman S, Arking R. A comparative biochemical and stress analysis of genetically selected Drosophila strains with different longevities. *Dev Genet*. 1995;17:340–351.
34. Dudas SP, Arking R. A coordinate upregulation of antioxidant gene activities is associated with the delayed onset of senescence in a long-lived strain of Drosophila. *J Gerontol Biol Sci*. 1995;50A:B117–B127.
35. Rose MR. *Evolutionary Biology of Aging*. New York: Oxford University Press; 1991.
36. Sambrook J, Fritsch EF, Maniatis T. *Molecular Cloning: A Laboratory Manual*. 2nd ed. Cold Spring Harbor: Cold Spring Harbor Laboratory Press; 1989.
37. Costlow N. [dissertation]. Ithaca, NY: Cornell University; 1984.
38. O'Connell P, Rosbash M. Sequence, structure, and codon preference of the Drosophila ribosomal protein 49 gene. *Nucleic Acids Res*. 1984; 12:5495–5513.
39. Kelley MR, Venugopal S, Harless J, Deutsch WA. Antibody to a human DNA repair protein allows for cloning of a Drosophila cDNA that encodes an apurinic endonuclease. *Mol Cell Biol*. 1989;9:965–973.
40. Grabowski DT, Deutsch WA, Derda D, Kelley MR. Drosophila AP3, a presumptive DNA repair protein, is homologous to human ribosomal associated protein PO. *Nucleic Acids Res*. 1991;19:4297.
41. Rozek CE, Davidson N. Drosophila has one myosin heavy-chain gene with three developmentally regulated transcripts. *Cell*. 1983;32:23–34.
42. Chalfie M, Tu Y, Euskirchen G, Ward WW, Prasher DC. Green flourescent protein as a marker for gene expression. *Science*. 1994; 263:802–805.
43. Wang S, Hazelrigg T. Implications for bcd mRNA localization from spatial distribution of exu protein in Drosophila oogenesis. *Nature*. 1994;369:400–403.
44. Larsen PL. Aging and resistance to oxidative stress in Caenorhabditis elegans. *Proc Natl Acad Sci USA*. 1993;90:8905–8909.
45. Vanfleteren JR. Oxidative stress and aging in Caenorhabditis elegans. *Biochem J*. 1993;292:605–608.
46. Lithgow GJ, White TM, Hinerfeld DA, Johnson TE. Thermotolerance of a long-lived mutant of *Caenorhabditis elegans*. *J Gerontol Biol Sci*. 1994;49:B270–B276.
47. Lithgow GJ, White TM, Melov S, Johnson TE. Thermotolerance and extended life span conferred by single-gene mutations and induced by thermal stress. *Proc Natl Acad Sci USA*. 1995;92:7540–7544.
48. Murakami S, Johnson TE. A genetic pathway conferring life extension and resistance to UV stress in *Caenorhabditis elegans*. *Genetics*. 1996; 143:1207–1218.

Received July 27, 1999
Accepted May 9, 2000
Decision Editor: John A. Faulkner, PhD

Electrophoretic Analysis of Methuselah Flies from Multiple Species

Denise J. Deckert-Cruz[1], Luciano M. Matzkin[2], Joseph L. Graves Jr.[3] and Michael R. Rose[4]

[1]*Dept. of Biology, Thomas Nelson Community College, Hampton, VA 23670-9407*
deckert-cruzd@tncc.vccs.edu

[2]*Dept. of Ecology and Evolutionary Biology, BioSciences West 310, University of Arizona, Tucson, AZ 85721-0088*

[3]*Dept. of Ecology and Evolutionary Biology, University of California, Irvine, Irvine, CA 92697-2525*
joseph.graves@cox.net

[4]*Dept. of Ecology and Evolutionary Biology, University of California, Irvine, Irvine, CA 92697-2525*
mrrose@uci.edu; also to whom correspondence should be sent

Key words: *Drosophila*, longevity, allozymes, phosphoglucomutase, glycogen

Abstract

We selected five-fold replicated lines of *Drosophila pseudoobscura*, *Drosophila hydei*, and *Drosophila arizonae* for postponed aging. After 10 generations of postponed reproduction, female mean longevity had increased in all *D. pseudoobscura* lines and all *D. hydei* lines, and male longevity had increased in all *D. pseudoobscura* lines and all *D. arizonae* lines. We assayed eight metabolic enzymes for differentiation in allozyme frequencies between control (B) lines and selected (O) lines. Frequencies of phosphoglucomutase allozymes differentiated in response to selection in both species in which this locus was electrophoretically polymorphic. This evolution at the *Pgm* locus may explain the increased glycogen content associated with postponed aging in *Drosophila*.

Introduction

A central problem facing the identification of the particular loci involved in selectively postponed aging is linkage disequilibrium between the alleles of the locus being observed and the alleles of other loci. When a locus is found to be statistically differentiated between selected and control populations, either the locus under study or the loci in linkage disequilibrium with it could be involved in the postponement of aging by selection. Comparisons among species can help to distinguish the loci with a causal role in postponed aging, as different species are likely to carry different adventitious linkage relationships among specific alleles.

We created five-fold replicated control (B) and selected (O) lines from outbred populations of *Drosophila pseudoobscura, Drosophila hydei* and *Drosophila arizonae,* three species phylogenetically distant from *Drosophila melanogaster. D. hydei* and *D. arizonae* are in the *repleta* group of the *Drosophila* subgenus, while *D. pseudoobscura* is in the *obscura* group of the *Sophophora* subgenus. *D. hydei* is a colonizing species (Johnston and Templeton, 1982), while *D. arizonae* breeds in cactus in southwestern North America (Heed, 1982). *D. pseudoobscura* is a temperate woodland species opportunistic in fruit orchards (Powell, 1997). The life history of each species was used to determine the larval rearing densities and larval rearing time, as well as the total generation time in the laboratory. Each O population was derived from a B population to facilitate pairwise comparisons. For the O populations, generation time was gradually increased.

Postponed aging in *Drosophila melanogaster* has been correlated with starvation resistance and desiccation resistance (Rose *et al.*, 1992), with tolerance to 15% ethanol vapor (Service *et al.*, 1985), and with tethered flight duration (Graves *et al.*, 1988). Hence we chose to focus our electrophoretic analysis on enzymes with positions in metabolic pathways relevant to these forms of stress resistance. We calculated allele frequencies from the frequencies of allozymes distinguishable by starch gel electrophoresis. We looked for parallel changes in allele frequencies across species.

Materials and Methods

Drosophila pseudoobscura and *D. hydei* were collected in orchards in Orange County, California, USA. The stocks were founded from at least 50 females. For *D. pseudoobscura,* only females with the standard chromosome inversion were included in the stock. The *D. arizonae* stock was founded from one-hundred gravid females collected from orange trees in Tempe, Arizona, USA. Each stock was maintained in the laboratory at a population size of 3000 flies for a year before beginning selection. This not only allowed these populations to adapt to the lab but gave them about 20 generations to break up adventitious linkage disequilibrium between alleles.

After a year of adaptation to the laboratory, each species {*Drosophila pseudoobscura* (PST), *D. hydei* (H) and *D. arizonae* (AZ)} was divided into five replicate B lines and five replicate O lines maintained on discrete generations. These lines represent baseline and "old reproduced" populations of flies established according to the techniques utilized to derive Rose's original *Drosophila melanogaster* B and O populations formed in 1980 (Rose, 1984). The larval rearing density was adjusted according to the body size of the species, and the larval rearing time was based on development time (Table 1). The adults of species with intrinsically longer life cycles spent more days in population cages. When the populations were first founded, they were divided into five independent populations. Each of these populations was later split into an O line and a B line which shared the same numerical subscript. The O lines were selected for postponed aging by starting each new generation with eggs from progressively older mothers, up to the generation times shown in Table 1.

At the beginning of each generation, *Drosophila* eggs were placed in vials of banana food. This medium contains agar (Difco bacto), ripe bananas, corn syrup (Karo), barley malt (Eden), yeast (Fleischmann's), ethanol, and *p*-hydroxybenzoic acid to retard fungal growth. Eggs were placed in a number of vials adequate to rear 3000 individuals per line at the standard density. After the allotted number of days for development, the adults were transferred to population cages, where they were fed with petri dishes of banana food changed three times per week. At the end of the generation time, the eggs were collected from the dish of banana food and placed in the vials.

Table 1 Rearing Conditions

Species	Lines	Density (Eggs/Vial)	Egg to Adult Time	Generation Time
D. pseudoobscura	PSTB1–PSTB5	40–60	19 days at 21°C	25 days
	PSTO1–PSTO5	40–60	19 days at 21°C	70 days
D. hydei	HB1–HB5	40–60	18 days at 25°C	28 days
	HO1 HO5	40 60	18 days at 25°C	100 days
D. arizonae	AZB1–AZB5	60–80	15 days at 25°C	25 days
	AZO1–AZO5	60–80	15 days at 25°C	100 days

Longevities were measured on individual flies maintained in groups of four flies of each sex per vial, without replacing dead flies. Deaths were recorded daily, and the flies were transferred to vials of fresh banana food three times per week. For each line, the longevities of 160 females and 160 males were measured, and the mean longevities were calculated separately for each sex. For starvation assays, four flies of the same sex were isolated in the bottom of a vial with a sponge, and the remainder of the vial contained 3 ml water in two cotton balls. Deaths were recorded every six hours. For desiccation assays, four single-sex flies were isolated from the bottom of a vial with a sponge, and the remainder of the vial contained 5 ml desiccant (Drierite). Deaths were recorded hourly. For each stress assay, the mean longevities were determined for 80 females and 80 males from each line, and the mean longevity was calculated for each line, for females and males separately.

Electrophoretic gels were prepared by cooking a 12% solution of electrolyzed potato starch (Sigma S-4501) in a pH 7.0 buffer of Tris (Sigma) and citric acid (Sigma), and ethylenediamine-tetraacetic acid. Flies were ground in the same buffer using a pellet pestle and cordless motor (Kontes). The gels were exposed to a constant voltage of 200V in a Tris-citrate-EDTA buffer at pH 7.0 for four to five hours.

The following enzymes were stained using mixtures applied as agar overlays:

Alcohol dehydrogenase (ADH): 0.1M Tris-HCl pH 8.6, 95% ethanol, NAD, MTT, and PMS.

Alpha-Glycerophosphate dehydrogenase (GPDH): 0.1M Tris-HCl pH 8.6, glycerophosphate, $MgCl_2$, NAD, MTT, and PMS.

Hexokinase (HEX): 0.1M Tris-HCl pH 8.0, glucose, $MgCl_2$, glucose-6-phosphate dehydrogenase, ATP, NADP, MTT, and PMS.

Malic enzyme (ME): 0.1M Tris-HCl pH 8.0, malic acid, $MgCl_2$, NADP, MTT, and PMS.

6-Phosphogluconate dehydrogenase (PGD): 0.1M Tris-HCl pH 8.0, 6-phosphogluconic acid, $MgCl_2$, NADP, MTT, and PMS.

Phosphoglucomutase (PGM): 0.1 M Tris-HCI pH 8.0, glucose-1-phosphate, $MgCl_2$, glucose-6-phosphate dehydrogenase, NADP, MTT, and PMS.

The following enzymes were stained using solutions in which the gel slice was submerged:

Glucose-6-phosphate dehydrogenase (G6PDH): 0.05M Tric-HCl pH 7.1, glucose-6-phosphate, $MgCl_2$, NADP, NBT, and PMS.

Superoxide dismutase (SOD): 0.1M pH 8.6, NAD, NBT, and PMS incubated in the dark.

Results

Ten generations of selection for postponed reproduction significantly increased longevity in at least one sex of all three species. For *Drosophila pseudoobscura*, the mean female longevity (Fig. 1) and the mean male longevity (Fig. 2) in each PSTO line were greater than in the corresponding PSTB line, with the differences between B and O line ranging from 13 to 20 days. Paired Student's t-tests of the mean longevities yielded values of $t = -7.95$ and $p = 0.001$ for the females, and $t = -13.74$ and $p = 0.0001$ for the males. Starvation resistance increased significantly for females ($t = -2.98$, $p = 0.041$) and marginally for males ($t = -2.16$, $p = 0.097$). Desiccation resistance did not increase consistently across the lines ($p = 0.175$ for females, $p = 0.395$ for males).

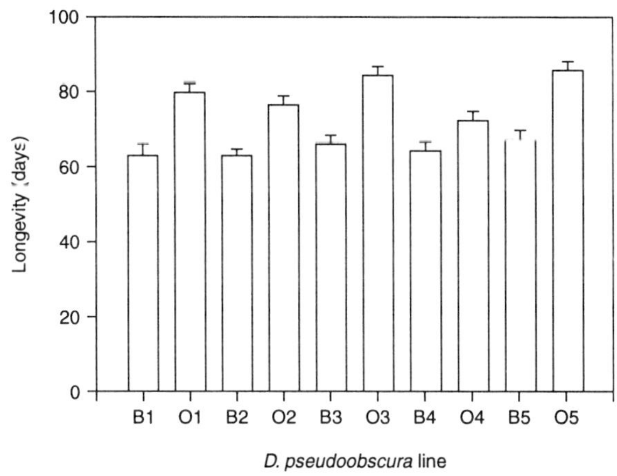

Fig. 1. Female longevity in *Drosophila pseudoobscura*. After ten generations of selection for postponed reproduction, mean female longevity was longer in each PSTO line than in the PSTB line from which it was derived.

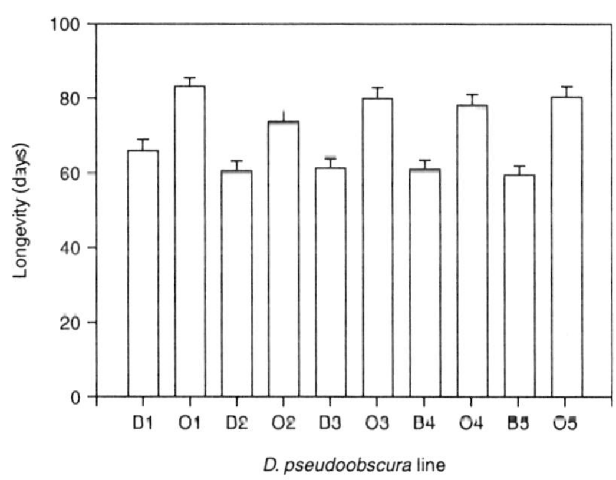

Fig. 2. Male longevity in *Drosophila pseudoobscura*. After ten generations of selection for postponed reproduction, mean male longevity was longer in each PSTO line than in the PSTB line from which it was derived.

For *Drosophila hydei*, the mean female longevity (Fig. 3) in each HO line was 18–29 days greater than in the corresponding HB line ($t = -8.81$, $p = 0.0009$). There was increased mean male longevity in four of the HO lines relative to their controls (Fig. 4), but line HO4 registered a decrease in mean male longevity, and the change across the five pairs of populations was not significant ($t = -2.00$, $p = 0.116$). There were no significant increases in starvation resistance or desiccation resistance.

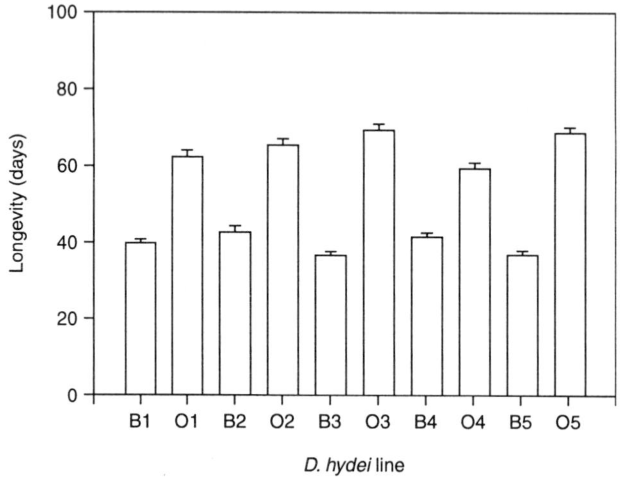

Fig. 3. Female longevity in *Drosophila hydei*. After ten generations of selection for postponed reproduction, mean female longevity was longer in each HO line than in the HB line from which it was derived.

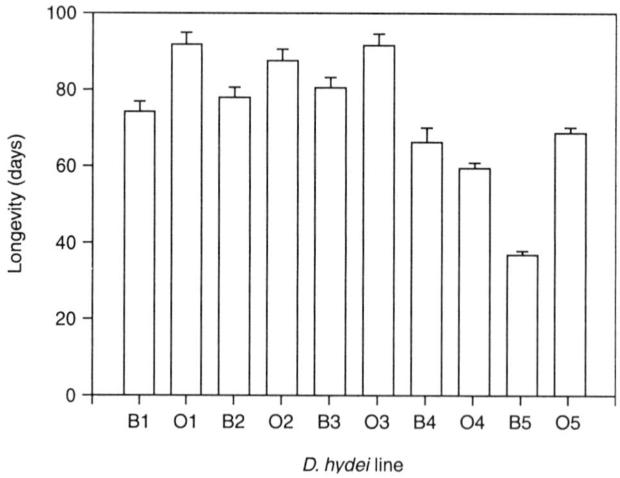

Fig. 4. Male longevity in *Drosophila hydei*. After ten generations of selection for postponed reproduction, mean male longevity was generally longer in the HO line than in the HB line from which they were derived.

For *Drosophila arizonae,* the mean female longevity was higher in three of the five AZO lines, relative to their corresponding AZB lines (Fig. 5), but the change across the five pairs of populations was not significant ($t = -1.53$, $p = 0.20$). Mean male longevity (Fig. 6) increased consistently across all five pairs of lines ($t = -4.33$, $p = 0.012$). Starvation resistance did not increase consistently ($p = 0.878$ for females, $p = 0.388$ for males). Desiccation resistance increased significantly for females ($t = -3.82$, $p = 0.012$) and marginally for males ($t = -2.51$, $p = 0.066$).

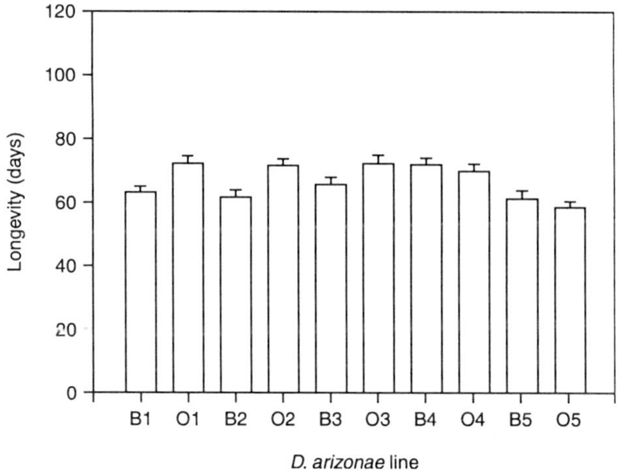

Fig. 5. Female longevity in *Drosophila arizonae.* After ten generations of selection for postponed reproduction, mean female longevity was generally longer in the AZO lines than in the AZB lines from which they were derived.

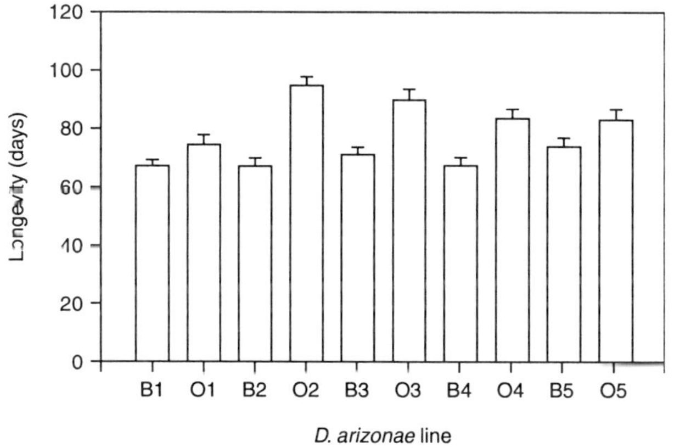

Fig. 6. Male longevity in *Drosophila arizonae.* After ten generations of selection for postponed reproduction, mean male longevity was longer in each AZO line than in the AZB line from which it was derived.

Of the enzymes assayed electrophoretically, only phosphoglucomutase was polymorphic in the *Drosophila pseudoobscura* lines and *Drosophila arizonae* lines. The *D. hydei* lines were monomorphic for phosphoglucomutase to the extent detectable by allozyme electrophoresis.

In *D. pseudoobscura*, *D. arizonae*, and *D. melanogaster*, there are three PGM allozymes distinguishable by electrophoresis, designated F, M, and S, for fast, medium, and slow migration in the gel. In *D. pseudoobscura*, the frequency of the allele corresponding to the slowest migrating (S) allozyme was higher in each of the five PSTO populations, compared with the PSTB population from which it had been derived (Table 2, Fig. 7). Paired *t*-tests were utilized to evaluate the differentiation in allele frequencies between the PSTB's and PSTO's. For the S allele, the frequency data are: $t = -6.500$ ($p < 0.01$); for the M allele, $t = 6.291$ ($p < 0.01$); and for the F allele, $t = 0.612$ ($p > 0.50$), although these are not strictly independent statistical tests. Apparently, during selection for postponed reproduction, the *Pgm* S allele increased in frequency, the *Pgm* M allele decreased in frequency, while the *Pgm* F allele remained rare.

The *Pgm* locus also responded to selection for postponed reproduction in the *D. arizonae* lines (Table 3). The frequency of the S allele was higher in four of the five AZO populations, relative to the AZB populations from which they were derived (paired *t*-test for four lines, $t = 4.07$, $p < 0.03$). The AZB5 population was exceptional in having a very high frequency of the S allele. Rather than increasing further, the frequency of this allele appears to have decreased over the ten generations of selection on the AZO line. Although there was no significant

Table 2 Allele Frequencies of *Phosphoglucomutase* in the *Drosophila pseudoobscura* lines

	F	M	S
PSTB1	0.03	0.39	0.58
PSTB2	0.03	0.34	0.63
PSTB3	0.00	0.52	0.48
PSTB4	0.02	0.52	0.48
PSTB5	0.00	0.46	0.54
PSTO1	0.01	0.31	0.68
PSTO2	0.01	0.24	0.75
PSTO3	0.00	0.33	0.67
PSTO4	0.00	0.36	0.64
PSTO5	0.03	0.35	0.62

The mean frequency of the S allele is 0.54 ± 0.03 in the PSTB's and 0.67 ± 0.02 in the PSTO's.

differentiation of *Pgm* when this line was included in the analysis (paired *t*-test, $t = 1.04, p > 0.350$), the other four *D. arizonae* lines still suggest a pattern of increase in the S allele as aging is postponed. The lines that appear to have the greatest increases in *Pgm* S frequency are those in which female longevity as well as male longevity increased.

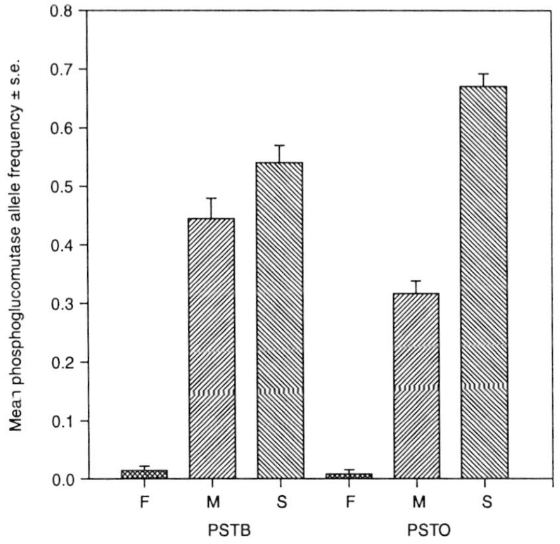

Fig. 7. Frequencies of *Pgm* alleles in *Drosophila pseudoobscura*. The frequency of the *Pgm* S allele has increased in the PSTO lines after 10 generations of selection for postponed reproduction.

Table 3 Allele Frequencies of *Phosphoglucomutase* in the *Drosophila arizonae* lines

	F	M	S
AZB1	0.730	0.084	0.186
AZB2	0.635	0.215	0.150
AZB3	0.638	0.183	0.179
AZB4	0.702	0.061	0.237
AZB5	0.545	0.101	0.354
AZO1	0.637	0.084	0.268
AZO2	0.613	0.093	0.294
AZO3	0.697	0.038	0.265
AZO4	0.587	0.138	0.276
AZO5	0.612	0.153	0.235

The mean frequency of the *Pgm* S allele is 0.22 ± 0.04 in the AZB's and 0.27 ± 0.01 in the AZO's.

Discussion

Pgm, the locus that encodes phosophoglucomutase, is of particular interest because of its striking differentiation in the original *Drosophila melanogaster* B and O lines. After 75 generations of selection for postponed reproduction, the O lines had much greater frequencies of the slowest migrating allozyme. Frequencies of the *Pgm* S allele expressed by this allozyme ranged from 0.39 to 0.65 for the O lines. Four of the five B lines had *Pgm* S frequencies no greater than 0.05; in the fifth line (B1) the frequency was 0.25 (Deckert-Cruz *et al.*, 1997).

In the current study, we show that in *D. pseudoobscura* and *D. arizonae*, the frequencies of the slowest migrating PGM allozyme also increase in response to selection for postponed reproduction. While it is the *Pgm* S allele that increases in frequency in each species, this allele might vary in nucleotide sequence across species, as well as within species. It is not necessarily the same allele.

Phosphoglucomutase is an enzyme with a key role in metabolism. It is located at a branch point in the metabolic pathways that lead to glycolysis on one hand and glycogen synthesis on the other. Flies depend on glycogen reserves to sustain flight, as they have no anerobic metabolism in the flight muscles (Wigglesworth, 1972; Chippendale, 1978). Verrelli and Eanes (2001) found a two-fold range in enzyme activity over 10 PGM protein haplotypes, and also found that the glycogen content of *Drosophila melanogaster* can vary over a three-fold range, depending on the PGM haplotypes present. Their haplotype 8, which shares an amino acid change with the S allele in *D. melanogaster*, confers high enzyme activity and glycogen content.

Graves *et al.* (1992) found that the glycogen content in the original O lines was higher than in the B lines. Increased glycogen content has been associated with desiccation resistance (Graves *et al.*, 1992) and may also be important for starvation resistance (Chippindale *et al.*, 1996). When *Drosophila melanogaster* were selected for stress resistance, responses to desiccation were partly independent of responses to starvation, although the two types of response were correlated (Rose *et al.*, 1992). In *D. pseudoobscura*, after 10 generations of selection in the PSTO lines, enhanced starvation resistance appears to be part of the postponed aging response, while enhanced desiccation resistance does not. The opposite is true for *D. arizonae*, in which enhanced desiccation resistance but not enhanced starvation resistance evolved. Glycogen content could have played a role in both responses, as they involve overlapping components of postponed aging.

The accumulation of glycogen stores has the potential to be a component of postponed aging across multiple species. It could be mediated by selection for PGM allozymes with high activity. However, elevated PGM activity alone does not significantly increase lifespan, as shown in *Drosophila melanogaster* with altered *Pgm* S expression (Allikian *et al.*, 2002). The PGM activity level, the actual structural form of phosphoglucomutase, and the genetic background all need to be considered in evaluating effects of *Pgm* alleles on lifespan.

We have shown that *D. pseudooscura* and *D. arizonae*, as well as *D. melanogaster*, respond to selection for postponed aging with evolution at the *Pgm* locus. Each of these species carries an allozyme class with a selective advantage in an environment of postponed reproduction. Demonstrating that the *Pgm* locus can respond to selection in species which have evolved separately for tens of millions of years minimizes the possibility that *Pgm* is adventitiously linked to another locus under selection. The presence of parallel *Pgm* allele frequency changes in three widely divergent species helps to confirm a causal role for this locus in postponed aging in *Drosophila*, as originally suggested by the work of Deckert-Cruz *et al.* (1997), using stocks derived from the B and O Methuselah flies of Rose (1984).

Acknowledgements

We are grateful for the contributions of the following individuals. Barbara Joos helped establish the populations. Fred Simmons determined the chromosome inversion of *D. pseudoobscura* individuals. Teri Markow collected the founders of the *D. arizonae* stock. Georgina Jimenez Matzkin, Jonathan Ilano, and many other students helped maintain flies and perform longevity and stress assays. Trung Nguyen helped with electrophoresis. This research was supported in part by grants from the NIA.

References

Allikian, M.J., D. Deckert-Cruz, M.R. Rose, G.N. Landis, and J. Tower. 2002. Doxycycline-induced expression of sense and inverted-repeat constructs modulates *phosphogluconate mutase* (*Pgm*) gene expression in adult *Drosophila melanogaster*. *Genome Biol* 3(5): research0021.1–0021.10.

Chippendale, G.M. 1978. The function of carbohydrates in insect life processes. In: *Biochemistry of the Insects*, M. Rockstein (ed.). Academic Press, New York, pp. 1–54.

Chippindale, A.K., T.J.F. Chu, and M.R. Rose. 1996. Complex trade-offs and the evolution of starvation resistance in *Drosophila melanogaster*. *Evolution* **50**: 753–766.

Deckert-Cruz, D.J., R.H. Tyler, J.E. Landmesser, and M.R. Rose. 1997. Allozymic differentation in response to laboratory demographic selection of *Drosophila melanogaster*. *Evolution* **51**: 865–972.

Graves, J.L. Jr., L.S. Luckinbill, and A. Nichols. 1988. Flight duration and wing beat frequency in long- and short-lived *Drosophila melanogaster*. *J Insect Physiol* **34**: 1021–1026.

Graves, J.L., Jr., E.C. Toolson, C. Jeong, L.N. Vu, and M.R. Rose. 1992. Desiccation, flight, glycogen, and postponed senescence, in *Drosophila melanogaster*. *Physiol Zool* **65**: 268–286.

Heed, W.B. 1982. The origin of *Drosophila* in the Sonoran Desert. In: *Ecological Genetics and Evolution: The Cactus-Yeast-Drosophila Model System*, J.S.F. Barker and W.T. Starmer (eds.). Academic Press, Sydney, pp. 65–80.

Johnston, J.S. and A.R. Templeton. 1982. Dispersal and clines in *Opuntia* breeding *Drosophila mercatorum* and *D. hydei* at Kamuela, Hawaii. In: *Ecological Genetics and Evolution: The Cactus-Yeast-Drosophila Model System*, J.S.F. Barker and W.T. Starmer (eds.). Academic Press, Sydney, pp. 241–256

Powell, J.R. 1997. *Progress and Prospects in Evolutionary Biology: The Drosophila Model*. Oxford University Press, Oxford.

Rose, M.R. 1984. Laboratory evolution of postponed senescence in *Drosophila melanogaster*. *Evolution* **38**: 1004–1010.

Rose, M.R., L.H. Vu, S.U. Park, and J. L. Graves, Jr. 1992. Selection on stress resistance increases longevity in *Drosophila melanogaster*. *Exper Gerontol* **27**: 241–250.

Service, P.M., E.W. Hutchinson, M.D. MacKinley, and M.R. Rose. 1985. Resistance to environmental stress in *Drosophila melanogaster* selected for postponed senescence. *Physiol Zool* **58**: 380–390.

Verrelli, B.C. and W.F. Eanes. 2001. The functional impact of *Pgm* amino acid polymorphism on glycogen content in *Drosophila melanogaster*. *Genetics* **159**: 201–210.

Wigglesworth, V.B. 1972. *The Principles of Insect Physiology*, 7th edition, Chapman and Hall, London.

V

REVERSE EVOLUTION OF METHUSELAH FLIES

Reverse Evolution of Aging

Michael R. Rose[1] and Hardip B. Passananti[2]

[1]*Department of Ecology and Evolutionary Biology, University of California, Irvine, CA 92697-2525, USA mrrose@uci.edu; also to whom correspondence should be sent*

[2]*Orrick, Herrington & Sutcliffe LLP, 4 Park Plaza, Suite 1600, Irvine, CA 92614 hpassananti@orrick.com*

One of the most important tools in the analysis of the evolution of aging in fruit flies is reverse evolution. Reverse evolution is a general evolutionary technique, one of the three basic tools in the evolutionary biologist's tool-box, the other two being forward evolution and the analysis of genetic variation. Reverse evolution can be defined as returning a population, or populations, to a previous selective regime (cf. Teotónio and Rose, 2001). Alternatively, reverse evolution can be defined as the effect of such an evolutionary manipulation. Large-scale experiments suggest that many characters may respond evolutionarily to the imposition of reverse evolution (e.g. Teotónio and Rose, 2000; reprinted here).

One of the early experiments with the Methuselah flies involved reverse evolution (Service *et al.*, 1988; reprinted here). Service *et al.* returned O-type Methuselah populations to early reproduction, the B-type regime. It is important to understand that the ancestor of the O populations had been maintained under B conditions for five years before the Methuselah flies were created. Thus the switching of O populations back to the B regime is a case of reverse evolution.

The main results of this early study were that early fecundity rapidly increased in the reversed O flies, relative to B flies, while starvation resistance fell quickly. Similar results were found by Teotónio and Rose (2000; reprinted here) in a larger collection of reverse lines, that is, fecundity rising as starvation resistance fell. These results suggested that starvation resistance trades off with early fecundity and possibly other early reproductive characters.

An interesting feature of the study by Service *et al.* is that two other characters associated with postponed aging, desiccation resistance and ethanol resistance, did **not** fall during reverse evolution. This suggested that these particular enhancements associated with postponed aging do not have a trade-off relationship with early fitness.

The reverse-selected lines of Service *et al.* were studied further by Graves *et al.* (1992; reprinted here). After more than 130 generations of reverse evolution, mean longevity fell to a level close to that of the B populations. [Note that there is an error in Fig. 1 of Graves *et al.*, in which the reverse-selected ("RU") and O coding is reversed.] For a number of physiological characters, such as starvation resistance, the reverse-selected flies remain intermediate between B and O populations. However, the large number of unmonitored generations between the Service *et al.* and Graves *et al.* studies make it difficult to interpret the results of the latter study.

Two additional studies by Passananti and others have further supported these initial findings. Passananti *et al.* (2004a, this volume) repeated the Service *et al.* study, sustaining reverse evolution for more than 50 generations, and studying longevity in addition to early fecundity, starvation resistance, and desiccation resistance. Longevity fell dramatically and immediately over the course of reverse evolution under B conditions, as did starvation resistance. Again, early fecundity rose with reverse evolution, while desiccation resistance did not generally respond. The longevity response to reverse evolution was particularly important, because it showed that longevity must be pleiotropically interconnected with early fitness. Genetic drift and inbreeding depression are extremely unlikely to account for the reverse evolution of longevity, because these genetic mechanisms are unlikely to reduce longevity by more than a few per cent over 50 generations. The actual reduction in longevity that was observed was about 25–30 per cent during this relatively brief period.

An additional point in favor of genetic trade-offs between early and late life was the increase in fitness characters detected in the reverse-selected O flies by Teotónio *et al.* (2002; reprinted here). During reverse evolution under B conditions, populations derived from O flies demonstrably increase in fitness **under B conditions.** That is, they must have lost adaptation to early life as a result of culture at later ages, adaptation which is regained during reverse evolution.

On the other hand, the work of Teotónio *et al.* (2000, 2002; both reprinted here) and a further study of Passananti *et al.* (2004b, this volume) support the notion that some of the changes that have arisen in the O flies do not involve pleiotropic interactions between early adult life and later adult life. Passananti *et al.* (2004a,b, this volume), like Service *et al.* (1988; reprinted here), found that desiccation resistance is generally unaffected by reverse evolution. This suggests that increased desiccation resistance evolves in the Methuselah O flies at little expense to early reproduction.

In evolutionary terms, the results of these studies of reverse evolution support both antagonistic pleiotropy and mutation accumulation as genetic mechanisms of aging, at least in fruit flies. There are few comparable studies of other organisms. One of these is the nematode study of Gordon Lithgow and colleagues (Walker *et al.*, 2000), who competed asexual nematode strains against each other. One of these strains was a longer-lived mutant, the other strain was its wild-type ancestor. Over time, the longer-lived mutant was eliminated, provided the culture regime involved nutritional deprivation. In this study, like ours, the shorter-lived stock had the evolutionary advantage when selection favored early reproduction under difficult conditions (cf. Leroi *et al.*, 1994). Evidently, slowed or postponed aging evolves only at a cost to reproductive fitness early in life, under difficult conditions.

References

Graves, J.L., E.C. Toolson, C. Jeong, L.N. Vu, and M.R. Rose. 1992. Desiccation, flight, glycogen, and postponed senescence in *Drosophila melanogaster*. *Physiol Zool* **65**: 268–286.

Leroi, A.M., A.K. Chippindale, and M.R. Rose. 1994. Long-term laboratory evolution of a genetic trade-off in *Drosophila melanogaster*. I. The role of genotype x environment interaction. *Evolution* **48**: 1244–1257.

Passananti, H.B., D.J. Deckert-Cruz, A.K. Chippindale, B.H. Le, and M.R. Rose. 2004a. Reverse evolution of aging in *Drosophila melanogaster*. This volume.

Passananti, H.B., K.A. Beckman, and M.R. Rose. 2004b. Relaxed stress selection in *Drosophila melanogaster*. This volume.

Service, P.M., E.W. Hutchinson, and M.R. Rose. 1988. Multiple genetic mechanisms for the evolution of senescence in *Drosophila melanogaster*. *Evolution* **42**: 708–716.

Teotónio, H. and M.R. Rose. 2000. Variation in the reversibility of evolution. *Nature* **408**: 463–466.

Teotónio, H. and M.R. Rose. 2001. Perspective: Reverse Evolution. *Evolution* **55**: 653–660.

Teotónio, H., M. Matos, and M.R. Rose. 2002. Reverse evolution of fitness in *Drosophila melanogaster*. *J Evol Biol* **15**: 608–617.

Walker, D.W., G. McColl, N.L. Jenkins, J. Harris, and G. J. Lithgow. 2000. Natural selection: Evolution of lifespan in *C. elegans*. *Nature* **405**: 296–297.

Evolution, 42(4), 1988, pp. 708–716

MULTIPLE GENETIC MECHANISMS FOR THE EVOLUTION OF SENESCENCE IN *DROSOPHILA MELANOGASTER*

P. M. Service,[1] E. W. Hutchinson,[2] and M. R. Rose[2]
Department of Biology, Dalhousie University, Halifax, NS B3H 4J1, CANADA

Abstract.—We present the results of selection experiments designed to distinguish between antagonistic pleiotropy and mutation accumulation, two mechanisms for the evolution of senescence. Reverse selection for early-life fitness was applied to laboratory populations of *Drosophila melanogaster* that had been previously selected for late-life fitness. These populations also exhibited reduced early-age female fecundity and increased resistance to the stresses of starvation, desiccation, and ethanol, when compared to control populations. Reverse selection was carried out at both uncontrolled, higher larval rearing density and at controlled, lower larval density. In the uncontrolled-density selection lines, early-age female fecundity increased to control-population levels in response to the reintroduction of selection for early-age fitness. Concomitantly, resistance to starvation declined in agreement with previous observations of a negative genetic correlation between these two characters and in accordance with the antagonistic-pleiotropy mechanism. However, resistance to stresses of desiccation and ethanol did not decline in the uncontrolled-density lines during 22 generations of reverse selection for early-life fitness. The latter results provide evidence that mutation accumulation has also played a role in the evolution of senescence in this set of *Drosophila* populations. No significant response in early-age fecundity or starvation resistance was observed in the controlled-density reverse-selection lines, supporting previous observations that selection on *Drosophila* life-history characters is critically sensitive to larval rearing density.

Received September 29, 1987. Accepted February 16, 1988

The modern evolutionary theory of senescence rests upon the argument that the force of natural selection decreases with advancing age (Medawar, 1946; Hamilton, 1966; Charlesworth, 1980). That is, increments or decrements in viability or fecundity that are confined to later ages have relatively little impact on lifetime fitness. Two genetic mechanisms of senescence, consistent with this evolutionary argument, have been proposed. Medawar (1946) suggested that senescence may result from mutations with deleterious effects confined to later ages. Such alleles will be maintained at relatively high equilibrium frequencies (Charlesworth, 1980 p. 218), and they will tend to accumulate in populations (Edney and Gill, 1968). We will refer to this as the mutation-accumulation mechanism (Rose and Charlesworth, 1980). A second mechanism of senescence, also suggested by Medawar (1946) and developed by Williams (1957), is antagonistic pleiotropy. According to this mechanism, senescence results from selection for alleles that have beneficial effects on fitness early in life but later, deleterious effects. Decline with age in the force of selection increases the likelihood that such alleles will have positive net effects on fitness.

If the evolutionary interpretation of senescence is correct and there is appropriate genetic variation, it should be possible to alter patterns of senescence by selection on experimental populations. If only old individuals are allowed to produce each subsequent generation, selection will favor alleles that enhance survivorship and late-age reproduction; that is, selection will act to forestall senescence. Conversely, if only the youngest individuals are allowed to produce subsequent generations, alleles contributing to senescence will be shielded from adverse selection.

These methods have been used successfully in a number of experiments to produce lines of *Drosophila* with divergent mean longevities (Wattiaux, 1968; Rose and Charlesworth, 1981*b*; Rose, 1984; Luckinbill et al., 1984; Luckinbill and Clare, 1985). However, other attempts to alter lifespan in *Drosophila* by selection in the laboratory have not been successful (Lints and Hoste, 1974; Lints et al., 1979; Luckinbill and

---

[1] Present address: T. H. Morgan School of Biological Sciences, University of Kentucky, Lexington, KY 40506.
[2] Present address: Department of Ecology and Evolutionary Biology, University of California, Irvine, CA 92717.

Clare, 1985). The absence of response to selection in these latter cases is almost certainly due to the fact that selection was applied to flies that were reared at a larval density of 10 per shell vial. In contrast, a response to selection has occurred when flies were reared at less controlled and, consequently, higher (approximately one order of magnitude) larval densities (Rose and Charlesworth, 1981*b*; Rose, 1984; Luckinbill and Clare, 1985). Phenotypic differentiation of major alleles affecting longevity is apparently suppressed at low rearing densities, making selection on those alleles difficult if not impossible. This interpretation is supported by additional observations of density effects. When longevity comparisons are made between populations that have been previously differentiated by selection at high density, the apparent difference between the populations is a function of the density at which assayed flies are reared (Clare and Luckinbill, 1985; Luckinbill and Clare, 1986). Specifically, the differences in longevity between differentiated lines are much less when assayed flies are grown at low density.

In those cases where selection on lifespan in *Drosophila* has been successful, there have been concomitant changes in female fecundity patterns. Increased longevity is associated with relatively lower fecundity early in life and increased fecundity late in life (Rose and Charlesworth, 1981*b*; Rose, 1984; Luckinbill et al., 1984, 1987). This tradeoff between early-life and late-life fitness components supports the antagonistic-pleiotropy mechanism for the evolution of senescence.

There have been several tests of mutation accumulation in *Drosophila*. Rose and Charlesworth (1980, 1981*a*) evaluated the additive genetic variance of daily fecundity as a function of age. If the mutation-accumulation mechanism were correct, additive genetic variance of a fitness component, such as daily fecundity, should have increased with age. They found no age-specific trend in genetic variance for daily fecundity. However, Clark (1987) suggests that tests of this sort may suffer from lack of sensitivity. Kosuda (1985) showed that variance in male mating activity among chromosome-substitution lines was much greater for old flies than for young. Furthermore, the reduction in mating activity due to second-chromosome homozygosity, relative to heterozygosity, was greater at older ages than at younger ages. Lastly, there was no correlation among lines between early-age and late-age mating activity. Taken together, these results are consistent with mutation accumulation. Mueller (1987) compared early-age and late-age fecundity of *D. melanogaster* populations that had been maintained either through reproduction by young females (*r*-selected), or through reproduction by females of indefinite age (*K*-selected). Although early-age fecundity did not vary between *r*- and *K*-selected lines, the *r*-selected lines were significantly less fecund than the *K*-selected lines at the later age. These results are also consistent with the notion that absence of natural selection at older ages in the *r*-selected lines permitted the accumulation of alleles with deleterious late-age effects.

In this paper, we propose an additional test of the mutation-accumulation mechanism and present the experimental results of such a test. This test involves selection for late-life fitness, followed by reverse selection for early-life fitness. The outcomes of reverse selection for early-life fitness are different under antagonistic pleiotropy and mutation accumulation (Table 1). For simplicity of illustration, we assume a single diallelic locus affecting any arbitrary trait. Under both genetic mechanisms, allele A_2 enhances late-life fitness relative to the effect of A_1. Under antagonistic pleiotropy, A_1 is relatively beneficial for early-life fitness, but under mutation accumulation, A_1 and A_2 are neutral with respect to early-life fitness. Selection for late-life fitness favors A_2 in both mechanisms. Subsequent reverse selection would increase the frequency of A_1 if the locus in question stands in an antagonistic-pleiotropic relationship with early-life and late-life fitness. However, under mutation accumulation, reverse selection favors neither allele, and we would expect to see no change in the population mean value of the trait in question.

Genetically increased longevity in *D. melanogaster* is associated with increased adult resistance to starvation, desiccation, and lethal concentrations of ethanol vapor

TABLE 1. Genetic models of senescence. Plus and minus signs indicate the effect of each allele relative to the other on either early-life or late-life fitness. Circles indicate neutrality.

	Antagonistic pleiotropy		Mutation accumulation	
	Allele		Allele	
Trait	A_1	A_2	A_1	A_2
Early-life fitness	+	−	o	o
Late-life fitness	−	+	−	+

(Service et al., 1985). If long-lived flies are subjected to selection for increased early-life fitness, different outcomes are expected, depending upon whether antagonistic pleiotropy or mutation accumulation is operating, as just explained. If alleles that increase adult stress resistance have antagonistic-pleiotropic effects on early-life fitness, then selection for increased early-life fitness will result in decreased stress resistance. On the other hand, under mutation accumulation, selection for increased early-life fitness should have negligible effects on stress resistance.

MATERIALS AND METHODS

Populations

Five of the populations used in this study had been maintained in discrete generations of two-weeks' duration for approximately five years. These populations, referred to as B_1–B_5, had, therefore, been subjected to strong selection for early-life fitness. The remaining five populations (O_1–O_5) had been cultured in discrete generations of ten-weeks' duration and had, consequently, been under selection for late-life fitness. These ten populations were derived simultaneously from a common ancestral population that had been maintained for the preceding five years under conditions favoring early-life fitness. The history and details of the selection procedures and culture conditions of the B and O populations are given by Rose (1984) and Service et al. (1985). The divergent selection pressures have resulted in a number of differences between B and O populations. For approximately the first two weeks of adult life, B females have greater fecundity than O females, after which the relative fecundities are reversed (Rose, 1984). O-population males and females have greater longevities than do B-population flies (Rose, 1984). In addition, O-population flies are more resistant than B-population flies to stresses of starvation, desiccation, and ethanol vapor (Service et al., 1985). Thus, increased longevity is associated with increased resistance to several potentially important causes of mortality.

Reverse Selection

Uncontrolled Larval Density.—Five reverse-selection populations, designated R_1–R_5, were started by obtaining approximately 2,000 eggs from O_1–O_5, respectively. The R populations were then handled in exactly the same way and at the same time as the B populations; that is, they were subjected to natural selection for early-life fitness. Every two weeks, adult flies were allowed to oviposit en masse in fresh 8-dram shell vials until an adequate number of eggs were laid. Egg densities were estimated to vary between 100 and 200 per vial. If too many eggs were laid, the excess was removed by scraping the surface of the medium. Other than that, no attempt was made to regulate larval density. After oviposition, adult flies were discarded or used for assays of selection response (see below). Twenty-two R generations were obtained (21 generations of selection). Population sizes were generally between 1,000 and 2,000 each generation.

Controlled Larval Density.—Other studies have shown that selection on *Drosophila* life-history characters can be strongly affected by larval rearing density (Clare and Luckinbill, 1985; Luckinbill and Clare, 1985). Therefore, we established ten additional experimental populations in order to assess the effects of rearing density on natural selection for early-life fitness. Five controlled-density reverse-selection populations were started by sampling 600 eggs or larvae from each O population and transferring them to fresh food in 8-dram shell vials at a density of 30 per vial. These five populations were designated RC_1–RC_5. In order to provide a standard against which to assess any selection response, five controlled-density B populations (BC_1–BC_5) were similarly established. The RC and BC populations were maintained in two-week discrete generations. Each subsequent generation was produced by sampling eggs from the current generation and transferring them to 20 rearing vials per population at a density of 30 per vial. Thus, each generation produced something less than 600 adult flies per population. Of these, 100 females per population were used for assays of selection response (see below), leaving somewhat fewer than 200 females as parents for each subsequent generation. The procedure was carried out for 12 generations.

Assays of Selection Response

Uncontrolled-Density Selection.—Four characters were tested for response to selection for early-life fitness in the R populations: early-age female fecundity and resistances to starvation, desiccation, and ethanol vapor. Fecundity was measured when flies were 4–5 days old. Fifty female/male pairs per population were placed in vials containing charcoal medium covered with a suspension of live yeast and allowed to oviposit for one 24-hr period. The procedures for testing starvation, desiccation, and ethanol resistance were similar to those used by Service et al. (1985). Forty females (6 days old) per population were used to assay each stress-resistance character.

Flies used for phenotypic assays were either taken directly from the experimental populations or were the second-generation descendants of flies sampled from the experimental populations. These latter flies were the product of two generations of controlled-density rearing (30 eggs per 8-dram shell vial) under conditions designed to forestall selection for early-life fitness components. The procedures for controlled-density assays are intended to reduce or eliminate unwanted environmental causes of variance, including possible maternal effects resulting from systematic differences in culture conditions between selected and unselected lines. All previous phenotypic assays of the B and O populations (Rose, 1984; Service et al., 1985) have used two generations of controlled-density rearing. However, Luckinbill et al. (1984) have demonstrated that response to selection for early-life and late-life fitness can be reliably detected when assayed flies are taken directly from experimental populations. In fact, phenotypic divergence between early- and late-selected lines is more marked in uncontrolled-density assays (Clare and Luckinbill, 1985). The crucial difference between controlled- and uncontrolled-density rearing of flies used for phenotypic assays is probably not control per se, but the fact that uncontrolled density generally means much higher density. Thus, while we shall adopt the practice of referring to our two types of assays as uncontrolled density (for flies taken directly from the experimental populations) or controlled density (for grandchildren of flies in the experimental populations), it should be borne in mind that the important distinction is probably high versus low density, respectively. Controlled-density assays were done on the grandchildren of flies from every other generation of selection, beginning with the first and ending with the eleventh. Uncontrolled-density assays were done on alternate generations, beginning with the second generation and continuing through the twenty-second.

Controlled-Density Selection.—Early-age fecundity and female starvation resistance were assayed for each generation, using flies taken directly from the experimental populations. Fifty flies were used per population for each character. Since the experimental populations (RC_1–RC_5 and BC_1–BC_5) were maintained at controlled density, these are controlled-density assays.

Method of Data Analysis

We analyzed our results by least-squares linear regression of population means on generation. For example, the mean starvation resistance level of the R_1 population in the eighth generation was taken as a single datum (R_{18}). In order to mitigate the effects of uncontrolled environmental variation, the means of individual reverse-selected lines in a generation were divided by the average level of the appropriate B controls for that generation. Therefore, one of our calculations involved the linear regression of the ratio $R_{1j}/B._j$ on generation, where subscript j refers to generation and $B._j$ is the mean of the five B populations in the jth generation. Tests of our hypotheses centered on tests of the significance of regression coefficients. However, because of our replication of re-

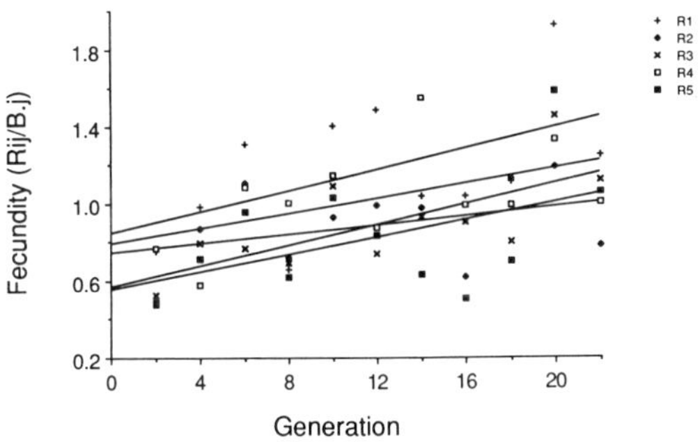

FIG. 1. Response of early-age fecundity to reverse selection for early-life fitness with uncontrolled larval rearing density and using uncontrolled-density assays. Each line is the linear least-squares regression on generation of a population mean value (R_{ij}) divided by the overall mean of the control populations ($B_{.j}$). Subscripts i and j refer to population and generation, respectively. The different symbols refer to the five reverse-selection populations.

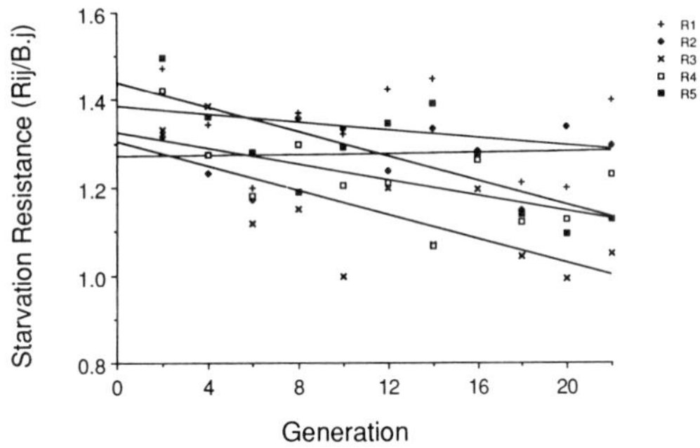

FIG. 2. Response of starvation resistance to reverse selection for early-life fitness with uncontrolled larval rearing density and using uncontrolled-density assays. Explanation of lines and symbols is as for Figure 1.

verse-selection lines and the problem of genetic sampling effects (Hill, 1971), we used each slope as a data point and performed t tests for the significance of deviation from zero of a set of five regression slopes. Since the appropriate alternative statistical hypothesis is for response to selection in the direction of the B-population controls, one-tailed P values are given.

One assumption of the regression model is that the error terms, ϵ_i, associated with the several values of the independent variable, X_i, are themselves independent random variables. This assumption can be violated in time-series data when the error terms become positively autocorrelated. Such autocorrelation is likely to arise if important variables that are correlated with time are omitted from the model (Neter and Wasserman, 1974 p. 352). In the present case, we could imagine some environmental trend in the laboratory. However, unless the control and selected populations reacted in different ways to such "missing" environmental factors, use of control lines to "standardize" the trait values in the selected lines should have eliminated or reduced the effects of such factors.

RESULTS

Fecundity and Starvation Characters

Uncontrolled-Density Selection.—Figures 1 and 2 show the results of 22 generations of uncontrolled-density reverse selection for the fecundity and starvation characters, respectively, using uncontrolled-density assays. Fecundity increased significantly over the course of selection, while starvation resistance decreased sig-

TABLE 2. Responses to reverse selection. Tabled values are mean regression coefficients with t statistics in parentheses, and significance levels are associated one-tailed probabilities (4 *d.f.*).

Line, assay, and generation	Fecundity	Starvation	Desiccation	Ethanol
Uncontrolled-density selection				
Uncontrolled-density assay				
22 generations	0.021 (7.63)[c]	−0.008 (2.99)[b]	−0.003 (1.24)	−0.005 (0.35)
12 generations	0.037 (6.06)[c]	−0.012 (2.35)[a]	0.012 (1.91)	0.025 (0.64)
Controlled-density assay				
11 generations	0.018 (1.65)	−0.035 (2.53)[a]	−0.001 (0.13)	−0.018 (1.79)
Controlled-density selection				
12 generations	−0.005 (0.90)	−0.0002 (0.03)	—	—

[a] $P \leq 0.05$.
[b] $P \leq 0.025$.
[c] $P \leq 0.005$.

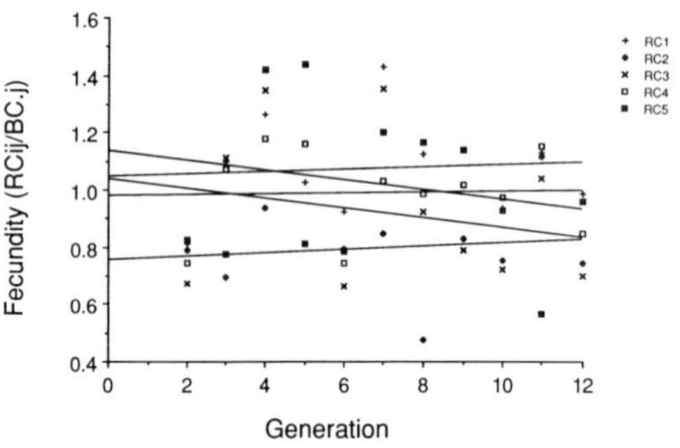

FIG. 3. Response of early-age fecundity to reverse selection for early-life fitness with controlled larval rearing density and using controlled-density assays. Each line is the linear least-squares regression on generation of a population mean value (RC_{ij}) divided by the overall mean of the control populations ($BC._j$). Subscripts *i* and *j* refer to population and generation, respectively. The different symbols refer to the five reverse-selection populations.

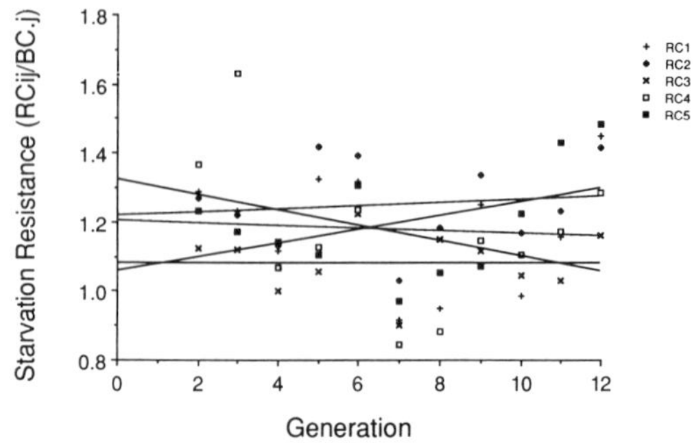

FIG. 4. Response of starvation resistance to reverse selection for early-life fitness with controlled larval rearing density and using controlled-density assays. Explanation of lines and symbols is as for Figure 3.

nificantly (Table 2). The former is only to be expected, given the reintroduction of selection for early fertility. The latter is evidently a reflection of antagonistic pleiotropy between starvation resistance and some early-life fitness component(s). Service and Rose (1985) estimated a large-magnitude negative genetic correlation between early fecundity and starvation resistance, suggesting that fecundity itself might be the key character whose rise depresses starvation resistance.

Controlled-Density Selection.—Figures 3 and 4 show the comparable results of reverse selection with controlled larval rearing densities over twelve generations for fecundity and starvation resistance, respectively. In contrast to the results for uncontrolled-density selection, there was no statistically significant change in either character over the generations studied (Table 2). This difference between the results for uncontrolled- and controlled-density selection cannot be attributed to the fewer generations in the latter experiment. Analysis for the corresponding 12 generations of reverse selection upon uncontrolled-density lines still indicates a statistically significant response to selection (Table 2). One interpretation of this is that it might reflect problems of assay sensitivity at lower rearing densities (cf. Clare and Luckinbill, 1985). However, the controlled-density assay of the uncontrolled-density lines reveals a statistically significant decrease in starvation resistance (Table 2). Therefore, the lack of response to reverse selection seems to reflect a difference in selection response, rather than an assay artifact.

Desiccation- and Ethanol-Resistance Characters

Figures 5 and 6 show the effects of 22 generations of reverse selection upon the desiccation- and ethanol-resistance characters in the uncontrolled-density selection lines, using uncontrolled-density assays. There was no consistent significant change over the period studied. Assays with controlled and uncontrolled larval rearing densities were not different (Table 2); in both cases, there was no significant change. These results do not accord well with an interpre-

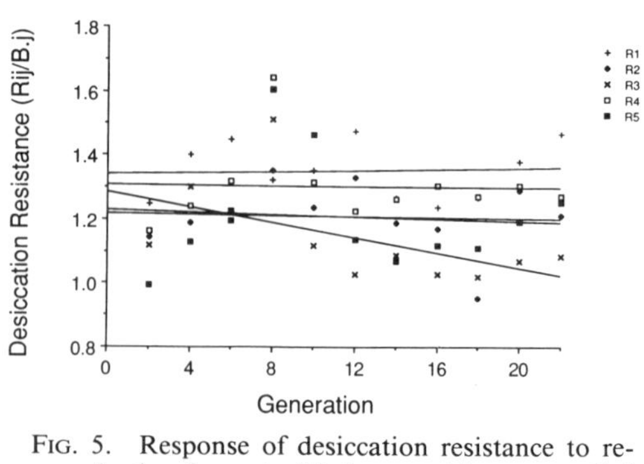

FIG. 5. Response of desiccation resistance to reverse selection for early-life fitness with uncontrolled larval rearing density and using uncontrolled-density assays. Explanation of lines and symbols is as for Figure 1.

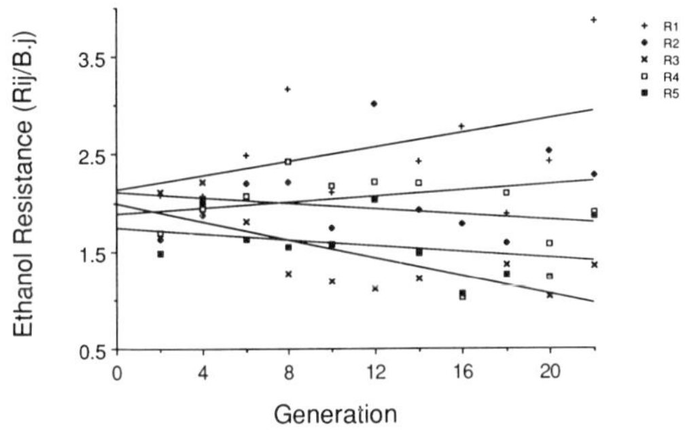

FIG. 6. Response of ethanol resistance to reverse selection for early-life fitness with uncontrolled larval rearing density and using uncontrolled-density assays. Explanation of lines and symbols is as for Figure 1.

tation of antagonistic pleiotropy, as we discuss below.

DISCUSSION

The foregoing results seem to us to be significant in four respects. Firstly, we have obtained independent evidence that the starvation-, desiccation-, and ethanol-resistance characters in our populations are not controlled by a single physiological mechanism. These results support the findings of Service (1987) and further suggest that the life-history differences between the B and O populations have a polygenic basis.

Secondly, we have corroborated the findings of Luckinbill and Clare (1985, 1986) and Clare and Luckinbill (1985): selection response in *Drosophila* life-history experiments can be critically dependent on larval rearing density. In particular, low rearing densities seem to reduce responses to selection (cf. Lints and Hoste, 1974; Lints et al., 1979).

Thirdly, we have still more evidence for antagonistic pleiotropy in the fall of starvation resistance upon a return to selection for early fertility and a relaxation of selection for later fertility. The rapid rise of fecundity upon reverse selection also suggests the existence of genetic antagonism between early reproduction and later reproduction, which must have been reducing early reproduction in our longer-lived late-reproduced stocks. These results are entirely consistent with previous work on these populations (Rose and Charlesworth, 1981*a*, 1981*b*; Rose, 1984; Service and Rose, 1985).

Fourthly, and of more interest, the results with respect to desiccation and ethanol resistance suggest that these characters can respond to selection for later reproduction in such a way that there is little or no antagonistic response in early-life fitness components. At the level of gene action, this suggests that there are alleles that enhance survival to, or reproduction at, later ages that do not have deleterious effects at early ages. This is a third line of evidence in support of mutation accumulation as a population-genetic mechanism for the evolution of senescence in *Drosophila* (see also Kosuda [1985] and Mueller [1987]). In the context of our experiments, we interpret these results to indicate that selection for early-life fitness in the ancestral fly population allowed alleles with adverse effects on desiccation and ethanol resistance to achieve relatively high frequencies. Subsequent selection in the O populations for late-life fitness reduced the frequencies of these deleterious alleles. Finally, during the relatively short period of reverse selection, allele frequencies in the R populations remained at O-population levels because of effective neutrality under conditions favoring early-life fitness.

There is one major problem with this line of interpretation: the possibility that the original longer-lived populations could have been fixed for the alleles enhancing resistance to desiccation and ethanol. There are two arguments against this alternative interpretation. Firstly, it is likely that there were many loci contributing to the genetic variance that underlay the response to selection for late-life fitness. In the large populations that we have used, it is unlikely that all the relevant alleles would have proceeded to fixation in the course of selection. This argument is not necessarily inconsistent with the estimate by Luckinbill et al. (1987) of a low number of effective factors controlling senescence in *D. melanogaster*. Their method of estimation (Wright, 1968; Lande, 1981) is based on the assumption that all factors have equal effect. Violation of this assumption will underestimate the number of effective factors, in some cases seriously (Wright, 1968 pp. 384–386). Secondly, if there were a few loci having fixed alleles of large effect that also exhibited antagonistic pleiotropy with respect to early- and late-life fitness, then they should have made their effects manifest by preventing a return to control levels of early-life fitness characters. The one such character that we observed, early-age fecundity, exhibited a return to the control level (Fig. 1), as would be expected if there were no fixed alleles that had depressed it in the longer-lived populations.

There may, of course, be other, unexamined early-life fitness components that did not return to control-population levels, so our argument against antagonistic pleiotropy and, by implication, for mutation accumulation as the mechanism for the evolution of desiccation and ethanol resistance is necessarily qualified. Also, the pattern of selection and reverse-selection responses that we have attributed to mutation accumulation could result from epistatic interactions in a system initially characterized by antagonistic pleiotropy. Lenski (1988) has shown that fitness costs associated with phage resistance in *E. coli* can be partly overcome by changes at loci not initially involved in the evolution of phage resistance. In the present context, the retention of enhanced desiccation and ethanol resistance in the R lines together with a return to control levels of early-life fitness (early-age fecundity) might have been due to changes at "modifier" loci. Pursuing this argument, the antagonism or neutrality, with respect to early-life fitness, of major alleles controlling desiccation and ethanol resistance would be a function of genetic background: ancestral versus O (or R) populations, respectively.

Despite these potential alternatives to the interpretation of mutation accumulation, we feel that the weight of evidence in the present experiments supports the conclusion that both mutation accumulation and antagonistic pleiotropy are important mechanisms for the evolution of senescence. Both now have multiple, independent lines of evidence for their action. However, both are attested to primarily by work with *Drosophila* and, therefore, stand in need of evaluation with other species (Rose and Service, 1985). It is worth commenting that the fecundity character used by Rose and Charlesworth (1980, 1981*a*) to test the mutation-accumulation mechanism appears, from the present results, to be a poor character for the evaluation of the mechanism. There is evidently a great deal of additive genetic variance for the character at early ages—genetic variance that is tied up with an antagonistic-pleiotropy mechanism or mechanisms (Rose, 1984; Service and Rose, 1985). Tests of mutation accumulation based on age-dependence of genetic variance in a fitness component require that antagonistic pleiotropy be virtually absent. The present type of test does not require this, nor does that of Mueller (1987). On the other hand, the present test is weakened in that the argument for mutation accumulation is based on a negative result. As antagonistic pleiotropy is undoubtedly common in the evolution of senescence (Rose and Service, 1985), coupled selection and reverse-selection experiments, when practical, should be considered in conjunction with overall tests of genetic variation and covariation (sib analyses) in future work on population-genetic mechanisms of senescence.

Acknowledgments

We thank L. Johnston, D. M. Lane, B. Musgrave, A. J. Shaw, B. Singh, and B. Tremblay for assistance in the laboratory. R. E. Lenski and K. C. Nishikawa provided helpful criticisms of the manuscript. This research was supported in part by Operating Grant U0178 to M.R.R. from the Natural Sciences and Engineering Research Council of Canada, in part by Fellowship 5 F32 AG05394 to P.M.S. from the National Institutes of Health of the United States, and in part by Grant AG06346 to M.R.R. from the National Institutes of Health.

Literature Cited

Charlesworth, B. 1980. Evolution in Age-Structured Populations. Cambridge Univ. Press, Cambridge, U.K.

Clare, M. J., and L. S. Luckinbill. 1985. The effects of gene environment interaction on the expression of longevity. Heredity 55:19–29.

Clark, A. G. 1987. Senescence and the genetic correlation hang-up. Amer. Natur. 129:932–940.

Edney, E. B., and R. W. Gill. 1968. Evolution of senescence and specific longevity. Nature 220:281–282.

Hamilton, W. D. 1966. The moulding of senescence by natural selection. J. Theoret. Biol. 12:12–45.

Hill, W. G. 1971. Design and efficiency of selection

experiments for estimating genetic parameters. Biometrics 27:293–311.

KOSUDA, K. 1985. The aging effect on male mating activity in *Drosophila melanogaster*. Behav. Genet. 15:297–303.

LANDE, R. 1981. The minimum number of genes contributing to quantitative variation between and within populations. Genetics 99:541–553.

LENSKI, R. E. 1988. Experimental studies of pleiotropy and epistasis in *Escherichia coli*. II. Compensation for maladaptive effects associated with resistance to virus T4. Evolution 42:433–440.

LINTS, F. A., AND C. HOSTE. 1974. The Lansing effect revisited. I. Life-span. Exp. Geront. 9:51–69.

LINTS, F. A., J. STOLL, G. GRUWEZ, AND C. V. LINTS. 1979. An attempt to select for increased longevity in *Drosophila melanogaster*. Gerontology 25:192–204.

LUCKINBILL, L. S., R. ARKING, M. J. CLARE, W. C. CIROCCO, AND S. A. BUCK. 1984. Selection for delayed senescence in *Drosophila melanogaster*. Evolution 38:996–1003.

LUCKINBILL, L. S., AND M. J. CLARE. 1985. Selection for life span in *Drosophila melanogaster*. Heredity 55:9–18.

———. 1986. A density threshold for the expression of longevity in *Drosophila melanogaster*. Heredity 56:329–335.

LUCKINBILL, L. S., M. J. CLARE, W. L. KRELL, W. C. CIROCCO, AND P. A. RICHARDS. 1987. Estimating the number of genetic elements that defer senescence in *Drosophila*. Evolutionary Ecology 1:37–46.

MEDAWAR, P. B. 1946. Old age and natural death. Modern Quart. 2:30–49.

MUELLER, L. D. 1987. Evolution of accelerated senescence in laboratory populations of *Drosophila*. Proc. Nat. Acad. Sci. USA 84:1974–1977.

NETER, J., AND W. WASSERMAN. 1974. Applied Linear Statistical Models. Irwin, Homewood, IL.

ROSE, M. R. 1984. Laboratory evolution of postponed senescence in *Drosophila melanogaster*. Evolution 38:1004–1010.

ROSE, M., AND B. CHARLESWORTH. 1980. A test of evolutionary theories of senescence. Nature 287: 141–142.

———. 1981*a*. Genetics of life history in *Drosophila melanogaster*. I. Sib analysis of adult females. Genetics 97:173–186.

———. 1981*b*. Genetics of life history in *Drosophila melanogaster*. II. Exploratory selection experiments. Genetics 97:187–196.

ROSE, M. R., AND P. M. SERVICE. 1985. Evolution of aging. Rev. Biol. Res. Aging 2:85–98.

SERVICE, P. M. 1987. Physiological mechanisms of increased stress resistance in *Drosophila melanogaster* selected for postponed senescence. Physiol. Zool. 60:321–326.

SERVICE, P. M., E. W. HUTCHINSON, M. D. MACKINLEY, AND M. R. ROSE. 1985. Resistance to environmental stress in *Drosophila melanogaster* selected for postponed senescence. Physiol. Zool. 58:380–389.

SERVICE, P. M., AND M. R. ROSE. 1985. Genetic covariation among life-history components: The effect of novel environments. Evolution 39:943–945.

WATTIAUX, J. M. 1968. Cumulative parental age effects in *Drosophila subobscura*. Evolution 22:406–421.

WILLIAMS, G. C. 1957. Pleiotropy, natural selection, and the evolution of senescence. Evolution 11:398–411.

WRIGHT, S. 1968. Evolution and the Genetics of Populations. I. Genetic and Biometric Foundations. Univ. Chicago Press, Chicago, IL.

Corresponding Editor: P. E. Smouse

268

Desiccation, Flight, Glycogen, and Postponed Senescence in *Drosophila melanogaster*

J. L. Graves[1]
E. C. Toolson[2]
C. Jeong[1,*]
L. N. Vu[1]
M. R. Rose[1]
[1]Department of Ecology and Evolutionary Biology, School of Biological Sciences, University of California, Irvine, California 92717; [2]Department of Biology, University of New Mexico, Albuquerque, New Mexico 87131

Accepted 8/22/91

Abstract
Drosophila melanogaster *selected for postponed aging have increased resistance to desiccation and increased flight duration. These and other populations were tested for longevity, flight duration, glycogen content, and stress resistance under a variety of conditions. It was found that desiccation resistance was positively associated with flight duration and glycogen content, while glycogen was exhausted in the course of desiccation, as it is in flight. Flies with postponed aging also have lower water-loss rates (WLRs) when dead, suggesting that factors other than glycogen content are partly responsible for their increased desiccation resistance. However, total epicuticular hydrocarbon does not appear to determine desiccation resistance. Starvation resistance does not vary in a manner that corresponds with desiccation resistance, under either selection or manipulation, but does vary in association with lipid level. This suggests at least two physiological mechanisms by which aging has been postponed in these flies: increased lipid content and increased glycogen content. These mechanisms are at least somewhat independent, evolutionarily, genetically, and physiologically.*

Introduction

Artificial selection experiments employing selection for late-life fitness in insects, mainly in *Drosophila melanogaster*, have succeeded in creating lines exhibiting postponed senescence (Wattiaux 1968*a*, 1968*b*, Rose and

* Present address: Medical College of Pennsylvania, Philadelphia, Pennsylvania 19129.

Physiological Zoology 65(2):268–286. 1992. 0031-935X/92/6502-90146$02.00

Charlesworth 1981; Rose 1984; Luckinbill and Clare 1985). These experiments confirm the modern evolutionary theory of senescence, that aging is the result of the declining force of natural selection with age of the adult soma (Hamilton 1966; Rose 1991).

However, the evolutionary theory does not, by itself, supply physiological mechanisms that could explain variation in longevity. A number of experimental studies have addressed this problem in *D. melanogaster* (Rose 1984; Rose et al. 1984; Luckinbill and Clare 1985; Service et al. 1985; Luckinbill et al. 1987; Service 1987; Service, Hutchinson, and Rose 1988; Graves, Luckinbill, and Nichols 1988; Luckinbill et al. 1988; Graves and Rose 1990; Rose, Graves, and Hutchinson 1990). These studies have found many physiological differences associated with evolutionarily postponed senescence: reduced early fecundity, decreased early metabolic rate, and decreased early locomotion. Senescence is countered by increased resistance to desiccation, ethanol vapor, starvation, increased tethered-flight duration, increased later locomotion, and increased lipid content.

These findings in turn raise additional questions concerning the action of the specific physiological mechanisms involved in postponed senescence and their relationships to each other. Service et al. (1988) attempted to address these questions using reverse selection. They placed lines previously selected for postponed aging (O's) on an early reproduction schedule to select for early fertility. The rationale for this was to separate genes that contributed to postponed senescence at a cost to early fitness, a situation known as antagonistic pleiotropy (Williams 1957; Rose 1985), from those genes that do not exhibit antagonistic pleiotropy. This study found that, after 22 generations of reverse selection (RU), lines exhibited an increase in early fecundity, a decrease in starvation resistance, and no change in desiccation or ethanol vapor tolerance. Service et al. (1988) proposed that these results indicated the presence of antagonistic pleiotropy between starvation and early fecundity, as had been found before (Service and Rose 1985). On the other hand, they concluded that desiccation and ethanol resistance are largely free of antagonistic pleiotropy because they did not respond to a reversal of selection intensities on early and late ages. In addition, these stress-resistance characters are at least somewhat independent from the resistance-to-starvation character.

This study continues the study of these reversed-selection stocks after more than 100 generations of further selection. Specifically, it answers two population genetics questions: (1) Where do reverse selection lines fall relative to the control and postponed aging populations after additional generations of selection, and do all the characters respond to reverse selection at the same rate? (2) Has inbreeding played any role in the evolution of the reversed-selection stocks?

But another level of interest is that of physiological mechanism. The mechanisms of desiccation resistance in insects, and in *Drosophila* particularly, have been studied intensively (Wigglesworth 1972; Clark and Doane 1983; Lighton 1988; Toolson and Kuper-Simbron 1989; Toolson et al. 1990). Given the increased desiccation resistance of postponed aging stocks (Service et al. 1985), an examination of mechanisms of desiccation resistance in these stocks should be of interest from the perspective of insect physiology. Thus, this study estimates rates of water loss in the B and O lines.

Another point of experimental concern is differences in surface hydrocarbons, which are known to play an important role in determining rates of water loss through the cuticle (Edney 1977; Hadley 1984). Another physiological character of interest to us is flight duration (see, e.g., Graves et al. 1988; Graves and Rose 1990). The metabolic basis of this character is reasonably well-known in *Drosophila.* Glycogen is known to be the exclusive reserve substance fueling flight, exhaustion of glycogen leading to cessation of flight (Williams, Barnes, and Sawyer 1943; Wigglesworth 1949). However, the glycogen used appears to be that in the fat body; females flown to exhaustion contain glycogen in the oocytes (Wigglesworth 1949). Lipids are not consumed for flight, nor is there any conversion of lipid into glycogen or vice versa.

Our results indicate the presence of at least two distinct physiological mechanisms involved in the postponement of aging, one bound up with the determination of starvation resistance, depending primarily on lipid level, and one involved in desiccation resistance and flight duration, depending primarily on glycogen level.

Material and Methods

Stocks

The lines used in this study were originally described in Rose (1984) and Service et al. (1985, 1988). Five replicate lines were subjected to selection for early- and late-life fitness (B and O lines, respectively). The O lines are currently maintained in discrete 10-wk generations while their control populations (B) are maintained on 2-wk generations. The 10 populations were derived simultaneously from a common ancestral population that had been maintained for the preceding 5 yr under conditions favoring early-life fitness. All populations used in this study have been maintained in cages at population sizes on the order of 10^3–10^4.

All flies used in these experiments developed at larval densities between 60 and 100 in standard molasses-banana medium, as described in Rose (1984). These populations were maintained at those densities as adults until the physiological assays were performed. The age of the flies used in the physiological assays was between 7 and 12 d after eclosion and only females were tested, unless otherwise specified. The flies were reared and maintained at 25°C on a 24-h light cycle.

Four replicate RU lines were created by placing O lines on a 2-wk reproduction schedule as described in Service et al. (1988). To test for inbreeding depression, a series of reciprocal crosses were made between all possible combinations of the replicate RU lines: RU1 × RU2, RU1 × RU3, RU1 × RU4, RU2 × RU3, RU2 × RU4, and RU3 × RU4. These reciprocal crosses were prepared by removing 20 replicate samples of 50 virgin females from the pure cultures and mating them in vials with 50 males. Each crossed population was reared as 20 vials of 60–100 eggs each. Individuals used in the physiological assays were randomly chosen from flies that emerged from these vials. The crossed populations were tested against control pure lines that had been produced in the same fashion.

In addition, flies selected for increased desiccation resistance, called D stocks, were compared with their controls, C stocks, that were otherwise handled in the same manner (Rose et al. 1990). These D and C stocks were derived from O stocks but were cultured at early ages. The longevity of D stocks is greater than that of C stocks (data not shown). Desiccation resistance is substantially increased in D stocks relative to C, O, or B stocks (data not shown).

Longevity, Flight Duration, Glycogen, and Stress-Resistance Assays

Longevity for females and males, flight duration, and resistance to desiccation, ethanol, and starvation were tested for the RU selection lines relative to B and O control lines.

Longevity of females and males was measured by placing one female and one male in a standard 25 × 95-mm shell vial with the bottom quarter filled with molasses-banana medium. The flies were transferred three times a week. Upon the death of one of the flies, a replacement fly of the same sex was added, so that all flies experienced the presence of a mate throughout the longevity assay. Forty pairs were assayed per replicate population.

The technique used to stimulate tethered flight is described in Graves et al. (1988) and Graves and Rose (1990). This involves lightly etherizing the fly and then tethering it to a light test piece of fishing line by use of Duco cement. The fly can be stimulated to fly by use of the tarsal reflex or by lightly passing an air current along the head-tail axis. *Drosophila* can be flown to exhaustion by this technique by an observer who immediately stimulates the fly at any momentary cessation of flight.

The desiccation, ethanol, and starvation-resistance assays are described in Service et al. (1985). In brief, these assays consist of four females being placed in a 25 × 95-mm shell vial. For the starvation assay, the females are confined to the bottom quarter of the vial, with a half sponge separating the flies from two cotton plugs that have been saturated with 3 mL of distilled water. The vial is then covered with Parafilm to prevent evaporation. In the case of the ethanol assay 3 mL of 15% ethanol solution are added to the cotton plug in place of water, and in the case of the desiccation assay, the cotton plugs are replaced with 3 mL of Drierite desiccant. The flies undergoing stress are then observed hourly in the case of desiccation and every 6 h in the case of starvation. The flies were recorded as dead when they no longer responded to physical stimulus.

Glycogen was assayed by microseparation of glycogen after the technique of Van Handel (1965) was performed on the five replicate B and O populations both before and after desiccation. Three females were used for each assay; the females were chosen that were of equal size distributions between the two lines. A total of five assays, thus containing 15 females, were performed for each population, in the two lines. The results are reported as micrograms of substance per three females.

Water Loss Rates and Surface Hydrocarbons

Virgin B and O line males and females were separated at eclosion and analyzed at 5 d of age. Flies were killed by exposure to cyanide vapor, and water-loss rates (WLRs) were determined gravimetrically on replicate samples of 20 flies as discussed in Toolson (1982); three replicate samples were used for each sex from each of the 10 experimental populations.

After WLR determination, epicuticular hydrocarbons (HCs) were extracted and isolated by placing each sample of flies on a Biosil-A minicolumn and washing the column with 8 mL hexane (Toolson et al. 1990). An internal stardard (5.0 μg n-docsane) was added to each extract, and the HCs were then analyzed by capillary gas-liquid chromatography.

The Effect of Tethered Flight on Desiccation and Starvation Resistance

To test the relationship of desiccation and starvation resistance to flight duration and hence to glycogen reserves in these lines, 40 females were flown to exhaustion for the five replicate B, O, and four replicate RU selection lines. Typical flight durations for these lines are shown in figure 1. Flies flown to exhaustion were untethered and placed in the desiccation and starvation assays as described above.

Two control groups were set up, the first group was not flown (NF), or exposed to ether (NE) or gluing (NG). The second control group was etherized, and then tethered with glue, but not flown (G,E). The second control was allowed to wake up and then released and placed in starvation and desiccation assays. The largest cause of immediate death in the (G,E) group was mechanical damage due to release from the tether. Flies killed in this way were not recorded in the stress assays; however, some flies with sublethal injuries may have been included in this group. To minimize this, flies were observed for several minutes for erratic behavior; suspect flies were removed from the (G,E) control. There is no reason to believe that there was any bias in the injuries due to release for any population.

The individual survival times were recorded for each assay from the replicate populations. Each population mean was pooled to get a mean value for the line in that treatment. The mean survival values after flight for each group and treatment were compared by use of a Student's t-test.

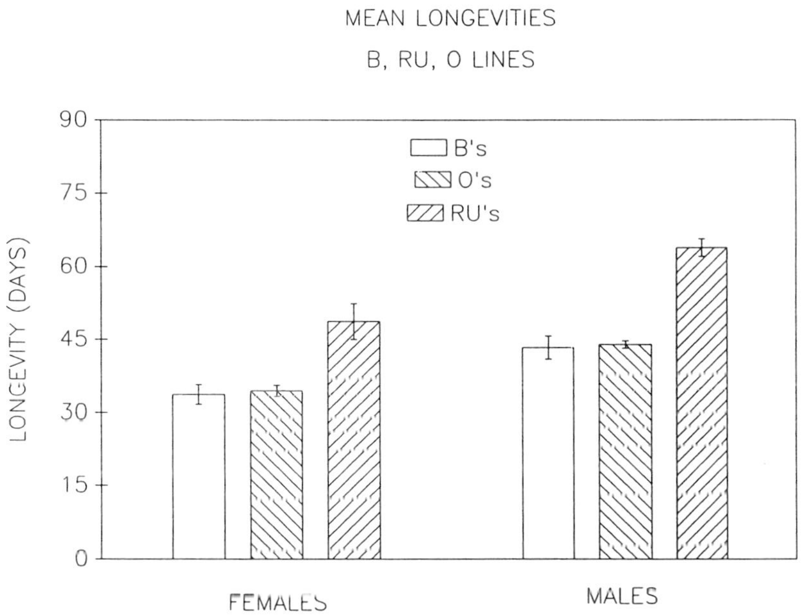

Fig. 1. The mean and SE for longevity from each line by sex is shown; n *= 4 for each line, and* n *= 40 for each population within a line. The mean value for the line is computed from the individual means for each of the replicate populations. Longevities are reported in days.*

Results

Physiological Assays of Pure and Crossed Reverse Selection Lines

Figure 1 shows the mean and standard error for longevities in females and males from four replicate populations of B, RU, and O lines at greater than 100 generations of reverse selection. A two-way ANOVA for line and sex found that both were highly significant ($F = 42.42, 40.92$). There was no significant interaction between line and sex. Pairwise t-tests revealed that males exhibited greater longevity than females, and O lines > RU = B lines, $P < 0.001$.

Figure 2 gives the mean and standard error for tethered flight duration of B, RU, O females between 7 and 12 d old. Pairwise t-tests revealed that O lines > RU = B lines, $P < 0.004$. The RU lines were slightly lower than B lines.

Figure 3 reports the mean and standard error for desiccation, ethanol, and starvation resistances for B, RU, and O females. Mean differences were tested by paired Student's t-test. The desiccation and ethanol results showed O > B = RU at $P < 0.006$, and $P < 0.01$, respectively. As in the flight duration results the RU lines were slightly lower than the B lines. Starvation resistance

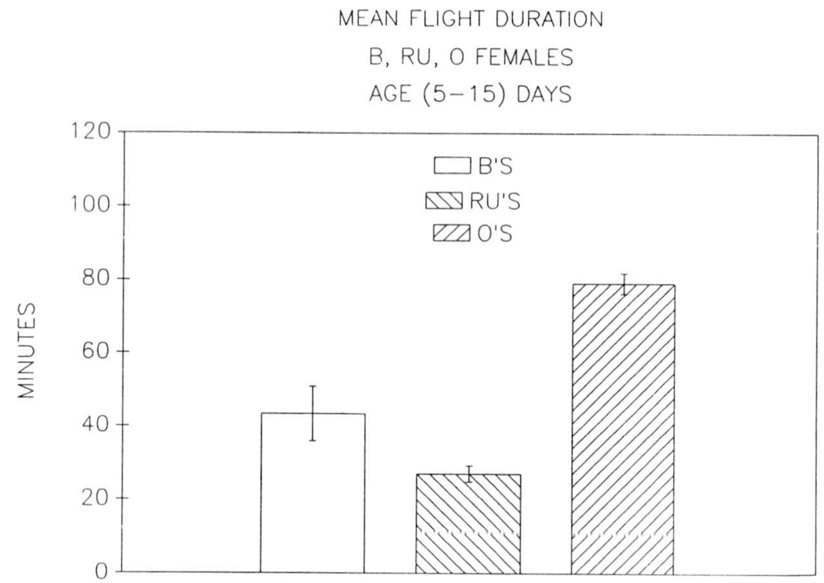

Fig. 2. The mean and SE for flight duration in females from each line is shown; n = *4 for each line, and* n = *40 for each population within a line. The mean value for the line is computed from the individual means for each of the replicate populations. Flight duration is reported in minutes.*

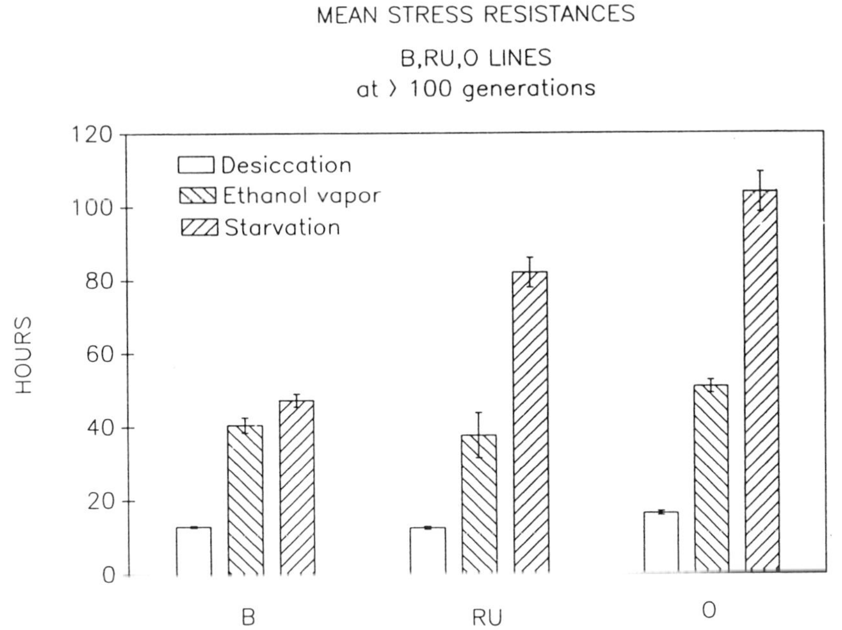

Fig. 3. The mean and SE of stress resistance in females from each line is shown; n = 5 *for the B and O lines,* n = 4 *for the RU line, and* n = 40 *for each population within a line. The mean value for the line is computed from the individual means for each of the replicate populations. Survival times are given in hours.*

ranked O > RU > B, with all significantly different from each other, at $P < 0.01$ and $P < 0.001$, respectively.

Figure 4 gives the mean and standard error for glycogen contents of B, RU, and O stocks when they are handled normally and under desiccation. Mean differences are tested by Student's *t*-test, unpaired except where comparisons were made within treatments. The nondesiccated between-line glycogen comparisons show that O = RU > B, at $P < 0.02$, while after desiccation B > O = RU, at $P < 0.004$. The within-line comparisons were all significantly less in glycogen after desiccation at $P < 0.003$. Desiccation dramatically exhausts glycogen, with the amounts that remain correlating negatively with the average time of survival per line. The B, RU, and O lines had 13%, 5%, and 3%, respectively, of the original glycogen content of the nondesiccated flies. These values are in the opposite order of their survival times.

Figure 5 gives the results of the comparison of longevity and flight duration from the pure versus crossed RU lines. Mean differences were tested by paired Student's *t*-test. No significant differences could be found for longevity

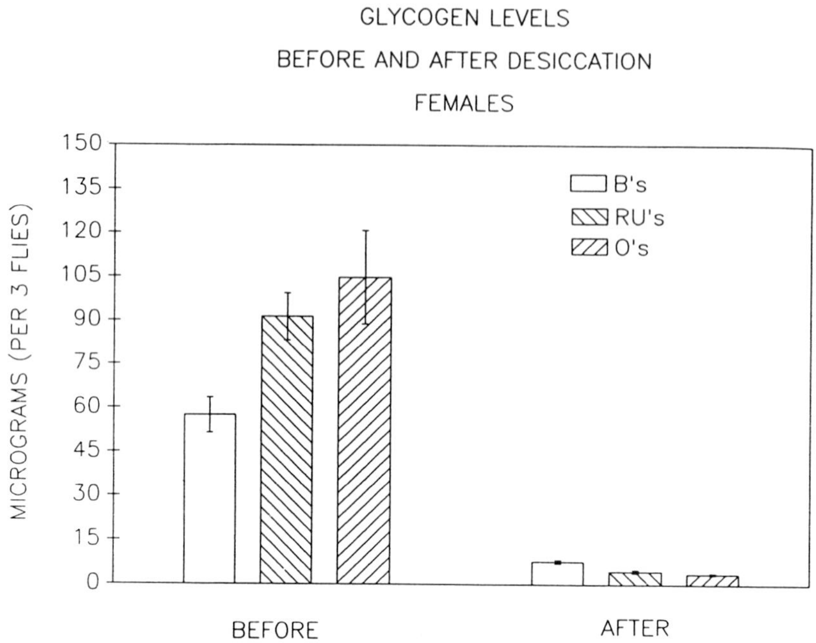

Fig. 4. The mean and SE of glycogen content in females from each line is shown before and after desiccation; n = *5 for B and O lines,* n = *4 for the RU line;* n = *8 replicates containing three flies were extracted for each population within a line for both treatments. The mean value for the line is computed from the individual means for each of the replicate populations. Glycogen is reported in micrograms per three flies.*

(female or male) or flight duration. Figure 6 gives the results for ethanol, desiccation, and starvation characters. Of all the characters examined only the desiccation assay showed a significant difference, favoring the pure line. This pattern is counter to the prediction of inbreeding depression, which would have found all characters in the crossed at higher values than the pure lines.

Water Loss Rates and Epicuticular Hydrocarbons

Results of WLR determination are presented in table 1. Using a two-way ANOVA for treatment and sex effects, male WLRs were significantly higher than WLRs of females of the same strain, and WLRs of O-strain males and females were significantly less than WLRs of their B-strain counterparts. Analysis of the epicuticular HCs yielded more than 20 individual components ranging from C_{20} to C_{33}. However, 12 components accounted for about 90%

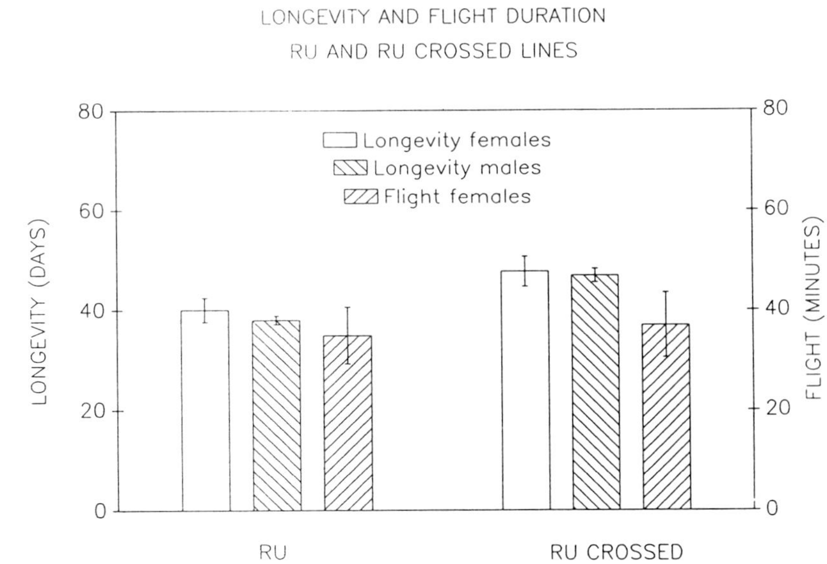

Fig. 5. The mean and SE for longevity and flight duration are shown; n *= 4 for the RU lines. Each population had 40 flies of each sex tested for the longevities, and 20 females per population were tested for the flight duration tests. Times are given in days for longevity and minutes for flight duration, respectively. The mean value for the line is computed from the individual means for each of the replicate populations.*

of the total HC fraction, and we restricted subsequent analyses to these HCs, which ranged in size from C_{21} to C_{29}. As shown in table 1, males yielded approximately twice as much total HC as females, a significant difference, but no statistically significant treatment differences were detected.

The Effect of Flight on Desiccation and Starvation Resistance

Figure 7 shows the within-line comparisons for the effect of flight on desiccation resistance. The B flown population exhibited significantly lower desiccation resistance than either the B(G,E) or B (NF,NG,NE) groups. The B (G,E) and B (NF,NG,NE) were not significantly different from each other, indicating that there was not a major effect of gluing or ethering on this line. The RU and O lines show the same response for flight on desiccation resistance. In these populations there was a significant effect of the tethering procedure, but both (G,E) treatments were significantly higher than the flown groups.

Figure 7 also shows the between-line comparisons for the effect of flight on desiccation-resistance experiments. The O (NF,NG,NE) group was sig-

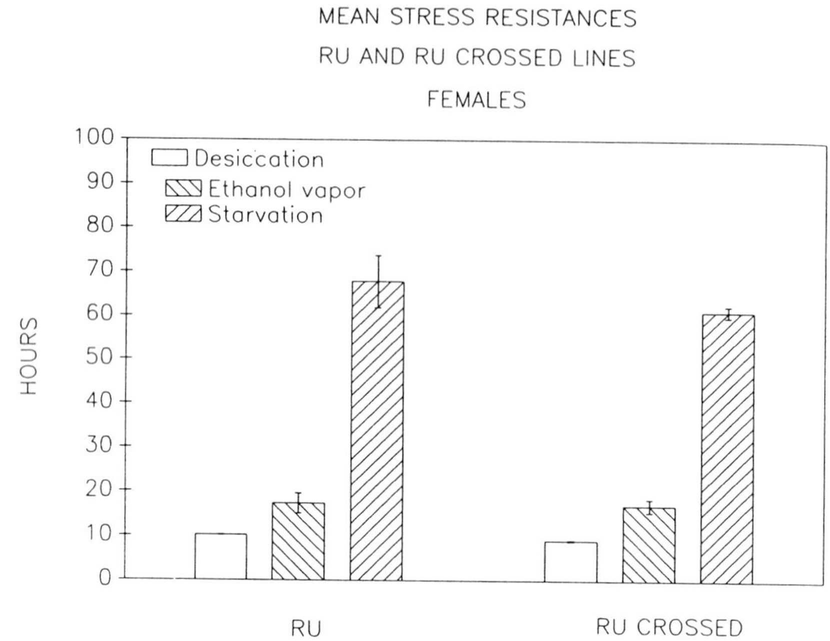

Fig. 6. The mean and SE for stress resistances are shown; n = *4 for the RU lines. Each population had 40 females tested. Times are given in hours. The mean value for the line is computed from the individual means for each of the replicate populations.*

nificantly higher than the B (NF,NG,NE) and RU (NF,NG,NE) groups, $P < 0.001$. The B and RU groups in this treatment were not significantly different from each other. The ranking of the (G,E) group is consistent with the (NF,NG,NE) group, indicating that the tethering procedure affects each line uniformly. The flown populations, however, are not significantly different from each other, a 5.5-h differential in the (G,E) populations is reduced to a 0.71-h difference, which is not statistically significant. Mean differences were tested by means of a Student's *t*-test for the replicate populations.

Figure 8 shows the within-line comparisons for the effect of flight on starvation resistance. The mean differences were tested by means of a Student's *t*-test for the replicate populations. Within-line comparisons showed that both control groups (NF,NG,NE) and (NF,G,E) were not significantly different from flies flown to exhaustion in the B and RU lines. In the O lines there was not a significant decrease in survival time due to flight (flown and NF,G,E). However, there was a slight effect of exposure to glue and ether on starvation resistance (NF,NG,NE and NF,G,E). In the between-line com-

TABLE 1

Mass, epicuticular hydrocarbons (HC), and water loss rate (WLR) of B and O flies

	B	O
Females:		
Mass (mg)	1.17 ± .05	1.26 ± .06
HC (μg)	1.01 ± .57	1.09 ± .23
WLR (μg cm^{-2} h^{-1}) ..	7.73 ± 1.4	6.49 ± 1.4
Males:		
Mass (mg)	.75 ± .04	.87 ± .02
HC (μg)	2.00 ± .47	1.93 ± .35
WLR (μg cm^{-2} h^{-1}) ..	10.01 ± 1.7	7.91 ± 1.4

Note. Three flies were used for each sex from each of the five replicate B and O populations.

parisons, O > RU > B, for the (NF,NG,NE) control at $P < 0.001$, however, this pattern is not repeated for the (NF,G,E) control. This group gives O > RU = B, indicating some differential effect of glue and ether on the survival time of the RUs.

The Effect of Selection for Desiccation Resistance on Glycogen

Figure 9 shows the relationship between B, O, C, and D stocks with regard to glycogen level for both females and males. A two-way ANOVA for line and sex revealed that both treatments were highly significant at $P < 0.01$. Pairwise *t*-tests between the B, O, C, and D lines found the rank D > O = C > B for both sexes at $P < 0.07$ and $P < 0.005$, respectively. Evidently, the most resistant populations, the D's, have the greatest glycogen levels. This fits the phenotypic effect of desiccation on glycogen, indicating that glycogen level is at least one factor limiting desiccation resistance in these populations.

Discussion

The decrease in longevity measured at over 100 generations of reversed selection (RU lines) is consistent with the increase in early fecundity observed in Service et al. (1988) at 22 generations in that it was found that early fecundity had decreased in conjunction with increased longevity (e.g.,

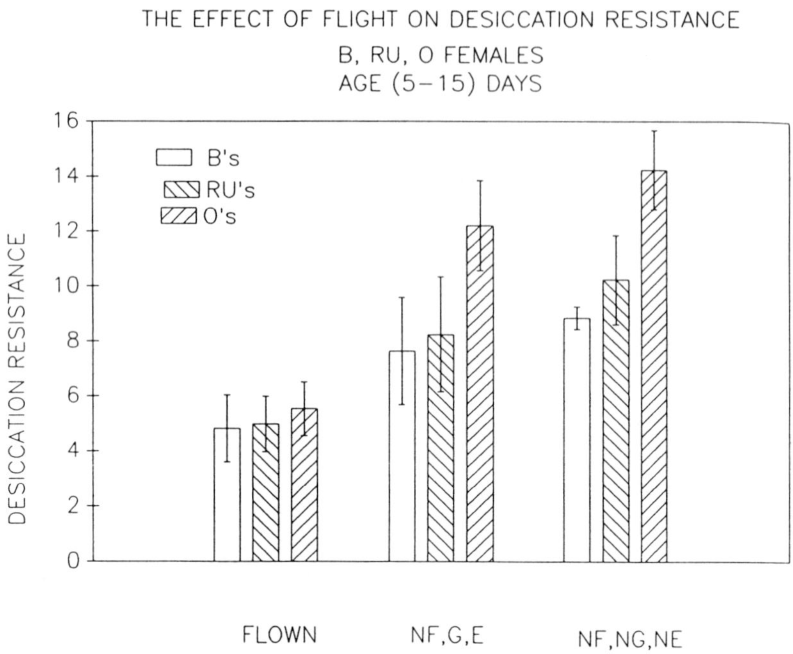

Fig. 7. The means and SEs for desiccation resistance in females flown to exhaustion (flown) and their controls are shown (NF, NG, NE and NF, G, and E). All times are in hours; n = 5, *for the B and O lines, and* n = 4 *for the RU lines. Each population within a line had 40 females tested. The mean value for the line is computed from the individual means for each of the replicate populations.*

Rose 1984; Luckinbill and Clare 1985). The rankings of the starvation resistances are also consistent with the earlier measurements at 22 generations of reversed selection, which found RU starvation resistance intermediate between O and B values. The desiccation and ethanol results conflict with the earlier rankings. Flight duration, not measured in the earlier study, has also declined in the RU lines. Nonetheless, glycogen levels appear to be only intermediate between B and O levels.

Longevity, flight duration, and stress resistance have all dropped in RU lines at 100 generations. There is no apparent difference between pure RU lines and their crosses. Thus, the crosses do not exhibit hybrid vigor, indicating an absence of inbreeding depression in the response to reversed selection. One possible explanation for the decreases in flight duration, desiccation resistance, and ethanol resistance observed in the RU stocks is that the flight duration, desiccation, and ethanol characters are also in some way pleiotropically linked to early fitness. However, this must be a different

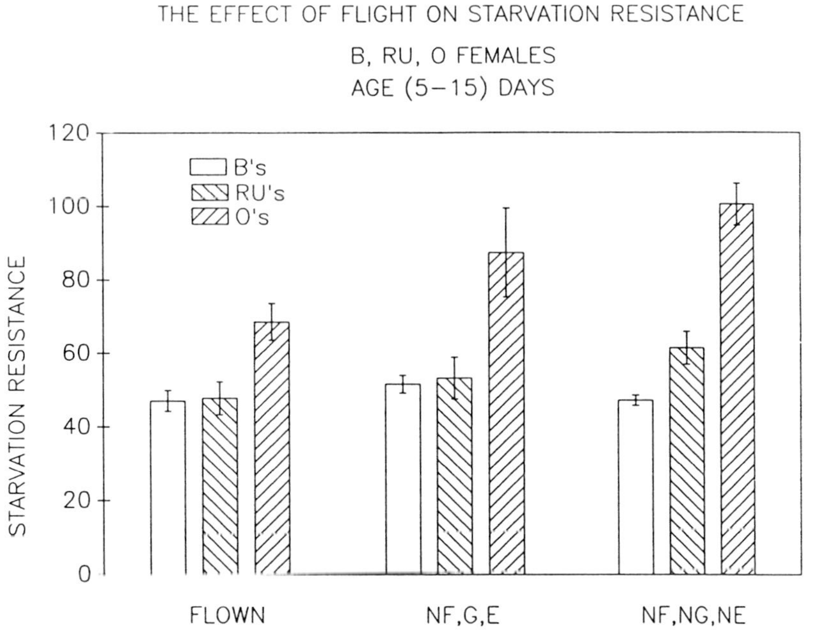

Fig. 8. The means and SEs for starvation resistance in females flown to exhaustion (flown) and their controls are shown (NF, NG, NE and NF, G, and E). All times are in hours; n = *5, for the B and O lines, and* n = *4 for the RU lines. Each population within a line had 40 females tested. The mean value for the line is computed from the individual means for each of the replicate populations.*

physiological genetic system from that involving starvation resistance, since at least two of these characters did not respond in parallel in earlier reverse-selection experiments (Service et al. 1988).

The WLRs for B and O lines in table 1 are considerably higher than those of other *Drosophila* species (Toolson 1982; E. C. Toolson, unpublished data) and reflect relatively high cuticular permeabilities. Although WLR data from young laboratory stocks of *Drosophila melanogaster* are not yet available, we attribute the high WLRs of the B and O strains to the fact the stocks have been maintained under conditions of high relative humidity for 10 years, as has been reported for *Drosophila pseudoobscura* (Toolson and Kuper-Simbron 1989). The WLR data indicate that there are physiological differences between dead B and O flies that follow the same qualitative pattern as desiccation resistance, the O flies losing water at a lower rate and surviving longer. Therefore, it may be the case that differences between the B and O stocks in desiccation resistance in part reflect some inert feature of the flies.

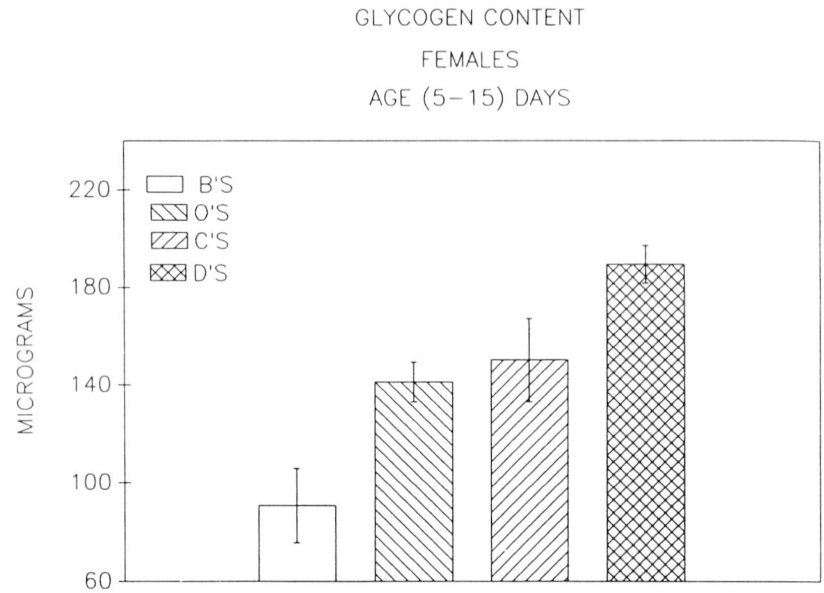

Fig. 9. The mean and SE of glycogen content in females from B, O, C, and D lines are shown; n = *5 for all lines;* n = *8 replicates containing three flies were extracted for each population within a line for both treatments. Amounts are reported in micrograms per three flies. The mean value for the line is computed from the individual means for each of the replicate populations.*

One candidate for an inert physiological difference between the B and O flies is a difference in epicuticular hydrocarbons, such that O flies with more hydrocarbon are relatively less permeable. The epicuticular HCs of the male and female flies differed significantly, both quantitatively and qualitatively. Males yielded nearly twice as much epicuticular hydrocarbons as females, and the male extracts were characterized by high proportions of tricosene (n-$C_{23:1}$); female HC extracts contained high proportions of heptacosadiene (n-$C_{27:2}$). Although some statistically significant differences in epicuticular HC profile between B's and O's were detected, these were small and of uncertain physiological or behavioral significance. Total HC levels were not statistically significant between B and O stocks. Therefore, increased levels of epicuticular hydrocarbons, in general, do not appear to be involved in resistance to desiccation. Moreover, males have about twice as much hydrocarbon as females but are less resistant to desiccation. No significant effect of relative amounts of individual HC components on cuticular permeability was detected. The lack of any effect of epicuticular HCs on cuticular permeability contrasts with findings from a number of arthropods

(Hadley 1984), although similar results have been found in a stock of *D. pseudoobscura* (Toolson and Kuper-Simbron 1989). Given the refutation of the epicuticular hydrocarbon hypothesis, an alternative possibility is alterations in the underlying cuticle. Unfortunately, this structure would be difficult to study in *D. melanogaster*, because of its small size.

Another possible mechanism underlying desiccation resistance is differences in glycogen reserves. This is suggested by the flight duration–stress resistance experiments and the glycogen content results. Flies flown to exhaustion from the B, RU, and O lines were indistinguishable in desiccation resistance, and glycogen is the principal fuel of flight in *D. melanogaster*. In addition, glycogen levels correspond to the ranking of the desiccation-resistance levels in B, RU, O, C, and D stocks. These results seem to indicate that the physiological difference in the ability to resist desiccation between these selected lines may be related to the amount of glycogen normally stored in their tissues at this age. By contrast, the ranking of the lines in the starvation assay after flight to exhaustion was not altered, which fits the evidence against the utilization of lipid in *Drosophila* flight (see Wigglesworth 1949; Service 1987).

Flight duration, which is powered by glycogen reserves, has been found to be correlated with longevity in lines selected for postponed senescence (Graves et al. 1988; Graves and Rose 1990). The results of the flight-followed-by-desiccation experiments in this article suggest that desiccation resistance and flight duration are physiologically correlated characters. At 100 generations of reverse selection, both characters had declined in RU lines. Roff (1977) showed a clear negative correlation between flights to exhaustion and early fecundity in *D. melanogaster*. Rose et al. (1990) have shown that selection for desiccation resistance enhances flight duration in lines derived from the O populations while selection for starvation resistance causes no increase in flight duration. Together, these studies indicate two physiologically and genetically linked suites of characters associated with postponed senescence in *D. melanogaster*, glycogen, desiccation resistance, flight duration, and ethanol tolerance versus lipid, starvation resistance, and early fecundity. The flight-duration suite is most likely linked to glycogen reserves, as shown here while the starvation suite is primarily linked to lipid reserves (Service 1987). Hoffman and Parsons (1989*a*, 1989*b*) partly confirm this analysis; they found no increase in lipid reserves associated with the increase in desiccation resistance in their selected lines. These studies did not check glycogen reserves.

The chief anomaly for this interpretation is the intermediate value of glycogen in the RU populations, given the absence of a difference in desiccation resistance and flight duration between B and RU stocks. It is possible that the cause of this problem is found in the action of genes that affect *both* allocation of calories to different reserve substances as well as acquisition of such caloric reserves. In particular, we do not want to assert that glycogen is the sole determinant of differentiation in desiccation or flight phenotypes. The WLR data from dead flies given here already indicate that this might not be a tenable interpretation, although one could image that the WLR might be slowed by the hydrophilic properties of residual glycogen reserves in dead flies.

Acknowledgments

This article is dedicated to George C. Williams on his retirement from the faculty at State University of New York, Stony Brook. We thank Natasha Parks, Phoung Tu Troung, and Quynh Ha for dedicated attention to these experiments. Teresa Tran and Vinh Vu aided in flight trials while Jorge Nieva and Phuong-Minh Tran assisted with the glycogen assays. This research was supported in part by a University of California President's Postdoctoral Fellowship to J.L.G. and in part by U.S. Public Health Service grant AG06346 to M.R.R.

Literature Cited

Clark, A. G., and W. W. Doane. 1983. Desiccation tolerance of the adipose mutant of *Drosophila melanogaster.* Hereditas 99:165–175.

Edney, E. B. 1977. Water balance in land arthropods. Springer, New York. 282 pp.

Graves, J. L., L. S. Luckinbill, and A. Nichols. 1988. Flight duration and wing beat frequency in long- and short-lived *Drosophila melanogaster.* J. Insect Physiol. 34: 1021–1026.

Graves, J. L., and M. R. Rose. 1990. Flight duration in *Drosophila melanogaster* selected for postponed senescence. Pages 57–63 in D. Harrison, ed. Genetic effects on aging. II. Telford, Caldwell, N.J.

Hadley, N. F. 1984. Cuticle: ecological significance. Pages 685–693 in J. Bereiter-Haha, A. G. Matoltsy, and K. S. Richards, eds. Biology of the integument. Vol. 1. Invertebrates. Springer, Heidelberg.

Hamilton, W. D. 1966. The moulding of senescence by natural selection. J. Theor. Biol. 12:12–45.

Hoffman, A., and P. A. Parsons. 1989*a*. An integrated approach to environmental stress tolerance and life-history variation desiccation tolerance in *Drosophila*. Biol. Bull. Linnean Soc. 37:117–136.

———. 1989*b*. Selection of increased desiccation resistance in *Drosophila melanogaster:* additive genetic control and correlated responses for other stresses. Genetics 122:837–845.

LIGHTON, J. R. B. 1988. Discontinous CO_2 emission in a small insect, the Formicine ant *Camponotus vicinus*. J. Exp. Biol. 134:363–376.

LUCKINBILL, L. S., and M. CLARE. 1985. Selection for life span in *Drosophila melanogaster*. Heredity 55:9–18.

LUCKINBILL, L. S., M. J. CLARE, W. L. KRELL, C. C. CIROCCO, and P. A. RICHARDS. 1987. Estimating the number of genetic elements that defer senescence in *Drosophila*. Evol. Ecol. 1:37–46.

LUCKINBILL, L. S., J. L. GRAVES, A. TOMKIW, and O. SOWIRKA. 1988. A qualitative analysis of some life history correlates of longevity in *Drosophila melanogaster*. Evol. Ecol. 2:85–94.

ROFF, D. 1977. Dispersal in dipterns: its costs and consequences. J. Anim. Ecol. 46: 443–456.

ROSE, M. R. 1984. Laboratory evolution of postponed senescence in *Drosophila melanogaster*. Evolution 38:1004–1010.

———. 1985. Life history evolution with antagonistic pleiotropy and overlapping generations. Theor. Popul. Biol. 28:342–358.

———. 1991. Evolutionary biology of aging. Oxford University Press, New York.

ROSE, M. R., and B. CHARLESWORTH. 1981. Genetics of life-history in *Drosophila melanogaster*. II. Exploratory selection experiments. Genetics 97:187–196.

ROSE, M. R., M. L. DOREY, A. M. COYLE, and P. M. SERVICE. 1984. The morphology of postponed senescence in *Drosophila melanogaster*. Can. J. Zool. 62:1576–1580.

ROSE, M. R., J. L. GRAVES, and E. W. HUTCHINSON. 1990. The use of selection to probe patterns of pleiotropy in fitness characters. Pages 29–42 in F. GILBERT, ed. Insect life cycles: genetics, evolution, and coordination. Springer, Berlin.

SERVICE, P. M. 1987. Physiological mechanisms of increased stress resistance in *Drosophila melanogaster* selected for postponed senescence. Physiol. Zool. 60:321–326.

SERVICE, P. M., E. W. HUTCHINSON, M. D. MACKINLEY, and M. R. ROSE. 1985. Resistance to environmental stress in *Drosophila melanogaster* selected for postponed senescence. Physiol. Zool. 58:380–389.

SERVICE, P. M., E. W. HUTCHINSON, and M. R. ROSE. 1988. Multiple genetic mechanisms for the evolution of senescence in *Drosophila melanogaster*. Evolution 42:708–716.

SERVICE, P. M., and M. R. ROSE. 1985. Genetic covariation among life-history components: the effect of novel environments. Evolution 39:943–945.

TOOLSON, E. C. 1982. Effects of rearing temperature on cuticle permeability and epicuticular lipid composition in *Drosophila pseudoobscura*. J. Exp. Zool. 222: 249–253.

TOOLSON, E. C., and R. KUPER-SIMBRON. 1989. Laboratory evolution of epicuticular hydrocarbon composition and cuticular permeability in *Drosophila pseudoobscura*: effects on sexual dimorphism and thermal acclimation ability. Evolution 43:468–473.

TOOLSON, E. C., T. M. MARKOW, L. L. JACKSON, and R. W. HOWARD. 1990. Epicuticular hydrocarbon composition of wild and laboratory-reared *Drosophila mojavensis* Patterson and Crow (Diptera: Drosophilidae). Ann. Entomol. Soc. Am. 83:1165–1176.

Van Handel, E. 1965. Microseparation of glycogen, simple sugars, and lipids. Anal. Biochem. 11:266–271.

Wattiaux, J. M. 1968*a*. Cumulative parental age effects in *Drosophila subobscura*. Evolution 22:406–421.

———. 1968*b*. Parental age effects in *Drosophila pseudoobscura*. Exp. Gerontol. 3: 55–61.

Wigglesworth, V. B. 1949. The utilization of reserve substances in *Drosophila* during flight. J. Exp. Biol. 26:150–163.

———. 1972. The principles of insect physiology. Chapman & Hall, London.

Williams, C. M., L. A. Barnes, and W. H. Sawyer. 1943. The utilization of glycogen by flies during flight and some aspects of the physiological aging of *Drosophila*. Biol. Bull. 84:263–272.

Williams, G. C. 1957. Pleiotropy, natural selection, and the evolution of senescence. Evolution 11:398–411.

Variation in the reversibility of evolution

Henrique Teotónio*† **& Michael R. Rose***

* *Department of Ecology and Evolutionary Biology, University of California, Irvine, California 92612, USA*

How reversible is adaptive evolution[1–6]? Studies of microbes give mixed answers to this question[6–11]. Reverse evolution has been little studied in sexual populations[12–14], even though the population genetics of sexual populations may be quite different. In the present study, 25 diverged replicated populations of *Drosophila melanogaster* are returned to a common ancestral environment for 50 generations. Here we show that reverse evolution back to the ancestral state occurs, but is not universal, instead depending on previous evolutionary history and the character studied. Hybrid populations showed no greater tendency to undergo successful reverse evolution, suggesting that insufficient genetic variation was not the factor limiting reverse evolution. Adaptive reverse evolution is a contingent process which occurs with only 50 generations of sexual reproduction.

† Present address: University Museum of Zoology, Downing Street, Cambridge CB2 3EJ, UK.

Over the past 20 years, we have created an artificial evolutionary radiation by imposing demographic and stress selection on large, outbred, replicated populations of *Drosophila melanogaster*. Part of this radiation is shown in Fig. 1. The five different groups of replicated populations are multifold differentiated from each other for several life-history, physiological and biochemical characters[15–21]. One of these groups has been maintained under the ancestral 'Ives' conditions since 1980 (the 'B' populations). One has been selected for reproduction late in life (the 'O' populations). One has been selected for survival under conditions of complete starvation (the 'SO' populations). One has been selected for reproduction in mid-life (the 'CO' populations). Finally, one group has been selected for reproduction very early in life (the 'ACO' populations) (see Fig. 1 legend and Methods). Most of the differentiation occurred within 50 generations of the start of selection.

At the start of the present study, a new population was derived from each of these 25 populations and then returned to the ancestral-Ives environmental conditions for 50 generations. The evolutionary response of these populations was assayed using eight characters which are related to fitness or were highly differentiated from the ancestral condition. By measuring these characters approximately every eight generations we were able to determine reverse evolutionary patterns and level of convergence.

The patterns of reverse evolution varied among evolutionary histories and among characters. Analysis of the evolutionary trajectories indicated that, except for fecundity, populations with different evolutionary histories had statistically heterogeneous responses to the reimposition of the ancestral environment (analysis of covariance, ANCOVA F-tests for evolutionary history and interaction term; $P < 0.05$). Results are shown in Fig. 2. Four kinds of evolutionary trajectories were observed. The first was reversion to ancestral character values with full convergence after little more than 20 generations. This was the case, for example, for IO and ISO female developmental time ('I' indicates a derived population; see Methods). The second kind of trajectory was a pattern of response without convergence within the 50 generations of the study, as in the case of IACO developmental time. The third kind of trajectory was

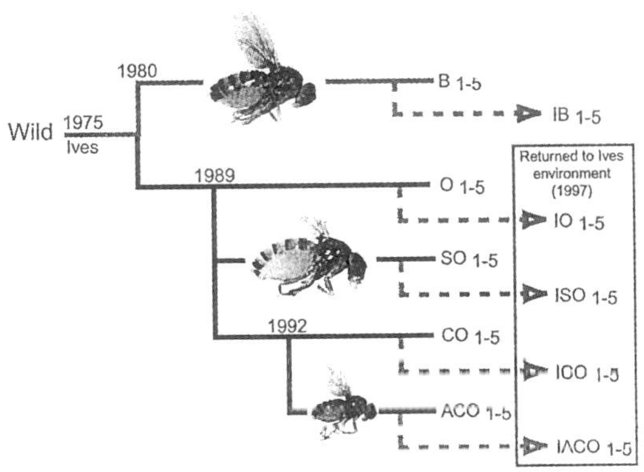

Figure 1 Laboratory radiation due to five selection treatments. Each group of populations was fivefold replicated (shown as subscript numbers). All populations descend from the Ives population, sampled from nature in 1975 (see Methods). We show female flies of the B, SO, and ACO treatments. These were cultured in identical conditions and are shown at the same relative scale. The direct descendants employed in our reverse evolution study are shown by dashed branches. Populations represented by yellow branches were maintained in the ancestral-Ives environment. Hybrid populations were derived either from crosses within or between different selection histories. A total of 44 populations were returned *de novo* to the ancestral-Ives environment and followed for fifty generations. Eight highly differentiated characters were measured during reverse evolution.

initial rapid reversion, followed by a stalling of evolution, without full convergence to ancestral character values, as for ISO starvation resistance or ICO developmental time. In this case, new stable states may have been attained. Finally, the fourth pattern was no significant change throughout the 50 generations of evolution, the best examples being ICO and IACO fecundity under high-density assay conditions.

The results show that reverse evolution back to an ancestral condition may or may not occur, at least over a 50-generation period of selection. Reverse evolution is neither inevitable nor impossible; it is contingent. This raises the question of the genetic mechanisms responsible for this contingency. Could populations that fail to return to the ancestral state lack genetic variation? Alternatively, could these populations be stuck in the vicinity of selective equilibria at which epistasis and linkage disequilibrium prevent their return to an earlier evolutionary state at which higher fitness would be attained? Both of these hypotheses can be tested by creating hybrids of the differentiated populations before reverse evolution. Hybrids should have more genetic variation (if the uncrossed experimental populations have become genetically impoverished and they should have undergone perturbation away from a strong evolutionary attractor (if that is their initial condition) through drastic gene frequency changes, subsequently coupled with recombination. If these genetic mechanisms are responsible for constraints on reverse evolution then we expect that hybrids should show more reverse evolution in the same 50 generations.

Nineteen hybrid populations were derived (see Methods). These underwent reverse evolution for 50 generations in parallel with the uncrossed derivatives already discussed. The appropriate test for improved convergence of the hybrids compared to the uncrossed must take into account the quantitative differentiation from the ancestral state in the two types of population. Many of the hybrids are less differentiated from the ancestral condition at the outset of reverse evolution, because of their hybridization. Therefore, we examined whether the hybrids and the uncrossed populations fall on the same regression of evolutionary rate on initial differentiation. Figure 3 shows this relationship among the 44 populations. There is no detectable difference between hybrid and uncrossed populations in their reverse evolution, once initial differentiation is allowed for. This indicates that the factors limiting reverse evolution are probably not related to a lack of genetic variation, epistasis, or linkage disequilibrium.

To account for these results we propose instead that the return to

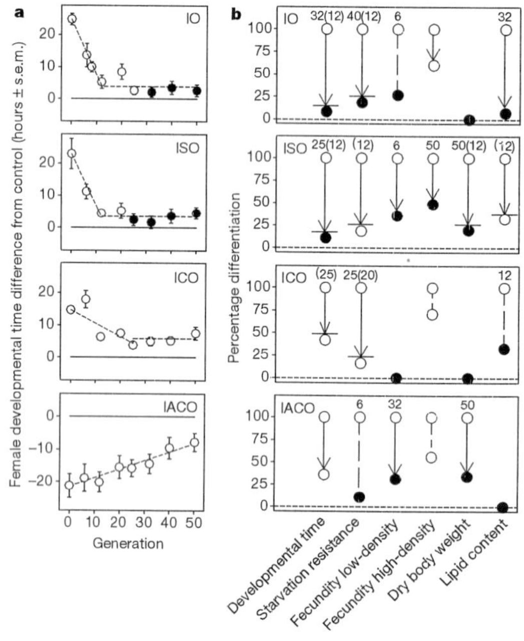

Figure 2 Reverse evolution during 50 generations. **a**, Response to selection in populations of different evolutionary history for female developmental time as a function of time (generations). Open circles show that the average difference from the control levels of the five replicate populations is significant (*t*-tests; $P < 0.05$). Black circles show non-significant differences from the controls, indicating convergence to ancestral levels. Dashed lines represent the best-fit regression models (IO, $P < 0.001$, adjusted $R^2 = 0.92$; ISO, $P < 0.001$, adjusted $R^2 = 0.95$; ICO, $P = 0.008$, adjusted $R^2 = 0.68$; IACO, $P < 0.001$, adjusted $R^2 = 0.93$). **b**, The categorized *x*-axis shows the response of each character separately, and the *y*-axis represents the percentage of differentiation relative to the control populations. Open circles indicate non-convergence to the control levels ($P < 0.05$), and black circles indicate convergence, as before. For all characters in all groups of populations, 100% differentiation was the initial starting condition, with the exception of a few characters which were undifferentiated from the beginning, these being represented by a single black circle at the zero differentiation line. Arrows between circles represent significant one-phase models, and arrow-lines indicate that a significant model with a first linear phase of convergence followed by a second no-response phase was the best fit to the evolutionary trajectory (see Methods). Broken lines show that no significant model was found, that is, that no significant change occurred during evolution. Numbers above the character for each group of populations show the generation of convergence to the control levels, and numbers in parentheses show the generation at which reverse evolution stalled. Reverse evolution patterns differ according to the character considered and according to selection history.

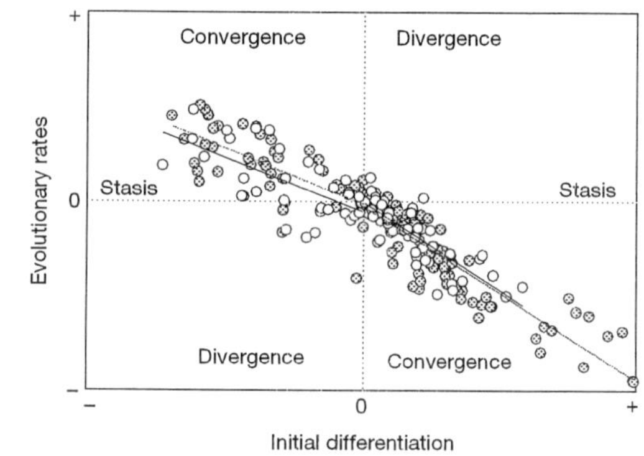

Figure 3 Reverse evolutionary rates as a function of differentiation at the start of the study. For each character and population, we estimated both the slope and the intercept of a one-phase linear regression, which were then used as the data points plotted, standardized to a common scale. Standardization involved dividing the estimated slopes and intercepts for each character by the largest slope or intercept for all data from that character, respectively. Data from fecundity under low adult density conditions were not used in this analysis. Uncrossed populations are shown by grey circles and hybrids by open circles. Out of 238 independent data points 209 fall in the two convergence quadrants, while only 29 fall in the two divergence quadrants. Assuming that stasis occurs when the evolutionary rate is less than 1% per generation, only 6 points fall in this class. The probability of chance occurrence of this pattern is much lower than 0.001 (sign test for convergence versus divergence). Hybrid and uncrossed evolutionary rate comparisons by ANCOVA (see Methods) did not reveal significant differences, when all characters were analysed either together or separately. Hybrids are not more likely to converge than uncrossed populations. For a significance level $\alpha = 0.05$, the statistical power to detect a difference between these groups, as estimated from the ANCOVA model, was larger than 95%, when global analysis was undertaken. Similar dependences of reverse evolutionary rates on initial differentiation are observed in the hybrid (solid lines) and uncrossed populations (dotted lines), in both global and separate linear regressions.

the ancestral-Ives environment did not produce uniform selection pressures among populations of different evolutionary histories. This can be attributed to variation in the relationship of each character with fitness in the ancestral environment according to the specific genetic background of each experimental population. Selection in the environments imposed before reverse evolution could have created different genetic architectures among groups of populations, in turn producing different genetic relationships between these characters and fitness in the ancestral environment. For example, populations which take longer to develop than the controls respond faster to natural selection on developmental time than populations which began by taking less time to develop. Thus, in the former populations, developmental time is intimately connected with fitness in the ancestral environment, whereas in the latter, the relationship with fitness is weaker in the ancestral environment (see Fig. 2).

Despite the cases in which reverse evolution fails to occur, or is stalled, our study shows an overwhelming tendency to evolutionary reversal during 50 generations of selection for adaptation to the ancestral environment. This demonstrates the power of selection in sexual populations to achieve adaptation[14,22–25]. History still plays a significant role in the observed evolutionary dynamics and outcomes[3,4,26,27], although this history is only a few hundred generations in duration. □

Methods

Experimental populations

We studied four different evolutionary histories, fivefold replicated, all derived from a common ancestor (Ives). These independent evolutionary groups included populations with 95 generations of selection history for increased late-life fertility (the O_{1-5})[15], with 100 generations of selection for increased starvation resistance (the SO_{1-5})[17], with 4–5-week intermediate generation times (the CO_{1-5})[17], and 190 generations of selection for decreased developmental time (the ACO_{1-5})[21]. Derived populations were designated by 'I' followed by the acronym for their evolutionary history. The five populations which had always been maintained in the ancestral-Ives conditions served as the control populations: IB_{1-5} (Fig. 1). These have been maintained for approximately 440 generations[15]. At the start of the experiment, hybrids within each selection treatment were created by five-way crosses of all five replicate populations. One of these five-way crosses used the B populations, which served as our control hybrid populations after replicating the cross five times. Hybrids were also created by crosses among all possible combinations of different selection treatments. Hybrid and uncrossed populations were followed in parallel during the 50 generations of reverse evolution. The ancestral-Ives environmental conditions include discrete two-week generations in vials, with egg and adult density controlled, and egg laying taking place during about two hours after mixing all adults of the same population. The amount of nutrients, temperature, humidity and light schedule were controlled. Census population sizes were always above 1,000 individuals per replicate population.

Assay designs

Eight characters were analysed in this study, most measured under conditions close to those of the ancestral-Ives environment. Developmental time in both genders was measured as the time from egg to adult emergence[21]. A total of 600 eggs per population were measured in each assay. Starvation resistance in both males and females was mesured as the time until death under starvation in high humidity conditions[17]. A total of 40 flies per gender and per population were measured in each assay. Early fecundity under high adult density conditions was measured as the number of eggs laid by 20 pairs of flies per vial during one hour. Sample sizes for thsi character were eight vials per population per assay. Early fecundity under low adult density conditions was measured as the number of eggs laid by a pair of flies per vial during 24 hours, the sample size being 40 such vials per population. Female dry body weight was calculated from the weight of 8 samples of 10 flies each per population, to the nearest 0.001 mg. Female whole-body lipid content was measured as the difference in dry weight before and after lipid extraction with a Soxhlet petroleum ether extractor.

Statistical analysis

All the starting populations employed in this study have been maintained independently of one another throughout their history and as such each one is an evolutionary experiment that is treated as a unique data point in our statistical analysis. Among the uncrossed populations, our experimental design allows a separation of genetic sampling effects from the main selection treatment effects. The IB_{1-5} are the control populations used to test the hypothesis of convergence to the ancestral character states. In each assay generation, each group of populations was independently compared for differences from the control populations by one-tailed unpaired Student *t*-tests. The evolutionary trajectories of the uncrossed populations were analysed with two general least-squares regression models to explain the response to selection: a one-phase linear model, with a constant rate of response throughout the experiment, or a two-phase linear model, with a first phase of response to selection followed by a plateau. The generation at which the second phase started was empirically determined[28]. Adjusted R^2, PRESS, Ellner and Turchin V_2, and Mallows C_p were the statistics used to choose the best regression model[29,30]. Heterogeneity among evolutionary trajectories was analysed using ANCOVA. These models used each separate character as the dependent variable, with generation defined as the covariate and evolutionary histories as a design factor with four categories: IO, ISO, ICO and IACO. Comparison of the rates of response under reverse evolution between uncrossed and hybrid populations were done by ANCOVA. In this model, the dependent variable was the slope estimated from one-phase linear regression models (using the response during reverse evolution) and the covariate was the estimated intercept of the regression model. Hybrid condition defined the factor in the ANCOVA, with uncrossed and hybrid categories. Power analyses were performed using the hybrid and uncrossed estimated least-squares means and standard deviations to estimate the power of *t*-tests for mean differences[30].

Received 25 May; accepted 18 September 2000.

1. Muller, H. J. Reversibility in evolution considered from the standpoint of genetics. *Biol. Rev.* **14**, 261–280 (1939).
2. Simpson, G. G. *The Major Features of Evolution* (Columbia Univ. Press, New York, 1953).
3. Wright, S. *Evolution and the Genetics of Populations. Vol 3 Experimental Results and Evolutionary Deductions* (Univ. Chicago Press, Chicago, 1977).
4. Bull, J. J. & Charnov, E. L. On irreversible evolution. *Evolution* **39**, 1149–1155 (1985).
5. Gayon, J. *Darwinism's Struggle for Survival: Heredity and the Hypothesis of Natural Selection* (Cambridge Univ. Press, Cambridge and New York, 1998).
6. Crill, W. D., Wichman, H. A. & Bull, J. J. Evolutionary reversals during viral adaptation to alternating hosts. *Genetics* **154**, 27–37 (2000).
7. Lenski, R. E. Experimental studies of pleiotropy and epistasis in *Escherichia coli*. II. Compensation for maladaptive effects associated with resistance to virus T4. *Evolution* **42**, 433–440 (1988).
8. Rainey, P. B. & Travisano, M. Adaptive radiation in a heterogeneous environment. *Nature* **394**, 69–72 (1998).
9. Burch, C. L. & Chao, L. Evolution by small steps and rugged landscapes in the RNA virus φ6. *Genetics* **151**, 921–927 (1999).
10. Moore, F. B. G., Rozen, D. E. & Lenski, R. E. Pervasive compensatory adaptation in *Escherichia coli*. *Proc. R. Soc. Lond. B* **267**, 515–522 (2000).
11. Levin, B. L., Perrot, V. & Walker, N. Compensatory mutations, antibiotic resistance and the population genetics of adaptive evolution in Bacteria. *Genetics* **154**, 985–997 (2000).
12. Clarke, C. A., Mani, G. S. & Wynne, G. Evolution in reverse: clean air and the peppered moth. *Biol. J. Linn. Soc.* **26**, 189–199 (1985).
13. Service, P. M., Hutchison, E. W. & Rose, M. R. Multiple genetic mechanisms for the evolution of senescence in *Drosophila melanogaster*. *Evolution* **42**, 708–716 (1988).
14. Cook, L. M., Dennis, R. L. & Mani, G. S. Melanic morph frequency in the peppered moth in the Manchester area. *Proc. R. Soc. Lond. B* **266**, 293–297 (1999).
15. Rose, M. R. Laboratory evolution of postponed senescence in *Drosophila melanogaster*. *Evolution* **38**, 1004–1010 (1984).
16. Graves, J. L., Toolson, E. C., Jeong, C., Vu, L. N. & Rose, M. R. Desiccation, flight, glycogen, and postponed senescence in *Drosophila melanogaster*. *Physiol. Zool.* **65**, 268–286 (1992).
17. Rose, M. R., Vu, L. N., Park, S. U. & Graves, J. L. Selection on stress resistance increased longevity in *Drosophila melanogaster*. *Exp. Gerontol.* **27**, 241–250 (1992).
18. Fleming, J. E., Spicer, G. S., Garrison, R. C. & Rose, M. R. Two-dimensional protein electrophoretic analysis of postponed aging in *Drosophila*. *Genetica* **91**, 183–198 (1993).
19. Tyler, R. H. *et al.* The effect of superoxide dismutase alleles on aging in *Drosophila*. *Genetica* **91**, 143–149 (1993).
20. Leroi, A. M., Chippindale, A. K. & Rose, M. R. Long-term laboratory evolution of a genetic life-history trade-off in *Drosophila melanogaster*. 1. The role of genotype-by-environment interaction. *Evolution* **48**, 1244–1257 (1994).
21. Chippindale, A. K., Alipaz, J. A., Chen, H.-W. & Rose, M. R. Experimental evolution of accelerated development in *Drosophila*. 1. Development speed and larval survival. *Evolution* **51**, 1536–1551 (1997).
22. Fisher, R. A. *The Genetical Theory of Natural Selection* (Dover, New York, 1958).
23. Reznick, D. N., Bryga, H. & Endler, J. A. Experimentally induced life-history evolution in a natural population. *Nature* **346**, 357–359 (1990).
24. Coyne, J. A., Barton, N. H. & Turelli, M. Perspective: a critique of Sewall Wright's shifting balance theory of evolution. *Evolution* **51**, 643–671 (1997).
25. Losos, J. B., Jackman, T. R., Larson, A., de Queiroz, K. & Rodriguez-Schettino, L. Contingency and determinism in replicated adaptive radiations of island lizards. *Science* **279**, 2115–2117 (1998).
26. Cohan, F. M. & Hoffmann, A. A. Uniform selection as a diversifying force in evolution: evidence from *Drosophila*. *Am. Nat.* **134**, 613–637 (1989).
27. Travisano, M., Mongold, J. A., Bennet, A. F. & Lenski, R. E. Experimental tests of the roles of adaptation, chance, and history in evolution. *Science* **267**, 87–90 (1995).
28. SPSS, I. *SigmaPlot 5.0. Programming Guide* (SPSS Inc., Chicago, 1998).
29. Glantz, S. A. & Slinker, B. K. *Primer of Applied Regression and Analysis of Variance* (McGraw-Hill Health Professions Division, New York, 1990).
30. Zar, J. H. *Biostatistical Analysis* (Prentice Hall, Upper Saddle River, 1996).

Acknowledgements

We thank P. Baldado, A. K. Chippindale, L. D. Mueller, E. Sucena and particularly M. Matos for advice throughout the project and for comments on the manuscript. H.T. was supported by the Gulbenkian Foundation and Program PRAXIS XXI/FCT under the "Programa Gulbenkian de Doutoramento em Biologia e Medicina".

Correspondence and requests for materials should be addressed to H.T.
(e-mail: ht227@cam.ac.uk).

Reverse evolution of fitness in *Drosophila melanogaster*

H. TEOTÓNIO,* M. MATOS† & M. R. ROSE*

*Department of Ecology and Evolutionary Biology, University of California Irvine, Irvine, CA, USA
†Centro de Biologia Ambiental, Department Zoologiae Antropologia, Faculdade de Ciências da Universidade de Lisboa, Campo Grande, Portugal

Keywords:

adaptation;
experimental reverse evolution;
fitness;
life-history;
plasticity.

Abstract

The evolution of fitness is central to evolutionary theory, yet few experimental systems allow us to track its evolution in genetically and environmentally relevant contexts. Reverse evolution experiments allow the study of the evolutionary return to ancestral phenotypic states, including fitness. This in turn permits well-defined tests for the dependence of adaptation on evolutionary history and environmental conditions. In the experiments described here, 20 populations of heterogeneous evolutionary histories were returned to their common ancestral environment for 50 generations, and were then compared with both their immediate differentiated ancestors and populations which had remained in the ancestral environment. One measure of fitness returned to ancestral levels to a greater extent than other characters did. The phenotypic effects of reverse evolution were also contingent on previous selective history. Moreover, convergence to the ancestral state was highly sensitive to environmental conditions. The phenotypic plasticity of fecundity, a character directly selected for, evolved during the experimental time frame. Reverse evolution appears to force multiple, diverged populations to converge on a common fitness state through different life-history and genetic changes.

Introduction

One of the most important topics in evolutionary biology is adaptation, particularly the selective and genetic mechanisms that lead to adaptation. Among the many empirical problems that face the study of adaptation is the inherent difficulty of inferring adaptation from the evolution of characters whose relationship with fitness may be questionable. Another important problem facing many empirical studies of adaptation is that they are frequently carried out in environmental conditions which are of unclear evolutionary relevance, if not actually a source of artefacts (vid. Lewontin, 1974; Wright, 1977; Haymer & Hartl, 1982; Hedrick & Murray, 1983; Service & Rose, 1985; Leroi *et al.*, 1994; Reznick & Travis, 1996; Rose *et al.*, 1996; Chippindale *et al.*, 2001). The obvious solution to these difficulties is to study the evolution of fitness itself in interpretable selective contexts, a solution that is also hard to achieve.

Historically, the direct estimation of fitness in sexually reproducing organisms has been difficult, because fitness measures of single genotypes in natural genetic and environmental contexts are usually impossible to obtain, unlike the situation for clonal organisms (e.g, Lenski *et al.*, 1991; Crill *et al.*, 2000). When isogenicity can be contrived, fitness may be estimated, but inbreeding effects can have devastating effects on the fitness of each line, and the whole set of inbred lines will often represent a biased sample of the genetic variation found in the original population. For a long time, intraspecific and interspecific competition experiments have been used in *Drosophila* to infer relative genotypic fitness (Latter & Robertson, 1962; Prout, 1971; Wright, 1977; Jungen & Hartl, 1979; Haymer & Hartl, 1982; Hedrick & Murray, 1983; Mackay, 1986; Wilton *et al.*, 1989; Joshi & Thompson, 1995; Weber, 1996; Fowler *et al.*, 1997; Barton & Partridge, 2000). Differences in relative fitness between wild-caught flies and their laboratory derivatives or between laboratory populations have usually

Correspondence: Henrique Teotónio, Department of Ecology and Evolutionary Biology, Princeton University, Princeton, NJ 08540–1003, USA
Tel.: +609 2585587; fax: +609 2581334;
e-mail: teotonio@princeton.edu

J. EVOL. BIOL. 15 (2002) 608–617 © 2002 BLACKWELL SCIENCE LTD been found. But inbreeding and genotype-by-environment interaction make it hard to establish a relationship between these fitness differences and the genetic and environmental circumstances in which the original populations evolved, which renders the scientific value of such findings highly uncertain (Rose *et al.*, 1996).

To address these problems in the study of adaptation one needs, at a minimum, an experimental system in which it is possible to estimate the fitness of members of genetically defined populations relative to the fitness of members of control populations that have already adapted to the experimental environment. If the experimental populations are allowed to evolve in the environment of the previously adapted control populations, their evolution has a well-defined point of reference. One such design is experimental reverse evolution, where the ancestral populations or their equivalent are available for comparison with derived differentiated populations (e.g. Lenski, 1988; Service *et al.*, 1988; Rainey & Travisano, 1998; Burch & Chao, 1999; Crill *et al.*, 2000; Teotónio & Rose, 2000, 2001). This was the experimental paradigm used in the present study. Note that this type of experimental paradigm is very different from both relaxed artificial selection and reverse artificial selection, most importantly with respect to the population sizes involved, the opportunity for natural selection to operate, and the degree of replication. Experimental evolution, forwards or reverse, involves more and larger populations compared with artificial selection, and natural selection is deliberately allowed to occur (see Wright, 1977; Rose *et al.*, 1996; Teotónio & Rose, 2001).

Reverse evolution in sexually outbreeding *Drosophila* populations appears to depend on previous evolutionary history, particularly with respect to the degree of convergence. Yet incomplete convergence to ancestral states could not be explained for most studied characters either by lack of genetic variation or the presence of gene-interaction systems in these *Drosophila* populations (Teotónio & Rose, 2000). However, this earlier study did not consider the reverse evolution of fitness itself, being limited to characters whose relationship to fitness is less obvious. Thus, it was not ascertained whether adaptation is itself dependent on previous evolutionary history. Also, the selective mechanisms that determine reverse evolution were not identified.

In this study we measure a total of 10 characters, including life-history characters and competitive fitness relative to a morphologically marked stock, for populations which have been kept in diverse environments for numerous generations, and their descendent populations selected in the environment of their common ancestor for 50 generations (vid. Teotónio & Rose, 2000). We specifically address: (1) *whether* heterogeneous populations derived from a common ancestor will attain similar degrees of adaptation during reverse evolution; (2) *whether* the evolutionary dynamics of fitness characters are similar to characters whose relation with fitness is less obvious; and (3) the relationship between the ancestral environment and the selective mechanisms behind reverse evolution.

Materials and methods

Experimental populations

All populations were derived from a common ancestor introduced to the laboratory in 1975, the 'Ives' population (Rose, 1984; Teotónio & Rose, 2000). In 1980, five populations were derived and ever-since maintained under Ives-ancestral conditions (B_{1-5}), which comprise a 2-week generation cycle. Another five populations were selected for increased late-life fertility (O_{1-5}), being maintained on a 10-week generation cycle (Rose, 1984). In 1989, populations derived individually from the O populations were selected for increased starvation resistance (SO_{1-5}), with another group maintained as the fed demographic control (CO_{1-5}) (Rose *et al.*, 1992). Both these population groups were maintained on a 3–4 week generation cycle. Lastly, from the CO populations, a set of five populations was selected for decreased developmental time and increased early fertility (ACO_{1-5}) (Chippindale *et al.*, 1997). These were maintained with generation lengths no longer than 9 days. All populations have been maintained at high population sizes ($n > 1500$), without systematic inbreeding or hybridization, throughout their history. By the time the experiments described here were performed, the O populations had undergone 110 generations of selection for late-life fitness, the SO and CO had undergone 130 generations in their selective environment, and the ACO populations 270 generations.

New populations were obtained from the four groups of selected populations (O, SO, CO and ACO), each new population derived from the same numbered replicate ancestor population (e.g. IO_1 derived from O_1, IO_2 derived from O_2, etc.), and cultured in the common ancestral Ives-environment for 50 generations (Teotónio & Rose 2000). This environment features discrete 2-week generations, with natural selection for early fertility in crowded conditions. Egg-laying occurs within 2 h, density being controlled at 50–100 eggs per vial. Each of the five IB populations was derived from a single B population. These were the control populations in the experimental design. Approximately 440 generations had elapsed since their formation by the time the reverse evolution experiment started (Teotónio & Rose, 2000).

A total of 45 populations were studied: the control IB_{1-5}, the 20 differentially selected populations (O_{1-5}, CO_{1-5}, ACO_{1-5}, and SO_{1-5}), and the 20 returned to the Ives environment (IO_{1-5}, ICO_{1-5}, $IACO_{1-5}$, and ISO_{1-5}). All characters were measured at generation 50 of reverse evolution.

Competitive fitness assays

Three competitive fitness assays were done: male, female, and population fitness. In each of them, morphologically marked populations were competed against the experimental populations in Ives-environment conditions.

For the male fitness assay, five vials were used per population comparison, each containing 25 eggs of the population to be tested and 75 eggs of a brown-eyed *bw* competitor outbred stock (supplied by A.K. Chippindale, UCSB). After 14 days, 20 *bw* females from each assay vial were isolated into separate vials. The proportion of these females with *wild-type* (*wt*) progeny (as a consequence of being fathered by males from the experimental population) was taken as the male fitness estimate. This includes male survivorship, mating success, and viability of offspring (see also Chippindale *et al.*, 2001). Although most vials had all *bw* or *wt* progeny, in approximately 20% of them the progeny were of varied parental origin, because of multiple mating. In such cases, each vial was given a score of 0.75 if the majority of the progeny had *wt* eye colour (65–95%), 0.5 if approximately half was *wt* (35–65%), and 0.25 if a minority of flies were *wt* (5–35%). This scoring system was adopted because the volume of data rendered exact counts impractical.

For the female fitness assay, five pairs of *wt* adults were placed together with 15 pairs of *bw* adults in eight replicate vials per population tested. The females were allowed to lay eggs for 2 h after which all adults were discarded. Egg density was controlled by removing excess eggs. After 14 days, female fitness was estimated as the proportion of *wt* progeny. This estimate includes female survivorship, fertility, and offspring viability.

The population fitness assay employed a two-generation competition contrived so that the marked population could not hybridize with the experimental populations. This was done using a compound autosome stock as the marked competitor because all hybrid progeny with the experimental population will be inviable (Jungen & Hartl, 1979; Novitski *et al.*, 1981). The marked stock no. 1113, with genotype *C(2)EN, b[1] bw[1]; st[1]*, was obtained from the Bloomington *Drosophila* Stock Center. In this assay, seven replicate vials were used per population. In each vial, three pairs of adults from the experimental population were placed together with 17 pairs of the marker stock. These adults were allowed to lay eggs for 24 h, after which egg density was controlled. After 14 days all emerged adult flies were scored for their genotype and allowed to lay for 24 h under crowded conditions for a second generation, egg density again being controlled. The proportion of *wt* adult progeny, after the first and second generation of competition, was used as an estimate of population fitness.

Life-history assays

For developmental time, 10 vials per population were collected with 60 eggs each and emerging adults were counted every 6 h. Population viability, regardless of sex, was taken as the egg to adult survivorship. In the starvation resistance assay, four same-sex flies were placed in each of 10 replicate vials under high humidity conditions but no food. Survivorship was scored every 6 h (Teotónio & Rose 2000).

Early fecundity under low crowding conditions was measured by placing one pair (one male and one female) of 14-day-old-flies in each of the 30 vials assayed per population. These flies were allowed to lay eggs for 24 h. For early fecundity under high crowding conditions, 20 pairs were used for each vial, with 1, 2 and 6 h for the laying period. Eight vials for the 1 h assay and five vials for the 2 and 6 h assays were collected from each population.

Data for developmental time, starvation resistance, and 1 h early fecundity under high crowding conditions for the reverse evolved populations IO_{1-5}, ISO_{1-5}, ICO_{1-5} and $IACO_{1-5}$ were previously reported in Teotónio & Rose (2000).

Data analysis

For each selection history, there are five independent replicate populations. Because of this, our observational units are the mean values of these populations. All group comparisons were made relative to the IB populations.

Two sets of hypotheses were tested. First, we tested whether particular independent groups of populations with different evolutionary histories converged to control IB values. For example, we tested whether O_{1-5} populations were differentiated from the IB populations, and whether the IO_{1-5} converged to the IB control levels. This was done with unpaired two-tailed Student's *t*-tests, with α set to 0.05. Sequential Bonferroni was used for multiple comparison corrections on each character separately, under the composite null hypothesis that all group differences to control are equal to zero (Rice, 1989). The number of tests for all characters were eight, with the exception of fecundity where the number of tests were 32.

Secondly, we tested the general question of convergence to the ancestral values, and the role that previous history has on reverse evolution, using a mixed analysis of variance (ANOVA) with a fixed effects 'selection' factor and a random effects 'history' factor. The historical factor was reflected in the division of populations into four groups according to their specific evolutionary history (the antecedent selection in O, CO, ACO or SO populations). For the selection factor, the populations were divided into two groups: those which had recently undergone Ives-culture and those which had not. A significant interaction would reveal historical effects on adaptation. For each population the data used were the absolute differences from the mean of the five IB populations. This was done because differences in sign might underestimate average differentiation among 'selection' categories and overestimate the interaction term between 'history' and 'selection', as a result of sign differences in initial differentiation.

For the analysis of differential plasticity in fecundity, analysis of covariance (ANCOVA) was performed on the 1, 2 and 6 h assays. A separate ANCOVA was performed for each of the four evolutionary histories (that is, one for the O ancestry group, including both IO_{1-5} and O_{1-5} populations, one for the CO ancestry group, etc.). Selection (Ives selection vs. non-Ives selection) was the fixed factor with number of hours as the covariate. A significant interaction between 'selection' and 'hours' would indicate a change with reverse evolution in the relationship between observed fecundity and the number of hours allowed for egg laying.

Results

Competitive fitness

The results for the male fitness assay are summarized in Fig. 1a and Table 1. Student *t*-tests revealed that the immediately ancestral O_{1-5}, SO_{1-5}, and ACO_{1-5} populations were differentiated from the IB populations, the CO_{1-5} being marginally differentiated ($P = 0.07$), whereas all reversely evolved populations, IO_{1-5}, ISO_{1-5}, ICO_{1-5}, and $IACO_{1-5}$ converged to control levels. ANOVA indicated that type of selection (populations cultured in Ives vs. non-Ives conditions) to be highly significant, although incidental evolutionary history (O, SO, CO, and ACO ancestry grouping) and the interaction between them were nonsignificant (see Table 1).

Female fitness results are shown in Fig. 1b. As tested with *t*-tests no group of populations was different from the control populations. However, ANOVA on female fitness data showed both a significant 'selection' and 'history' factor (Table 1).

For the population fitness assay, in the first generation, the SO populations were heteroscedastic, as measured by Bartlett's tests. Angular transformation partially corrected this and transformed data were used in subsequent analysis of all populations. Comparisons using *t*-tests revealed that ACO_{1-5} ($P = 0.02$) were significantly less fit in the assay's first generation. All other populations were undifferentiated (see Fig. 1c). However, the CO_{1-5} ($P = 0.07$), IO_{1-5} ($P = 0.07$), and O_{1-5} ($P = 0.08$) were marginally differentiated from control levels. By the second generation of the population fitness assay, most of the populations undergoing assay had cleared the experimental vials of morphologically marked competitor flies (not shown). For this reason, multigenerational statistical analyses could not be pursued. ANOVA did not reveal any significant effects of any factor, when each generation was analysed separately.

Life-history characters

Results from population viability, female developmental time, and female starvation resistance are shown in Fig. 2 and Table 1.

For female developmental time, all but the IO populations were differentiated from the control populations, by *t*-tests. Male developmental time results are similar to female results, with the exception of the ISO populations, which for differences were not significant. ANOVA shows that both selection and interaction between selection and history are significant.

The CO, ICO and O populations were different from the IB populations with respect to viability (Fig. 2). The interaction between selection and history in the ANOVA for population viability was significant (Table 1).

Analysis of both female and male starvation resistance showed that all but the ICO and IACO populations were differentiated from the IB populations. Also for both sexes, only the interaction term in the ANOVA was significant (Table 1). As the SO treatment has disproportionately higher starvation resistance values than those of any other group of populations (Fig. 2), we re-analysed the data using the absolute differences of the assayed populations from the control populations divided by the value of the immediately ancestral differentiated population group. This analysis was undertaken in order to minimize any effects of initial differentiation on the degree of convergence, which could adversely affect the estimation of the interaction term. The ANOVA in this case showed for females that both type of selection ($F_{1,3} = 17.5$; $P = 0.03$) and the interaction term ($F_{3,32} = 3.35$; $P = 0.03$) are significant. In males, selection is marginally significant ($F_{1,3} = 6.41$; $P = 0.09$), whereas the interaction is highly significant ($F_{3,32} = 10.3$; $P < 0.01$).

The early fecundity results are presented in Fig. 3 and also in Table 1. At low adult density with 24 h of egg-laying, ACO and SO are different from the IB values as tested with *t*-tests. The interaction term was the only factor found significant in the ANOVA.

All populations but ISO are differentiated from the control for early fecundity at high density with 1 h egg laying. In the 2 h assay all but ISO and ICO groups are different from the control. Finally, analysis of the 6-h assay showed that only O, SO, ACO were significantly different from the control. The fecundity data indicate a return to the ancestral fecundity character with increased time of egg-laying in the reverse evolution populations (Fig. 3). ANOVA revealed that type of selection had significant effects in the 1- and 2-h assays, although the interaction was significant for the 1, 2 and 6 h assays (for the 1 h at the 6% level; Table 1).

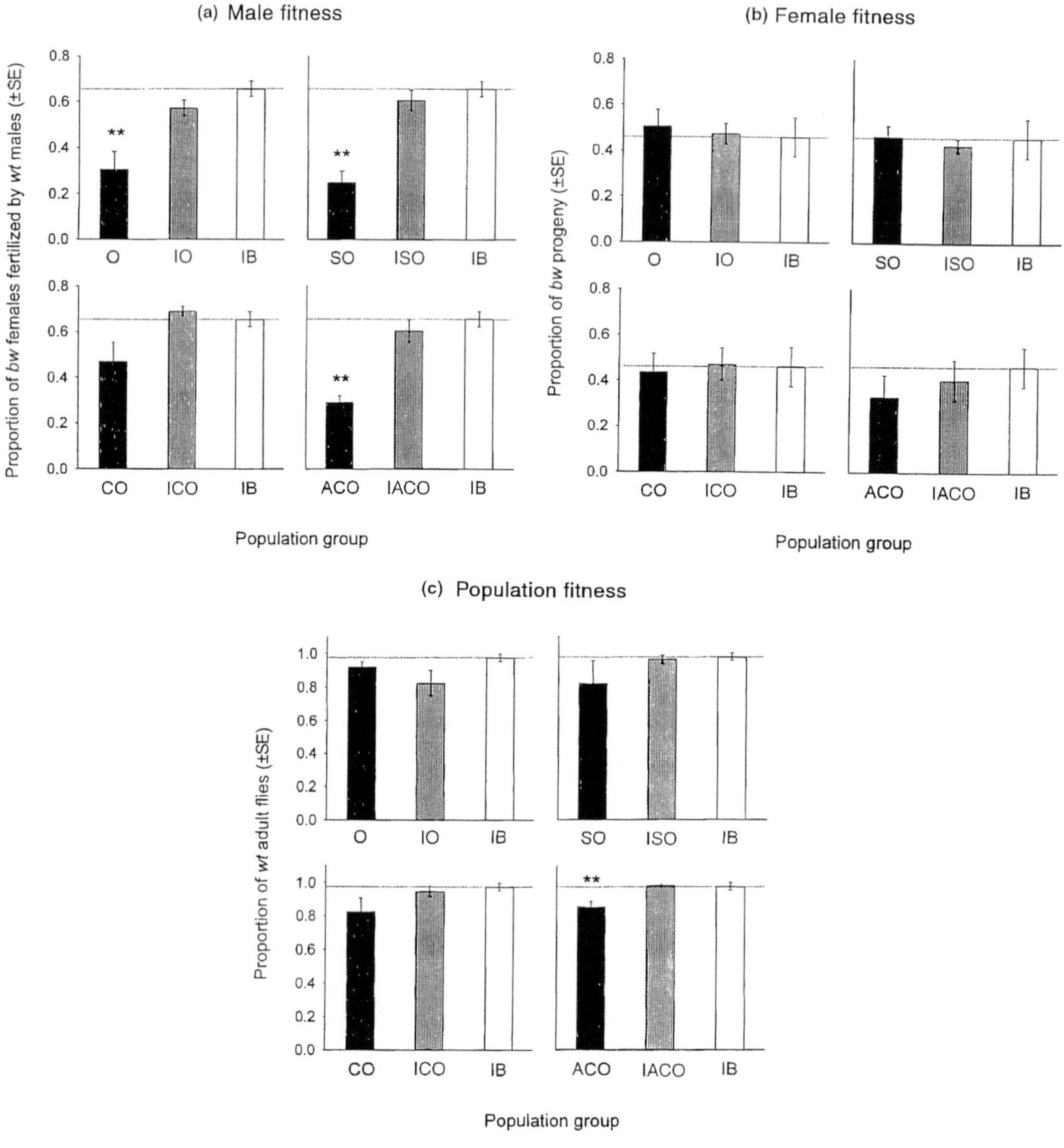

Fig. 1 In (a) Male fitness results are shown. Male fitness was measured as the proportion of tester marked stock females fertilized by males from experimental populations (see Materials and methods). Differentially selected populations are shown by black bars, whereas populations derived from them after 50 generations of reverse evolution are shown by grey bars. Control populations are represented by white bars and their mean value by a horizontal grey line. In (b), female fitness measured as the number of offspring mothered by the tester populations is shown. In (c), generation one of the population fitness assay is shown as the untransformed proportion of experimental flies that survived in competition to marked tester flies; angular transformation was used for the analysis. Error bars represent the standard error of the five replicate populations, for each group. Two-tailed Student's *t*-tests: $*0.01 < P \le 0.05$; $**P \le 0.01$. Sequential Bonferroni with number of tests equal to eight: †-tests that lose significance with $\alpha = 0.05$.

Table 1 Mean Squares for a two-factor mixed ANOVA table with selection factor fixed and history factor random.

	Fitness			Developmental Time			Starvation Resistance		Fecundity			
	♀	♂	Population †	♀	♂	Viability	♀	♂	1-hour	2-hour	6-hour	low-density
Selection df=1	0.013*	0.615 **	0.035	1518.8**	1309.5**	2.2E-4	21317	12156	127478*	461081*	816896	793.3
History df=3	0.028**	0.035	0.004	49.7	77.1	2E-3	17105	8773	2814	43050	125887	159.3
Selection × History df=3	9E-4	0.014	0.03	37.6*	36.5*	2.2E-3*	8808**	5778**	6698	46371**	168943**	259.5*
Error df=32	5.3E-3	0.012	0.021	8.64	8.72	6.2E-4	390	122	2412	4238	19921	74.3

significance testing of F statistics: *$0.01 < p \leq 0.05$; **$p \leq 0.01$.
† generation 1 of angular transformed population fitness assay.

The ANCOVA performed on the fecundity data shows that the fecundity characters of all populations responded to 50 generations of Ives-culture, as the selection factor was significant, irrespective of history (not shown). A change of the linear relationship of fecundity with hours was found for the populations with O and SO ancestry, shown by a significant interaction term ($F_{1,26} = 9.2$, $P < 0.01$ for O ancestry; $F_{1,26} = 5.8$, $P = 0.02$ for the SO ancestry), but it was not found for populations with CO and ACO ancestry.

Discussion

Reverse evolution and measures of fitness

Our findings show that the evolutionary dynamics of reverse evolution depend in part on previous selective history, as many of the ANOVA interaction terms reflecting selective ancestry are significant. Our results also show that phenotypic convergence is not universal, particularly for characters whose relationship to fitness

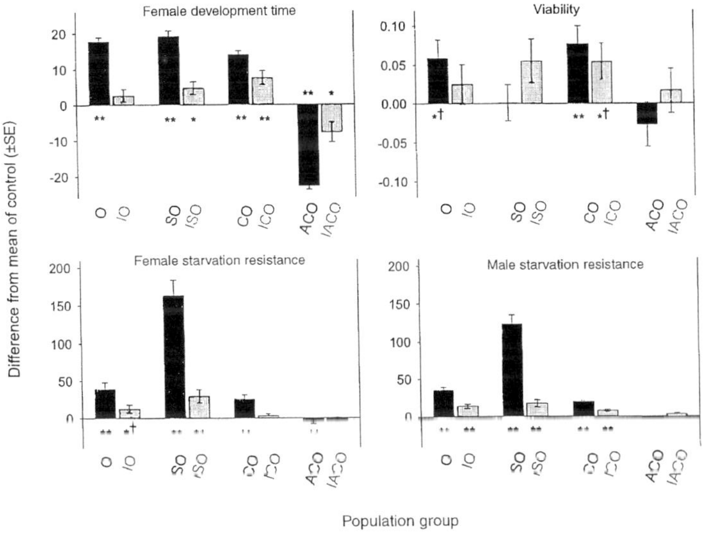

Fig. 2 The upper-left panel shows female developmental time as a difference from control in number of hours from egg to adult. Black bars show the ancestor populations, whereas grey show the reverse evolved populations. Similar results were found for male developmental time data, except that ISO populations were not different from controls. Upper-right panel shows population viability as a difference from control levels in the probability of survival to adulthood. The bottom panels show female and male starvation resistance in number of hours surviving until death. Error bars represent the standard error of the difference between a particular group of populations from the control populations. The development time and starvation resistance data presented for the reverse evolved populations was previously reported in Teotónio & Rose (2000). Student's t-tests: *$0.01 < P \leq 0.05$; **$P \leq 0.01$. Sequential Bonferroni with number of tests equal to eight: †tests that lose significance with $\alpha = 0.05$.

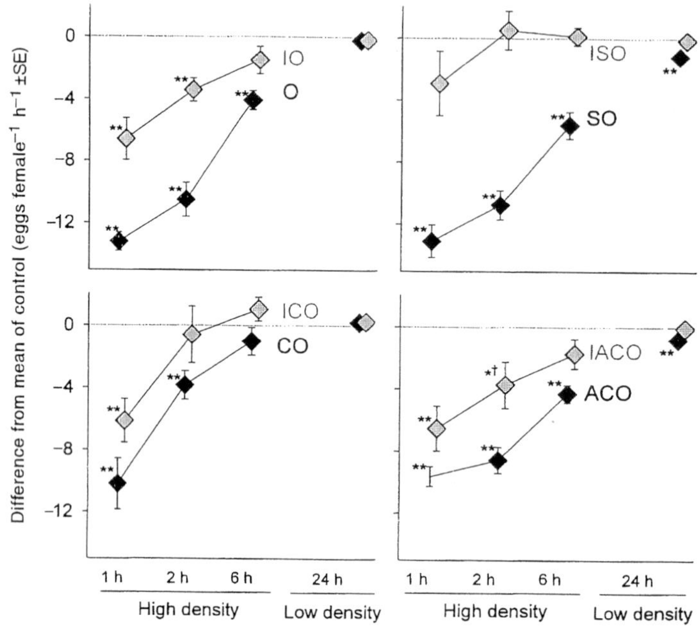

Fig. 3 Early fecundity mean replicate values ± standard error of the difference, with adult density and egg-laying time as environmental variables. All points show the mean difference of each group of populations from the control. The error bars in the low-density environment are too small to be visualized. Student's *t*-tests: *0.01 < $P \le 0.05$; **$P \le 0.01$. Sequential Bonferroni with number of tests equal to 32: †tests that lose significance with $\alpha = 0.05$.

is less clear. Similar results have been obtained in experimental studies of adaptation in bacteria (e.g. Travisano *et al.*, 1995). These conclusions are in agreement with several other studies of reverse evolution in both sexual and asexual species (Lenski, 1988; Service *et al.*, 1988; Rainey & Travisano, 1998; Crill *et al.*, 2000; Teotónio & Rose 2000, 2001).

Among the three competitive fitness estimates, the one that gave the clearest results was male fitness. This character was differentiated considerably in the starting populations, converging to a level not significantly different from that of the common ancestral control by generation 50, for all populations selected in the ancestral environment. The particular components of male fitness which are being selected during reverse evolution is unknown and may be different from population to population. For example, rapidity to achieve sexual maturity might be important in populations that take more time to develop and mature than the ancestral populations. In populations selected for late-life fertility, increased starvation resistance, and intermediate age of reproduction, individuals may not attain their peak mating ability by the time that mating usually occurs in the ancestral environment (Rose, 1984; Leroi *et al.*, 1994). In populations selected for accelerated development, on the other hand, selection should be different (Chippindale *et al.*, 1997). Because the generation length of these particular populations is set to 9 days, a male fitness decline after this day may have evolved as a result of a trade-off with very early mating success, producing radically accelerated male senescence. But these possibilities are only speculative at this point.

The total fitness assays did not give much useful information. Differences in total population fitness between the differentiated populations and the controls were negligible, with the exception of the competitive fitness of those selected for accelerated development during the first generation of assay. Taking these results separately from those of the second assay generation they reveal that significant female competitive fitness differences were present, despite the fact that we did not observe those differences in the female fitness assay. However, multigeneration population competition did not reveal any fitness differentiation involving the accelerated development stocks and thus we must conclude that all populations had similar competitive population fitness.

The multigenerational competition did not reveal significant population differences because of the very poor performance of the mutant stock employed. The use of this type of assay is attractive in strictly sexually reproducing organisms because it can directly assess a full generation of competitive success for a genetically unmanipulated experimental population, by contrast with techniques involving the use of inbreeding to estimate the fitness of single chromosomes (e.g. Fowler *et al.*, 1997; Barton & Partridge, 2000). In our case however, it proved a poor experimental technique.

Although there was a significant selection and historical effect in female competitive fitness estimates, these revealed little differentiation, before or after reverse evolution. This is surprising, given that the ancestral environment imposes selection for reproduction at day 14 after egg collection, and these conditions were closely replicated. The lack of differentiation for females may be due to one of two causes. First, perhaps the test employed is not sensitive enough to population differences because the competitor population has such a good performance in this environment that it masks any differences (but see Chippindale *et al.*, 2001). Secondly, the lack of differentiation may be real. The question is then what component of female fitness is actively being selected, if any? The results for fecundity characters measured directly reveal extensive differentiation of populations. An explanation for the lack of female fitness differentiation might have to wait for experiments which can test the extent to which fecundity plasticity reveals female fitness differences (see below).

It is interesting to note that although male fitness was highly differentiated and responsive to selection, female fitness was neither, As in previous studies with *Drosophila*, the two characters are apparently able to evolve separately from each other (Rice, 1996; Chippindale *et al.*, 2001). The fact that the Pearson correlation coefficient between male and female fitness is equal to -0.002 ± 0.023 reinforces this conclusion.

Alternative routes to adaptation

The specific life-history mechanisms behind reverse evolution were tackled by measuring life-history characters which are closely related to fitness in the ancestral environment: developmental time, starvation resistance and particularly viability and fecundity. Generally the life-history data do not indicate a widespread return to ancestral values after 50 generations of reverse evolution. Our results are not due to a lack of relevant genetic variation or presence of an epistatic genetic architecture (Teotónio & Rose, 2000). In contrast to this, and to the best of our resolution, the significant results from the male competitive fitness assays show a more pronounced convergence to ancestral levels.

These two patterns together suggest that similar degrees of adaptation were achieved through different underlying life-history and genetic mechanisms (cf. Prout, 1971; Wright, 1977; Cohan, 1984; Cohan & Hoffman, 1989; Travisano & Lenski, 1996; Crill *et al.*, 2000). The populations previously selected for increased starvation resistance reverted to ancestral character values in fecundity, remaining differentiated for starvation resistance and developmental time. The populations with accelerated development did not return to ancestral values for fecundity or developmental time. The late-life fertility populations also did not completely return to ancestral fecundity or starvation resistance levels. Nevertheless, all these groups of populations returned to ancestral male fitness levels.

Alternative routes to adaptation have been experimentally observed in *Drosophila* through asymmetry of correlated responses to selection in alternative environments (Shiotsugu *et al.*, 1997) or similar environments (e.g. Cohan & Hoffman, 1989; see also Bohren *et al.*, 1966; Gromko, 1995). Our study however, is one of the first to attempt a direct measure of fitness. Microbial systems have long permitted the simultaneous estimation of fitness and life-history characters. It is well established in the microbial literature that different genetic, physiological and life-history mechanisms are implicated in similar adaptive responses. Generally, this is supported by numerous studies on the evolution of bacterial resistance to phages and antibiotics (e.g. Lenski, 1988; Cohan *et al.*, 1994; review in Lenski, 1998), bacterial adaptation to minimal media (e.g. Travisano *et al.*, 1995; Travisano & Lenski, 1996), viral adaptation to alternative hosts (Crill *et al.*, 2000), and also by the adaptive recovery of deleterious mutations in both viruses and bacteria (Burch & Chao, 1999; Moore *et al.*, 2000).

The role that sexual recombination might have in these evolutionary patterns and their genetic basis has however, not been adequately determined. For example, it has been experimentally determined that bacterial evolution during antibiotic exposure and recovery involve compensatory genetic changes, thus revealing an epistatic mode of action. In sexual populations, because of recombination, such cases might not be so common (Teotónio & Rose, 2001).

Environmental sensitivity of adaptation

Our study illustrates the importance of characterizing adaptation in evolutionarily relevant environments. In particular it shows some of the pitfalls that can occur when not describing the environment properly. For example, if we had measured early fecundity only at low density, no differences between selected populations and controls would have been detected. Furthermore, if early fecundity at high density with 1-h of egg-laying had been the only fecundity character studied, it might have been concluded that reverse evolution did not lead to convergence for fitness characters (Teotónio & Rose, 2000). Nonetheless, with more time for egg-laying, measurable evolutionary convergence occurs. The rapidity of egg-laying might be what is selected for in the ancestral environment, as well as being the evolutionary constraint involved during convergence. The reverse evolution of fitness might then be explained by different rates of response to an environmental transition at the time of egg-laying, such as the contact with fresh food, CO_2, or the presence of laid eggs. Both amount of nutrients and adult crowding have been previously implicated in the evolution of early fecundity in some of our laboratory populations (Leroi *et al.*, 1994). Here the data indicate that the speed of response to an environmental transition was the focus of selection for early fecundity. This is a striking example of the environmental specificity and dependence of adaptation (Travisano & Lenski, 1996). The findings on fecundity also indicate that the change in plasticity for a fitness character during reverse evolution depended on previous selection history.

Despite the intricacy of these details, it is nonetheless clear that plasticity is eminently evolvable, even in an evolutionary time frame as short as 50 generations.

Complexity in experimental evolution

Overall our findings illustrate the complexity of studying the evolutionary processes of adaptation. Clear expectations of the outcome of reverse evolution were defined *a priori*, but were not fully met. Presumably this reflected the effects of previous selection on the populations analysed, the variable associations between life-history characters and fitness, and the subtle environmental dependence of the expression of these characters. This complexity should not be underestimated, both when designing studies involving experimental evolution and when interpreting their results.

Acknowledgments

For essential technical help we thank Y. Chau, T. Vu, N. Vu, C. Hammerle and E. Gass. A. K. Chippindale and L. D. Mueller have given helpful advice throughout the project. C. L. Burch, G.A.C. Bell, and two anonymous reviewers improved the manuscript with their suggestions. M. Matos received a travel grant from FLAD. H. Teotónio was supported by the Gulbenkian Foundation, FLAD, and PRAXIS / FCT (Portugal).

References

Barton, N.H. & Partridge, L. 2000. Measuring fitness by means of balancer chromosomes. *Genet. Res. Camb.* **75**: 297–313.

Bohren, B.B., Hill, W.G. & Robertson, A. 1966. Some observations on asymmetrical correlated responses to selection. *Genet. Res. Camb.* **7**: 44–57.

Burch, C.L. & Chao, L. 1999. Evolution by small steps and rugged landscapes in the RNA virus φ6. *Genetics* **151**: 921–927.

Chippindale, A.K., Alipaz, J.A., Chen, H.-W. & Rose, M.R. 1997. Experimental evolution of accelerated development in *Drosophila*. 1. Development speed and larval survival. *Evolution* **51**: 1536–1551.

Chippindale, A.K., Gibson, J.R. & Rice, W.R. 2001. Negative genetic correlation for adult fitness between sexes reveals ontogenetic conflict in Drosophila. *Proc. Natl. Acad. Sci. USA* **98**: 1671–1675.

Cohan, F.M. 1984. Can uniform selection retard random genetic divergence between isolated conspecific populations. *Evolution* **38**: 495–504.

Cohan, F.M. & Hoffman, A.A. 1989. Uniform selection as a diversifying force in evolution: evidence from *Drosophila*. *Am. Nat.* **134**: 613–637.

Cohan, F.M., King, E.C. & Zawadzki, P. 1994. Amelioration of the deleterious pleiotropic effects of an adaptive mutation in *Bacillus subtilis*. *Evolution* **48**: 81–95.

Crill, W.D., Wichman, H.A. & Bull, J.J. 2000. Evolutionary reversals during viral adaptation to alternating hosts. *Genetics* **154**: 27–37.

Fowler, K., Semple, C., Barton, N.H. & Partridge, L. 1997. Genetic variation for total fitness in *Drosophila melanogaster*. *Proc. R. Soc. Lond. B* **264**: 191–199.

Gromko, M.H. 1995. Unpredictability of correlated response to selection: pleiotropy and sampling interact. *Evolution* **49**: 685–693.

Haymer, D.S. & Hartl, D.L. 1982. The experimental assessment of fitness in *Drosophila*. I. Comparative measures of competitive reproductive success. *Genetics* **102**: 455–466.

Hedrick, P.W. & Murray, E. 1983. Selection and measures of fitness. In: *Genetics and Biology of Drosophila* (M. Ashburner, H. L. Carson & J. N. Thompson, Jr, eds), pp. 61–104. Academic Press, Orlando.

Joshi, A. & Thompson, J.N. 1995. Alternative routes to the evolution of competitive ability in two competing species of *Drosophila*. *Evolution* **49**: 616–625.

Jungen, H. & Hartl, D.L. 1979. Average fitness of populations of *Drosophila melanogaster* as estimated using compound-autosome strains. *Evolution* **33**: 371–380.

Latter, B.D.H. & Robertson, A. 1962. The effects of inbreeding and artificial selection on reproductive fitness. *Genet. Res. Camb.* **3**: 110–138.

Lenski, R.E. 1988. Experimental studies of pleiotropy and epistasis in *Escherichia coli*. II. Compensation for maladaptive effects associated with resistance to virus T4. *Evolution* **42**: 433–440.

Lenski, R.E. 1998. Bacterial evolution and the cost of antibiotic resistance. *Intern. Microbiol.* **1**: 265–270.

Lenski, R.M., Rose, R., Simpson, S.C. & Tadler, S.C. 1991. Long-term experimental evolution in *Escherichia coli*. I. Adaptation and divergence during 2,000 generations. *Am. Nat.* **138**: 1315–1341.

Leroi, A.M., Chippindale, A.K. & Rose, M.R. 1994. Long-term laboratory evolution of a genetic life-history trade-off in *Drosophila melanogaster*. 1. The role of genotype-by-environment interaction. *Evolution* **48**: 1244–1257.

Lewontin, R.C. 1974. *The Genetic Basis of Evolutionary Change*. Columbia University Press, New York.

Mackay, T.F.C. 1986. A quantitative genetic analysis of fitness and its components in *Drosophila melanogaster*. *Genet. Res. Camb.* **47**: 59–70.

Moore, F.B.-G., Rozen, D. & Lenski, R.E. 2000. Pervasive compensatory adaptation in *Escherichia coli. Proc. R. Soc. Lond. B.* **267**: 515–522.

Novitski, E., Grace, D. & Strommen, C. 1981. The entire compound autosomes of *Drosophila melanogaster. Genetics* **98**: 257–273.

Prout, T. 1971. The relation between fitness components and population prediction in *Drosophila*. I: the estimation of fitness components. *Genetics* **68**: 127–149.

Rainey, P.B. & Travisano, M. 1998. Adaptive radiation in a heterogeneous environment. *Nature* **394**: 69–72.

Reznick, D. & Travis, J. 1996. The empirical study of adaptation in natural populations. In: *Adaptation* (M. R. Rose & G. D. Lauder, eds), pp. 243–289. Academic Press, San Diego.

Rice, W.R. 1989. Analysing tables of statistical tests. *Evolution* **43**: 223–225.

Rice, W.R. 1996. Sexually antagonistic male adaptation triggered by experimental arrest of female evolution. *Nature* **381**: 232–234.

Rose, M.R. 1984. Laboratory evolution of postponed senescence in *Drosophila melanogaster. Evolution* **38**: 1004–1010.

Rose, M.R., Vu, L.N., Park, S.V. & Graves, J.L. Jr. 1992. Selection of stress resistance increases longevity in *Drosophila melanogaster. Exp. Geront.* **27**: 241–250.

Rose, M.R., Nusbaum, T.J. & Chippindale, A.K. 1996. Laboratory evolution: the experimental wonderland and the Cheshire Cat syndrome. In: *Adaptation* (M. R. Rose & G. V. Lauder, eds), pp. 221–241. Academic Press, San Diego.

Service, P.M., Hutchinson, E.W. & Rose, M.R. 1988. Multiple genetic mechanisms for the evolution of senescence. *Evolution* **42**: 708–716.

Service, P.M. & Rose, M.R. 1985. Genetic covariation among life history components: the effect of novel environments. *Evolution* **39**: 943–945.

Shiotsugu, J., Leroi, A.M., Yashiro, H., Rose, M.R. & Mueller, L.D. 1997. The symmetry of correlated selection responses in adaptive evolution: an experimental study using *Drosophila. Evolution* **163–172**: 163–172.

Teotónio, H. & Rose, M.R. 2000. Variation in the reversibility of evolution. *Nature* **408**: 463–466.

Teotónio, H. & Rose, M.R. 2001. Perspective: reverse evolution. *Evolution* **55**: 653–660.

Travisano, M. & Lenski, R.E. 1996. Long-term experimental evolution in *Escherichia coli*. IV. Targets of selection and the specificity of adaptation. *Genetics* **143**: 15–26.

Travisano, M., Mongold, J.A., Bennett, A.F. & Lenski, R.E. 1995. Experimental tests of the roles of adaptation, chance, and history in evolution. *Science* **267**: 87–90.

Weber, K.E. 1996. Large genetic change at small fitness cost in large populations of *Drosophila melanogaster* selected for wind tunnel flight: Rethinking fitness surfaces. *Genetics* **144**: 205–213.

Wilton, A.N., Joseph, M.G. & Sved, J.A. 1989. Can chromosomal heterosis in *Drosophila* be explained by deleterious recessive genes? Negative results from a dichromosomal population test. *Genet. Res. Camb.* **53**: 129–140.

Wright, S. 1977. Evolution and the Genetics of Populations, **Vol. 3**. Experimental Results and Evolutionary Deductions. University of Chicago Press, Chicago.

Received: 31 October 2001; revised: 4 February 2002; accepted: 2 March 2002

Reverse Evolution of Aging in *Drosophila melanogaster*

Hardip B. Passananti[1], Denise J. Deckert-Cruz[2], Adam K. Chippindale[3], Brian H. Le[4] and Michael R. Rose[5]

[1]*Orrick, Herrington & Sutcliffe LLP,*
4 Park Plaza, Suite 1600, Irvine, CA 92614
hpassananti@orrick.com

[2]*Dept. of Biology, Thomas Nelson Community College,*
Hampton, VA 23670
deckert-cruzd@tncc.vccs.edu

[3]*Dept. of Biology, Queen's University, Kingston,*
ON K 7L 3N6, Canada
chippind@biology.queensu.ca

[4]*Dept. of Pathology H179, Milton S. Hershey Medical Center,*
Pennsylvania State University, 500 University Drive, Hershey, PA 17033
brianle@psu.edu

[5]*Dept. of Ecology and Evol. Biology, University of California,*
Irvine, Irvine, CA 92697-2525
mrrose@uci.edu; also to whom correspondence should be sent

Key words: Aging, reverse selection, life-history, allele frequencies, mutation accumulation, antagonistic pleiotropy

Abstract

This study used reverse selection on populations of *Drosophila melanogaster* to test the evolutionary theory of aging, including antagonistic pleiotropy and mutation accumulation, the two non-exclusive population genetic mechanisms of aging. Specifically, reversed demographic selection was imposed on five populations selected for late-life fertility for 83 generations ($O_{1\text{-}5}$), returning them to an ancestral demographic schedule of 14 days. The five ancestral populations ($B_{1\text{-}5}$) were assayed each generation to serve as a control for environmental fluctuations over time. Relaxing selection for late-fecundity and imposing selection for early fecundity resulted in a rapid drop in longevity, and an increase in early fecundity, suggesting that longevity and some early life fitness component(s) are subject to antagonistic pleiotropy. As longevity fell, the frequency of the S allele of *Pgm* also decreased. Starvation resistance fell dramatically in reverse-selected males, and remained unchanged in females, suggesting that different physiological genetic mechanisms control resistance to starvation in the two sexes. Desiccation resistance remained unchanged under reverse demographic selection, implicating mutation accumulation as the primary mechanism for its evolution. Overall, these results provide some support for evolutionary theories of aging.

Introduction

Laboratory evolution of *Drosophila* species has been used repeatedly to test both the general evolutionary theory of aging and the specific population genetic mechanisms that are thought to underlie the evolution of aging (Rose, 1991). Many stocks with postponed aging have been produced using selection for later fertility (Rose and Charlesworth, 1980; Rose, 1984; Luckinbill *et al.*, 1984; Luckinbill and Clare, 1985; Partridge and Fowler, 1992; Patridge *et al.*, 1999), and a wide range of functional characters are differentiated in these stocks, including mean longevity, maximum longevity, later fertility, stress resistance, flight duration, and so on (Service and Rose, 1985; Rose *et al.*, 1992a,b; Graves *et al.*, 1992). Recently, differentiation at allozymic loci has also been associated with selectively postponed aging (Deckert-Cruz *et al.*, 1997).

Another technique that has been used to study the evolution of aging is laboratory reverse evolution (cf. Teotónio and Rose, 2000, 2001; Teotónio *et al.*, 2002). Reverse selection allows the detection of mutation accumulation and antagonistic pleiotropy, the two genetic mechanisms by which aging is thought to evolve. Service *et al.* (1988) and Graves *et al.* (1992) used reverse selection on *D. melanogaster* populations that had initially been selected for early reproduction, but were then switched to a regime of delayed reproduction (Rose, 1984). Reverse selection took the form of returning late-reproduced populations to early reproduction. When reverse selection causes a character to revert to its ancestral condition quickly, this character or its correlates must be subject to strong selection under early reproduction (Teotónio and Rose, 2001). When this occurs, selection for late reproduction will lead to a correlated reduction in some fitness component(s) at early ages, because of negative genetic correlations between late characters and early characters. Although linkage disequilibrium is a theoretical possibility, the usual population genetic interpretation of this situation is antagonistic pleiotropy between fitness-components at early and late ages (Service *et al.*, 1988; Rose, 1991), on the grounds that incidental linkage disequilibrium over multiple loci is not expected to be consistent with respect to the pattern of the genetic correlation produced. On the other hand, if the character in question never returns to its ancestral condition during reverse selection, or does so very slowly, then it could be governed by mutation accumulation, because of the presumptive absence of antagonistic pleiotropy. But these two mechanisms are not mutually exclusive in the evolution of aging in any one species, even if they interfere with each other in the evolution of individual loci.

In the experiments of Service *et al.* (1988), the reversed-selection lines rapidly lost their enhanced starvation resistance, but gained in early fecundity, thereby returning to their ancestral condition. This suggested antagonistic pleiotropy connecting starvation resistance and early fecundity, a result confirmed in independent experiments with different designs, including direct measurement of the genetic correlation between the two characters (Service and Rose, 1985; Leroi *et al.*, 1994b). Desiccation resistance did not respond to reversed selection, suggesting that this trait is governed by the mutation accumulation mechanism of evolution. This study has been criticized, however, as follows: Service *et al.* (1988)

> "did not measure how late fecundity or lifespan responded to reverse selection, although the original improvement of these demographic traits under selection for late fitness was cited as the primary evidence for postponed senescence. Clearly, to understand the effects of reverse selection on senescence one should measure the return rate of the demographic traits assayed originally. In particular, did lifespan rapidly return to the level of the control population?"
>
> *Promislow and Tatar*, 1998, p. 301

An additional problem is that Service *et al.* (1988) only observed their study populations for 22 generations. While Graves *et al.* (1992) returned to those populations some 70 generations later, the intervening period was missed. Another failing was one of omission. We now know that there is differentiation of the allozymic loci in longer-lived populations. While such loci are not necessarily themselves targets of selection in these experimental populations, they can be used as markers for their surrounding chromosomal regions. If these loci exhibit quick reversions in allele frequency under reverse selection, then they are likely to be near the loci that are subject to antagonistic pleiotropy, if they are not such loci themselves.

The present study attempts to respond to these criticisms. We imposed reverse demographic selection on five populations with postponed aging, and then followed the response to selection over 67 generations. Allele frequencies at three electrophoretic loci were also observed during reverse selection. As before, the evolutionary theory of aging is strikingly corroborated. In addition, we find clear evidence for both antagonistic pleiotropy and mutation accumulation as contributing mechanisms to the evolution of aging.

Materials and Methods

Populations

All 15 *Drosophila melanogaster* populations used in this study descend from a single ancestral population derived from a sample obtained in 1975 from a South Amherst, Massachusetts, population studied by P.T. Ives (1970). This population has since been maintained on two-week discrete generations at 25°C with constant light. In 1980, ten populations were derived from this IV stock. Five of these populations, designated $B_{1\text{-}5}$ ("baseline"), were handled in the same manner as the IV's, and the other five populations, designated $O_{1\text{-}5}$ ("old-reproduced") were selected for late reproduction, as described in Rose (1984). The five replicate O populations have spent most of their evolutionary history on a ten-week generation cycle, with the first two weeks spent in 8 dram vials with banana-molasses medium and the remaining eight weeks in Plexiglas cages. Three days prior to egg collection (the end of the 10 week cycle), the O populations are given a dietary yeast paste in order to increase the female egg-laying capacity. The five B replicate populations, on the other hand, spend the entire two weeks in 8 dram vials with banana molasses medium. In addition, the B populations do not receive any dietary yeast supplements to boost their fecundity.

The five replicate RO ("reverse O's") populations were derived from the corresponding five replicate O's in October of 1994 (see Fig. 1) at generation 83 of O selection. The RO's are demographically identical to the B's in that they too are on a discrete two week generation cycle. However, there are some important differences between the two selection regimes. The RO's are dumped into Plexiglas cages on day 11 and given a plate of banana molasses medium coated with the dietary yeast paste. In order to prevent inadvertent selection for accelerated development, all the rearing vials are saved and dumped into the cages again on days 12 and 13. The populations are given fresh yeasted plates on days 12 and

13, and the eggs are collected for the next generation on day 14. This last phase of the RO life cycle, where they experience a cage phase with three days of fresh dietary yeast, is identical to the last three days of the O selection regime. The five replicate RO populations experience direct, reverse selection for early fecundity, but no direct selection on longevity. With respect to longevity, the RO's are thus "relaxed selection" populations rather than "reverse selection" populations.

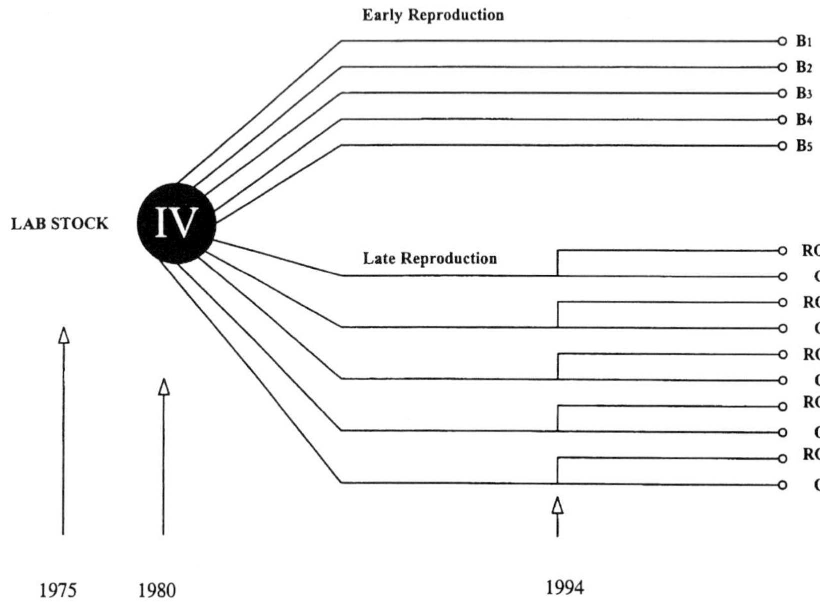

Fig. 1. Phylogeny of populations used in experimental study. All 15 populations were ultimately derived from the IV stock. $O_{1\text{-}5}$ represent the "old" populations selected for late-fertility and increased longevity. $B_{1\text{-}5}$ represent the "baseline" populations and are selected for early fecundity. $RO_{1\text{-}5}$ represent the "reverse selected" populations that were derived from the O's, and then subjected to selection for early fertility.

Experimental Assays

In each assay, samples from all 15 populations ($B_{1\text{-}5}$, $O_{1\text{-}5}$, $RO_{1\text{-}5}$) were placed on a common two week cycle for two generations in order to synchronize the sampled flies as well as to remove any parental effects which might have been created by their different selection regimes. Egg collections for the experimental flies were made from these synchronized samples, and the larvae were reared at 25°C at a density of approximately 60 eggs per 8 dram vial. When the adults began to eclose, the earliest to emerge were discarded; adults were collected 11 days after the egg collection. This sampling procedure was repeated at generations 7, 9, 12, 16, 20, 25, 30, 42, 50, and 67 of RO selection. 51 600 individuals were assayed in this experiment.

Assay Conditions

Two different nutritional levels were employed in the assays. The high yeast solution is made up of 5 grams of Fleischmann's yeast, 50 mL of distilled water, and 2 mL of 1% acetic acid. The low yeast solution is made by taking 1 mL of the high yeast solution and diluting it with 63 mL of distilled water. These nutritional levels correspond to levels 1 and 5 from Chippindale *et al.* (1993), and will hereafter be called "low" and "high." One hundred ml of this yeast solution was pipetted into a vial with banana molasses medium and allowed to dry overnight.

Flies employed in the early fecundity, starvation resistance, and desiccation resistance assays were conditioned for four days on either the low or high nutritional levels. On day 11 after egg collection, four mated pairs were placed in high and low conditioning vials with 40 flies of each sex being employed per character. Two hundred and forty flies per yeast level were conditioned for each population, giving a total of 7200 flies per generation. These experimental flies were transferred to fresh yeasted vials after two days. On day 15 after egg collection, the flies were placed in the appropriate experimental vial.

Characters Studied

Longevity. Four males and four females were placed in a vial of banana molasses medium within 24 hours of emergence from the rearing vials. Eighty individuals per population were assayed for a total of 1200 adults (15 different populations). Flies were transferred to fresh medium every Monday, Wednesday, and Friday and scored for mortality daily. Longevity assays were performed at generations 9, 20, 30, 42, and 50 of the RO selection, utilizing 6000 individuals.

Early fecundity. A single female, aged 4–5 days, was transferred from the conditioning vials using CO_2 to a vial with charcoal-colored medium coated with 100 μl of either the high or low yeast for 24 hours. Charcoal-colored medium was used to facilitate the counting of eggs. Forty females per population, each in a separate vial, were tested at each yeast level for a total of 1200 observations per generation. Early fecundity assays were performed at generations 7, 9, 12, 16, 20, 25, 42, and 50 of RO selection on 9600 female flies that were 4–5 days of adult age.

Starvation resistance. Following Service and Rose (1985), four males and four females, 4–5 days of age, were placed in 8 empty dram vials from the conditioning vials. A half sponge was then placed approximately three-fourths of the way into the vial so as to keep the flies in the top fourth of the vial. Two pieces of cotton moistened with water were then placed in the vial, which was sealed with Parafilm (the half sponge separated the flies from the moist cotton). Mortality checks were performed every six hours, and deaths noted accordingly. Forty males and 40 females from each population were assayed at each yeast level for a total of 2400 individuals per generation. Starvation resistance assays were performed at generations 7, 9, 12, 16, 20, 25, 50, and 67 of the RO selection, for a total of 19 200 individuals.

Desiccation resistance. Desiccation resistance assays were set up in a fashion similar to that of starvation resistance, except that the flies were not given a water source. In lieu of moistened cotton, 5 mL of Drierite desiccant was placed in the vial in order to remove all humidity from the air. See Service and Rose (1985) for details. Mortality checks were performed every hour until all flies had died. Two thousand four hundred 4–5 day old adults were assayed at generation 7, 9, 12, 16, 20, 25, and 50 of the RO selection, sacrificing 16,800 fruitflies.

Electrophoretic Allele Frequencies

Superoxide dismutase. One selectively differentiated locus, Cu,Zn Superoxide dismutase (Tyler *et al.*, 1993), has produced increased longevity in some transformation experiments (Reveillaud *et al.*, 1991; Orr and Sohal, 1994), but not others (Orr and Sohal, 1993). An association between the S allele of SOD and increased life-span has already been established in the B and O populations (Deckert-Cruz *et al.*, 1997). As a free radical scavenger, SOD has long been speculated to play a role in aging (Harman, 1956) by catalyzing the conversion of superoxide anion radicals to hydrogen peroxide, which is subsequently converted to water by catalase.

Alcohol dehydrogenase. Other studies have measured the response to selection (Deckert-Cruz *et al.*, 1997) of allele frequencies at the alcohol dehydrogenase (Adh) locus in the B and O populations. Adh allele frequencies do not respond to this type of selection, which makes the Adh protein interesting as a guide to the response of unselected loci to reverse evolution.

Phosphoglucomutase. Three electrophoretically distinguishable allozymes of the phosphoglucomutase enzyme (PGM) are found in the B and O stocks and designated F (fast), M (medium), and S (slow), according to their electrophoretic migration. Phosphoglucomutase is important for the metabolism of polysaccharides, as it converts glucose-1-phosphate to glucose-6-phosphate, an intermediate in glycolysis. In stocks reproduced early in life, the M allozyme is more abundant, and presumably the allele(s) encoding it is/are the most abundant allele(s) at the *Pgm* locus (Deckert-Cruz *et al.*, 1997). The S allozyme is rarely found in early-reproduced (B) stocks. Selection for postponed reproduction has resulted in substantial increases in the frequency of the *Pgm* S allele in each of the five replicate O populations. After 75 generations of selection, the mean frequency of the *Pgm* S allele was 51% (Deckert-Cruz *et al.*, 1997). In the B populations, only 5% of *Pgm* alleles, on average, encode the S allozyme.

Gel preparation. Electrophoretic gels were prepared by cooking a 12% solution of electrolyzed potato starch (Sigma S-4501) in a pH 7.0 buffer of Tris, citric acid, and ethylenediaminetetraacetic acid. Allozyme frequencies were recorded immediately upon development of gels. Allele frequencies were calculated from approximately 200 alleles per population.

Statistical Analysis

As the trajectories of population means are the focus of the scientific hypotheses under test, all data analyses proceeded using these means. Every RO population was paired with its ancestral O population, giving five independent contrasts of a pair of populations for each analysis. RO–O differences, rather than absolute values, were used in the analyses to minimize the importance of both environmental fluctuations over time and the shared genetic background inherited from each O ancestral population.

For each trait, two steps of statistical analysis were performed. The first step determined whether the reverse-selected populations differed from their ancestral controls, the long-lived populations, for any of the characters examined. For longevity and early fecundity, a two-factor ANOVA was performed, with replicate-pair a random factor and selection regime a fixed factor. The character under study (e.g. RO–O longevity, RO–O fecundity) was the dependent variable. Generation and sex were factors in the longevity analysis, while generation and nutrition were factors in the fecundity analysis. A three-factor ANOVA was performed for starvation resistance and desiccation resistance, with generation as a random factor, and nutrition and sex as fixed-effect factors. Only significant interaction terms are reported. Significance tests were performed first using the residual mean square error term, produced by variation between populations within treatments. Additional ANOVA's were performed using the interaction term of population (selection treatment) and generation as an error term. There was no difference in the results with either choice of error terms. There were also no differences in the results using either the populations grouped as separate treatments or grouped according to RO–O differences. Only the RO–O statistical results are reported. Data analyses also proceeded both for individuals independently and nested within vials, in order to take into account any statistical effect due to the microenvironment shared within individual vials. There were no differences in the results of any statistical tests between these two types of analyses. Only the vial-independent analyses are reported here.

The second step in the statistical analysis was least-squares linear regression against generation number. Independent regressions were made for each population and for the difference between the five sets of paired (O and RO) populations. Generation number was always the independent variable, while the trait of interest in a particular population, or the difference between the two populations for the trait, was the dependent variable. Hypothesis testing always employed the variation between replicates as the error term. Thus each regression was employed as the equivalent of a single datum, an extremely conservative method of analysis.

Separate regressions were performed for the difference between each population pair, sex, and nutritional level. In each case, the initial difference between the control and reverse-evolved population was fixed as zero, because the two populations began the experiment as the same population. The estimated regression slopes were then treated as raw data to test if (i) the five regression slopes were significantly different from zero; and (ii) they were affected by sex or nutrition.

Results

Longevity

Figure 2 gives the mean difference between RO and O longevity, in days, as a function of RO generation for both males and females. The solid line represents the mean difference of the five replicate populations. The patterns of mean longevity for $B_{1\text{-}5}$, $O_{1\text{-}5}$, and $RO_{1\text{-}5}$ populations are summarized in Table 1.

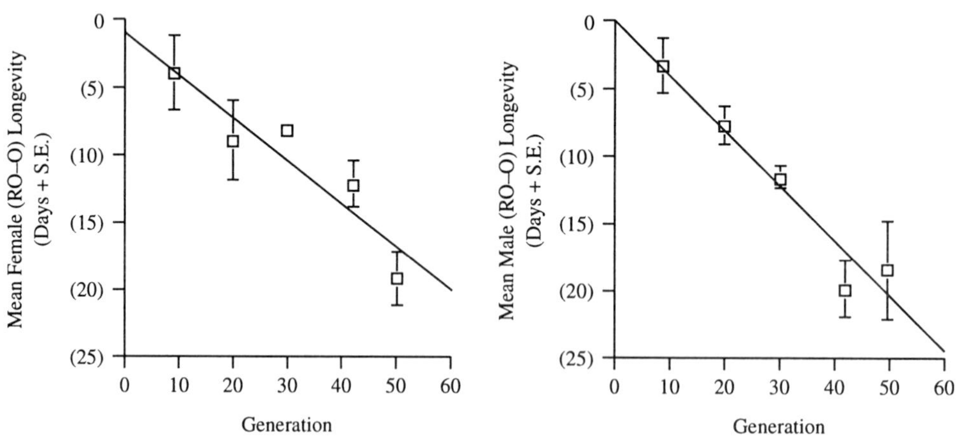

Fig. 2. Mean RO–O longevities for females and males plotted against generation (9, 20, 30, 42, 50) of RO selection. The vertical line represents the difference of the 5 replicate populations, for a total of 800 observations per data point (generation). Linear regression analysis shows the slope to be significantly different from zero for both males (p = 0.0005) and females (p = 0.0027).

Table 1 B, O, and RO Longevity (in days) for Males and Females

Population		B		O		RO	
Gen	Replicate	Male	Female	Male	Female	Male	Female
	1	27.21	31.91	59.16	42.19	49.61	47.69
	2	30.25	28.30	54.92	44.14	53.98	43.08
9	3	27.32	32.47	66.51	48.50	62.08	40.94
	4	23.92	26.83	62.33	50.95	57.74	40.79
	5	35.58	26.97	56.36	51.24	58.82	44.46
	mean	28.86	29.30	59.86	47.40	56.45	43.39
	s.e.	1.96	1.21	2.09	1.82	2.14	1.27
	1	24.90	19.56	49.97	49.73	40.23	30.69
	2	25.70	25.76	49.41	43.82	40.49	41.29
20	3	27.03	26.03	53.68	51.26	42.57	44.28
	4	22.69	22.95	54.09	47.68	49.16	43.00
	5	29.00	25.39	57.21	57.32	53.17	45.68
	mean	25.86	23.94	52.88	49.96	45.12	40.99
	s.e.	1.05	1.22	1.44	2.22	2.58	2.67
	1	36.08	26.25	57.63	51.05	46.49	43.3
	2	29.25	35.67	58.3	48.45	46.49	40.73
30	3	27.46	33.68	64.72	54.98	55.93	48.08
	4	31.90	36.84	60.83	54.00	48.13	43.34
	5	37.78	36.26	63.23	54.5	49.66	46.49
	mean	32.49	33.74	60.94	52.60	49.34	44.39
	s.e.	1.96	1.95	1.37	1.24	1.75	1.30
	1	31	25.97	59.87	59.38	41.27	41.36
	2	29.5	32.63	58.38	50.46	45.53	37.44
42	3	28.65	28.73	69.69	52.23	47.73	42.79
	4	24.29	28.36	61.06	51.51	35.74	39.26
	5	31.77	30.33	64.41	53.9	43.88	45.68
	mean	29.04	29.20	62.68	53.50	42.83	41.31
	s.e.	1.31	1.10	2.01	1.57	2.06	1.42
	1	27.37	22.51	57.49	51.86	41.99	33.74
	2	25.06	30.18	57.39	50.38	41.22	24.79
50	3	24.22	25.35	50.67	48.23	43.12	34.77
	4	23.22	24.96	64.85	51.89	38.28	33.5
	5	31.03	24.4	66.49	52.13	38.45	31.37
	mean	26.18	25.54	59.38	50.90	40.61	31.63
	s.e.	1.39	1.33	2.86	0.74	0.97	1.80

Data from the longevity experiments conducted at generations 9, 20, 30, 42, and 50 of RO selection. Each replicate entry represents the mean of 40 individuals, for a total of 6000 observations. The analysis is a two-factor ANOVA of RO–O longevity with population means as replicates. Analysis of variance shows that the difference between RO and O paired replicate populations differs significantly ($p < 0.0001$) across generations for both sexes.

ANOVA Table (RO–O Longevity)

Source	df	MS	F	P
Generation	4	367.33	15.408	< 0.0001
Sex	1	34.778	1.459	0.2342
Interaction	4	36.853	1.546	0.2075
Residual	40	23.841		

Statistical analysis used a two-factor ANOVA of RO–O longevity with population pairs as replicates, and sex and generation as factors (categorical variables). Analysis of variance shows that the mean difference in longevity between the five RO and O paired replicate populations differ significantly across generations ($p < 0.0001$). Males and females differ from each other in longevity. There are no significant interaction effects between generation time and sex.

Linear regression of RO–O longevity shows that the slope of the paired longevity difference also differs significantly from zero for both males ($p = 0.0005$) and females ($p = 0.0027$), analyzed separately. The significant effect of generation in the RO–O ANOVA suggests that RO–O longevity varies among generations, but it could be increasing one generation and decreasing the next. The linear regression analysis, however, shows the RO–O generational effect is, on average, significantly different from zero, with the five RO populations significantly falling in longevity relative to the five O populations.

Early Fecundity

Figure 3 gives the early fecundity difference between RO and O females, at both low and high yeast levels. The solid line represents the mean RO–O difference of the five replicate populations. Linear regression of this data on generation number shows that the slope of the line is significantly different from zero at the low yeast level ($p = 0.0002$), and marginally distinguishable from zero at the high yeast treatment ($p = 0.0596$). Also, the positive slope for the RO–O regression line shows that the difference in egg-laying between the two populations is increasing over generations, with the RO populations having a greater egg output than the O populations.

Analysis of variance of RO–O early fecundity, summarized in Table 2, shows that this difference is significant across generations. There is no significant interaction effect between generation and dietary treatment.

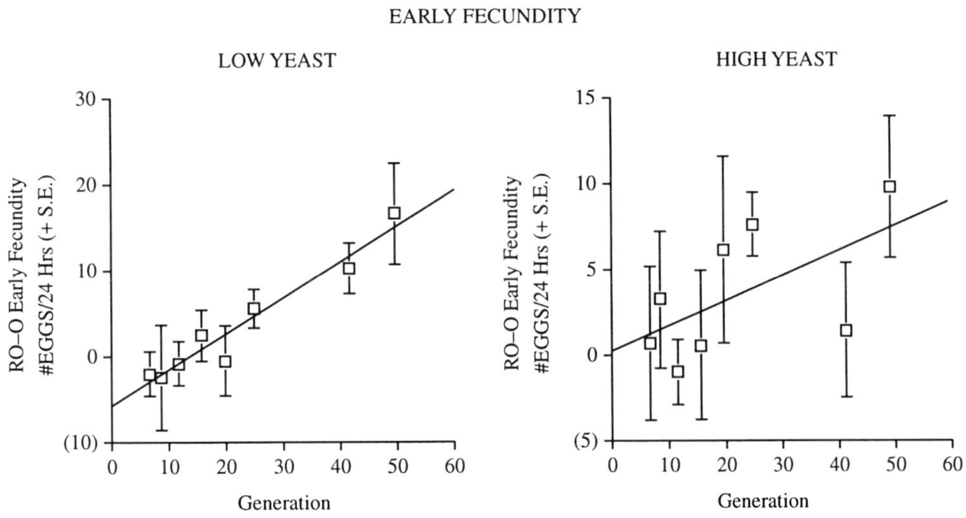

Fig. 3. Mean early fecundity difference between RO and O females, at both low and high yeast levels. The vertical line represents the difference of the 5 replicate populations. Linear regression analysis shows the slope of the line to be significantly different from zero at the low nutritional level ($p = 0.0002$) and marginally distinguishable from zero at the high nutritional level ($p = 0.0596$).

Table 2 B, O, and RO Early Fecundity under Low and High Nutritional Levels

Population		B		O		RO	
Gen	**Replicate**	**Low**	**High**	**Low**	**High**	**Low**	**High**
	1	19.71	73.95	18.93	75.33	19.50	69.58
	2	14.08	63.97	20.31	78.49	8.71	88.17
7	**3**	21.91	47.48	10.19	76.39	11.26	67.79
	4	17.98	56.98	10.67	54.33	8.19	67.80
	5	25.15	64.85	7.29	73.00	10.03	68.15
	mean	19.77	61.45	13.48	71.51	11.54	72.30
	s.e.	1.86	4.41	2.58	4.38	2.06	3.98
	1	32.57	78.94	33.03	100.68	25.79	102.54
	2	26.16	61.29	36.20	99.83	47.70	91.08
9	**3**	32.45	76.72	41.45	102.41	21.86	111.39
	4	35.44	59.20	32.13	105.40	44.30	105.50
	5	42.85	77.32	37.35	103.11	28.52	117.64
	mean	33.89	70.69	36.03	102.29	33.63	105.63
	s.e.	2.70	4.29	1.66	0.97	5.19	4.46
	1	31.97	67.48	15.11	94.36	17.05	91.79
	2	28.24	63.62	19.79	98.83	20.13	99.91
12	**3**	29.81	69.62	22.53	96.48	12.35	89.9
	4	28.73	71.65	10.66	84.2	16.31	83.2
	5	36.07	78.15	11.49	93.55	10.09	98.35
	mean	30.96	70.10	15.92	93.48	15.19	92.63
	s.e.	1.43	2.41	2.31	2.50	1.78	3.02

Table 2 (*continued*)

	1	22.08	62.46	11.35	83.83	18.48	88.03
	2	14.70	57.10	8.85	81.83	14.38	66.85
16	**3**	19.38	61.05	25.13	82.09	21.18	90.79
	4	22.28	66.77	13.75	73.95	8.85	81.68
	5	11.90	68.56	10.80	84.65	20.05	82.55
	mean	18.07	63.19	13.98	81.27	16.59	81.98
	s.e.	2.06	2.05	2.90	1.90	2.25	4.14
	1	37.00	67.65	27.95	72.75	15.26	87.00
	2	40.93	69.48	17.78	98.88	27.55	101.40
20	**3**	30.95	79.95	18.50	85.76	18.45	86.03
	4	32.95	77.48	17.90	87.33	12.30	79.18
	5	26.85	74.92	18.95	87.15	25.83	109.43
	mean	33.74	73.90	20.22	86.37	19.88	92.61
	s.e.	2.43	2.34	1.94	4.15	2.96	5.55
	1	26.38	70.85	9.43	78.53	19.18	86.38
	2	24.43	34.39	7.83	82	17.98	89.6
25	**3**	19.25	64.05	6.75	76.46	11.25	79.75
	4	29.75	64.92	4.95	68.03	11.78	82.35
	5	24.25	62.08	11.28	69.1	8.93	74.65
	mean	24.81	59.26	8.05	74.82	13.82	82.55
	s.e.	1.71	6.39	1.09	2.71	2.01	2.59
	1	14.76	62.8	12.13	91.25	27.63	90.41
	2	19.54	71.89	13.55	96.82	13.87	91.50
42	**3**	10.25	66.05	13.48	94.43	28.64	90.08
	4	22.18	68.76	7.55	89.23	21.44	91.53
	5	16.64	67.55	12.93	82.05	20.68	98.39
	mean	16.67	67.41	11.93	90.76	22.45	92.38
	s.e.	2.04	1.50	1.12	2.54	2.67	1.53
	1	27.93	68.18	31.48	92.50	34.49	90.25
	2	31.80	62.20	17.95	79.58	44.31	100.25
50	**3**	42.26	65.10	24.00	91.13	26.47	95.00
	4	37.08	78.45	12.90	75.03	41.60	88.80
	5	33.95	66.23	9.28	76.53	32.98	89.93
	mean	34.60	68.03	19.12	82.95	35.97	92.85
	s.e.	2.42	2.78	3.96	3.70	3.18	2.13

Data from the early fecundity experiment utilizing 1200 different females at two nutritional levels, per generation. Each replicate entry represents the mean of 40 females. Early fecundity was measured at generations 7, 9, 12, 16, 20, 25, 42, and 50 of RO selection, for a total of 9600 observations. The analysis was a two-factor ANOVA with population means as replicates.

ANOVA Table (RO–O Early Fecundity)

Source	df	MS	F	P
Generation	7	231.463	2.998	0.0087
Nutrition	1	0.248	0.003	0.955
Interaction	7	78.034	1.011	0.4323
Residual	64	77.204		

Starvation Resistance

Figure 4 shows the difference in starvation resistance between the five paired RO and O populations of both sexes, at both low and high yeast levels. The patterns of starvation resistance among $B_{1\text{-}5}$, $O_{1\text{-}5}$, and $RO_{1\text{-}5}$ populations are summarized in Table 3. The mean RO–O difference is becoming greater and more negative, indicating a significant drop in RO starvation time relative to the O's.

The analysis of variance used was a three-way ANOVA of RO–O starvation resistance with the five population pairs as replicates. The results indicate that this difference in hours starved significantly varies across generations ($p < 0.0001$), as well as across nutritional levels ($p = 0.0075$). There is a significant interaction effect between the dietary treatment effect and the generation time effect.

Linear regression of RO–O starvation resistance shows that the slope of the line is significantly different from zero for males at both low ($p = 0.0073$) and high ($p = 0.0039$) nutritional levels. The slope of the line is also significantly different from zero for females at both the low nutritional level ($p = 0.0449$), and the high nutritional regime ($p = 0.0451$).

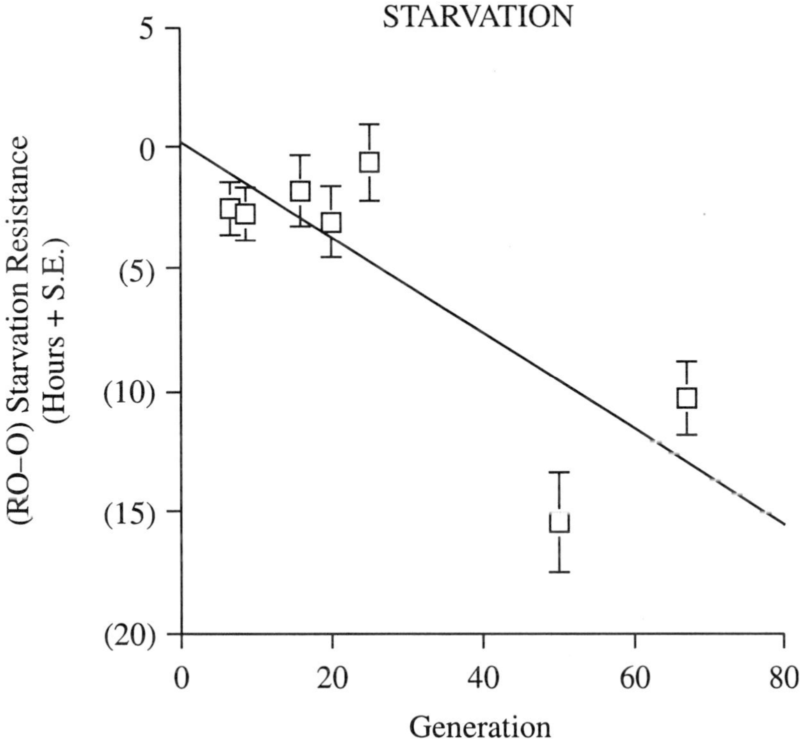

Fig. 4. (RO–O) average starvation resistance in hours for both males and females under low and high nutritional levels. When data from males and females are analyzed together, the slope of the line is significantly different from zero.

Table 3 B, O, and RO Starvation Resistance in hours for Males and Females under Low and High Nutritional Levels

Population		B				O				RO			
		Low Yeast		High Yeast		Low Yeast		High Yeast		Low Yeast		High Yeast	
Gen	Rep.	Male	Female	Male	Female	Male	Female	Male	Female	Male	Female	Male	Female
	1	18.15	33.15	16.70	18.80	44.70	54.30	26.50	26.92	31.57	47.95	23.83	23.67
	2	17.05	38.00	10.95	20.85	40.55	55.70	32.65	33.55	40.75	52.25	29.68	28.15
7	3	15.90	41.25	10.45	24.70	39.80	54.20	27.50	29.75	41.75	57.75	24.22	29.50
	4	19.55	42.00	15.70	23.75	32.60	57.67	32.10	25.50	39.25	49.75	26.15	25.10
	5	15.75	42.30	19.90	21.95	42.63	57.25	32.30	31.70	38.10	64.20	27.90	26.60
	Mean	17.28	39.34	14.74	22.01	40.06	55.82	30.21	29.48	38.28	54.38	26.36	26.60
	s.e.	0.71	1.73	1.79	1.05	2.05	0.72	1.32	1.48	1.79	2.96	1.10	1.04
	1	14.40	30.45	12.40	22.15	35.25	45.28	25.55	17.60	24.95	39.08	24.85	20.80
	2	9.60	29.95	8.40	16.50	36.48	48.55	31.75	26.05	23.90	38.28	25.25	21.05
9	3	14.55	29.60	7.60	18.70	33.06	43.50	22.30	23.65	26.20	45.25	20.35	21.45
	4	12.00	25.30	12.50	17.60	28.30	42.75	25.42	21.55	30.25	38.95	21.05	20.60
	5	14.95	32.35	14.40	20.25	33.05	47.75	28.75	22.90	35.18	49.20	27.30	26.10
	Mean	13.10	29.53	11.06	19.04	33.23	45.57	26.75	22.35	28.10	42.15	23.76	22.00
	s.e.	1.02	1.16	1.31	0.99	1.40	1.14	1.61	1.39	2.07	2.17	1.32	1.03
	1	15.60	30.45	13.70	21.05	23.95	39.10	22.45	19.15	20.85	38.55	16.80	21.75
	2	11.95	33.40	9.05	19.10	25.40	43.85	25.35	23.40	23.65	41.35	21.00	23.25
16	3	28.05	46.35	11.20	24.55	12.10	32.65	25.00	25.50	27.55	46.00	18.45	19.65
	4	13.81	27.50	12.56	20.06	23.85	40.65	24.35	23.90	20.60	38.75	19.90	22.00
	5	13.30	35.35	17.75	24.05	27.40	47.35	28.00	28.75	20.40	34.65	23.05	29.50
	Mean	16.54	34.61	12.85	21.76	22.54	40.72	25.03	24.14	22.61	39.86	19.84	23.23
	s.e.	2.94	3.22	1.45	1.08	2.69	2.46	0.90	1.56	1.37	1.87	1.07	1.67
	1	10.25	33.65	6.80	18.50	19.40	51.05	20.85	30.45	15.35	39.35	13.00	28.60
	2	8.65	25.75	10.33	17.35	16.50	56.10	24.30	35.85	26.65	41.35	23.25	26.67
20	3	8.20	39.85	10.40	28.90	19.50	48.90	13.20	33.60	16.70	32.88	18.15	34.05
	4	10.65	30.15	18.15	22.05	20.85	46.30	18.90	27.15	14.20	40.15	19.70	28.25

Table 3 (*continued*)

Population		B				O				RO			
		Low Yeast		High Yeast		Low Yeast		High Yeast		Low Yeast		High Yeast	
Gen	Rep.	Male	Female	Male	Female	Male	Female	Male	Female	Male	Female	Male	Female
	5	8.05	29.20	9.90	21.60	14.30	44.30	20.65	32.35	17.05	49.45	17.60	31.25
	Mean	9.16	31.72	11.12	21.68	18.11	49.33	19.58	31.88	17.99	40.64	18.34	29.76
	s.e.	0.54	2.39	1.88	2.01	1.19	2.04	1.82	1.47	2.22	2.65	1.66	1.30
	1	9.70	32.05	12.45	17.55	18.80	28.80	17.15	30.05	9.10	34.70	18.30	24.90
	2	11.06	35.85	14.00	21.20	28.75	69.10	34.05	46.05	26.45	66.80	30.75	42.40
25	3	7.20	36.50	7.95	26.35	8.90	48.65	25.05	44.85	22.85	60.20	25.15	38.50
	4	14.40	48.35	19.60	30.35	27.25	61.55	28.25	42.05	30.33	70.25	26.85	36.00
	5	12.60	41.10	19.45	26.35	24.65	71.00	31.70	41.75	25.25	57.20	24.60	46.20
	Mean	10.99	38.77	14.69	24.36	21.67	55.82	27.24	40.95	22.80	57.83	25.13	37.60
	s.e.	1.23	2.79	2.21	2.24	3.62	7.81	2.95	2.84	3.63	6.23	2.02	3.62
	1	15.15	36.25	13.25	27.65	40.45	57.25	34.55	41.30	23.05	43.03	20.20	29.65
	2	11.80	35.40	11.00	26.45	51.15	77.85	31.75	36.40	26.85	46.05	26.30	34.40
50	3	8.55	17.00	12.30	26.25	44.45	63.20	36.25	35.65	27.95	50.45	22.15	31.45
	4	14.20	40.90	14.30	25.25	48.00	76.50	34.75	34.30	20.85	40.20	26.95	28.75
	5	14.80	37.30	12.45	25.35	45.35	74.90	36.65	42.50	28.50	55.10	24.00	30.75
	Mean	12.90	33.37	12.66	26.19	45.88	69.94	34.79	38.03	25.44	46.97	23.92	31.00
	s.e.	1.23	4.20	0.55	0.44	1.79	4.10	0.86	1.63	1.49	2.65	1.26	0.97
	1	15.25	36.15	9.80	18.35	36.05	49.25	25.55	23.45	17.20	35.33	15.00	22.65
	2	13.50	33.65	11.00	17.45	38.40	41.85	25.85	24.05	17.65	35.00	14.50	22.90
67	3	13.30	31.55	7.45	25.30	39.70	45.10	25.65	26.75	17.55	35.10	17.00	22.10
	4	16.35	38.50	13.80	22.40	29.60	45.20	28.55	24.20	17.35	30.90	14.50	17.50
	5	15.35	34.10	11.30	17.80	29.10	44.40	19.20	24.30	12.20	32.75	16.40	26.60
	Mean	14.75	34.79	10.67	20.26	34.57	45.16	24.96	24.55	16.39	33.82	15.48	22.35
	s.e.	0.58	1.18	1.04	1.54	2.21	1.19	1.54	0.57	1.05	0.87	0.52	1.45

Data from the starvation resistance experiments conducted on 19 200 fruit flies at generation 7–67 at two different nutritional levels. Each replicate entry represents the mean of 40 individuals. The analysis was a three-factor ANOVA with population means as replicates. Only significant interaction terms are reported.

ANOVA Table (RO–O Starvation Resistance)

Source	df	MS	F	P
Generation	6	601.274	17.531	< 0.0001
Sex	1	43.848	1.278	0.2606
Nutrition	1	254.529	7.421	0.0075
Gen*Nut	6	197.771	5.766	< 0.0001
Residual	112	34.298		

Desiccation Resistance

Table 4 summarizes the mean desiccation resistance of $B_{1\text{-}5}$, $O_{1\text{-}5}$, and $RO_{1\text{-}5}$ populations. The difference between the five RO and the five O selection treatments, even after 50 generations of relaxed selection, is negligible for both sexes at both high and low nutritional levels.

The analysis is a three-way ANOVA with population means as replicates. None of the factors (generation, sex, nutrition) are statistically significant (p = 0.431, p = 0.881, and p = 0.095, respectively).

Linear regression analysis shows the slopes of the lines to be indistinguishable from zero for both males and females at both nutritional levels (males: p = 0.4957 low yeast, p = 0.0701 high yeast; females: p = 0.5565 low yeast, p = 0.1609 high yeast) (see Fig. 5).

Electrophoretic Allele Frequencies

The reversal of selection which formed the five RO populations led to a decrease in the frequency of the *Pgm* S allele. In each RO population, the frequency of the *Pgm* S allele fell within ten generations of selection for early reproduction. Following ten generations of selection, the mean frequency was 49%; in the following ten generations, it was 43%. Following 20 generations of selection for early reproduction, the mean *Pgm* S allele frequency was 37%, reflecting another 6% decrease. A paired *t*-test comparing the generation 20 *Pgm* S allele frequencies with the generation 5 *Pgm* S allele frequencies yields a p value of 0.02. After declining precipitously during the first 20 generations of RO selection, the *Pgm* S allele frequency changed more slowly between generations 20 and 30. Following generation 30, the mean allele frequency was 34% (see Fig. 6). All decreases in the *Pgm* S allele frequency represent increases in the *Pgm* M allele frequency, as the mean *Pgm* F allele frequency remained relatively constant.

Table 4 B, O, and RO Desiccation Resistance in hours

Population		B				O				RO			
		Low Yeast		High Yeast		Low Yeast		High Yeast		Low Yeast		High Yeast	
Gen	Rep.	Male	Female	Male	Female	Male	Female	Male	Female	Male	Female	Male	Female
	1	5.40	12.18	4.93	9.88	5.65	11.45	6.18	12.03	6.88	12.45	5.81	10.43
	2	5.80	11.15	5.05	9.69	7.15	12.95	7.03	12.67	6.78	12.05	6.88	11.25
7	3	4.98	10.45	5.06	10.83	7.83	12.28	6.88	12.60	6.63	11.75	6.53	11.63
	4	6.20	12.50	5.52	11.43	8.08	12.00	7.28	13.45	7.08	14.08	7.15	12.10
	5	7.45	12.63	5.65	11.00	8.05	11.25			6.68	10.88	6.38	12.00
	Mean	5.97	11.78	5.24	10.57	7.35	11.99	6.84	12.69	6.81	12.24	6.55	11.48
	s.e.	0.42	0.42	0.14	0.33	0.46	0.30	0.24	0.29	0.08	0.53	0.23	0.30
	1	5.45	10.38	4.38	9.75	7.18	12.62	6.45	9.27	6.05	11.55	5.53	7.93
	2	4.50	9.70	3.70	7.10	7.38	13.08	6.00	10.23	7.03	12.05	6.48	10.70
9	3	5.03	8.78	4.40	8.63	5.93	11.43	5.88	10.43	7.68	10.55	6.80	10.98
	4	4.60	8.55	4.25	8.15	6.89	12.28	6.10	8.53	6.33	11.48	6.53	9.75
	5	5.63	10.43	5.46	10.30	6.18	10.95	6.40	12.90	6.93	13.18	6.25	12.28
	Mean	5.04	9.57	4.44	8.79	6.71	12.07	6.17	10.27	6.80	11.76	6.32	10.33
	s.e.	0.22	0.39	0.29	0.57	0.28	0.39	0.11	0.74	0.29	0.43	0.22	0.72
	1	5.05	10.13	4.78	8.14	6.46	11.35	6.33	10.56	7.08	11.95	6.69	10.56
	2	5.22	11.53	5.08	8.03	8.38	12.36	7.20	9.15	7.18	10.73	6.65	8.53
12	3	5.90	9.00	5.53	8.68	8.18	11.78	6.35	11.65	7.39	12.08	6.63	10.70
	4	5.10	11.08	4.88	8.55	7.03	12.43	6.54	11.38	6.93	12.98	6.45	11.45
	5	5.31	11.93	5.36	9.25	7.78	12.39	8.05	13.06	8.25	13.70	7.02	9.81
	Mean	5.12	10.73	5.13	8.53	7.57	12.06	6.89	11.16	7.37	12.29	6.69	10.21
	s.e.	0.07	0.53	0.14	0.22	0.36	0.21	0.33	0.64	0.23	0.50	0.09	0.49
	1	5.35	8.85	5.18	9.10	6.63	10.00	5.90	9.68	5.80	10.08	5.60	9.63
	2	5.38	9.30	4.33	7.13	6.13	9.93	6.13	9.90	5.68	9.50	6.15	9.80
16	3	5.35	9.38	5.20	9.10	4.48	7.93	6.28	10.68	6.65	11.38	7.15	10.98
	4	5.38	9.56	5.13	9.69	5.38	9.98	5.95	10.58	5.25	9.58	6.25	10.10
	5	5.08	9.25	5.53	10.00	5.50	8.13	6.08	11.38	6.23	10.08	5.88	10.93

Table 4 (*continued*)

Population		B				O				RO			
		Low Yeast		High Yeast		Low Yeast		High Yeast		Low Yeast		High Yeast	
Gen	Rep.	Male	Female	Male	Female	Male	Female	Male	Female	Male	Female	Male	Female
	Mean	5.51	9.27	5.07	9.00	5.62	9.19	6.07	10.44	5.92	10.12	6.21	10.29
	s.e.	0.22	0.12	0.20	0.50	0.36	0.48	0.07	0.30	0.24	0.34	0.26	0.28
	1	6.20	9.98	5.93	10.65	7.30	10.75	7.05	12.73	5.83	6.60	6.20	13.08
	2	4.48	6.10	4.80	9.73	6.63	9.15	6.70	15.15	6.70	11.55	6.30	13.48
20	3	5.23	9.13	4.92	11.40	6.23	9.73	7.10	13.40	6.30	10.20	5.45	12.05
	4	5.90	9.10	5.28	10.20	6.07	9.45	6.18	12.43	5.50	10.90	6.15	14.70
	5	5.30	8.68	4.70	8.25	5.83	9.20	6.33	15.17	5.03	8.33	6.75	14.93
	Mean	5.42	8.60	5.13	10.05	6.41	9.66	6.67	13.78	5.87	9.52	6.17	13.65
	s.e.	0.30	0.66	0.22	0.53	0.26	0.29	0.19	0.59	0.29	0.91	0.21	0.53
	1	8.40	15.78	8.65	18.05	10.43	18.30	9.50	22.40	8.15	14.68	8.23	14.70
	2	4.68	7.90	5.20	7.61	4.93	8.05	5.83	11.45	5.58	12.55	5.85	11.95
25	3	4.15	10.15	5.05	13.60	5.55	7.88	4.83	10.53	3.45	4.93	5.20	8.17
	4	2.85	5.38	4.08	8.53	5.03	7.30	5.08	8.90	4.75	8.80	5.85	12.33
	5	2.95	4.80	5.45	11.00	3.59	7.35	4.93	11.78	3.00	7.30	4.38	8.05
	Mean	4.61	8.80	5.69	11.76	5.91	9.78	6.03	13.01	4.99	9.65	5.90	11.04
	s.e.	1.01	1.99	0.78	1.89	1.18	2.14	0.88	2.40	0.91	1.76	0.64	1.29
	1	6.90	12.75	6.83	12.90	8.95	16.15	8.98	14.78	8.05	13.40	7.28	15.10
	2	6.40	12.10	6.40	12.60	8.15	13.65	8.08	14.78	8.15	16.50	8.23	14.75
50	3	6.58	11.88	5.43	13.48	7.60	13.08	7.85	14.98	8.25	14.50	6.85	12.98
	4	6.38	12.43	6.03	11.05	8.05	14.48	7.80	17.50	7.03	13.35	7.60	15.28
	5	6.15	11.80	6.25	13.03	7.45	13.00	8.30	18.33	7.33	14.78	7.15	14.98
	Mean	6.48	12.19	6.19	12.61	8.04	14.07	8.20	16.07	7.76	14.51	7.42	14.62
	s.e.	0.12	0.18	0.23	0.42	0.26	0.58	0.21	0.76	0.24	0.57	0.24	0.42

Data from the desiccation resistance experiments conducted on 16 800 fruitflies at generations 7–50 at two different nutritional levels. Each replicate entry represents the mean of 40 individuals. The analysis was a three-factor ANOVA with population means as replicates. None of the factors was statistically significant. Only significant interaction terms are reported.

ANOVA Table (RO–O Desiccation Resistance)

Source	df	MS	F	P
Generation	6	2.413	0.997	0.4312
Sex	1	0.055	0.023	0.8806
Nutrition	1	6.882	2.843	0.0946
Residual	110	2.420		

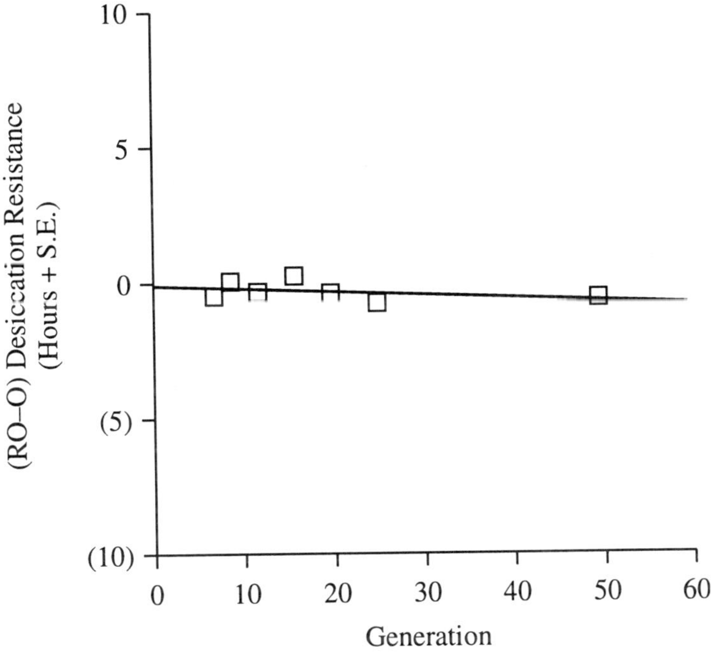

Fig. 5. (RO–O) average desiccation resistance in hours for both males and females under low and high nutritional levels. When analyzed separately, the slope of the line is indistinguishable from zero for both males and females at both nutritional levels (males: $p = 0.4957$ low yeast, $p = 0.0701$ high yeast; females: $p = 0.5565$ low yeast, $p = 0.1609$ high yeast).

After 30 generations of relaxed selection, the SOD locus has not responded to reverse selection. The frequency of the S allele of SOD is 15% in the five reverse selected populations, compared with 16–17% in the five replicate O populations. The allele frequencies of the F allele of ADH also remained unchanged in the five reverse selected populations (see Fig. 7).

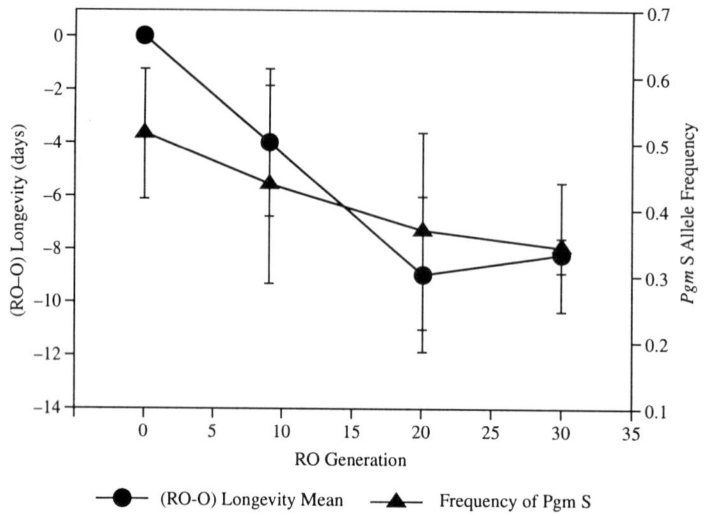

Fig. 6. *Pgm S* allele frequencies fell within 30 generations of the reversal of selection from late to early reproduction. Each solid point represents the mean of the five replicate RO populations (± S.E.).

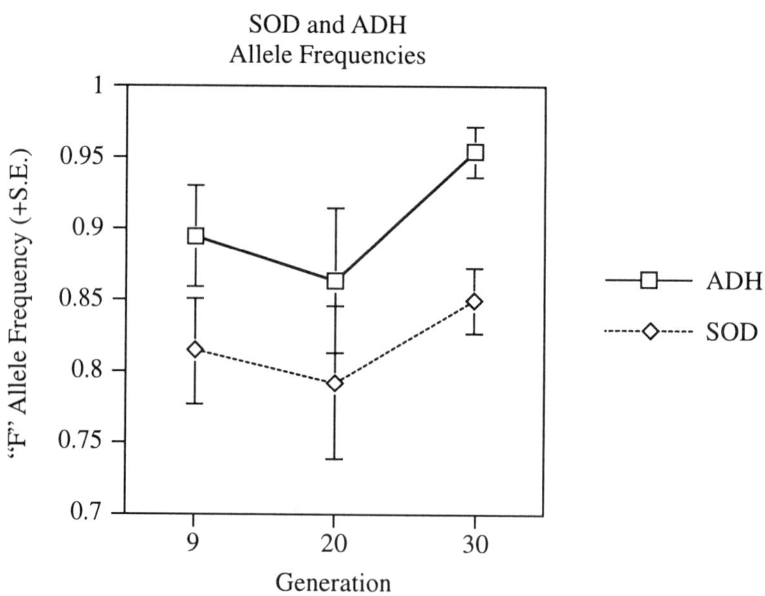

Fig. 7. SOD and ADH "F" allele frequencies plotted against generation of RO selection. There is no significant change in the frequency of the "F" allele for either SOD or ADH in the five replicate reverse selected populations.

Discussion

Summary of Results

Relaxing selection for late-fertility and imposing selection for early fertility causes reverse evolution not only in obviously important life-history characters, like longevity, but also in correlated characters, like starvation resistance. Longevity responded very rapidly to the reversed selection treatment. By generation 9 of reverse-selection, the longevity difference between the experimental population and its ancestral control was at the threshold of significance ($p = 0.08$), and by generation 20 they had significantly differentiated ($p = 0.0021$). The goodness of fit of the linear regression to the data is remarkably good — an R^2 of 0.859 for females and an R^2 of 0.938 for males. Over just 50 generations, average longevity fell by 18.4 days in males and 19.3 days in females, while early fecundity increased by 9.9 and 16.9 eggs per day, for high and low nutrition respectively. Also, as longevity fell, the frequency of the S allele of *Pgm* decreased 15% in frequency over 30 generations. Starvation resistance fell by 9.5 hours and 18.2 hours in the males in high and low nutritional environments, respectively. Desiccation resistance remained statistically unchanged in both sexes. We now turn to the significance of these results, with respect to earlier research and with respect to more general issues.

Analysis of Genetic Mechanisms

It is reasonably easy to sort out the responses to reverse selection into those that must involve some type of pleiotropy, or genetic correlation between ages, and those that might not. For example, the rapid decrease in RO longevity is too quick to be accounted for by mutation accumulation in the reverse selected populations. The effective population size of these cultures is about 1000 individuals (Mueller, pers. comm.). Drift fixes mutations on a time-scale of $4N_e$ generations, or 4000 generations, which is two orders of magnitude greater than the time-scale of our experiments. An alternative calculation is to assume a linear impact of mutations that occur at a rate μ per locus per generation, with N loci of average effect Δ, over 50 generations. If we suppose that 100 loci are involved in the selection regime that created the long-lived O's, they must have an average effect of 0.4, since the differentiation of B and O longevities was about 40 days. [Fleming *et al.* (1993) estimated 100–300 loci in the differentiation of B and O flies.] The rate of loss of differentiation due to mutation accumulation would therefore be about μ times 2000. A mutation rate per locus affecting a quantitative character is unlikely to be greater than 10^{-8}, so that the total decrease in longevity due to mutation accumulation is likely to be 10^{-4} or less, over the course of reverse evolution. Even if we take the very high mutation rate of 10^{-6}, the impact of mutation accumulation is not likely to exceed a few percent over 50 generations. Perhaps an additional scaling factor is that the response of early fecundity to direct selection was of the same magnitude as the indirect response of longevity. It is extremely unlikely that mutation accumulation affecting longevity would be on the same scale as a direct response to selection by early fecundity.

The dramatic drop in the longevity of the reverse-selected populations is correlated with a decrease in the frequency of the S allele of *Pgm*, and an increase in the frequency of the M allele of *Pgm*. An earlier survey of metabolic enzymes (Deckert-Cruz *et al.*, 1997) revealed that a PGM allozyme rare in the early reproduced populations was consistently more common in the longer lived populations, implicating the S allele, or a closely-linked allele, in the evolution of aging. With the re-imposition of an early fecundity selection regime, the falling of the S allele frequency indicates the presence of antagonistic pleiotropy in the effects of the locus for which PGM electromorphs serve as markers, perhaps *Pgm* itself.

From such arguments, we conclude that the reverse evolution of male and female longevity, PGM electromorph frequencies, and starvation resistance must have involved some type of genetic correlation between these characters and some early-age fitness-component. One might argue that the five long-lived populations, which are not selected for early fecundity, may have a substantial load of accumulated mutations acting on early fecundity. The dramatic increase in early fecundity in the five reverse long-lived populations (which are directly selected for early fecundity) may have been due to the removal of early-acting mutations that accumulated in the long-lived populations. This increase in reverse-selected fecundity may, through increased costs of reproduction, have caused the decrease in lifespan in the five reverse long-lived populations. Note that while this is an effect due to the cost of reproduction (cf. Sgrò and Partridge, 2000), it is still a type of antagonistic pleiotropy. It is plausible that desiccation resistance generally did not involve substantial early-late genetic correlation. The lack of evolution of desiccation resistance over the short period of reverse selection indicates that it may normally evolve by mutation-selection balance. In other words, the lack of a response in desiccation resistance under reverse selection suggests that the alleles affecting this character are age-specific in their effects, lacking significant antagonistic pleiotropy between early and late ages. It might be argued that the lack of response of desiccation resistance to reverse selection is due to fixation for the alleles that enhance desiccation resistance. This is unlikely, however, because of other selection lines that have been derived from these populations and successfully selected for large increases in desiccation resistance (Rose *et al.*, 1992b).

If all genetic variation had been exhausted for this trait in the postponed senescence populations, this increase in desiccation resistance in the desiccation-selected populations would not have been possible.

Genotype by Environment Interactions

The environment plays a crucial role in the response to selection (cf. Teotónio *et al.*, 2002). In the present experiments, two different assay environments were employed with some characters: "low yeast" and "high yeast." Fecundity responded to reverse selection detectably only when food was limited. For instance, under the low nutrition assay, reverse-selected females clearly increased their early egg-laying significantly. Under the high yeast conditions, the response of early fecundity is more sporadic, with much greater variance between lines, obviating statistical significance. On the other hand, starvation resistance and desiccation resistance do not show qualitatively different results between high and low nutrition assays. This is somewhat anomalous, given the close association between starvation resistance and early fecundity that has been found for these populations; the two characters inversely vary with nutrition (Chippindale *et al.*, 1993). However, Leroi *et al.* (1994a) showed that the genetic antagonism between early fecundity and longevity was suppressed under conditions of high, rather than low, nutrition. This offers a plausible explanation for the significant response of RO early fecundity only with low nutrition: at higher nutrition there is less of a physiological trade-off, presumably because there are more calories and nutrients available for both survival and fecundity.

There is also heterogeneity of the selection responses for the various traits studied. For example, longevity responded very rapidly to the reversed selection treatment. By generation 9 of reverse selection, the longevity difference between the experimental population and its ancestral control was at the threshold of significance, and by generation 20 they had significantly differentiated. Early fecundity in the low-yeast nutritional regime responded after 25 generations of selection, while starvation resistance took many more generations to respond to selection. This temporal heterogeneity of the selection responses illustrates the caution one must take in making general statements about selection responses.

Conclusion

This is the first experiment to show that the joint evolution of early and late life-history characters fully meets the expectations of evolutionary theory. When the force of natural selection acting on late-fertility is reduced to zero, longevity falls rapidly. Other experiments have uncovered the joint evolutionary patterns of early fecundity and stress resistance under reverse evolution (e.g. Service *et al.*, 1988; Teotónio and Rose, 2000), but such findings did not provide a direct test of the evolutionary theory of aging. We provide such a direct test here.

While the fundamental evolutionary theory of aging is supported by our results, there are a number of important complexities that impinge on our findings. As has been found in a number of related publications, assay conditions affect the specific correlations during reverse evolution. An important point is that assay conditions tend to be different from the conditions imposed during laboratory evolution (vid. Leroi *et al.*, 1994a; Teotónio *et al.*, 2002). Another important limitation is that we have not exhaustively identified all the early life characters that trade-off with longevity. While early fecundity recovers during reverse evolution, early male fertility has not been examined here. Indeed, the full complex of pleiotropic interrelationships among survival and fertility characters remain to be delineated for O-type Methuselah flies.

The rapidity of the fall in mean longevity in our experiment is remarkable, about 0.4 days of adult life being lost per generation. Adult life-history can evolve with great speed. Furthermore, this result was not obtained by artificial selection. Natural selection in the laboratory was the agent of evolutionary change in these fruit flies, just as it was in the original derivation of the O Methuselah flies. Natural selection on life-history is thus demonstrably powerful in a contrived setting, raising the possibility that it can be comparably powerful in nature.

Acknowledgements

We thank W. Chen, M. Le, A. Ngo, S. Sen, D. Templin, and K. Yee for assistance with laboratory work. We also thank L.D. Mueller, M. Matos and two anonymous reviewers for insightful comments on the manuscript. Special thanks to B. Yool for initiating the RO stocks, and to R.A. Passananti for indispensable support. This research was supported by NIH grant AG09970 to M.R. Rose.

References

Chippindale, A.K., T.J.F. Chu, and M.R. Rose. 1996. Complex trade-offs and the evolution of starvation resistance in *Drosophila melanogaster*. *Evolution* **50**: 753–766.

Chippindale, A.K., A.M. Leroi, S.B. Kim, and M.R. Rose. 1993. Phenotypic plasticity and selection in *Drosophila* life-history evolution. I. Nutrition and the cost of reproduction. *J Evol Biol* **6**: 171–193.

Deckert-Cruz, D.J., R.H. Tyler, J.E. Landmesser, and M.R. Rose. 1997. Allozymic differentiation in response to laboratory demographic selection of *Drosophila melanogaster*. *Evolution* **51**: 865–872.

Djawdan, M., A.K. Chippindale, M.R. Rose, and T.J. Bradley. 1998. Metabolic reserves and evolved stress resistance in *Drosophila melanogaster*. *Physiol Zool* **71**: 584–594.

Fleming, J.E., G.S. Spicer, R.C. Garrison, and M.R. Rose. 1993. Two–dimensional protein electrophoretic analysis of postponed aging in *Drosophila*. *Genetica* **91**: 183–198.

Graves, J.L., E.C. Toolson, C. Jeong, L.N. Vu, and M.R. Rose. 1992. Desiccation, flight, glycogen, and postponed senescence in *Drosophila melanogaster*. *Physiol Zool* **65**: 268–286.

Harman, D. 1956. Aging: A theory based on free radical and radiation chemistry. *J Gerontol* **11**: 298–300.

Ives, P.T. 1970. Further studies of the South Amherst population of *Drosophila melanogaster*. *Evolution* **38**: 507–518.

Leroi, A.M., A.K. Chippindale, and M.R. Rose. 1994a. Long-term laboratory evolution of a genetic life-history trade-off in *Drosophila melanogaster*. I. The role of genotype-by-environment interaction. *Evolution* **48**: 1244–1257.

Leroi, A.M., W.R. Chen, and M.R. Rose. 1994b. Long-term laboratory evolution of a genetic life-history trade-off in *Drosophila melanogaster*. II. Stability of genetic correlations. *Evolution* **48**: 1258–1268.

Luckinbill, L.S., M.J. Arking, M.J. Clare, W.C. Cirocco, and S. Buck. 1984. Selection for delayed senescence in *Drosophila melanogaster*. *Evolution* **38**: 996–1003.

Luckinbill, L.S. and M.J. Clare. 1985. Selection for life span in *Drosophila melanogaster*. *Heredity* **55**: 9–18.

Manchenko, G.P. 1994. *Handbook of Detection of Enzymes on Electrophoretic Gels*. CRC Press, Inc., Boco Raton, FL.

Orr, W.C. and R.S. Sohal. 1993. Effects of copper zinc superoxide dismutase overexpression on life span and resistance to oxidative stress in transgenic *Drosophila melanogaster*. *Arch Biochem Biophys* **301**: 34–40.

Orr, W.C. and R.S. Sohal. 1994. Extension of life-span by overexpression of superoxide dismutase and catalase in *Drosophila melanogaster*. *Science* **263**: 1128–1130.

Partridge, L. and A.K. Fowler. 1992. Direct and correlated responses to selection on age at reproduction in *Drosophila melanogaster*. *Evolution* **46**: 76–91.

Partridge, L., N. Prowse, and P. Pignatelli. 1999. Another set of responses and correlated responses to selection on age at reproduction in *Drosophila melanogaster*. *Proceedings of the Royal Society of London B* **266**: 255–261.

Pasteur, N., G. Pasteur, F. Bonhomme, J. Catalan, and J. Britton-Davidian. 1988. *Practical Isozyme Genetics*. Ellis Horwood Ltd., Chichester, UK.

Promislow, D.E.L. and M. Tatar. 1998. Mutation and senescence: Where genetics and demography meet. *Genetica* **102/103**: 299–314.

Reveillaud, I., A. Niedzwiecki, and J.E. Fleming. 1991. Expression of bovine superoxide dismutase in *Drosophila melanogaster* augments resistance to oxidative stress. *Mol Cell Biol* **11**: 632–640.

Rose, M.R. 1984. Laboratory evolution of postponed senescence in *Drosophila melanogaster*. *Evolution* **38**: 1004–1010.

Rose, M.R. 1991. *Evolutionary Biology of Aging*. Oxford University Press, New York.

Rose, M.R. and B. Charlesworth. 1980. A test of evolutionary theories of senescence. *Nature* **287**: 141–142.

Rose, M.R., T.J. Nusbaum, and J.E. Fleming. 1992a. *Drosophila* with postponed aging as a model for aging research. *Lab Animal Sci* **42**: 114–118.

Rose, M.R., L.N. Vu, S.U. Park, and J.L. Graves. 1992b. Selection for stress resistance increases longevity in *Drosophila melanogaster*. *Exper Gerontol* **27**: 241–250.

Service, P.M., E.W. Hutchinson, and M.R. Rose. 1988. Multiple genetic mechanisms for the evolution of senescence in *Drosophila melanogaster*. *Evolution* **42**: 708–716.

Service, P.M. and M.R. Rose. 1985. Genetic covariation among life-history components: The effects of novel environments. *Evolution* **39**: 943–945.

Sgrò, C.M. and L. Partridge. 2000. Evolutionary responses of the life history of wild-caught *Drosophila melanogaster* to two standard methods of laboratory culture. *Am Nat* **156**: 341–353.

Teotónio, H., M. Matos, and M.R. Rose. 2002. Reverse evolution of fitness in *Drosophila melanogaster*. *J Evol Biol* **15**: 608–617.

Teotónio, H. and M.R. Rose. 2000. Variation in the reversibility of evolution. *Nature* **408**: 463–466.

Teotónio, H. and M.R. Rose. 2001. Perspective: Reverse evolution. *Evolution* **55**: 653–660.

Tyler, R.H., H. Brar, M. Singh, A. Latorre, J.L. Graves, L.D. Mueller, M.R. Rose, and F.J. Ayala. 1993. The effect of superoxide dismutase alleles on aging in *Drosophila*. *Genetica* **91**: 141–149.

Relaxed Stress Selection in *Drosophila melanogaster*

Hardip B. Passananti,[1] Kelly A. Beckman, and Michael R. Rose[2]

[1]*Orrick, Herrington & Sutcliffe LLP,*
4 Park Plaza, Suite 1600, Irvine, CA 92614
hpassananti@orrick.com

[2]*497 Steinhaus Hall, University of California,*
Irvine, Irvine, CA 92697-2525
mrrose@uci.edu; also to whom correspondence should be sent

Key words: Aging, reverse selection, stress-selection, life-history, Drosophila melanogaster

Abstract

Typically, a selection response is lost when a selection regime is relaxed or terminated. Some possible explanations for this result are a negative genetic correlation between selected characters and fitness, antagonistic pleiotropy and linkage involving alleles of opposed effect, or a side-effect of relaxed inbreeding. The current study relaxes selection on stocks selected for resistance to starvation and desiccation stress, and the response to relaxed selection is then observed. Contrary to many findings, we found maintenance of the selection response with respect to some characters upon relaxation of the stress selection regime. Specifically, after 35 generations of relaxed selection, longevity, early fecundity, and desiccation resistance have not changed significantly in the relaxed desiccation-selection populations, suggesting that the alleles affecting these characters lack significant antagonistic pleiotropy. Starvation resistance, on the other hand, rose significantly in the relaxed desiccation populations. After 20 generations of relaxed selection, starvation resistance fell dramatically in the relaxed starvation-selection populations relative to their ancestral populations. Longevity, however, has not dropped significantly from that of its ancestral population. When early fecundity in the relaxed starvation-selection populations is analyzed separately at each generation, there is a significant increase in this character. This increase in early-life fecundity in association with the decrease in starvation resistance may reflect antagonistic pleiotropy between these two characters in the relaxed starvation resistance system.

Introduction

Ever since the publication of the *Origin of Species*, evolutionary biologists have drawn useful inferences about the genetic basis of evolution from the study of selection. For example, there are the pioneering studies of Mather and Harrison (1949) and the more recent studies of Yoo (1980) on the number of abdominal bristles in *Drosophila melanogaster*. Mather and Harrison (1949) scored bristles on several segments of the abdomen, while Yoo scored the bristles on a single segment. Yoo then imposed selection for increased bristle number for 86 generations, and after this time, the mean bristle number had increased by 316%. However, several of the populations stopped responding to selection, having reached a selection plateau. When selection was later terminated or relaxed, the mean bristle number declined quickly. The decline resulting from relaxing selection shows that alleles for low bristle number were still present, and hence the original selection plateaus were not caused by a loss of genetic variation. These results are fairly typical of quantitative genetic experiments in that selection plateaus are usually reached after many generations of an increase in the selected trait. Relaxing the selection regime usually results in a loss of the selection response (Yoo, 1980; Blows and Hoffman, 1993; and others).

Why does relaxing or terminating a selection regime result in an abrupt loss of the selection response? One possibility is a negative genetic correlation between selected character(s) and fitness. In Yoo's (1980) experiment, the decline when selection was relaxed indicates that bristle number was negatively correlated with fitness; presumably the selection plateau was similarly caused by natural selection acting in opposition to artificial selection. It is likely that linkage disequilibrium played some role in lowering fitness, because when Mather and Harrison (1949) resumed selection on lines in which they relaxed selection for a few generations, the bristle number increased without as great a diminution of viability and fecundity. This suggests that during the period of relaxation, recombination dissociated deleterious alleles from the alleles for high bristle number, and that chromosomes free of deleterious alleles then increased in frequency when selection was resumed.

These experiments and others like them confirm the important role of antagonistic pleiotropy and linkage involving alleles of opposed effect. Bohren *et al.* (1966) have shown that selection can generate the latter, especially in small populations. In selection experiments, the selected character does not change at a steady rate, probably because new mutations occur and because recombination sporadically rearranges closely linked loci into new combinations that are favorable material for selection. Replicate populations vary in response (Yoo, 1980 and others), probably because patterns of linkage disequilibrium differ among the samples from which the populations are initiated. The frequency of polygenic combinations may also vary because of genetic drift as selection proceeds. A change in the character lowers some components of fitness because of pleiotropy and/or linkage disequilibrium, and the decrement in fitness eventually becomes great enough to reduce the response to selection, even while the population retains genetic variation.

Another possibility as to why relaxation of selection leads to a loss of selection response is simply a side-effect of relaxed inbreeding. In any practical selection regime, the number of adult parents used is more or less restricted, with the result that some inbreeding inevitably takes place, and its effects are superimposed on those of the selection. Furthermore, most artificial selection experiments start off with small populations, usually with an N_e of approximately 30 individuals, which generates inbreeding depression. Suppose that relaxation of selection proceeds with a substantial increase in population size, an increase that allows natural selection to once again ameliorate components of fitness. With this scenario, the loss of an indirect, deleterious response to selection following the relaxation of the selection regime is then just a manifestation of regained vigor once population size has increased.

This loss of selection response on relaxed selection is a fairly trivial, that is commonplace, result. But what if there is no response to relaxed selection? In other words, what are the implications of the maintenance of the selection response upon relaxation? One could argue that there was no inbreeding in the experimental system, and no negative genetic correlations with fitness in the control environment.

It is important to distinguish between relaxed and reverse selection. In reverse evolution, populations are returned to earlier selection regimes from their evolutionary history, i.e. to an antecedent state (Teotónio and Rose, 2000, 2001; Teotónio *et al.*, 2002). Such reverse evolution or selection should be distinguished from relaxed selection. Relaxed selection, when performed correctly, involves the complete removal of selection in the experimental populations. This relaxation of selection is unlikely to have characterized any experimental population's evolutionary history for many generations, due to the physical difficulty of preventing selection from occurring. The complicated interrelationship between reverse and relaxed selection is a major issue in the interpretation of the results of this study. We will return to this issue in the Discussion section.

The present study utilizes experimental populations of *Drosophila melanogaster* derived from stocks selected for postponed senescence (Rose, 1984). A wide range of functional characters are differentiated in these stocks, including mean longevity, maximum longevity, later fertility, starvation resistance, desiccation resistance, flight duration, and so on (Rose *et al.*, 1992a; Rose *et al.*, 1992b; Graves *et al.*, 1992). Differentiation at allozymic loci has also been associated with these postponed aging populations (Deckert-Cruz *et al.*, 1997). Replicate populations were derived from these long-lived populations and selected for either starvation resistance or desiccation resistance (Rose *et al.*, 1992b). The current study relaxes stress selection on these stocks with enhanced resistance to starvation and desiccation stress. The response to relaxed selection is observed over 20 and 35 generations, respectively. Contrary to most other findings, we found maintenance of the selection response with respect to some characters upon relaxation of the stress selection regime. Other characters, such as starvation resistance, responded strongly to relaxed selection.

Materials and Methods

Extant Populations

Source Population: Ives

All 30 *Drosophila melanogaster* populations used in this study descend from a single ancestral population derived from a sample obtained in 1975 from a South Amherst, Massachusetts, population studied by P.T. Ives (1970). All 30 populations (see Fig. 1) are five-fold replicated, and are large and outbred so that the confounding effects of inbreeding can be minimized. The populations are maintained using discrete-generation culture methods, at 25°C with constant light.

B and O Populations

In 1980, ten populations were derived from this IV stock. Five of these populations, designated $B_{1\text{-}5}$ ("baseline"), were handled in the same manner as the IV's, and the other five populations, designated $O_{1\text{-}5}$ ("old reproduced") were selected for late reproduction, as described in Rose (1984). The five replicate O populations have spent most of their evolutionary history on a ten-week generation cycle, with the first two weeks spent in 8 dram vials with banana-molasses medium and the remaining eight weeks in Plexiglas cages. Three days prior to egg collection (the end of the 10-week cycle), the O populations are given a dietary yeast paste in order to increase the female egg-laying capacity. The five B replicate populations,

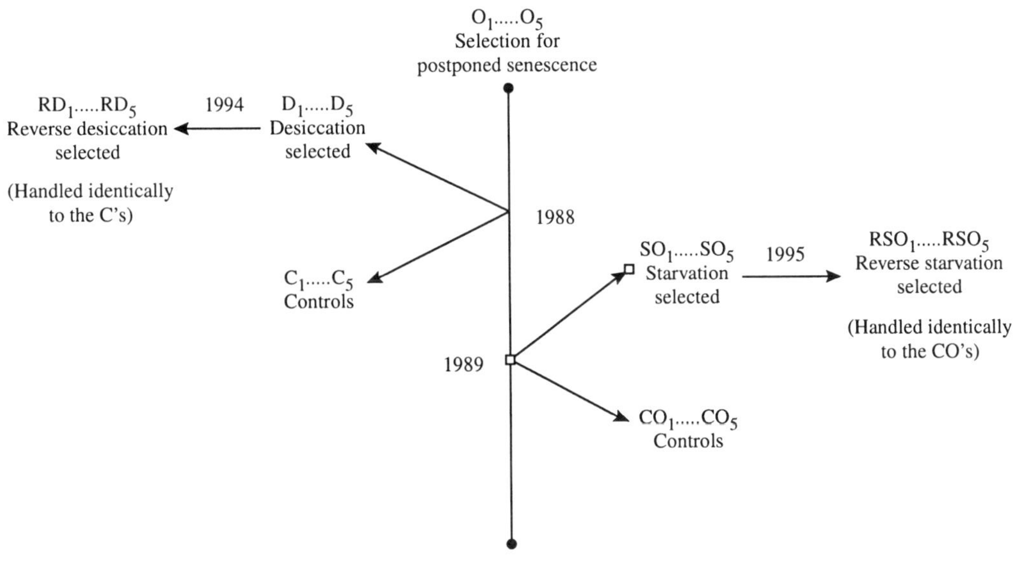

Fig. 1. Phylogeny of populations used in experimental study.

on the other hand, spend the entire two weeks in 8 dram vials with banana molasses medium. In addition, the B populations do not receive any dietary yeast supplements to boost their fecundity.

C and D Populations

The five replicate D ("desiccation") populations and their controls, the C's, were derived from the long-lived O populations in 1988 (Rose *et al.*, 1990; Rose *et al.*, 1992b). The five-fold replicated group of D populations was selected for resistance to desiccation, a character related to longevity in the ancestral O populations (Service *et al.*, 1985). On day 14 of the life-cycle, each of the C and D replicate populations is placed in a population cage. In the case of the D populations, each cage is prepared for selection by placing bags of desiccant inside. The cages are sealed off with plastic wrap so that atmospheric moisture is excluded. At the same time, the control C populations are placed in population cages provisioned with non-nutritive agar plates to provide moisture. Since this experiment is designed to isolate the effects of desiccation selection from those of starvation selection, the control populations, C, are mildly starvation selected while the D's are being desiccated and mildly starvation selected. When 70–80% of the D flies in each replicate population have died, the survivors of desiccation selection, as well as the appropriate control population survivors, are provided with standard banana-molasses medium. Three days after the last replicate D population has been taken off selection, eggs are collected from all the C and D populations to produce the next generation.

SO and CO Populations

In 1989, the $SO_{1\text{-}5}$ ("starvation O's") and the $CO_{1\text{-}5}$ ("control O's") populations were derived from the O's, after the O's had been under selection for 60 generations (Rose *et al.*, 1992b). The SO populations are selected for increased resistance to starvation, and the CO's are their controls. For the first 14 days of their life-cycles, the control and starvation selected populations are handled in an identical manner. Each generation, flies are raised at controlled densities of 60–80 per vial in 8 dram vials filled with approximately 5 ml food, and kept at 25°C in an incubator, as in the B and O culture (Rose 1984). At day 14 after oviposition, the eclosed flies from all the populations are placed in plexiglass chambers. The CO populations are given food in the form of petri dishes of banana-molasses medium replaced every two days. The SO's, on the other hand, are not given food, but instead receive petri dishes containing non-nutritive agar. After approximately 70–80% mortality due to starvation, each SO population, and its paired control, are given yeasted banana-molasses medium on which to recover and lay eggs for the following generation.

Creation of RD Populations

The five replicate RD ("relaxed D's") populations were derived from the D's in October of 1994 at generation 132 of D selection. The RD's are handled identically to the C's as outlined above. The $RD_{1\text{-}5}$ populations are placed in cages after they have spent 14 days in 8-dram vials. When 70–80% of a D replicate population have died, the corresponding RD population (as well as the corresponding C population) is provided with standard banana molasses medium.

Creation of RSO Populations

The $RSO_{1\text{-}5}$ ("relaxed starvation") populations were derived from the SO's in 1996 at generation 94 of SO selection. The RSO populations are handled identically to the CO's, as described above.

Experimental Assays

Samples from all 30 populations ($RD_{1\text{-}5}$, $D_{1\text{-}5}$, $C_{1\text{-}5}$, $RSO_{1\text{-}5}$, $SO_{1\text{-}5}$, $CO_{1\text{-}5}$) were placed on a common two-week cycle for two generations in order to synchronize the sampled flies as well as to remove any parental effects which might have been created by their different selection regimes. Egg collection for the experimental flies was done from these synchronized samples, and the larvae were reared at

25°C at a density of approximately 60 eggs per 8 dram vial. When the adults began to eclose, the earliest to emerge were discarded; adults were collected 11 days after the egg collection. This sampling procedure was repeated at generations 7, 9, 12, 16, 20, 25, and 35 of RD selection, and at generations 9, 12, 15, and 20 of RSO selection.

Assay Conditions

Two different nutritional levels were employed in the RD assays. The high yeast solution is made up of 5 grams of Fleischmann's yeast, 50 mL of distilled water, and 2 mL of 1% acetic acid. The low yeast solution is made by taking 1 mL of the high yeast solution and diluting it with 63 mL of distilled water. These nutritional levels correspond to levels 1 and 5 from Chippindale *et al.* (1993), and will hereafter be called "low" and "high." 100 uL of this yeast solution was pipetted into a vial with banana molasses medium and allowed to dry overnight.

Flies employed in the early fecundity, starvation resistance, and desiccation resistance assays were conditioned for four days on either the low or high nutritional levels. On day 11 after egg collection, four mated pairs were placed in high and low conditioning vials with 40 flies of each sex being employed per character. Two hundred and forty flies per yeast level were conditioned for each population, giving a total of 7200 flies per assay. These experimental flies were transferred to fresh yeasted vials after two days. On day 15 after egg collection, the flies were placed in the appropriate experimental vial.

No such conditioning phase was employed in the RSO assays. Instead, adults from the experimental egg collection were either subjected to longevity assays within 24 hours of eclosion, or desiccation, starvation, or early fecundity assays four days after eclosion.

Characters Studied

Longevity. Four males and four females were placed in a vial of banana molasses medium within 24 hours of emergence from the rearing vials. Eighty individuals per population were assayed for a total of 2400 adults. Flies were transferred to fresh medium every Monday, Wednesday, and Friday and scored for mortality daily. Longevity assays were performed at generations 9, 20, and 35 of RD selection, and at generations 9, 12, and 20 of RSO selection.

Early fecundity. A single female, aged 4–5 days, was transferred from the conditioning vials using CO_2 to a vial with charcoal-colored medium coated with 100 uL of either the high or low yeast for 24 hours. Charcoal-colored medium was used to facilitate the counting of eggs. Forty females per population, each in a separate vial, were tested at each yeast level for a total of 1200 observations per generation. Early fecundity assays were performed at generations 7, 9, 12, 16, 25, and 35 of RO selection on female flies that were 4–5 days of adult age.

In the RSO assays, there was no conditioning phase, and only the high yeast level was used. Early fecundity assays were performed at generations 12, 15, and 20 of RSO selection.

Starvation resistance. Following Service *et al.* (1985), four males and four females, 4–5 days of age, were placed in empty 8 dram vials from the conditioning vials. Note that in the RSO experiments, experimental flies were taken directly from the rearing vials and placed in the experimental starvation vial. A half sponge was then placed approximately three-fourths of the way into the vial so as to keep the flies in the top fourth of the vial. Two pieces of cotton moistened with water were then placed into the vial, which was sealed with Parafilm. Mortality checks were performed every six hours, and deaths noted accordingly. Forty males and forty females from each population were assayed at each yeast level for a total of 4800 individuals per generation. Starvation resistance assays were performed at generations 7, 9, 12, 16, 25, and 35 of RD selection, and at generations 9, 12, and 20 of RSO selection.

Desiccation resistance. Desiccation resistance assays were set up in a fashion similar to that of starvation resistance, except that the flies were not given a water source. In lieu of the moistened cotton, five mL of Drierite desiccant was placed in the vial in order to remove all humidity from the air. See Service *et al.* (1985) for details. Mortality checks were performed every hour until all flies had died. Two thousand four hundred 4–5 day old adults were assayed at generations 7, 9, 12, 16, 25, and 35 of RD selection. Desiccation resistance experiments were not performed on any RSO populations.

Statistical Analysis

As the trajectories of population means are the focus of the scientific hypotheses under test, all data analyses proceeded using these means. Every RD population was paired with its ancestral D population, giving five independent contrasts for each analysis. Also, each RSO population was paired with its ancestral SO population. When longitudinal patterns were to be compared, independent regressions were made on each population or on the difference between paired (RD and D; RSO and SO) populations. Hypothesis testing always employed the variation between replicates as the error term.

All statistical analyses employed Statview for the MacIntosh.

RD Results

Longevity

Figure 2 gives the RD–D longevity, in days, as a function of the RD generation for males and females in each of the five replicate populations. Statistical analysis used a two-factor ANOVA of RD–D longevity with populations as replicates. Analysis of variance shows that the mean difference in longevity between the RD and D paired replicate populations does not change significantly across generations ($p = 0.569$). The RD populations are not changing in longevity relative to their ancestral group, the D populations.

This lack of a response to relaxed selection in longevity is seen again in the linear regression analysis of RD–D longevity. The regression analysis shows that the slope of the line for the mean of the five replicate populations is not significantly different from zero for either males ($p = 0.329$) or females ($p = 0.923$), analyzed separately (Fig. 2). Also see Table 1 for the mean longevity in days for males and females.

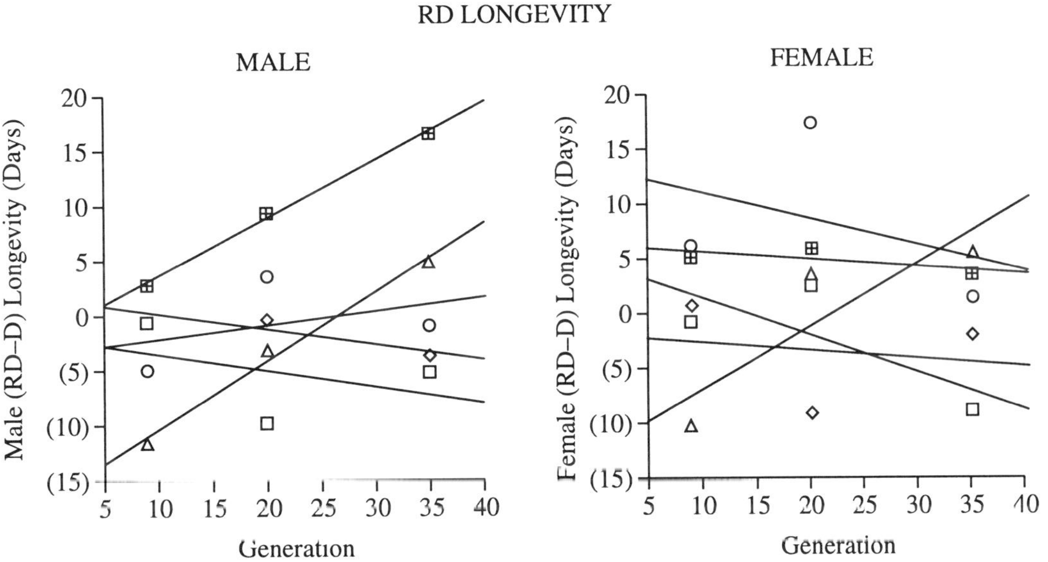

Fig. 2. RD–D longevity in days for all five replicate populations of males and females plotted against generation of RD selection. Linear regression analysis of the mean for all five replicate populations shows that the slope of the line is not significantly different from zero for either males ($p = 0.329$) or females ($p = 0.923$).

Table 1 C, D, and RD Longevities (in days) for Males and Females

		C		D		RD	
Gen	**Replicate**	**Male**	**Female**	**Male**	**Female**	**Male**	**Female**
	1	52.71	53.68	57.44	56.55	56.91	55.86
	2	60.42	56.53	65.20	54.55	65.02	55.33
9	3	70.00	49.71	73.44	48.13	68.51	54.50
	4	56.18	45.88	79.33	55.37	67.76	45.15
	5	67.11	48.56	52.94	42.17	55.84	47.44
	mean	**61.28**	**50.87**	**65.67**	**51.35**	**62.81**	**51.66**
	s.e.	3.24	1.89	4.88	2.72	2.70	2.23
	1	58.23	47.81	54.73	30.47	54.59	33.07
	2	49.61	44.68	64.33	45.72	64.06	36.71
20	3	61.95	48.86	62.35	32.63	66.08	50.19
	4	47.26	40.95	67.39	34.11	64.38	37.83
	5	64.15	43.98	54.08	37.69	63.58	43.78
	mean	**56.24**	**45.26**	**60.58**	**36.12**	**62.54**	**40.32**
	s.e.	3.34	1.41	2.65	2.67	2.03	3.01
	1	51.03	48.05	50.98	45.74	45.91	36.95
	2	55.50	42.95	51.33	45.62	47.82	43.81
35	3	54.66	47.31	58.29	44.03	57.53	45.68
	4	38.78	42.87	62.53	43.98	67.60	49.90
	5	55.83	49.61	48.39	37.88	65.16	41.72
	mean	**51.16**	**46.16**	**54.30**	**43.45**	**56.80**	**43.61**
	s.e.	3.21	1.38	2.63	1.44	4.39	2.14

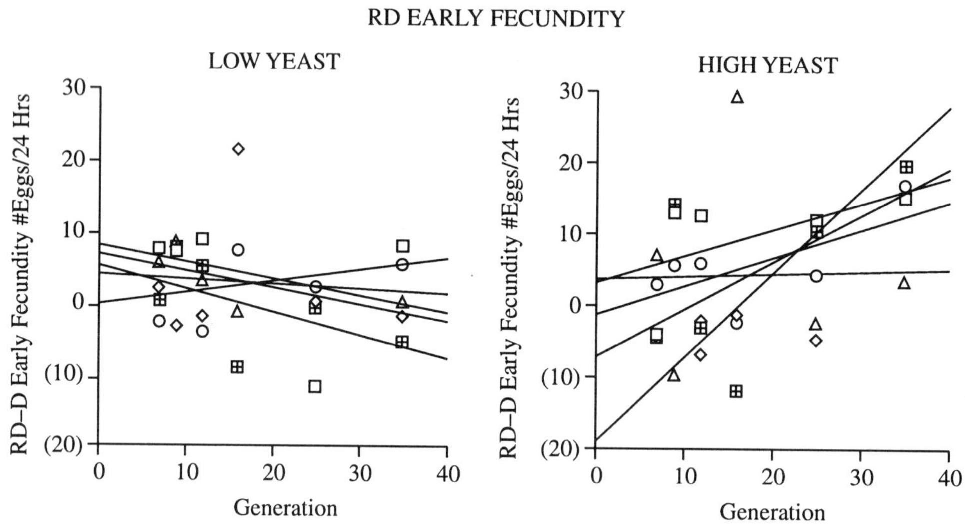

Fig. 3. Early fecundity difference between RD and D females for all five replicate populations against RD generation time, at both the low and high yeast level. Linear regression analysis of the mean shows the slope of the line to be indistinguishable from zero at both the low yeast level ($p = 0.219$) and at the high yeast level ($p = 0.06$).

Table 2 C, D, and RD Early Fecundity under Low and High Nutritional Levels

Gen	Population Replicate	C		D		RD	
		Low	High	Low	High	Low	High
	1	13.90	82.43	11.50	61.44	19.30	67.67
	2	10.40	83.05	3.38	68.09	5.64	64.23
7	3	19.54	66.79	11.13	49.71	8.69	52.77
	4	29.14	71.00	7.21	49.07	13.10	56.23
	5	19.90	81.05	8.84	58.41	9.34	54.31
	mean	**18.58**	**76.86**	**8.41**	**57.34**	**11.21**	**59.04**
	s.e.	3.19	3.34	1.48	3.61	2.34	2.92
	1	26.15	88.85	4.13	46.41	11.50	59.59
	2	34.33		9.15		6.06	
9	3	23.31	94.93	10.86	73.96	18.88	79.79
	4	28.00	103.52	3.75	75.36	12.57	65.77
	5	33.73	31.90	5.71	55.79	13.92	70.15
	mean	**29.10**	**79.80**	**6.72**	**62.88**	**12.59**	**68.83**
	s.e.	2.15	16.25	1.41	7.07	2.06	4.25
	1	61.63	80.00	19.18	59.38	28.21	72.00
	2	32.38	78.78	27.43	61.13	25.63	54.40
12	3	40.90	85.95	33.55	62.15	29.63	68.27
	4	50.95	77.09	20.25	51.28	23.72	49.61
	5	50.80	95.78	19.20	61.85	24.46	58.85
	mean	**47.33**	**83.52**	**23.92**	**59.16**	**26.33**	**60.63**
	s.e.	4.97	3.41	2.86	2.03	1.12	4.19
	1	31.33	87.48			16.75	89.48
	2	29.73	95.10	5.80	67.69	27.53	66.55
16	3	29.08	84.68	11.48	67.90	18.97	65.58
	4	27.31	95.00	2.78	39.56	1.54	68.97
	5	23.98	85.65	24.08	70.15	15.15	58.43
	mean	**28.29**	**89.58**	**11.04**	**61.33**	**15.99**	**69.80**
	s.e.	1.25	2.28	4.71	7.28	4.20	5.22
	1	22.46	76.45	5.08	61.15	10.38	70.78
	2	10.30	90.13	13.78	60.44	16.28	68.18
20	3	11.58	89.28	5.83	68.52	13.85	71.00
	4	37.10	103.13	2.45	61.58	6.90	58.95
	5	5.92	76.53	2.11	63.14	3.80	58.92
	mean	17.47	87.10	5.85	62.97	10.24	65.57
	s.e.	5.61	4.98	2.11	1.46	2.26	2.75
	1	19.68	84.54	19.68	63.71	8.08	75.95
	2	10.55	88.21	3.28	47.93	4.03	43.51
25	3	15.65	86.65	9.55	66.72	11.83	71.10
	4	21.78	94.18	1.42	53.25	2.48	51.05
	5	20.31	89.72	2.73	49.89	2.10	60.53
	mean	**17.59**	**88.66**	**7.33**	**56.30**	**5.70**	**60.43**
	s.e.	2.03	1.62	3.39	3.77	1.86	6.04
	1	34.53	102.82	17.10	67.21	25.30	82.62
	2	45.05	112.33	15.47	40.03	13.80	72.31
35	3	38.65	89.68	17.00	67.29	22.73	84.46
	4	42.20	98.70	4.15	56.95	4.48	60.58
	5	30.80	119.13	10.15	60.85	4.80	80.73
	mean	**38.25**	**104.53**	**12.77**	**58.47**	**14.22**	**76.14**
	s.e.	2.56	5.15	2.50	5.01	4.35	4.41

Early Fecundity

Figure 3 gives the early fecundity difference between the RD and D females, at both high and low yeast levels. Linear regression analysis of the mean of the five replicate populations shows the slope of the line to be indistinguishable from zero at the low yeast level ($p = 0.219$). Nor is it significantly different from zero at the high yeast level ($p = 0.06$). The lack of a slope for the RD–D regression line shows that the difference in egg-laying between the two populations is not changing over generations. It should be noted, however, that at every time point except one (generation 25, low yeast), the RD populations are laying more eggs in absolute numbers than their paired D population.

Analysis of variance of RD–D early fecundity shows that this difference is not significant across generations ($p = 0.231$) nor across nutritional levels ($p = 0.230$). There is also no significant interaction effect between generation time and dietary treatment. See also Table 2 for the mean early fecundity under both low and high nutritional levels.

Desiccation Resistance

The mean desiccation resistance difference between the RD and D populations, even after 35 generations of relaxed selection, is negligible for both sexes at both high and low nutritional levels.

A three-way ANOVA with population means as replicates was used to compare the RD and D desiccation resistance times. There is no statistically significant effect of generation ($p = 0.114$), or nutritional level ($p = 0.08$). There is a significant effect of sex ($p = 0.0005$), however.

Linear regression analysis shows the slopes of the lines to be indistinguishable from zero for males (see Fig. 4) at both nutritional levels ($p = 0.708$ low yeast, $p = 0.249$ high yeast), and also indistinguishable from zero for females (see Fig. 5) at both the high yeast level ($p = 0.360$) and at the low nutritional regime ($p = 0.066$). Neither the RD males nor the RD females, under high or low nutritional regimes, have deviated significantly from their founding source population with respect to desiccation resistance. Also see Table 3 for the mean desiccation resistance for males and females under low and high nutritional levels.

Starvation Resistance

Figures 6 and 7 show the starvation resistance difference between the RD and D populations of both sexes, at both high and low yeast levels for the five replicate populations. The mean RD–D difference is becoming greater and more positive, indicating a significant increase in RD starvation resistance time relative to the D's.

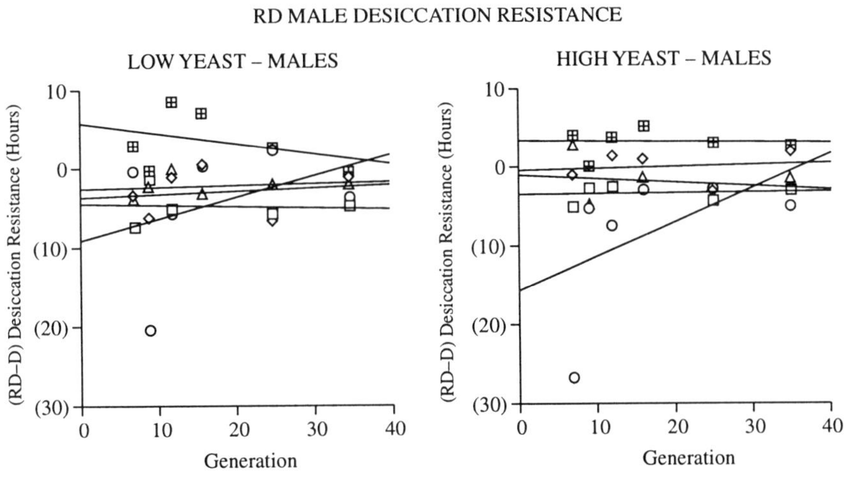

Fig. 4. (RD–D) male desiccation resistance for all five replicate populations in hours under low and high nutritional levels. The slope of the line is indistinguishable from zero for males at both nutritional levels ($p = 0.708$ low yeast, $p = 0.249$ high yeast).

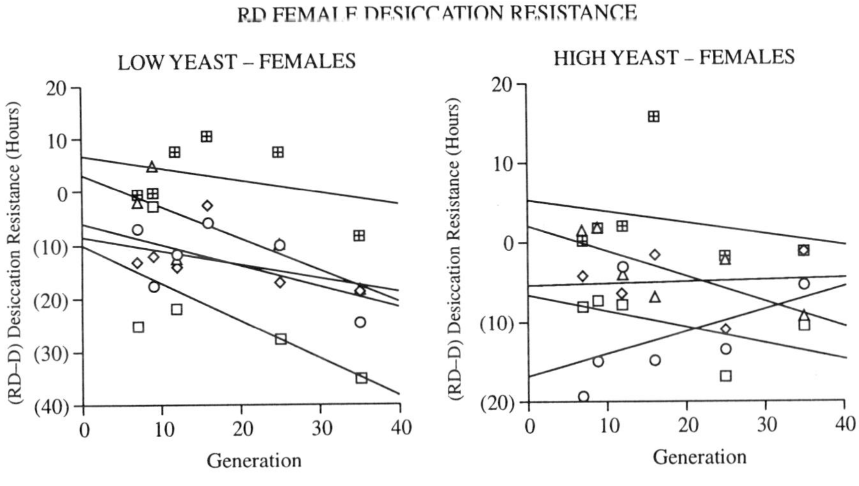

Fig. 5. (RD–D) female desiccation resistance for all five replicate populations in hours under low and high nutritional regimes. The slope of the line for the mean of the replicate populations is indistinguishable from zero for females under both food levels ($p = 0.066$ low yeast, $p = 0.360$ high yeast).

The analysis is a three-way ANOVA of RD–D starvation resistance with population means as replicates. The results indicate that this difference in hours starved changes significantly across generations ($p < 0.0001$), but not across nutritional levels ($p = 0.812$). There is also a significant effect of sex ($p = 0.018$).

Linear regression of RD–D starvation resistance shows that the slope of the line for the mean of the five replicate populations is significantly different from zero for males at both low ($p = 0.029$) and high ($p = 0.05$) nutritional levels, as

Table 3 C, D, and RD Desiccation Resistance under Low and High Nutritional Levels

Population		C				D				RD			
		Low Yeast		High Yeast		Low Yeast		High Yeast		Low Yeast		High Yeast	
Gen	Rep.	Male	Female	Male	Female	Male	Female	Male	Female	Male	Female	Male	Female
	1	9.11	15.76	9.43	17.97	31.70	87.65	27.25	70.83	24.30	62.57	22.10	63.08
	2	7.11	14.03	7.79	15.65	36.80	88.55	32.95	74.00	33.10	75.50	31.95	70.20
7	3	9.13	17.63	9.29	21.14	48.00	81.44	55.00	84.79	47.43	74.58	28.36	65.43
	4	9.17	19.13	8.35	19.46	30.50	62.28	22.61	64.64	26.50	60.25	25.29	66.38
	5	8.23	13.25	8.46	15.23	30.72	68.15	27.90	66.15	33.60	67.69	32.17	66.61
	mean	**8.55**	**15.96**	**8.66**	**17.89**	**35.54**	**77.61**	**33.14**	**72.08**	**32.99**	**68.12**	**27.97**	**66.34**
	s.e.	0.40	1.09	0.31	1.12	3.32	5.29	6.38	3.59	4.04	3.08	1.94	1.15
	1	8.48	17.75	7.75	18.60	19.25	60.95	18.10	59.35	17.85	58.35	15.50	52.36
	2	8.10	13.45	6.92	15.01	28.85	69.10	–	–	22.30	57.15	21.55	59.85
9	3	7.70	12.43	6.46	18.83	42.05	66.60	20.40	62.11	21.55	49.00	15.20	47.25
	4	9.80	11.78	7.78	19.65	20.75	55.55	22.50	50.20	18.25	60.55	17.60	52.55
	5	6.13	10.10	6.55	12.23	19.65	53.70	17.70	49.42	19.35	53.51	17.90	51.50
	mean	**8.04**	**13.10**	**7.09**	**16.86**	**26.11**	**61.18**	**19.68**	**55.27**	**19.86**	**55.71**	**17.55**	**52.70**
	s.e.	0.59	1.28	0.29	1.41	4.35	3.00	1.11	3.21	0.89	2.03	1.14	2.03
	1	8.55	15.28	8.28	17.95	21.60	75.90	17.85	51.90	16.40	54.05	15.45	44.45
	2	8.08	12.10	6.68	14.93	24.85	72.40	22.40	52.55	23.50	58.45	23.75	46.55
12	3	8.03	15.13	6.83	18.85	35.45	74.15	26.45	54.80	29.60	62.45	19.05	52.05
	4	7.63	12.45	7.65	17.13	20.15	67.10	18.55	52.65	19.80	54.70	15.95	48.80
	5	7.88	12.28	7.28	14.05	20.45	63.20	18.60	53.25	29.05	70.75	22.60	55.60
	mean	**8.03**	**13.45**	**7.34**	**16.58**	**24.50**	**70.55**	**20.77**	**53.03**	**23.67**	**60.08**	**19.36**	**49.49**
	s.e.	0.15	0.72	0.29	0.91	2.86	2.36	1.63	0.49	2.57	3.06	1.68	1.98
	1	7.55	13.95	8.55	21.25	–	–	–	–	11.95	40.00	15.00	37.80
	2	6.83	12.08	7.06	14.78	23.90	63.35	26.05	65.80	24.30	60.90	26.90	64.55
16	3	7.80	13.85	7.38	20.08	17.05	47.20	21.20	51.80	17.15	41.45	18.30	37.05
	4	6.50	11.40	6.65	15.55	20.60	61.25	19.95	61.05	17.20	55.60	18.55	54.25
	5	7.85	13.33	7.56	17.40	15.40	49.30	19.55	46.70	22.45	59.80	24.90	62.85

Table 3 (*continued*)

		C				D				RD			
Population		Low Yeast		High Yeast		Low Yeast		High Yeast		Low Yeast		High Yeast	
Gen	Rep.	Male	Female	Male	Female	Male	Female	Male	Female	Male	Female	Male	Female
	mean	**7.31**	**12.92**	**7.44**	**17.81**	**19.24**	**55.28**	**21.69**	**56.34**	**18.61**	**51.55**	**20.73**	**51.30**
	s.e.	0.27	0.51	0.32	1.25	1.90	4.10	1.50	4.33	2.19	4.51	2.22	5.93
	1	3.70	4.34	4.83	6.58	3.80	6.80	5.05	6.25	4.45	5.65	4.65	7.20
	2	4.65	9.20	5.48	9.90	10.10	24.50	11.65	41.05	10.80	32.10	13.95	33.60
20	3	6.03	10.80	5.85	11.33	9.95	30.20	11.70	23.55	9.65	22.10	8.90	24.95
	4	3.23	4.73	3.89	5.10	7.35	19.80	8.30	26.00	3.05	5.60	6.10	8.25
	5	2.48	2.65	3.35	4.98	2.45	5.15	6.85	8.80	4.05	4.65	5.70	9.45
	mean	**4.02**	**6.34**	**4.68**	**7.58**	**6.73**	**17.29**	**8.71**	**21.13**	**6.40**	**14.02**	**7.86**	**16.69**
	s.e.	0.61	1.55	0.47	1.29	1.57	4.91	1.32	6.32	1.59	5.57	1.68	5.33
	1	11.85	21.38	10.90	24.48	25.65	82.65	23.30	71.15	19.85	54.95	19.10	54.50
	2	11.13	13.03	10.65	18.88	33.80	85.70	29.60	73.55	26.95	68.65	26.55	62.85
25	3	10.00	17.68	10.55	24.58	34.80	77.50	26.05	69.25	37.10	67.50	23.45	56.00
	4	10.60	17.78	11.07	23.50	27.70	70.00	25.65	60.15	25.50	60.15	21.95	58.28
	5	9.40	14.50	10.35	20.05	28.35	64.95	24.30	70.35	31.00	72.50	27.45	69.00
	mean	**10.60**	**16.87**	**10.70**	**22.30**	**30.06**	**76.16**	**25.78**	**68.89**	**28.08**	**64.75**	**23.70**	**60.13**
	s.e.	0.43	1.45	0.13	1.19	1.79	3.86	1.07	2.30	2.88	3.16	1.52	2.63
	1	10.00	17.50	8.45	17.03	17.66	72.50	16.80	49.95	12.80	37.10	13.85	39.65
	2	7.85	12.13	8.55	15.30	21.35	75.95	18.30	46.20	20.00	57.05	20.15	45.35
35	3	8.68	16.03	9.20	20.63	21.25	65.95	21.55	38.80	17.50	41.35	16.70	33.80
	4	8.55	13.33	8.50	16.40	15.55	56.55	16.25	42.50	13.40	38.45	14.70	33.45
	5	8.00	11.55	8.23	14.73	20.40	60.45	16.80	45.90	19.20	52.20	19.65	45.15
	mean	**8.62**	**14.11**	**8.59**	**16.82**	**19.24**	**66.28**	**17.94**	**44.67**	**16.58**	**45.23**	**17.01**	**39.48**
	s.e.	0.38	1.15	0.16	1.03	1.14	3.61	0.96	1.88	1.48	3.97	1.27	2.60

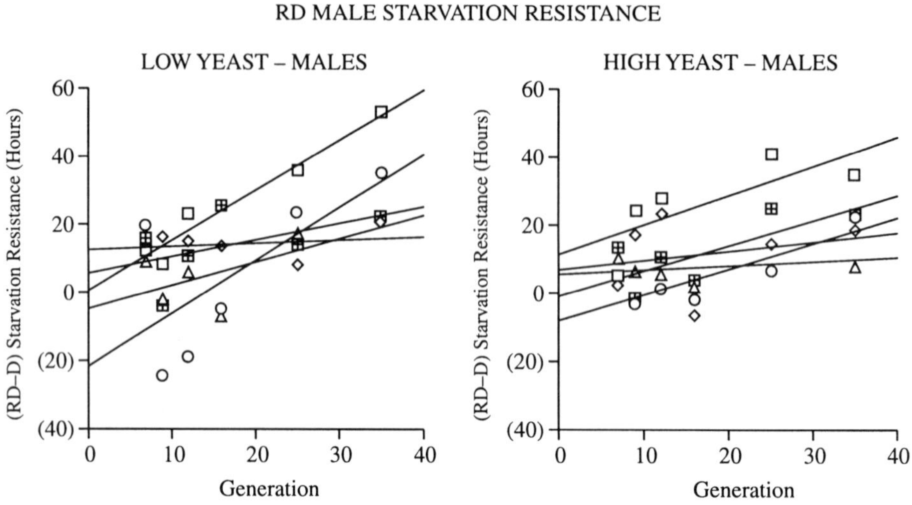

Fig. 6. (RD–D) starvation resistance in hours for all five replicate populations of males under low and high nutritional levels. The slope of the line for the mean of these five replicate populations is significantly different from zero for males at both yeast levels ($p = 0.029$ low yeast, $p = 0.05$ high yeast).

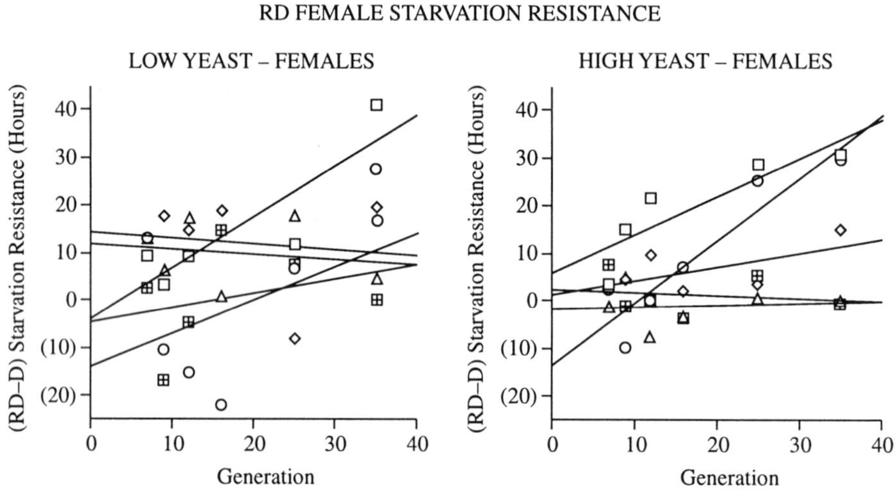

Fig. 7. (RD–D) female starvation resistance in hours for all five replicate populations under low and high nutritional levels. The slope of the line for the mean of these five replicate populations is significantly different from zero for females at both nutritional level ($p = 0.044$ low yeast, $p = 0.0004$ high yeast).

well as for females at both food levels ($p = 0.044$ low yeast, $p = 0.0004$ high yeast). Again, the positive slope for the RD–D regression line shows that the difference in starvation resistance time between the two populations is increasing over generations, with the RD populations being better adapted to deal with a bout of starvation. See also Table 4 for the mean starvation resistance for males and females under low and high nutritional levels.

Table 4 C, D, and RD Starvation Resistance in hours for Males and Females under Low and High Nutritional Levels

Population		C				D				RD			
		Low Yeast		High Yeast		Low Yeast		High Yeast		Low Yeast		High Yeast	
Gen	Rep.	Male	Female	Male	Female	Male	Female	Male	Female	Male	Female	Male	Female
	1	79.31	100.11	73.40	65.15	58.75	102.75	52.30	65.00	70.90	112.00	56.95	68.88
	2	93.25	112.50	87.08	72.10	53.55	105.75	56.85	69.75	66.50	115.75	58.88	73.38
7	3	87.00	117.75	87.56	95.67	67.63	98.25	58.25	76.00	87.45	111.25	62.06	78.68
	4	107.00	121.53	94.50	84.68	55.00	93.08	51.57	67.00	63.79	106.00	61.31	65.89
	5	75.25	90.70	60.90	55.80	52.67	100.79	53.20	70.42	68.50	103.50	66.19	78.42
	mean	**88.36**	**108.52**	**80.69**	**74.68**	**57.52**	**100.12**	**54.43**	**69.63**	**71.43**	**109.70**	**61.08**	**73.05**
	s.e.	5.60	5.74	6.01	7.05	2.73	2.15	1.32	1.87	4.17	2.20	1.57	2.54
	1	80.70	121.85	68.05	68.05	53.65	121.30	47.00	61.10	61.75	124.60	70.80	76.65
	2	91.35	135.70	80.10	75.75	57.95	105.05	55.05	72.10	74.25	122.75	71.70	76.95
9	3	110.30	156.20	114.00	92.08	79.56	132.73	51.60	70.05	55.25	122.30	48.15	60.30
	4	102.25	155.05	86.30	76.55	53.50	110.50	43.00	54.55	51.45	116.70	48.80	59.90
	5	37.60	123.30	74.90	60.90	58.80	118.60	47.45	63.95	54.45	101.80	45.50	63.20
	mean	**84.44**	**138.42**	**84.67**	**74.67**	**60.69**	**117.64**	**48.82**	**64.35**	**59.43**	**117.63**	**56.99**	**67.40**
	s.e.	12.73	7.43	7.92	5.20	4.84	4.75	2.07	3.16	4.07	4.17	5.85	3.88
	1	97.40	104.15	81.75	65.85	54.20	114.05	44.85	71.10	77.20	123.25	72.60	93.30
	2	85.40	121.40	78.15	74.25	54.60	100.05	41.15	59.15	69.55	114.70	64.10	69.20
12	3	107.40	136.10	85.80	93.90	80.05	124.30	52.10	74.45	61.40	109.10	53.10	75.00
	4	101.10	126.00	92.15	80.15	51.55	97.75	40.25	65.60	57.15	114.90	45.30	58.15
	5	88.90	106.15	67.35	60.70	54.45	112.13	45.20	70.55	65.15	107.67	55.65	71.10
	mean	**96.04**	**118.76**	**81.04**	**74.97**	**58.97**	**109.66**	**44.71**	**68.17**	**66.09**	**113.92**	**58.15**	**73.35**
	s.e.	4.00	6.05	4.14	5.80	5.30	4.87	2.09	2.66	3.45	2.75	4.70	5.72
	1	55.90	123.40	46.70	67.25	-	-	-	-	56.80	106.45	51.50	97.53
	2	38.55	97.88	44.75	71.15	38.80	103.45	43.75	69.10	52.40	122.30	36.85	71.65
16	3	59.00	108.20	66.10	94.00	37.00	78.25	32.90	60.02	32.10	100.20	30.80	67.70
	4	67.75	140.35	61.60	77.65	53.30	123.42	40.15	72.55	46.25	123.65	41.35	69.70

Table 4 (*continued*)

		C				D				RD			
Population		**Low Yeast**		**High Yeast**		**Low Yeast**		**High Yeast**		**Low Yeast**		**High Yeast**	
Gen	**Rep.**	**Male**	**Female**	**Male**	**Female**	**Male**	**Female**	**Male**	**Female**	**Male**	**Female**	**Male**	**Female**
	5	49.85	81.50	56.60	61.85	31.92	108.75	43.40	71.30	57.35	123.50	46.85	68.10
	mean	**54.21**	**110.27**	**55.15**	**74.38**	**40.26**	**103.47**	**40.05**	**68.24**	**48.98**	**115.22**	**41.47**	**74.94**
	s.e.	4.87	10.16	4.14	5.54	4.59	9.41	2.52	2.83	4.67	4.96	3.64	5.69
	1	6.25	8.20	6.10	10.75	5.85	5.55	6.20	8.45	5.65	10.75	5.70	7.95
	2	9.45	49.95	14.70	30.15	14.68	50.75	12.25	37.45	14.00	68.45	24.85	34.15
20	3	3.80	6.20	4.15	6.40	6.05	15.35	5.80	11.50	5.50	6.40	5.25	9.60
	4	6.30	49.95	7.30	21.40	11.60	33.35	9.45	35.25	5.85	12.75	7.83	17.58
	5	4.35	16.05	6.31	16.58	5.58	5.40	7.15	8.40	5.75	6.50	6.95	11.35
	mean	**6.03**	**26.07**	**7.71**	**17.06**	**8.75**	**22.08**	**8.17**	**20.21**	**7.35**	**20.97**	**10.12**	**16.13**
	s.e.	0.99	9.89	1.82	4.15	1.86	8.79	1.20	6.62	1.66	11.93	3.71	4.79
	1	122.15	150.50	105.10	94.30	67.23	146.55	61.80	85.80	103.15	158.50	103.05	115.35
	2	123.10	160.30	96.30	87.25	82.20	146.60	68.85	83.40	90.25	138.70	83.45	87.50
25	3	151.33	175.70	111.85	120.95	76.90	143.05	60.55	86.95	100.45	149.80	67.05	112.95
	4	127.75	172.05	95.30	87.05	77.95	153.40	49.95	77.55	95.15	171.20	65.30	78.83
	5	94.00	128.56	84.70	75.10	71.05	126.25	43.95	71.60	85.35	133.95	68.75	77.60
	mean	**123.67**	**157.42**	**98.65**	**92.93**	**75.07**	**143.17**	**57.02**	**81.06**	**94.87**	**150.43**	**77.52**	**94.45**
	s.e.	9.12	8.48	4.62	7.65	2.65	4.55	4.45	2.87	3.25	6.73	7.15	8.23
	1	105.70	125.95	83.90	52.40	53.83	112.05	45.50	57.10	107.35	153.25	80.88	89.07
	2	106.50	135.90	76.70	65.30	47.55	96.95	41.70	47.55	68.15	116.75	60.65	63.35
35	3	116.25	154.15	106.85	87.65	33.90	97.95	36.35	51.35	69.30	114.90	59.15	81.95
	4	82.40	129.50	72.20	54.35	44.70	114.15	35.45	44.15	65.45	118.85	43.45	44.80
	5	88.20	109.20	70.25	47.15	49.10	107.30	43.45	49.15	71.70	107.79	67.10	49.14
	mean	**99.81**	**130.94**	**81.98**	**61.37**	**45.82**	**105.68**	**40.49**	**49.86**	**76.39**	**122.31**	**62.25**	**65.66**
	s.e.	6.28	7.29	6.65	7.20	3.33	3.54	1.97	2.16	7.81	7.96	6.07	8.74

RSO Results

Longevity

Figure 8 gives the RSO–SO longevity, in days, as a function of RSO generation for both males and females. A two-factor analysis of variance with population means as replicates was used to compare RSO and SO longevities. The analysis shows that there is no significant effect of generation ($p = 0.328$), but there is a significant effect of sex ($p = 0.018$). When the sexes are analyzed separately, there is still no significant effect of generation for either males ($p = 0.508$) or females ($p = 0.619$). The life-span of RSO males and RSO females is not changing significantly from that of their ancestral populations, the SO's.

This lack of a response to relaxed selection is reiterated in the linear regression analysis of RSO–SO longevity which shows that the slope of the line for the mean of the five replicate populations is not significantly different from zero for either males ($p = 0.334$) or females ($p = 0.552$) (Fig. 8).

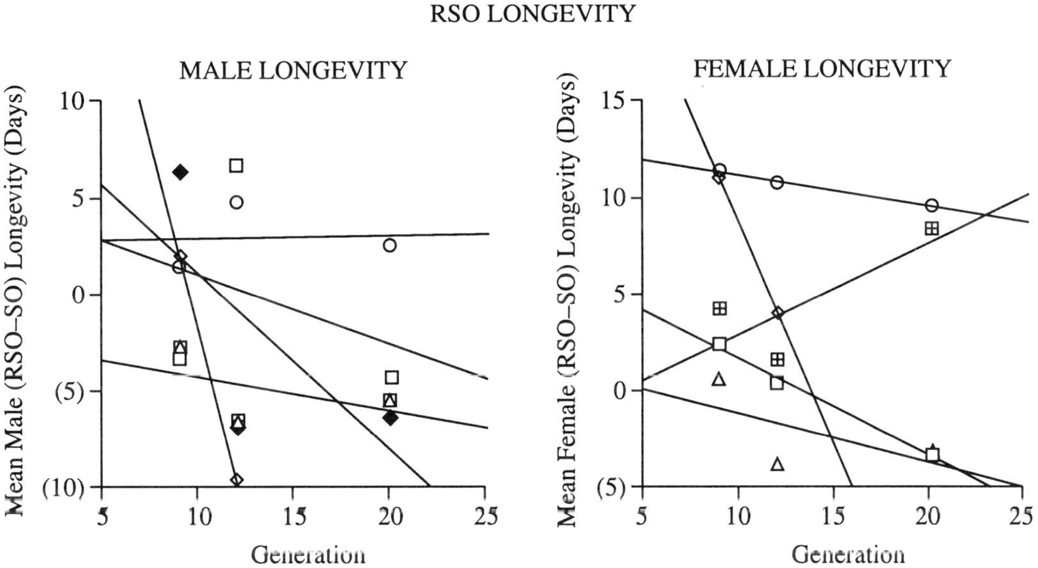

Fig. 8. (RSO–SO) longevity in days for all five replicate populations of males and females plotted against generation of RSO selection. Linear regression analysis of the mean of the five replicate populations shows that the slope of the line is not significantly different from zero for either males ($p = 0.334$) or females ($p = 0.552$)

Early Fecundity

Linear regression analysis of the early fecundity difference between RSO and SO females on generation number shows the slope of the line to be indistinguishable from zero at the high yeast level ($p = 0.134$) (Fig. 9). This lack of a significant slope for RSO–SO mean fecundity shows that early fecundity in the RSO populations is not responding to relaxed selection because the RSO's egg-laying capacity has not changed from that of its ancestral population, the SO's. It should be noted that the RSO early fecundity has less replication than the RD fecundity analysis (RD fecundity has six mean data points, while RSO fecundity has three mean data points; each data point consists of the mean of five populations). RSO–SO generation zero data was included in the linear regression analysis since each RSO population is strictly derived from its paired SO population.

Analysis of variance of RSO–SO mean early fecundity also shows that this difference is not significant across generations ($p = 0.093$). When early fecundity is analyzed separately at each generation, however, there is a significant difference between the RSO and SO early fecundity at generation 15 ($p = 0.04$), and again at generation 20 ($p = 0.02$). Therefore, it appears that although the RSO's had not differentiated from the SO's with respect to early fecundity by generation 12 of relaxed selection, the RSO females are beginning to increase their egg-laying capacity relative to the SO's.

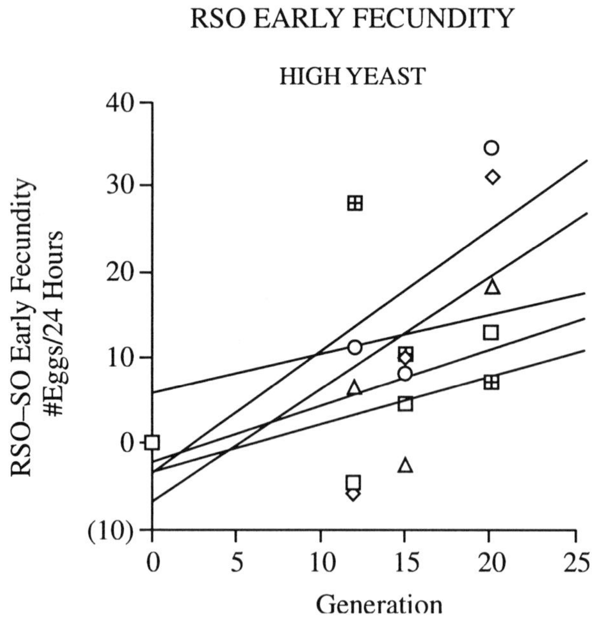

Fig. 9. Early fecundity difference between RSO and SO females at the high nutritional regime for all five replicate populations shows the slope of the line to be indistinguishable from zero ($p = 0.134$).

Starvation Resistance

Figure 10 gives the paired RSO–SO starvation resistance for all five replicate populations, in hours, as a function of RSO generation. Again, since the RSO starvation analysis has less replication than the RD starvation data, generation zero RSO–SO data was included in the starvation resistance analysis.

The statistical analysis is a two-way ANOVA with population means as replicates. Analysis of variance shows that the mean difference in starvation resistance between RSO and SO paired replicate populations differ significantly across generations ($p < 0.0001$), with the RSO populations significantly falling in starvation resistance relative to the SO populations. Males and females differ from each other in their ability to resist a bout of starvation ($p = 0.0014$). When the sexes are analyzed separately, there is still a significant effect of generation for both males ($p = 0.003$) and females ($p = 0.002$).

Linear regression of RSO-SO starvation resistance shows that the slope of the line is significantly different from zero for both males ($p = 0.01$) and females ($p = 0.007$). The mean RSO-SO difference is becoming greater and more negative, indicating a significant decrease in RSO starvation resistance time relative to the SO's (Fig. 10).

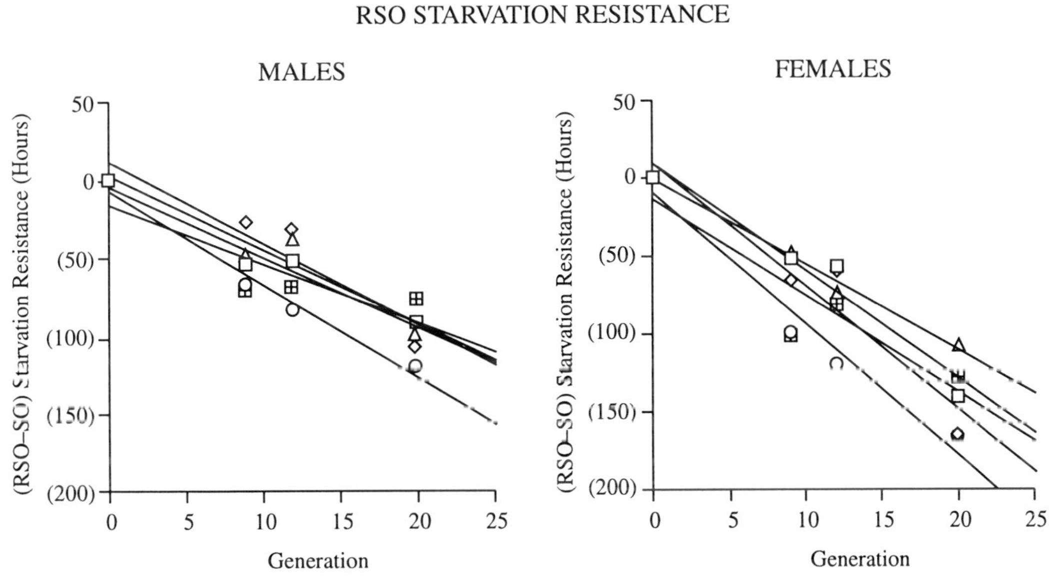

Fig. 10. (RSO–SO) male and female starvation resistance for all five replicate populations in hours. The slope of the line of the mean of all five replicate populations is significantly different from zero for both males ($p = 0.01$) and females ($p = 0.007$).

Discussion

Summary of Results

After 35 generations of relaxed selection, longevity, early fecundity, and desiccation resistance have not changed significantly in the relaxed desiccation (RD) populations, suggesting that the alleles affecting these characters lack significant antagonistic pleiotropy. A mitigation to this interpretation of the results is that early fecundity with high yeast had a linear regression coefficient with probability 0.06, just above our predetermined significance threshold of 0.05. Similarly, at low nutrition levels, the regression coefficient of female desiccation resistance on generation number had a probability of 0.07. Starvation resistance, on the other hand, rose significantly in the relaxed desiccation populations in both males and females, at both nutritional levels, indicating that they had responded to the new selection regime. The effect on starvation resistance is thus a robust one, unlike the borderline and mixed effect of relaxed selection on desiccation resistance. We conclude that only the effect on starvation resistance is meaningful.

Somewhat similar results are observed in the relaxed starvation (RSO) populations. After 20 generations of relaxed selection, RSO longevity has not dropped significantly from that of its ancestral population, the SO's. When RSO early fecundity is analyzed separately at each generation, there is a significant increase in RSO early fecundity at generation 15 and again at generation 20. Also, starvation resistance fell dramatically in the relaxed starvation-selection populations relative to their ancestral populations, the SO's. The increase in early-life fecundity in association with this decrease in starvation resistance is presumably a reflection of antagonistic pleiotropy between these two characters in the RSO system, as found originally by Service *et al.* (1988).

Evidence against Antagonistic Pleiotropy for Stress Resistance

Laboratory evolution of the Drosophila species has been used repeatedly to test both the general evolutionary theory of aging and the specific population genetic mechanisms that are thought to underlie the evolution of aging, antagonistic pleiotropy and mutation accumulation (Rose 1991). One method of differentiating between these population genetic mechanisms of aging is through reverse selection. These two genetic mechanisms are not mutually exclusive, as both could play a role in the evolution of aging. Service *et al.* (1988) and Graves *et al.* (1992) used reverse selection on *D. melanogaster* populations that had initially been selected for delayed reproduction, but were then switched to a regime of early reproduction. Reverse selection took the form of returning late-reproduced populations to early reproduction. The experimental issue was the pattern of response to reverse selection among various characters that had responded to selection for later reproduction. For instance, if reversing the previously imposed selection causes a differentiated character to quickly revert back to its ancestral condition, this would suggest that this character is subject to strong selection under the ancestral regime of early reproduction. Thus its change under selection for late reproduction probably caused a reduction in some fitness component(s) at early ages. The population genetic interpretation of this situation is that it involves some form of antagonistic pleitropy between fitness-components at early and late ages (Service *et al.*, 1988; Rose, 1991). On the other hand, if the character in question never returns to its ancestral condition, or does so very slowly, then it may be governed by mutation accumulation (Rose, 1991). In this case, the differentiation of the character under selection for late reproduction is effectively neutral at early ages.

After 35 generations of relaxed selection, longevity, early fecundity, and desiccation resistance have not changed significantly in the RD populations. For instance, the RD males are living, on average, 57 days, and the D males are living 54 days after 35 generations of relaxed selection. Similarly, there is no distinction between the RD and D females at generation 35, with the RD females surviving 44 days, and the D females surviving 43 days, on average. Neither has desiccation resistance changed in the five RD populations. The RD males are surviving 17 hours without water, and the D males are desiccating for one additional hour, on average. There is no significant difference between the early fecundity output of the RD and D females at generation 35 of relaxed selection either. This would suggest that strong antagonistic pleiotropy is not involved in the evolution of these characters in the original D populations.

Starvation resistance apparently does not trade off with longevity in the relaxed starvation selected (RSO) system. For instance, the RSO longevity has not dropped significantly from that of its ancestral population, the SO's. Specifically, after 20 generations of relaxed selection, the RSO males are living 52 days, on average, and the SO males are surviving 55 days. The RSO females are surviving, on average, three days longer than the SO's from which they were derived (RSO = 51 days, SO = 48 days). Other results, however, indicate that continued selection for starvation will lead to a drop in longevity (Phelan, unpublished data). This may be due to the starvation resistance flies accumulating detrimental levels of lipid, leading to a drop in longevity simply because the flies are too heavy.

The lack of a response in longevity, early fecundity in the RD's, and desiccation resistance under relaxed selection suggests that the alleles affecting these characters are age-specific in their effects, lacking significant antagonistic pleiotropy. It might be argued that the lack of response of these characters to relaxed selection is due to fixation for the alleles that enhance either longevity, fecundity, or desiccation resistance. This is unlikely, however, because of other selection lines that have been derived from these populations and successfully selected for these characters. For instance, if all genetic variation had been exhausted for desiccation resistance in the O populations, the substantial increase in desiccation resistance in the D populations would not have been possible (Rose *et al.*, 1992b). Yet, desiccation in the O populations did not respond to reverse selection. It should also be noted that our selection regimes will not necessarily pick up on all the antagonisms, just those involving the antecedent selection and fitness as it is operationally established in the control (original selection) regimes.

Another important factor to note is that neither longevity nor early fecundity, the characters involved in the classic trade-off that has been demonstrated in numerous other studies (Rose, 1984; Luckinbill, *et al.*, 1984; Luckinbill and Clare, 1985; Partridge and Fowlerk, 1992), were directly selected for in either the original stress selected populations or in the relaxed regime. Since they were not directly selected upon, this may then shed some light on why neither one of these characters, when analyzed across time, changed significantly from its founding populations in the relaxed selected populations.

Evidence for Antagonistic Pleiotropy

Other studies have implicated a trade-off between longevity and stress resistance, on the one hand, and early fecundity on the other (Rose, 1984; Luckinbill *et al.* 1984; Service and Rose, 1985; Rose *et al.*, 1990, 1992b; Zwann *et al.*, 1991; Chippindale *et al.*, 1993; Leroi *et al.*, 1994a,b). For example, selection for postponed senescence in *Drosophila melanogaster* has repeatedly been found to increase mean longevity and late-life fertility, and depress early-life fertility (Rose and Charlesworth, 1980; Rose, 1984; Luckinbill *et al.*, 1984). These longer-lived populations have exhibited increases in resistance to an array of environmental stresses, such as starvation, heat, toxic ethanol vapor, and desiccation stress (Service *et al.*, 1985). When stress resistance was the target of selection, longevity increased in selected populations (Rose *et al.*, 1992b). These results suggested a general trade-off between survival-related traits and early reproduction.

Starvation resistance fell dramatically in the RSO's relative to their founding population, the SO's. For instance, assayed females in the five RSO populations lasted without food for only 96 hours, on average, at generation 20 of the relaxed selection, whereas their founding population was starving for 235 hours, on average. The RSO males, also at generation 20, were starving 95 hours, while the SO males survived 192 hours on average without food.

When early fecundity in the RSO populations is analyzed separately at each generation, there is a significant difference between RSO and SO early fecundity at generation 15 ($p = 0.04$), and again at generation 20 ($p = 0.02$). By generation 20 of the RSO selection, the RSO females are laying, on average, 21.18 more eggs than the starvation selected flies from which they were derived (RSO = 62.20 eggs, SO = 41.02 eggs). Since the starvation stress is removed in the relaxed populations, it appears that the flies are re-directing resources from starvation into early life fitness components. This increase in early-life fecundity in association with a decrease in starvation resistance is evidently a reflection of antagonistic pleiotropy between the two characters, a result that has been confirmed in independent experiments with different designs (Service and Rose, 1985; Leroi *et al.*, 1994b).

Replacing one Selection Pressure with Another

The relaxed-selected RD populations exhibited a dramatic increase in starvation resistance relative to their controls. The RD response represents a new selection regime, in that they were originally in an environment in which the precursors to the RD stocks were selected for desiccation resistance, and were then put into an environment where they were selected for starvation resistance, like the C controls for the D stocks. Even though there may have been interference between selection for starvation resistance and selection for desiccation resistance, the selected flies predominantly died from stress associated with lack of water before they died from stress inflicted by lack of food, because desiccation is a much more severe stress than starvation. Given that the D selection regime exposed the D flies to mild selection for starvation resistance, and that the RD's were then directly selected for starvation resistance alone, one might expect them to surpass their ancestors in their ability to resist a bout of starvation. [However, we did not monitor the level of mortality due to starvation selection in the RD cultures.] That is indeed the case. By generation 35 of the RD selection, the RD females were starving 122 hours, while the D females survived only 106 hours without food. The RD males showed a similar increase in starvation resistance with the RD males starving 76 hours and the D males starving 16 hours without food. This may occur because the selection for starvation resistance in the RD lines may be more efficient in the absence of selection for desiccation resistance.

Genotype-by-Environment Interactions involving Nutrition

The environment typically plays a role in the expression of trade-offs (e.g. Chippindale *et al.*, 1996). This was not the case in the present study, however. In the RD experiment, two different assay environments, low yeast and high yeast, were used. Early fecundity, starvation resistance and desiccation resistance were measured on both of these nutritional regimes. Nutrition did not play a significant role in the differentiation of any of these characters (p values for effect of nutrition: $p = 0.23$ early fecundity; $p = 0.08$ desiccation; $p = 0.812$ starvation). No significant deviation in early fecundity in the RD_{1-5} populations relative to their ancestors, D_{1-5}, was found in either the high or low food level. Similarly, desiccation resistance did not respond to relaxed selection in the RD's, regardless of whether it was tested under the high or low nutritional regime. Even though starvation resistance in the RD's increased dramatically in response to relaxed selection, this was not done as a function of nutrition ($p = 0.812$). In other words, trade-offs between early-life fitness components and starvation resistance were expressed regardless of yeast level. This lack of genotype-by-environment interactions is in contrast to earlier work by Service *et al.* (1988) and Luckinbill and Clare (1985), but see Nusbaum and Rose (1999).

Relaxed and Reverse Selection

This study is unlike studies of reversed selection in that we have identified characters that do not apparently respond to relaxed selection (cf. Service *et al.*, 1988; Yoo, 1980; Graves *et al.*, 1992; Blows and Hoffman, 1993; additional literature cited in Falconer and Mackay, 1996). After 35 generations of relaxed selection, longevity, early fecundity, and desiccation resistance have not changed significantly in the relaxed desiccation-selection (RD) populations. Similarly, the relaxed starvation-selection populations have not changed significantly in longevity after 20 generations of relaxed selection.

The relaxed starvation populations are an unusually good example of relaxed selection because this five-fold replicated system involves the complete removal of the previous starvation selection in the experimental populations. The populations went from being in an environment in which they were starved, to one in which this selection pressure has been removed. It is important to keep in mind that this new selection regime is not reverse selection because the populations are not being returned to an earlier selection regime from their evolutionary history. Instead, the preceding starvation selection is removed in the new RSO populations, a selection treatment that was not imposed on these populations before.

The relaxed desiccation populations, on the other hand, cannot be as easily characterized as either strictly relaxed selection or reverse selected populations. The relaxed desiccation populations were taken from a desiccation resistance selection regime in which they were also mildly selected for starvation resistance to one in which they were only mildly selected for starvation resistance. Hence, they cannot be characterized as reverse selected populations because they are not being returned to an antecedent state. Nor can they be characterized as true relaxed populations because all selection is not removed.

This discussion illustrates the general difficulty of the distinction between relaxed and reverse selection (cf. Teotónio and Rose, 2000, 2001; Teotónio *et al.*, 2002). In some respects, we can not know with certainty whether selection has ever been relaxed. Recondite selection can be expected to occur with almost any experimental design. On the other hand, reverse selection requires the exact reinstatement of a prior selection regime from a population's history. This too is unlikely to be achieved completely, if only because the action of selection depends in part on a population's phenotypic features, which will have changed in the generations since the ancestral selective regime was last imposed. Nonetheless, inasmuch as any experimentally-imposed selection regime is fraught with potential confounds and ambiguities, relaxed and reverse selection are just as reasonable as categories for distinguishing experimental regimes as artificial selection and natural selection in the laboratory.

Conclusions

There are several conclusions of general interest that can be extracted from our results. One is that significant erosion of selection responses upon relaxation is not a general phenomenon. The many earlier instances of such erosions, especially in the quantitative genetics literature (cf. Falconer and Mackay, 1996), may be due to inbreeding and linkage effects, effects that are less likely to arise with the much larger population sizes employed here. On the other hand, many of these earlier cases of eroded response occur once the selected populations have reached plateaus in selection response. In the present study, the selected populations were not known to have reached a plateau. Yet a response to relaxed selection would be fostered by the presence of segregating genetic variation associated with a lack of a selection plateau, so the lack of an established plateau does not weaken the significance of our results.

A second point is that the patterns of correlated response to upward selection may not be preserved in the response to relaxed selection. There may be curvilinear relationships between functional characters, whereby selectively produced increases in character A give increases in character B only over part of the (A,B) phenotypic space, not over all of it. Thus characters that tend to respond together to upward selection on one of these characters may not exhibit the same degree, or even sign, of correlation on relaxed selection. A curvilinear relationship has recently been shown for stress resistance and longevity in a variety of our *Drosophila* stocks (Phelan *et al.*, 2003; Archer *et al.*, 2003). Extreme increases in stress resistance apparently no longer increase longevity, contrary to Rose *et al.* (1992). This illustrates a more general point that the evolutionary and genetic relationships between characters may not be fixed even over short periods of evolution, a point of salience for evolutionary models that assume such fixity.

Acknowledgements

Many hours of technical expert assistance were provided by B. Le, Y. Chau, W. Chen, A. Ngo, S. Sen, D. Templin, K. Yee, and countless other 199 students in the lab. Special thanks to B. Yool and A.K. Chippindale for creating the RD stocks, and R.A. Passananti for computer and graphical assistance. Helpful comments on the manuscript were provided by L.D. Mueller and T. Long. This research was supported by NIH Grant AG09970 to M.R. Rose.

References

Archer, M.A., J.P. Phelan, K.A. Beckman, and M.R. Rose. 2003. Breakdown in correlations during laboratory evolution. II. Selection on stress resistance in *Drosophila* populations. *Evolution* **57**: 536–543.

Blows, M.W. and A.A. Hoffman. 1993. The genetics of central and marginal populations of *Drosophila serrata.* I. Genetic variation for stress resistance and species borders. *Evolution* **47**: 1255–1270.

Bohren, B.B., W.G. Gill and A. Robertson. 1966. Some observations on asymmetrical correlated responses to selection. *Gene Res* **7**: 44–57.

Chippindale, A.K., A.M. Leroi, S.B. Kim, and M.R. Rose. 1993. Phenotypic plasticity and selection in *Drosophila* life-history evolution. I. Nutrition and the cost of reproduction. *J Evol Biol* **6**: 171–193.

Chippindale, A.K., T.J.F. Chu, and M.R. Rose. 1996. Complex trade-offs and the evolution of starvation resistance in *Drosophila melanogaster. Evolution* **50**: 753–766.

Deckert-Cruz, D.J., R.H. Tyler, J.E. Landmesser, and M.R. Rose. 1997. Allozymic differentiation in response to laboratory demographic selection of *Drosophila melanogaster. Evolution* **51**: 865–872.

Falconer, D.S. and T.F.C. Mackay. 1996. *Introduction to Quantitative Genetics.* Addison Wesley Longman Limited, England.

Graves, J.L., E.C. Toolson, C. Jeong, L.N. Vu, and M.R. Rose. 1992. Desiccation, flight, glycogen, and postponed senescence in *Drosophila melanogaster. Physiol Zool* **65**: 268–286.

Ives, P.T. 1970. Further studies of the South Amherst population of *Drosophila melanogaster. Evolution* **38**: 507–518.

Leroi, A.M., A.K. Chippindale, and M.R. Rose. 1994a. Long-term laboratory evolution of a genetic life-history trade-off in *Drosophila melanogaster*. I. The role of genotype-by-environment interaction. *Evolution* **48**: 1244–1257.

Leroi, A.M., W.R. Chen and M.R. Rose, 1994b. Long-term laboratory evolution of a genetic life-history trade-off in *Drosophila melanogaster*. 2. Stability of genetic correlations. *Evolution* **48**: 1258–1268.

Luckinbill, L.S., M.J. Arking, M.J. Clare, W.C. Cirocco. and S. Buck. 1984. Selection for delayed senescence in *Drosophila melanogaster*. *Evolution* **38**: 996–1003.

Luckinbill, L.S. and M.J. Clare. 1985. Selection for life span in *Drosophila melanogaster*. *Heredity* **55**: 9–18.

Mather, K. and B.J. Harrison. 1949. The manifold effect of selection. *Heredity* **3**: 1–52; 131–162.

Nusbaum, T. and M.R. Rose. 1999. The effects of nutritional manipulation and laboratory selection on lifespan in *Drosophila melanogaster*. *J Gerontol (Biol Sci)* **54A**: B192–B198.

Partridge, L. and A.K. Fowler. 1992. Direct and correlated responses to selection on age at reproduction in *Drosophila* L *melanogaster*. *Evolution* **46**: 76–91.

Phelan, J.P., M.A. Archer, K.A. Beckman, A.K. Chippindale, T.J. Nusbaum, and M.R. Rose. 2003. Breakdown in correlations during laboratory evolution. I. Comparative analyses of *Drosophila* populations. *Evolution* **57**: 527–535.

Rose, M.R. 1984. Laboratory evolution of postponed senescence in *Drosophila melanogaster*. *Evolution* **38**: 1004–1010.

Rose, M.R. 1991. *Evolutionary Biology of Aging*. Oxford University Press, New York.

Rose, M.R. and B. Charlesworth. 1980. A test of evolutionary theories of senescence. *Nature* **287**: 141–142.

Rose, M.R., J.L. Graves, and E.W. Hutchinson. 1990. The use of selection to probe patterns of fitness characters. In: *Insect Life Cycles: Genetics, Evolution and Coordination*. Gilbert, F. (ed.), Springer-Verlag, London, pp. 91–105.

Rose, M.R., T.J. Nusbaum, and J.E. Fleming. 1992a. *Drosophila* with postponed aging as a model for aging research. *Lab Ani Sci* **42**: 114–118.

Rose, M.R., L.N. Vu, S.U. Park, and J.L. Graves, 1992b. Selection for stress resistance increases longevity in *Drosophila melanogaster*. *Exper Gerontol* **27**: 241–250.

Service, P.M., E.W. Hutchinson, and M.R. Rose. 1988. Multiple genetic mechanisms for the evolution of senescence in *Drosophila melanogaster*. *Evolution* **42**: 708–716.

Service, P.M. and M.R. Rose, 1985. Genetic covariation among life-history components: The effects of novel environments. *Evolution* **39**: 943–945.

Service, P.M., E.W. Hutchinson, M.D. Macinley, and M.R. Rose. 1985. Resistance to environmental stress in *Drosophila melanogaster* selected for postponed senescence. *Physiol Zool* **58**: 380–389.

Teotónio, H., M. Matos, and M.R. Rose. 2002. Reverse evolution of fitness in *Drosophila melanogaster*. *J Evol Biol* **15**: 608–617.

Teotónio, H. and M.R. Rose. 2000. Variation in the reversibility of evolution. *Nature* **408**: 463–466.

Teotónio, H. and M.R. Rose. 2001. Perspective: Reverse Evolution. *Evolution* **55**: 653–660.

Yoo, H.H. 1980. Long-term selection for a quantitative character in large replicate populations of *Drosophila melanogaster*. I. Response to selection. *Gene Res* **35**: 1–17.

Zwaan, B.J., R. Bijlsma, and R.F. Hoekstra. 1991. On the developmental theory of ageing. I. Starvation resistance and longevity in *Drosophila melanogaster* in relation to pre-adult breeding conditions. *Heredity* **66**: 29–39.

VI

AGING, DEVELOPMENT, AND CROWDING

Aging, Development, and Crowding

Laurence D. Mueller[1] and Michael R. Rose[1,2]

[1]Department of Ecology and Evolutionary Biology,
University of California, Irvine, CA 92697-2525, USA

[2]mrrose@uci.edu; to whom correspondence should be sent

The stages of the life cycle of *Drosophila* are quite distinct from an ecological perspective. The first part of the pre-reproductive life is spent as a crawling larva with limited ability to disperse. This is followed by an immobile pupal stage in which metamorphosis takes place without feeding or dispersal. The adult is a highly mobile flying animal able to disperse several km in 24 hours (Coyne *et al.*, 1982). A natural question that arises is the extent to which the developmental processes of larvae and pupae affect adult phenotypes. As a matter of basic fly culture, it has been known for some time that the amount of food consumed by a larva affects the size of the adult. Larval nutrition in turn depends on crowding. In principle, larval ecology should impinge on the biology of the adult, possibly even the aging of the adult.

There are many interesting questions concerning the relationship between the larval and adult life cycles. One question that has been pursued in some depth is the extent to which the food that larvae are raised on affects the oviposition site-choice of adult females. When the female lays eggs, she effectively chooses the food resource the larvae will have to develop on. But it turns out there is little effect of larval food regime on adult female choice (Jaenike, 1982).

The first paper in this section (Mueller *et al.*, 1993; reprinted here) looks at the consequences of age-specific selection, such as the timing of reproduction, for characters that respond to density-dependent selection, and *vice versa*. Age-specific selection is enforced by controlling the age at which adults reproduce; there is no direct selection on larvae. This type of adult selection has no apparent effect on two larval phenotypes, pupation height and larval feeding rates (Mueller *et al.*, 1993; reprinted here). This result is consistent with previous work on oviposition choice. But does this mean there are no important connections between selection during the adult life stage and larval development?

Chippindale *et al.* (1994; reprinted here) provide some of the first evidence for such connections. They find development time differences between populations selected for postponed senescence (O) and their controls (B). The B-populations develop more rapidly than the O-populations. However, this decreased development time comes at a cost in reduced egg-to-adult survival. Some of the results in Chippindale *et al.* (1994) are at odds with related work by Partridge and Fowler (1992). However, Chippindale *et al.* argue convincingly that these differences can be understood in terms of uncontrolled larval densities and inbreeding in the Partridge and Fowler stocks. Chippindale *et al.* suggest that the evolution of more rapid development in the B-populations may be a response to the demands of early reproduction. These ideas are pursued in the next paper.

Chippindale *et al.* (1997; reprinted here) develop a new set of populations that have been selected for rapid development. This type of selection goes beyond the type of selection that the B-populations experienced, and for that reason gives a clearer picture of the evolution of developmental speed and related characters. The accelerated development lines produced by Chippindale *et al.* achieved most of their reductions in development time by shortening the larval period rather than pupal duration. In line with the previous study (Chippindale *et al.*, 1994), there was a marked reduction in egg-to-adult survival in the accelerated development lines that was not due to inbreeding depression. The accelerated lines also showed a reduction in pupation height. This may have been a means of conserving energy.

Borash *et al.* (2000) continued the examination of the phenotypic differentiation in the accelerated stocks. Their study showed that the accelerated stocks are more sensitive to larval stress. The stressful conditions used were additional ammonia and urea in the larval food. The reductions in viability suffered by the accelerated stocks were greater than the viability reductions experienced by other fly populations, which served as controls . Borash *et al.* also reported that the larval feeding rates of the accelerated stocks are higher than those of their controls. However, this result has recently been questioned by Prasad *et al.* (2001), who found no differences in feeding rates in their own accelerated development populations. These differences may be due to Borash *et al.* using larvae of the same absolute age vs. Prasad *et al.* using larvae at about the same physiological age.

The last paper in this section, by Chippindale *et al.* (2004; this volume), continues the examination of the accelerated development populations. They show that paradoxically the populations selected for accelerated development have lower growth rates than controls. That is, even though they reach maturity more rapidly, their weight gain per unit time is less. These results are also consistent with published results from Prasad *et al.* (2001). It appears that by shortening the larval growth period, the larvae put on less weight than their controls, but probably lose about the same amount of weight as controls during the metabolically active pupal phase.

Some of the features of the results connecting the evolution of development to the evolution of aging illustrate general themes in this field. First, selection can readily alter life-history patterns, whether they involve aging or development. Second, while there are no global interconnections that necessarily link life-cycle stages, there are some evolutionary interactions between these stages. Third, specific physiological genetic mechanisms shape the evolution of life-history, often in ways that have not been anticipated by conventional theories. Fruit flies, unfortunately, do not read papers in theoretical biology.

References

Borash, D.J., H. Teotónio, M.R. Rose, and L.D. Mueller. 2000. Density-dependent natural selection in *Drosophila:* Correlations between feeding rate, development time, and viability. *J Evol Biol* **13**: 181–187.

Chippindale, A.K., D.T. Hoang, P.M. Service and M.R. Rose. 1994. The evolution of development in *Drosophila melanogaster* selected for postponed senescence. *Evolution* **48**: 1880–1899.

Chippindale, A.K., J.A. Alipaz, H-W. Chen, and M.R. Rose. 1997. Experimental evolution of accelerated development in *Drosophila*. 1. Larval development speed and survival. *Evolution* **51**: 1536–1551.

Chippindale, A.K., J.A. Alipaz, and M.R. Rose. 2004. Experimental evolution of accelerated development in *Drosophila*. 2. Adult fitness and the fast development syndrome. This volume.

Coyne, J. A., I.A. Boussy, T. Prout, S.H. Bryant, J.S. Jones, and J.A. Moore. 1982. Long distance migration of *Drosophila*. *Am Nat* **119**: 589–595.

Jaenike, J. 1982. Environmental modification of oviposition behavior in *Drosophila*. *Am Nat* **119**: 784–802.

Mueller, L.D., J.L. Graves, and M.R. Rose. 1993. Interactions between density-dependent and age-specific selection in *Drosophila melanogaster*. *Funct Ecol* **7**: 469–479.

Partridge, L. and K. Fowler. 1992. Direct and correlated responses to selection on age at reproduction in *Drosophila melanogaster*. *Evolution* **46**: 76–91.

Prasad, N.G., M. Shakarad, D. Anitha, M. Rajamani, and A. Joshi. 2001. Correlated responses to selection for faster development and early reproduction in *Drosophila*: The evolution of larval traits. *Evolution* **55**: 1363–1372.

Functional Ecology 1993
7, 469–479

Interactions between density-dependent and age-specific selection in *Drosophila melanogaster*

L. D. MUELLER, J. L. GRAVES and M. R. ROSE
Department of Ecology and Evolutionary Biology, University of California, Irvine, California 92717, USA

Summary

1. Density-dependent natural selection and age-specific natural selection are important determinants of life-history evolution. A variety of laboratory populations of *Drosophila melanogaster* have been created to study the effects of these selection mechanisms.
2. Two types of populations have been selected for reproduction early (B) and late (O) in life. These have exhibited changes in life span and resistance to stresses, such as desiccation, starvation, ethanol vapours and flying to exhaustion.
3. Similarly, two types of populations have been selected at high adult and larval densities (K) and low adult and larval densities (r). These have exhibited changes in characters like larval feeding rates, pupation height and minimum food required for successful pupation.
4. To study whether age-specific and density-dependent selection act on the same traits either directly or through indirect effects, such as pleiotropy or linkage, we have examined the B and O populations for the traits that have become differentiated in the r and K populations and vice versa.
5. In general, there is a lack of similar response, except for starvation resistance which is greater in the K populations than the r populations.
6. We have tested, for the first time, longevity in all four types of populations as a function of adult density. The O populations show greater longevity than the B populations at all densities and this difference does not depend on density. In contrast, the K populations are able to resist the decline in longevity caused by increasing density much more effectively than are the r populations.
7. Lastly, a new set of populations, called CU, has been derived from the B populations and is maintained by crowding the larval life stage but raising adults under low densities. The CU populations have evolved increased feeding rates, pupation height and larval viability at high density relative to the B populations. These changes parallel the changes seen in the r and K populations and demonstrate the importance of crowding in the larval stages for much of the evolution seen in the r and K populations.

Key-words: Ageing, evolutionary physiology, fruit flies, life-history evolution, r and K selection
Functional Ecology (1993) 7, 469–479

Introduction

A number of *Drosophila* laboratory evolution studies have now been conducted focusing on age-specific or density dependent selection (e.g. Mueller & Ayala 1981; Rose 1984; Luckinbill & Clare 1985). A major concern in the studies of age specific selection has been the dependence of the response to selection on density (Clare & Luckinbill 1985; Service, Hutchinson & Rose 1988); it is now apparent that larval rearing densities can be critical in such experiments. On the other hand, 'r-*vs*-K' selection also gives rise to varying patterns of selection on different adult life-cycle stages, and thus age-specific selection (see Mueller 1987). Inadvertently, it seems as if research on these two different types of selection has been probing at their interactions, albeit unsystematically. A further problem in these experiments is that there is little in the way of theoretical expectations to guide the interpretation of those experimental results that indicate a possible interaction between age-specific and density-dependent selection.

Age-specific and density-dependent mechanisms of selection infuse discussions of such specific problems as the cost of reproduction, r and K selection, the evolution of ageing, etc. Implicitly, these modes of selection are also related to questions concerning patterns of pleiotropy between fitness components, the maintenance of genetic variation for fitness characters, and response to selection on fitness characters.

One of the features that has characterized the literature on life-history evolution from the start is ambiguity about patterns of results. Many of the original ideas about r and K selection also seem to entail different conceptions about the age structure of selection associated with r and K selection (MacArthur & Wilson 1967; Pianka 1970; discussed further in Mueller 1988a). This way of thinking is in fact quite natural; different ecological settings are expected to give rise to characteristic associations between age-dependent and density-dependent selection. Unfortunately, our theoretical understanding of the combined effects of the two forms of selection is limited to some local results in age-structured population genetics theory (Charlesworth 1980). Indeed, there are no general, theoretically motivated, expectations as to the effects of interactions between age-specific and density-dependent selection. However, recent experimental work on selection for different ages at reproduction has suggested that the outcome of evolution subject to this age-specific treatment may interact with the density under which larvae are raised (Luckinbill & Clare 1985; Service *et al.* 1988; Partridge & Fowler 1992).

Before the 1980s, most of the available evidence bearing on life-history evolution was circumstantial, chiefly consisting of comparative or phenotypic evidence, of little direct significance for the evolutionary genetics of life history. There were a few relevant experiments, but these either gave inconsistent results (e.g. Mertz 1975) or were interpreted in odd ways (e.g. Wattiaux 1968; Lints & Hoste 1974, 1977; these authors being neo-Lamarckians).

At the end of the 1970s, there was a burgeoning of evidence bearing on age specificity in selection. Much of this work has employed *Drosophila* (e.g. Rose & Charlesworth 1980; Rose 1984; Luckinbill & Clare 1985; Service *et al.* 1988). Other organisms have also been used: meadow grass (Law 1979), petunia (Roach 1986), and frogs (Travis, Emerson & Blouin 1987) being examples. There is now abundant evidence for antagonistic pleiotropy between life-history characters (e.g. Rose & Charlesworth 1981a,b; Luckinbill *et al.* 1984; Roach 1986; Service *et al.* 1988). There is also evidence for mutation accumulation affecting later life-history characters (Kosuda 1985; Mueller 1987; Service *et al.* 1988). The relationship between theory and experiment in this area has been one of useful interaction, in that theory has generated clear expectations that have been tested experimentally.

Much less work has been done on density-dependent selection. Luckinbill (1978, 1984) performed an early study on the covariation of r and K parameters in populations of bacteria and protozoa. He did not find the trade-off between r and K that had been suggested by proponents of the verbal theory (e.g. Pianka 1970). Mueller & Ayala (1981), Mueller & Sweet (1986), Mueller (1988b), Joshi & Mueller (1988), Bierbaum, Mueller & Ayala (1989) and Mueller, Guo & Ayala (1991) report a series of experiments involving sustained culture of populations at high and low densities. More in keeping with density-dependent selection theory, populations had relatively higher fitness at those densities at which they had been cultured for some time. Mueller (1988a) provides a discussion of the relationship between theory and experiment in this area.

One of the striking features of experimental work on life-history evolution is how difficult it is to separate age-specific and density-dependent factors in evolutionary–genetic experiments. In the experiments on *Drosophila* life-history genetics, a critical factor has been larval rearing densities (Clare & Luckinbill 1985; Luckinbill & Clare 1985; Service *et al.* 1988). At lower larval rearing densities, there is relatively less additive genetic variance for age-specific life-history characters. On the other hand, the density-dependent selection experiments of Mueller and colleagues have involved alterations in age structure in their experimental populations, giving rise to changes in age-specific selection patterns (see Mueller 1987). Thus, it seems evident that age-specific and density-dependent patterns of selection are difficult to separate, even under controlled laboratory conditions.

In the present study, we compare populations of *Drosophila melanogaster* Meigen that have been subjected to age-specific selection and density-dependent selection. In particular, we evaluate age-specific selected lines for their differentiation for characters that have been found to be involved in the response to density-dependent selection and vice versa. In addition, we report the creation and initial analysis of *D. melanogaster* lines selected specifically for adaptation to conditions of larval crowding.

Materials and methods

STOCKS

Age-selected stocks

Five *D. melanogaster* stocks, called 'O', have been selected for postponed ageing since February 1980 (Rose 1984). These have also been paired with five control stocks, 'B', which have been maintained from young females only, also since Feburary 1980. Larvae in the B and O cultures have always been maintained at low larval densities (60–100 vial^{-1}). This selection has continued for a total of 12 years. A number of observations have been made about the biology of

471
Density and age-dependent selection

the B–O difference. Firstly, life span is increased in the O lines by reductions in rates of mortality at every adult age (Rose 1984). Secondly, later fertility is increased somewhat, but earlier fecundity is considerably depressed (Rose 1984), a depression which is not relieved by crossing (Hutchinson & Rose 1991). These findings had been obtained earlier in a comparable experiment with less replication (Rose & Charlesworth 1981b). Perhaps more importantly, they were presaged by quantitative genetic analysis of the base population from which B and O stocks were derived. The base population exhibited pronounced negative genetic correlations between early fecundity and later survival and egg laying (Rose & Charlesworth 1981a).

Thirdly, a variety of functional characters are altered.

1. Overall body size is not affected by postponed senescence, but early ovary weight is reduced (Rose *et al.* 1984), at the same time as early fecundity is reduced.
2. O adult survival times under stress are enhanced for starvation, desiccation, and 15% ambient ethanol, but not for high temperatures or high levels of ambient ethanol (Service *et al.* 1985).
3. Early O metabolic rate and locomotion are depressed, but later locomotion is enhanced (Service 1987).
4. O lipid levels are increased in a strain-, age- and gender-dependent fashion that mimics that of starvation time (Service 1987). This suggests that the starvation resistance of O stocks is a result of increased lipid levels.
5. Flight duration is increased in the O stocks at every adult age (Graves & Rose 1990).

Density-selected stocks

The basis of this experimental programme was the creation of populations subjected to different density regimes. Six replicate population cultures of *D. melanogaster* were initiated in November 1978. Three were kept at high population densities in a serial transfer system (Mueller & Ayala 1981), and therefore were designated 'K' populations, while three were kept at low larval and adult densities, and so were designated 'r' populations. The r and K populations also differed in age of reproduction (Mueller 1987). The r populations reproduced during the first 6 days of adult life, while the K populations reproduced indefinitely, although the bulk of reproduction probably occurred during the first 2 weeks of adult life.

Relative to the r populations, the K populations showed:

1. Elevated population growth rates at high densities, but depressed growth rates at low densities (Mueller & Ayala 1981; Mueller *et al.* 1991).
2. Increased size and viability under crowded larval conditions (Bierbaum *et al.* 1989).
3. Increased pupation height (Mueller & Sweet 1986; Guo, Mueller & Ayala 1991).
4. Increased larval competitive ability, due to increased larval feeding rates (Joshi & Mueller 1988; Mueller 1988b).
5. Increased minimum larval food requirements for pupation (Mueller 1990).

ASSAY METHODS

Prior to initiating the assays described below, all test populations were passed through two generations of common conditions. Specifically, for two generations adults from the selected populations lay eggs that are then placed at low larval densities and raised under identical conditions to those of their controls. The flies that emerge from these cultures were then used to produce larvae or adults that were used in the assay procedure. These protocols ensure that differences observed between control and selected populations were due to genetic differences and not to environmental or maternal differences.

Starvation resistance

This was assayed by measuring the amount of time it took a fly to become completely unresponsive to mechanical stimulation under conditions of high humidity and no nutrition. The starvation environment was created by placing flies in groups of four into standard shell vials, and then inserting half a sponge stopper to seal in the flies, such that the lower edge of the sponge was 4–5 cm above the bottom of the vial. A cotton-ball was inserted into the part of the vial away from the flies on the other side of the sponge. Three millilitres of water was squirted into the cotton wool and the open end of the vial was sealed with Parafilm. The flies were then observed at regular intervals, every 4–8 h, the vial being agitated if a fly was motionless. Death due to starvation was inferred when a fly did not respond to agitation on three consecutive checks. The time of first inanition was then used as the time of death (see also Service *et al.* 1985).

Desiccation resistance

This was assayed in a fashion analogous to that used to assay starvation resistance. The main difference was that 3 g of Dri-Rite desiccant replaced the water-soaked cotton wool in the vial. Flies were checked more often, every 1–2 h, because they die much faster due to desiccation than starvation (Service *et al.* 1985).

Ethanol resistance

This was assayed like starvation resistance, except that the cotton wool was soaked with a 15% ethanol solution. Flies were checked more often, as for desiccation resistance, because they die much faster (see Service *et al.* 1985).

Glycogen

This was assayed by microseparation of glycogen after the technique of van Handel (1965a,b) and was performed on the five replicate B and O populations both before and after desiccation. Three females were used for each assay; the females were chosen to be of equal size distributions between the two lines. A total of five assays, each consisting of 15 females, was performed for each population, in the two lines.

Longevity

This was assayed using groups of mated flies, where the groups varied in size, as discussed below. All longevity assays were conducted on banana–molasses medium, at 25°C, in shell vials, with transfer to new vials every 2–3 days without anaesthesia.

Tethered flight duration

Flies were lightly etherized and then individually tethered to a light test piece of fishing line by use of Duco cement. The flies were stimulated to fly by passing light air current along the head–tail axis. Flies were flown to exhaustion by immediate stimulation upon cessation of flight.

Larval feeding rate

Newly hatched first-instar larvae were placed in a watch glass containing medium. Larvae were collected for assay 48h later. Larvae were placed individually on a Petri dish containing 3% agar and a thin solution of 2% live yeast. The larva was given 1·5 min to acclimate and then cephalopharyngeal retractions were recorded for 1 min using a dissecting microscope and a hand counter.

Pupation height

Fifty newly hatched first-instar larvae were collected and placed in 6-dram vials with 10ml of standard medium. After pupation, the height of each pupa was recorded as the distance from the top of the food medium to the spiracles of the pupal case. Any pupa touching the surface was given a height of 0. Experiments were run in 12:12 L:D.

Minimum food requirements

This was determined by placing larvae individually in vials with measured amounts of food (Mueller 1990). For each population, five food levels were used and 50 larvae per food level were tested. The three pairs of populations B_2–B_4 and O_2–O_4 were tested at 2, 3, 4, 5 and 6mg of yeast per larva. The fraction of larvae surviving on 2mg of yeast was not uniformly low (range 0·24–0·94). Subsequently, for B_1 and O_1 and B_5 and O_5, the food levels were 0·1, 0·5, 1, 2 and 5mg of yeast. While testing the O_2 and B_2 populations, a number of vials at all food levels inadvertently had no larvae placed in them. Thus, the total sample size for these experiments ranged from 39 to 49.

SURVEY OF EXTANT STOCKS

B and O populations

These stocks were compared with each other for the following larval characters: minimum food requirements, pupation height and feeding rate, as discussed above.

r and K populations

These stocks were compared with each other for the following adult characters: starvation, ethanol and desiccation resistance; tethered flight duration; glycogen levels; and longevity with one mated pair per vial.

Comparison of adult density effects in r, K, B and O populations

All 16 of the extant stocks were compared for the effect of varying adult density on longevity. The density treatments used with the r and K stocks were 2, 4, 20, 64, 80, 128 and 200, with 50:50 sex ratios initially. For B and O stocks, the densities were 4, 16, 20, 32, 50, 64, 80, 128, 160 and 200, with sex ratios as before. Vials were changed every 2 days after the first 5 days. Dead flies were not replaced.

CREATION OF CU LINES

One of the major problems facing the comparison of B–O with r–K stocks is that they were founded from different wild populations of *D. melanogaster*, making background effects potentially obscure. In addition, the r and K stocks were difficult to interpret in terms of selection mechanisms, because there were several important differences between the selection regimes imposed on these stocks: larval density, adult density and time of reproduction (Mueller & Ayala 1981). Accordingly, we derived five replicate populations from the B populations of Rose (1984), and selected these replicates under conditions of larval crowding (500 larvae/8-dram vial for 12 gener-

473
Density and age-dependent selection

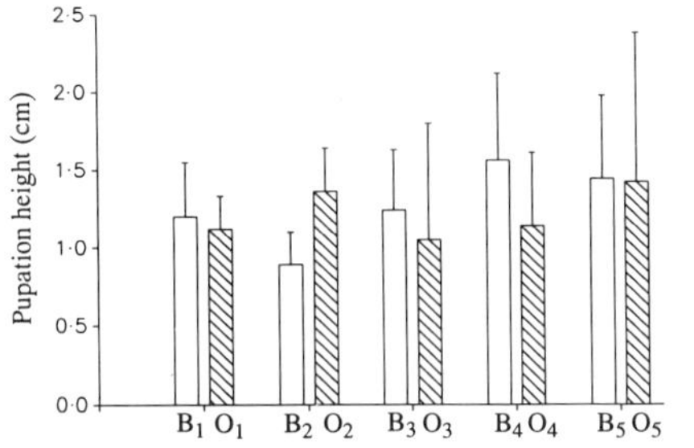

Fig. 1. Pupation heights with 95% confidence intervals in the five B (denoted by subscripts 1–5) and five O populations.

ations; 1000+ larvae/6-dram vial since) for 30 generations. These stocks were thus selected specifically for fitness under high larval densities, with no adult crowding.

Similarly, a series of populations has been derived from B called UC which are uncrowded as larvae (50–80 larvae per 8-dram vial) and crowded as adults (~150 adults per vial). These populations are cultured from young adults (1 week old) and a generation is competed in 3 weeks.

These stocks have been compared with B stocks for pupation height, feeding rate, longevity and resistance to larval crowding. The last assay involved rearing larvae at densities of 50 and 500 larvae per 8-dram vial, with the number of flies emerging at 12-h periods being recorded. These data allowed estimation of both average development for a stock, as well as its viability.

Results

COMPARISON OF LARVAE FROM AGE-SELECTED STOCKS

Four B populations have greater pupation heights than their matched O population but none of those differences is significant (Fig. 1). The one B population that has a lower pupation height than its matched O population (B_2 *vs* O_2) does show a significant difference (Fig. 1). There is clearly no consistent difference in pupation height between the B and O populations.

The results from the feeding rate study show that four B populations have lower feeding rates than their matched O populations (Fig. 2). One of those differences is statistically significant (B_5 *vs* O_5). If all five pairs of populations are compared by the non-parametric Wilcoxon signed-ranks test, the differences are not significant either ($P = 0{\cdot}0625$). The average O population has a feeding rate which is 5·9 counts greater than the average B population. While it is not possible to conclude that this difference is statistically significant, it is clearly much less than the difference between the average r and K populations (21 c.p.m.; Mueller & Sweet 1986).

The minimum food requirement results were analysed with a log-linear model. For the pairs of populations with exactly 50 samples at each food level, there are two dimensions in the table of counts of surviving larvae: food level (five values) and population (B *vs* O). Differential ability to survive on these various food levels would be indicated by a significant two-way effect in the model. This has been tested by comparing the goodness of fit to a model without the two-way interaction (food by population) to the fully saturated model (no degrees of freedom). The increase in the likelihood ratio χ^2 statistic for the reduced model and its significance is shown in Table 1. The B_2 and O_2 experiment had numerous food levels which had different numbers of vials. For these data there are three factors: food level, population and status (alive or dead). As a result, there are three different two-way interaction terms. The significance of the food by population interaction is tested by computing the difference in likelihood ratio χ^2 between the model with all two-way interactions to the one without the food by population interaction. It is clear from these results (Table 1) that there are no significant differences between the B and O populations in minimum food requirements.

COMPARISON OF ADULTS FROM r AND K STOCKS

The low-density assays are shown in Figs. 3 and 4. There was no significant difference in longevity in either sex between the r and K lines, and there was no

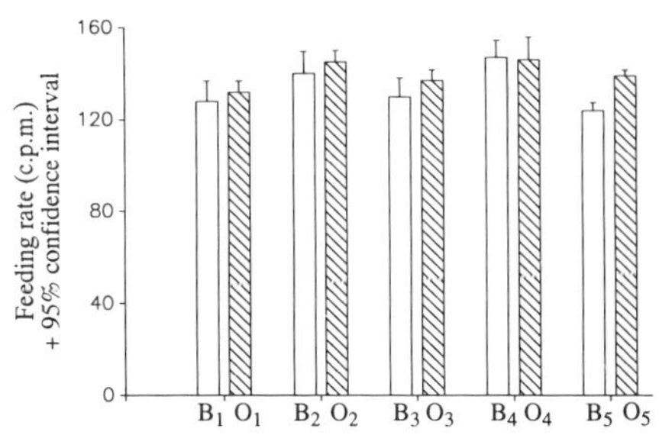

Fig. 2. Feeding rates and 95% confidence intervals in the five B and O populations.

Table 1. Results of log-linear analysis on the minimum food requirement tests

Populations	df	Test statistic	Probability
B_1 and O_1	4	2·686*	0·61
B_2 and O_2	4	2·805†	0·59
B_3 and O_3	4	2·202*	0·70
B_4 and O_4	4	0·883*	0·93
B_5 and O_5	4	0·496†	0·97

* Likelihood ratio χ^2.
† Partial χ^2.

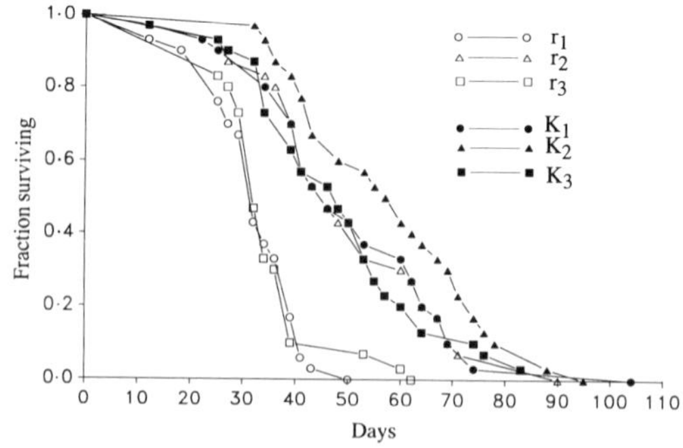

Fig. 3. The fraction of adult females that survive to various ages in the three r and three K populations.

sex difference within lines. As discussed below, this absence of a difference is not characteristic of these lines when adults are crowded.

Figure 5 shows the stress resistance and flight duration results for r and K females. Only the starvation resistance character is significantly differentiated. The males are not differentiated for starvation or desiccation resistance. Likewise the glycogen content of the r and K adults are not significantly different (Table 2).

EFFECT OF ADULT CROWDING ON LONGEVITY

Figure 6 shows the effect of adult density on longevity in the B and O stocks, while Fig. 7 shows the corresponding results for the r and K stocks. The B and O results involve a consistent differentiation of treatments that does not depend on density. By contrast, there is a substantial density dependence of differentiation of r and K stocks. Using a Newman–Kuels multiple comparison test, we find that longevities in the r populations are not significantly different from each other at 2, 4, and 20 adults, but this group has significantly greater longevity than those at 50. This non-parametric test is not sensitive to the distribution of longevity, which may be skewed. The longevity at 50 is also greater than the longevity at a density of 80 adults in the r populations. By comparison the longevities at 2, 4, 20, 50 and 80 adults are not significantly different from each other in the K populations. These results strongly suggest that K stocks have evolved resistance to the deleterious effects of adult crowding. It is noteworthy that when B and r populations are compared over varying densities, they exhibit the same sensitivity to adult density. This suggests that the K populations differ because of their selection history, not some feature specific to their ancestry relative to the other stocks.

Table 2. The glycogen content ($\mu g\ ml^{-1}$, three flies per 5 ml) of adult r and K males and females

	Sex			
	Male		Female	
	Population			
	K	r	K	r
Mean	8·5	6·6	17·2	14·9
95% confidence interval	±2·8	±7·5	±6·3	±3·3

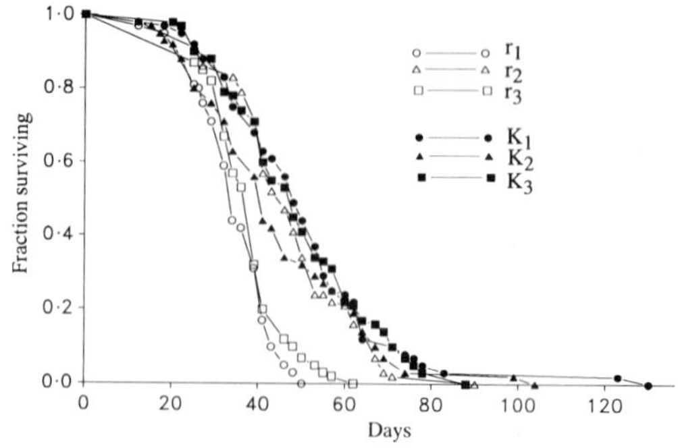

Fig. 4. The fraction of adult males that survive to various ages in the three r and three K populations.

475
Density and age-dependent selection

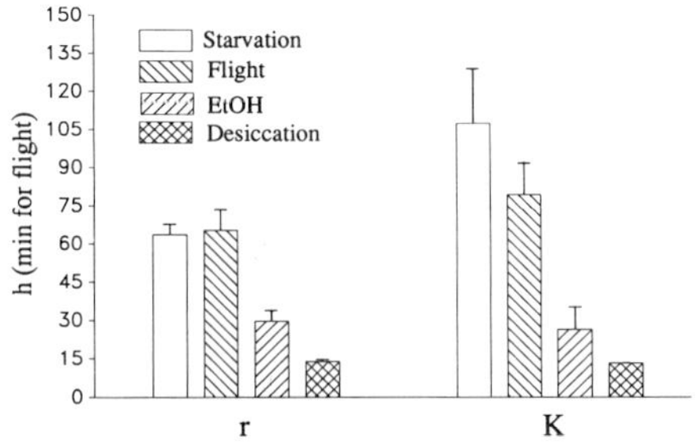

Fig. 5. The average (±1 SE) of females from the three r and three K populations for resistance to various stresses as described in the text.

SELECTION WITH LARVAL CROWDING: CU STOCKS

No differences in the pupation heights and feeding rates of the B and CU populations after 12 generations of selection in the CU environment were found (Figs. 8 and 9). At this time there were no consistent differences between the two groups of populations. Since one possible explanation for this observation is that the intensity of selection was too weak, the level of larval crowding was increased at that time (see Materials and methods). The differences in pupation height and feeding rate in the ensuing generations show a rapid differentiation of the two groups of lines (Figs. 8 and 9). The pupation height and feeding rates have increased in the CU populations relative to the B populations.

Larval development times indicate that the B populations develop more rapidly than CU populations at both densities (Table 3). The larval viabilities show a density by population interaction (Table 3). At the low larval density the populations have similar viabilities, but at high larval density the CU larvae are 33% more likely to survive than B larvae.

To test for significant differences between UC and B or CU and B we have used the Wilcoxon signed-rank test for paired observations, these tests do not require longevities to be normally distributed. In these cases CU_1 or UC_1 is paired with B_1 since B_1 is the parental population. The longevities of the CU, UC and B populations show no significant differences (Figs. 10 and 11). However, the median longevities of four of the UC populations are greater than their controls while the longevities of the fifth pair was equal. Nevertheless the differences between the mean UC and B longevities was small — 3 days.

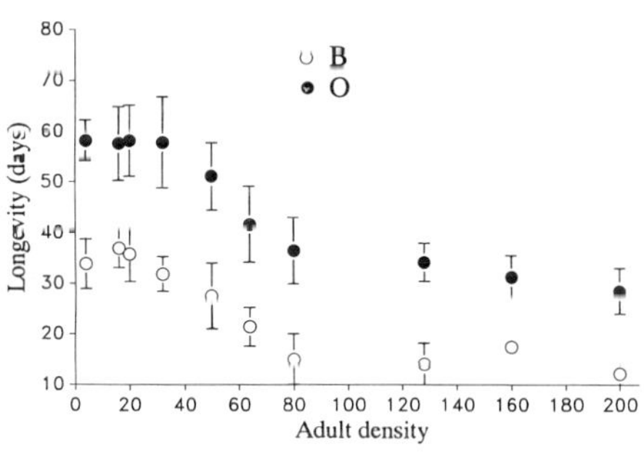

Fig. 6. The average longevity (±1 SE) of the B and O populations at 10 different adult densities.

Discussion

AGE-SPECIFIC AND DENSITY-DEPENDENT SELECTION REGIMES GIVE RISE TO DIFFERENT RESPONSES FOR MOST CHARACTERS

None of the characters that responded to density-dependent selection was found to respond in a statistically significant manner in stocks subjected to different age-specific selection regimes only. This conclusion might be qualified to the extent that the feeding rate data suggest that there has been a small change in the feeding rates of the O populations, a change that might be statistically detectable with massive replication. However, even if the effect does prove to be significant, its magnitude (a difference of about six cephalopharyngeal contractions between the B and O populations, compared with a difference of 21 contractions between r and K) is small. Therefore, the general pattern seems to be one in which the response to age-specific selection is distinct from that to density-dependent selection.

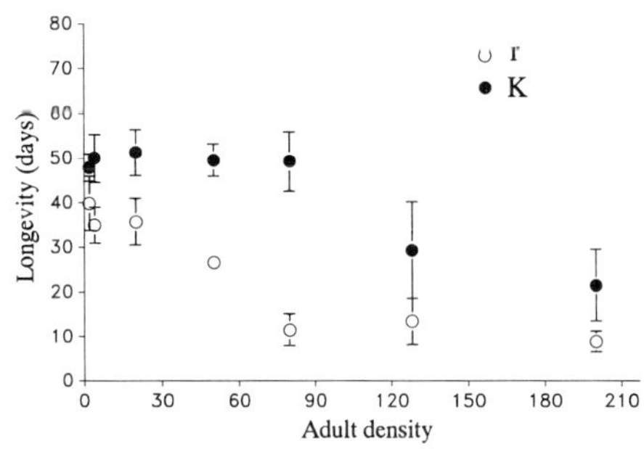

Fig. 7. The average longevity (±1 SE) of the r and K populations at seven different adult densities.

It remains true that K selection does not affect some of the characters involved in the response to age-specific selection: desiccation resistance, ethanol resistance and flight duration. However, starvation resistance is enhanced in the K lines, relative to the r lines. Interestingly, longevity at increased adult densities is also enhanced in the K populations (Fig. 7). Perhaps the lipid resources that presumably

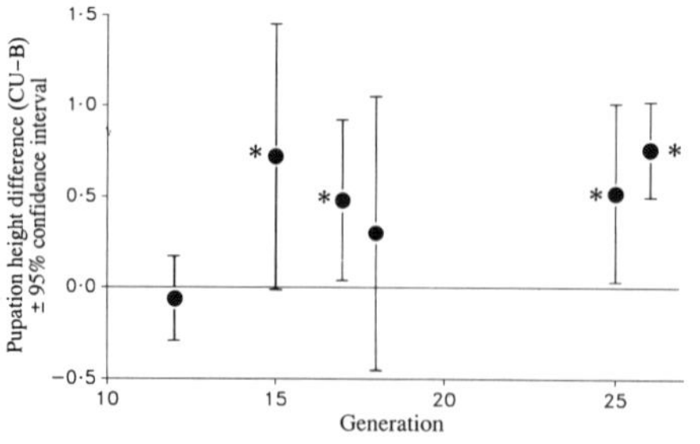

Fig. 8. The difference in mean pupation height (±1 SE) of the five CU and five B populations over 14 generations of selection.
* Significant differences between the populations were assessed by a random blocks design ANOVA.

underlie enhanced starvation resistance (cf. Service 1987) are used by K flies to enhance their adult survival under conditions of crowding. A problem with this interpretation is that the O flies have enhanced starvation resistance, but no specific enhancement in their resistance to adult crowding, compared with the B or r flies. Therefore, it might be more reasonable to regard the enhanced starvation resistance of K flies as a response to selection distinct from their increased resistance to adult crowding.

Another possible interpretation of the differences between the r and K populations is that the r populations have been inbred and hence show reduced values of the phenotypes in question. There are several reasons why we think this alternative is unlikely for the characters examined here. The B populations have been kept at very large population sizes and show no evidence of inbreeding depression (Hutchinson & Rose 1991). The value of longevity at various densities (Figs. 6 and 7) and starvation resistance in the B populations are quite similar to the r populations. The only character in the r populations that has shown signs of inbreeding depression are traits which act late in life (Mueller 1987). Thus, starvation resistance in early life would not be expected to be depressed. Although longevity differences are seen between the r and K populations, they appear most dramatically at higher population densities, not at low population densities, where flies live the longest.

There is a fundamental confusion in the comparison of the r–K differentiation with the B–O differentiation, that arises from the age-specific selection imposed on the K flies, which can lay eggs at later ages than the r flies. Thus the r–K stocks have both density-dependent and age-specific selection differences. The B and O cultures have been handled so as to minimize the role of crowding effects upon both larvae and adults. The results presented here suggest that the B and O populations indeed have not differentiated from each other with respect to adaptations arising from density-dependent selection.

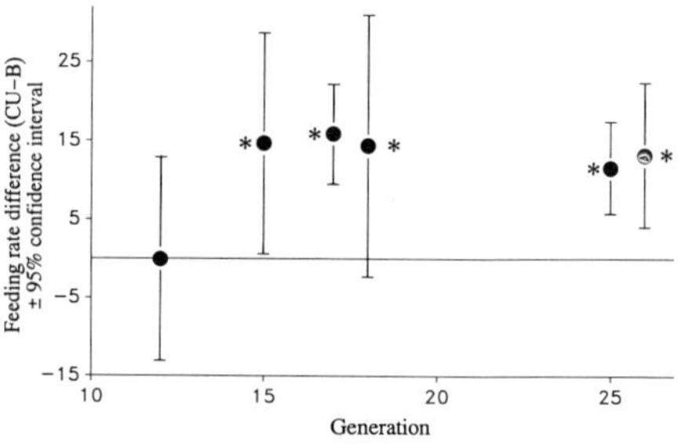

Fig. 9. The difference in mean feeding rate (±1 SE) of the five CU and five B populations over 14 generations of selection. * Significant differences between the populations were assessed by a random blocks design ANOVA.

Table 3. Viability (measured as the fraction surviving), male and female development times (DT; measured in days) ±95% confidence intervals, for the CU and B populations and two different larval densities

	Density 50			Density 500		
Population	Viability	Male-DT	Female-DT	Viability	Male-DT	Female-DT
B_1	0·78 ± 0·09	8·25 ± 0·30	8·07 ± 0·27	0·45 ± 0·11	15·3 ± 0·7	14·4 ± 0·9
B_2	0·75 ± 0·13	8·34 ± 0·39	8·14 ± 0·27	0·19 ± 0·35	11·1 ± 10·2	10·8 ± 8·8
B_3	0·73 ± 0·16	8·37 ± 0·26	8·14 ± 0·27	0·32 ± 0·04	12·9 ± 3·5	12·7 ± 3·2
B_4	0·81 ± 0·12	8·29 ± 0·27	8·20 ± 0·29	0·36 ± 0·25	11·7 ± 3·3	11·8 ± 3·1
B_5	0·87 ± 0·05	8·32 ± 0·33	8·21 ± 0·33	0·46 ± 0·09	14·9 ± 1·8	14·2 ± 1·2
B_{ave}	0·79 ± 0·07	8·31 ± 0·06	8·15 ± 0·07	0·36 ± 0·13	13·2 ± 2·3	12·8 ± 2·0
CU_1	0·76 ± 0·16	8·67 ± 0·08	8·59 ± 0·08	0·42 ± 0·12	13·5 ± 2·2	13·3 ± 1·5
CU_2	0·75 ± 0·07	8·59 ± 0·12	8·53 ± 0·19	0·50 ± 0·07	14·9 ± 0·4	13·9 ± 0·6
CU_3	0·74 ± 0·18	8·65 ± 0·14	8·61 ± 0·14	0·41 ± 0·07	14·7 ± 0·3	14·0 ± 0·9
CU_4	0·77 ± 0·12	8·66 ± 0·20	8·50 ± 0·22	0·51 ± 0·10	13·8 ± 1·0	13·7 ± 1·5
CU_5	0·74 ± 0·13	9·02 ± 0·47	8·81 ± 0·25	0·50 ± 0·11	13·3 ± 1·7	13·3 ± 1·7
CU_{ave}	0·75 ± 0·02	8·72 ± 0·21	8·61 ± 0·15	0·47 ± 0·06	14·0 ± 0·9	13·7 ± 0·4

477
Density and age-dependent selection

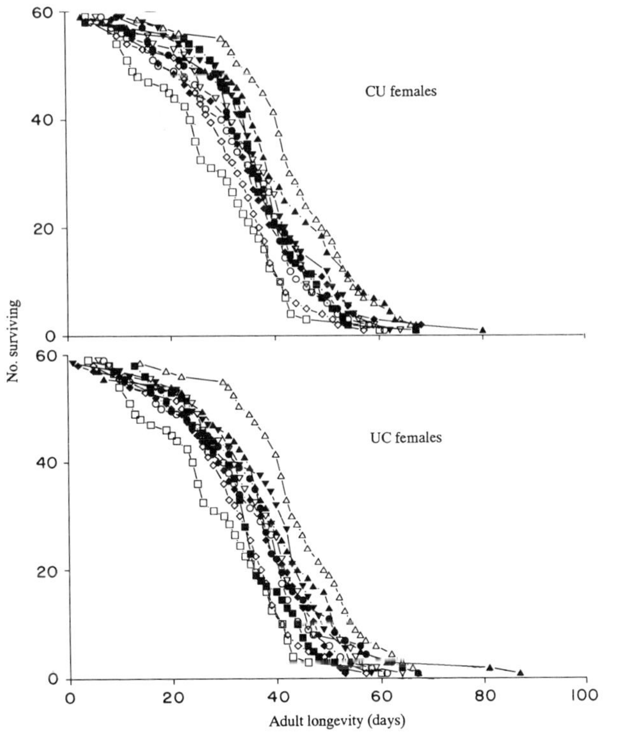

Fig. 10. The number (from 60 initially) of adult females that survive to various ages in the five B (open symbols) and five CU or UC (closed symbols).

Thus the B–O contrast appears to involve age-specific selection primarily, free of any significant admixture of density-dependent selection.

However, the comparison of the two selection regime differences, B–O *vs* r–K, is nonetheless deficient with respect to the clear articulation of the effects of density-dependent selection *vs* age-specific selection, given the confusion of selection mechanisms acting to differentiate K from r stocks. This was the motivation for the creation of the CU stocks, which have been selected specifically for adaptations to larval crowding. Further research devoted to the clarification of the multiple selection mechanisms in life-history evolution requires stocks of this kind, rather than the r–K stocks.

SOME OF THE RESPONSES TO THE K SELECTION REGIME ARISE WITH SELECTION UNDER LARVAL CROWDING ALONE

Notwithstanding these strictures, it is important to note that the CU stocks give similar differentiation of larval characters to those observed in the comparison of r and K stocks. Pupation height and larval feeding rate were both increased in CU stocks relative to B stocks, as they were in K stocks relative to r stocks (Mueller & Sweet 1986; Joshi & Mueller 1988). Another interesting result is the test of the dependence of larval viability on larval density in CU *vs* B stocks. CU stocks have greater resistance to viability depression resulting from crowding, suggesting that they are indeed adapted to more crowded larval conditions. This result was obtained for r and K stocks also (Bierbaum *et al.* 1989), which again strengthens the connection between some of the K differentiation and selection under conditions of larval crowding.

Conclusion

The responses of *Drosophila* populations to age-specific selection and density-dependent selection do not appear to be generally similar. It is not the case that K and O populations, both of which have been selected for adaptations to ecologically demanding regimes, compared to their matched r and B controls, normally respond by equivalent mechanisms. This is not to preclude the possibility that some of the characters associated with adaptation to age-specific selection or density-dependent selection are related,

478
L. D. Mueller
et al.

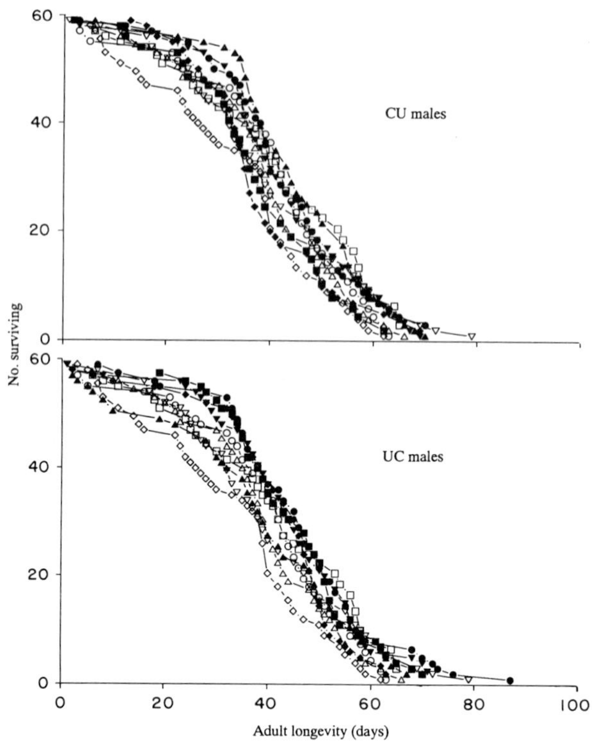

Fig. 11. The number (from 60 initially) of adult males that survive to various ages in the five B (open symbols) and five CU or UC (closed symbols).

only to assert that the overall adaptive syndromes are not similar.

Acknowledgements

We thank an anonymous referee for helpful comments and M. Duarte, M. Duong, K. Grimm, C. M. Jeong, L. Hoang, L. McPhee, L. Mena, J. Nelson, K. Nguyen, M. Nguyen, P. Nguyen, N. Parks, S. Smith, L. Vu and V. Vu for help in completing the experiments. This study was supported in part by a University of California Presidents' fellowship to J.L.G., US-PHS grant AG06346 to M.R.R. and US-PHS grant AG09970 to all authors.

References

Bierbaum, T.J., Mueller, L.D. & Ayala, F.J. (1989) Density-dependent evolution of life history characteristics in *Drosophila melanogaster*. *Evolution* **43,** 382–392.

Charlesworth, B. (1980) *Evolution in Age-Structured Populations*. Cambridge University Press, London.

Clare, M.J. & Luckinbill, L.S. (1985) The effects of gene-environment interaction on the expression of longevity. *Heredity* **55,** 19–29.

Graves, J.L. & Rose, M.R. (1990) Flight duration in *Drosophila melanogaster* selected for postponed senescence. *Genetic Effects on Aging*, vol. 2 (ed. D. E. Harrison), pp. 59–65. Telford Press, Caldwell, New Jersey.

Guo, P.Z., Mueller, L.D. & Ayala, F.J. (1991) Evolution of behavior by density-dependent natural selection. *Proceedings of the National Academy of Science USA* **88,** 10905–10906.

van Handel, E. (1965a) Estimation of glycogen in small amounts of tissue. *Analytical Biochemistry* **11,** 256–265.

van Handel, E. (1965b) Microseparation of glycogen, sugars, and lipids. *Analytical Biochemistry* **11,** 266–271.

Hutchinson, E.W. & Rose, M.R. (1991) Quantitative genetics of postponed aging in *Drosophila melanogaster*. I. Analysis of outbred populations. *Genetics* **127,** 719–727.

Joshi, A. & Mueller, L.D. (1988) Evolution of higher feeding rate in *Drosophila* due to density-dependent natural selection. *Evolution* **42,** 1090–1093.

Kosuda, K. (1985) The aging effect on male mating activity in *Drosophila melanogaster*. *Behavioral Genetics* **15,** 297–303.

Law, R. (1979) The cost of reproduction in annual meadow grass. *American Naturalist* **113,** 3–16.

Lints, F.A. & Hoste, C. (1974) The Lansing effect revisited. I. Life-span. *Experimental Gerontology* **9,** 51–69.

Lints, F.A. & Hoste, C. (1977) The Lansing effect revisited. II. Cumulative and spontaneously reversible parental age effects on fecundity in *Drosophila melanogaster*. *Evolution* **31,** 387–404.

Luckinbill, L.S. (1978) r- and K-selection in experimental populations of *Escherichia coli*. *Science* **202,** 1201–1203.

479
Density and age-dependent selection

Luckinbill, L.S. (1984) An experimental analysis of a life history theory. *Ecology* **65,** 1170–1184.

Luckinbill, L.S. & Clare, M.J. (1985) Selection for life span in *Drosophila melanogaster*. *Heredity* **55,** 9–18.

Luckinbill, L.S., Arking, R., Clare, M.J., Cirocco, W.C. & Buck, S.A. (1984) Selection for delayed senescence in *Drosophila melanogaster*. *Evolution* **38,** 996–1003.

MacArthur, R.M. & Wilson, E.O. (1967) *The Theory of Island Biogeography*. Princeton Unviersity Press, Princeton, New Jersey.

Mertz, D.B. (1975) Senescent decline in flour beetles selected for early adult fitness. *Physiological Zoology* **48,** 1–23.

Mueller, L.D. (1987) Evolution of accelerated senescence in laboratory populations of *Drosophila*. *Proceedings of the National Academy of Science USA* **84,** 1974–1977.

Mueller, L.D. (1988a) Density-dependent population growth and natural selection in food limited environments: the *Drosophila* model. *American Naturalist* **132,** 786–809.

Mueller, L.D. (1988b) Evolution of competitive ability in *Drosophila* due to density-dependent natural selection. *Proceedings of the National Academy of Science USA* **85,** 4383–4386.

Mueller, L.D. (1990) Density-dependent natural selection does not increase efficiency. *Evolutionary Ecology* **4,** 290–297.

Mueller, L.D. & Ayala, F.J. (1981) Trade-off between *r*-selection and *K*-selection in *Drosophila* populations. *Proceedings of the National Academy of Science USA* **78,** 1303–1305.

Mueller, L.D. & Sweet, V.F. (1986) Density-dependent natural selection in *Drosophila*: evolution of pupation height. *Evolution* **40,** 1354–1356.

Mueller, L.D., Guo, P.Z. & Ayala, F.J. (1991) Density-dependent natural selection and trade-offs in life history traits. *Science* **253,** 433–435.

Partridge, L. & Fowler, K. (1992) Direct and correlated responses to selection on age at reproduction in *Drosophila melanogaster*. *Evolution* **46,** 76–91.

Pianka, E.R. (1970) On *r*- and *K*-selection. *American Naturalist* **104,** 952–956.

Roach, D.A. (1986) Life history variation in *Geranium carolinianum*. I. Covariation between characters at different stages of the life cycle. *American Naturalist* **128,** 47–57.

Rose, M.R. (1984) Laboratory evolution of postponed senescence in *Drosophila melanogaster*. *Evolution* **38,** 1004–1010.

Rose, M.R. & Charlesworth, B. (1980) A test of evolutionary theories of senescence. *Nature* **287,** 141–142.

Rose, M.R. & Charlesworth, B. (1981a) Genetics of life history in *Drosophila melanogaster*. I. Sib analysis of adult females. *Genetics* **97,** 173–186.

Rose, M.R. & Charlesworth, B. (1981b) Genetics of life history in *Drosophila melanogaster*. II. Exploratory selection experiments. *Genetics* **97,** 187–196.

Rose, M.R., Dorey, M.L., Coyle, A.M. & Service, P.M. (1984) The morphology of postponed senescence in *Drosophila melanogaster*. *Canadian Journal Zoology* **62,** 1576–1580.

Service, P.M. (1987) Physiological mechanisms of increased stress resistance in *Drosophilia melanogaster* selected for postponed senescence. *Physiological Zoology* **60,** 321–326.

Service, P.M., Hutchinson, E.W., MacKinley, M.D. & Rose, M.R. (1985) Resistance to environmental stress in *Drosophila melanogaster* selected for postponed senescence. *Physiological Zoology* **58,** 380–389.

Service, P.M., Hutchinson, E.W. & Rose, M.R. (1988) Multiple genetic mechanisms for the evolution of senescence in *Drosophila melanogaster*. *Evolution* **42,** 708–716.

Travis, J., Emerson, S.B. & Blouin, M. (1987) A quantitative genetic analysis of larval life-history traits in *Hyla crucifer*. *Evolution* **41,** 145–156.

Wattiaux, J.M. (1968) Cumulative parental age effects in *Drosophila subobscura*. *Evolution* **22,** 406–421.

Received 5 May 1992; revised 17 November 1992; accepted 1 February 1993

Evolution, 48(6), 1994, pp. 1880–1899

THE EVOLUTION OF DEVELOPMENT IN *DROSOPHILA MELANOGASTER* SELECTED FOR POSTPONED SENESCENCE

ADAM K. CHIPPINDALE[1], DAT T. HOANG[1], PHILIP M. SERVICE[2], AND MICHAEL R. ROSE[1]
[1]*Department of Ecology and Evolutionary Biology, University of California, Irvine, California 92717*
E-Mail: AKC@TOAD.BIO.UCI.EDU
[2]*Department of Biological Sciences, Northern Arizona University, Flagstaff, Arizona 86011-5640*

Abstract.—The role of development in the evolution of postponed senescence is poorly understood despite the existence of a major gerontological theory connecting developmental rate to aging. We investigate the role of developmental rate in the laboratory evolution of aging using 24 distinct populations of *Drosophila melanogaster*. We have found a significant difference between the larval developmental rates of our *Drosophila* stocks selected for early (B) and late-life (O) fertility. This larval developmental time difference of approximately 12% (O > B) has been stable for at least 5 yr, occurs under a wide variety of rearing conditions, responds to reverse selection, and is shown for two other O-like selection treatments. Emerging adults from lines with different larval developmental rates show no significant differences in weight at emergence, thorax length, or starvation resistance. Long-developing lines (O, CO, and CB) have greater survivorship from egg to pupa and from pupa to adult, with and without strong larval competition. Crosses between slower developing populations, and a variety of other lines of evidence, indicate that neither mutation accumulation nor inbreeding depression are responsible for the extended development of our late-reproduced selection treatments. These results stand in striking contrast to other recent studies. We argue that inbreeding depression and inadvertent direct selection in other laboratories' culture regimes explain their results. We demonstrate antagonistic pleiotropy between developmental rate and preadult viability. The absence of any correlation between longevity and developmental time in our stocks refutes the developmental theory of aging.

Key words.—Aging, developmental rate, *Drosophila*, life-history evolution, natural selection, pleiotropy, survival, trade-offs.

Received August 23, 1993. Accepted April 22, 1994.

At least 45 trade-offs between life-history characters are known (Stearns 1992), most of which can be classified as costs of current reproduction for future reproduction or survival. The trade-off between early fecundity and longevity in outbred laboratory populations of *Drosophila melanogaster* is one especially well-documented and well-replicated result (Rose and Charlesworth 1981a,b; Rose 1984; Luckinbill et al. 1984; Tucić et al. 1988; Zwaan 1993), lending support to Williams' (1957) pleiotropy theory for the evolution of senescence and much other life-history theory based on a cost to reproduction (Stearns 1992).

Recently, Partridge and Fowler (1992) appeared to have found an exception to this result. Partridge and Fowler selected for later-life fertility and found no cost in early fertility. Finding slower development, larger body size, and lower viability, they suggested that there is a trade-off between larval viability and developmental time, on one hand, and adult body size and survival, on the other. However, other reports on the size of flies selected for postponed senescence have revealed no consistent weight differences between selection treatments and controls (Rose et al. 1984; Luckinbill et al. 1988; Zwaan 1993). To further muddy the waters, upward selection on fresh weight at eclosion by Hillesheim and Stearns (1991, 1992) produced flies that were *shorter*-lived than unselected controls. The heavier flies also had higher early-life fecundity and lower late-life fecundity, with no advantage in lifetime fecundity. This baffling array of findings raises some potentially severe problems for the study of life-history trade-offs, problems that call out for empirical clarification.

Another point of interest in the relationship between development and adult life history is the developmental theory of aging. This theory posits that aging is a part of development, after the fashion of differentiation or growth (Lints 1978; 1988). The main experimental evidence for the developmental theory comes from a study by Lints and Lints (1969), who observed increases in both adult lifespan and larval developmental time when *Drosophila* larvae were reared at high densities. From this observation, they suggested that developmental time is important in determining the rate of senescence of the adult. Recent work has failed to corroborate this finding (Zwaan et al. 1991,1992; Zwaan 1993). Zwaan et al. suggested that the effects of larval crowding on adult aging are more likely to be the result of viability selection than any meaningful connection between development and longevity. In any case, such phenotypic manipulations skirt around the critical issue of genetic correlation. If developmental time and longevity were positively genetically correlated, there might be some evolutionary rationale for the developmental theory of aging. But laboratory selection experiments on developmental time and longevity by the same group (Zwaan 1993) failed to reveal evidence for a genetic correlation between developmental time and longevity. It would be useful if the developmental theory of aging could be supported by some reproducible experiments, or disposed of by further experimental refutation.

Nonetheless, there is, in general, the potential for trade-offs to arise between distinct life-cycle stages, such as larva and adult in holometabolous insects. Selection for adult survival and late-life fertility thus could, in principle, elicit extended larval growth to increase the size and provisions available to the emerging imago. But later maturation will reduce fitness, all other things being equal (Charlesworth 1980). Insects with holometabolous life cycles, like *Drosophila* may, therefore, face trade-offs among fitness components imposed by the timing of metamorphosis. In particular, even when direct selection is restricted to one life stage, there is potential for trade-offs to impinge on other stages.

The principal goal of the present communication is an intensive investigation of the role of development in the laboratory evolution of aging, with a view to clarifying or refuting extant confusions. Using 24 distinct populations of *D. melanogaster* selected for different adult demographic parameters, we examine stage-specific developmental time, viability, size and weight, starvation resistance, and response to limited larval food. We test experimentally contradictory findings from other laboratory selection experiments, and argue that inbreeding depression and inadvertent selection can explain these findings. We determine the preadult response to selection for postponed senescence in our stocks, and demonstrate antagonistic pleiotropy between developmental rate and viability. We also use our selected populations to refute the developmental theory of aging.

Materials and Methods

Origin and Handling of Selection Treatments

The experimental *Drosophila melanogaster* populations used in these experiments were all derived from a South Amherst, Massachusetts, population first studied by P. T. Ives (Ives 1970). The population was sampled in 1975 and has been maintained as an outbred population in 2-wk discrete generations with abundant food to the present time. In 1980, after over 120 generations of laboratory culture, 10 populations were derived from the basal IV stock. Five replicate B populations were maintained on the same 2-wk schedule and thus selected for early-life net fertility, and 5 replicate O populations were selected for late-life fertility by progressively extending the total generation time to 10 wk, as outlined in Rose (1984). Additionally, a set of populations called RU were reverse-selected from the O populations back to the B-type culture regimen starting in 1985 (Service et al. 1988). At the time of most experiments reported here (1992), the Bs had undergone about 250 generations, the Os about 60 generations, and the RUs had been reverse selected for about 120 generations. RU replicate number 5 was lost during selection; thus, only 4 populations exist at present. Replicates are maintained independently at population sizes greater than 1000 adults per generation.

Two other selection treatments are relevant to this paper. In 1989, two starvation-resistance selected treatments, SB and SO, were created, each consisting of five replicate populations founded from the same-numbered replicate of each B and O population (Rose et al. 1992). As controls, two lines, named CB and CO, were also derived. The CB and CO lines are maintained in demographic parallel to the S lines, though provided with food throughout all phases of the life cycle. The life cycles of CB and CO lines are identical to the O life cycle in all ways except for the duration of the selection phase. The total (egg-to-egg) generation time of these lines is about 5 wk at writing: intermediate between B and O, though slightly variable in length depending upon the duration of selection of the starvation lines. The S lines and their controls had undergone about 40 generations of selection when these assays were done.

Although the adult selection treatments of stocks discussed in this paper differ radically in timing of reproduction, the larval rearing and early adult environments are identical. Populations are reared at regulated moderate larval densities of 50–150 eggs per 8-dram vial on approximately 5 ml (excess) banana-molasses medium and transferred to their selection treatment at 2 wk from egg collection. The larval culture environment is 25°C with a 24:0 light-dark cycle. The lack of a light-dark cycle may be a particularly important feature of our system, because it means that the evolution of differences in developmental time has been freed from a modality that might either mute or exaggerate the response to selection if emergence is rhythmically dependent on light conditions (Brett 1955).

All experiments reported on in this communication were performed in our laboratory at the University of California, Irvine, with the exception of the first (1988) assay, which was done at the University of Kentucky.

Statistical Methods

The selection scheme employed by our laboratory entails identical treatment of five independent replicate populations per selection treatment (four in the case of the RU treatment). For statistical comparisons, the population mean is the primary datum. This technique is conservative, because even large experiments reduce to a small number of degrees of freedom in the analysis, as tests involving treatment means are conducted using *only* the five replicate population means. Distributional nonnormality of data for individuals within populations is therefore not a factor in our analysis. Tests for the normality in distribution of the replicate means were not conducted; with four or five data points, it is very unlikely that any test could demonstrate nonnormality. Normality of the replicate means is expected, particularly as our sample sizes and replication levels are high. For this reason, we have also not performed any transformations on the data, as is often done, particularly with viability data.

The B and O lines were founded en masse from the IV population and therefore have no special pairwise associations of numbered replicate populations caused by ancestry (e.g., B_1 to O_1). Statistical comparisons of B and O lines are therefore always unpaired with respect to replicate number. For all other selection treatments described, there is a special historical relationship between replicates of the same number: subsequently derived stocks have been founded in a paired manner (e.g., CB_1 and SB_1 were both derived from B_1); analyses have been pairwise by population number for these selection lines.

All statistical analyses were performed using Statview II and SuperAnova on Macintosh computers.

Standard Procedures for Developmental Time Assays

Before experimentation, all selection treatments were put through two full generations of 2–3 wk common culture as a standardizing procedure. Standardization was performed to synchronize selection treatments on previously different schedules, bolster population sizes, and remove any parental and grandparental age effects potentially arising from the selection itself.

Standard developmental time experiments were performed with exactly 60 eggs per 8-dram vial and excess banana-molasses food (5 ml/vial), except where noted. About 2000 adult flies of each replicate were given approximately 2 d of fresh, yeasted food in a cage to increase egg output before experimental collection. This procedure was employed to minimize the egg-laying period (1–4 h, depending on the experiment) and to ensure that eggs laid were not advanced in embryonic development owing to retention of eggs by the females. Ten replicate vials were collected for each population sample by gentle brush-transfer of eggs to thin squares of standard medium, which were then placed in the vial. Exactly 60 eggs were counted into each vial. When several people assisted in setting up an assay, each population sample of ten vials was split between two egg collectors (five vials each), so that no whole sample was taken by one person, and each egg collector handled a range of selection treatments.

Flies were collected at 6-h intervals (at 6:00, 12:00, 18:00, and 24:00 h each day). Checks were continued until no adults had emerged for 72 h. A week after the termination of emergence checks, the vials were reexamined. In the postmortem analysis we checked for very late emergent flies, total numbers of pupae, and numbers of flies dead in puparium (these are easily discerned from empty pupal cases using transmitted light), as well as adults that had died during emergence from pupa. Inevitably one or two adults were found stuck in the culture medium. These flies were counted in the postmortem check, but not considered in the egg-to-adult viability estimate (i.e., they were considered inviable in the transition from pupa to adult; the food medium being a natural hazard in their laboratory environment).

Because small differences in incubator position and proximity to the light can have an effect on developmental time, in all experiments controlled interspersion of selection treatments with random daily repositioning of blocks was employed. For example, in the B/O/RU experiment, $B_1/O_1/RU_1$ were housed together in the same experiment rack as one block. Racks were randomly repositioned vertically and horizontally on a daily basis within the same incubator.

All assays of egg-to-adult developmental time and viability followed the standard procedures outlined above, except where noted and in the earliest study of B versus O developmental time. The first assay, conducted in the summer of 1988, differed substantially in several ways from the protocol later adopted. This pilot experiment was smaller, using only four of the five replicates (B_{2-5} and O_{2-5}) for each selection treatment and four vials for each population. It employed a cornmeal/brewer's yeast medium with dextrose and sucrose sweeteners, and each vial was inoculated with a live yeast suspension the day following the collection of eggs for the experiment.

Hatching Time and Hatchability of B and O Treatments

In the same experiment and at the same time that eggs were collected for the assay of egg-to-adult developmental time in the Bs, Os, and RUs, a subset of eggs was separated for an egg hatching-time assay on B and O lines. Eggs were collected from population cages over one hour and exactly 100 eggs were counted onto a food-colored agar plate for each of the populations. Hatching was scored every hour for 37 h, when it appeared that the hatching period was over. As a precaution, the assay plates were recounted about two days later to make sure that unhatched eggs were indeed inviable; no eggs scored as unhatched at the end of the timed checks hatched subsequently.

Pupation Time and Height of B and O Treatments

To measure pupation height (Mueller and Sweet 1986) and time from egg to pupa, eggs were collected in the standard manner. The vials were wrapped in clear plastic sleeves 48 h after egg collection, and the distance between the surface of the medium and the bottom of the sponge stopper was standardized to ±1 mm. Pupae were distinguished from larvae according to the criteria of Ashburner (1989) for a P1 pupa: immobile, without visible larval segments, and with a smooth tanning of the pupal coat visible to the unaided eye. Each 6-h check was scored with a different color of marking pen. When all pupae had formed, the plastic sleeves were removed from the vials and data on spatial and temporal formation of the pupae were taken. We used a Summagraphics digitizing pad and Macintosh computer driven by MacMeasure II to measure pupation height. Adults emerging from this assay were sorted and sexed as in the standard assay (see above).

Crosses within the O-Selection Treatment

Crosses among O populations were performed to test for mutation-accumulation as a mechanism for the evolution of extended developmental time of the O populations. We performed crosses between different population replicates and compared them to pure-breeding samples to try to detect F_1 heterosis in developmental time. To balance the design, we did not run reciprocal crosses (e.g., O_1 male × O_2 female and O_2 male × O_1 female), and each true-breeding sample was run twice from independently derived and sampled flies. Accordingly, ten crossed samples were compared to ten true-breeding samples handled identically but independently. Egg-to-adult developmental time was scored in the standard manner (see above).

Developmental Time and Viability with Varied Food Levels

Two experiments involving (1) B and O selection treatments, and (2) B, CB, and CO treatments were conducted to determine developmental time and viability at different larval food levels. We varied the food level while keeping the number of eggs constant. Exactly 60 eggs were counted into each vial on a small piece of non-nutritional agar. Food levels of 1.0, 1.5, 2.0, and 4.0 ml food per vial were employed in the first experiment; these translate into a standardized density series of 16.7, 25.0, 33.3, and 66.7 μl of food per larva. In the second experiment, a fifth food level of 0.5 ml (8.3μl/larva) was added. The 4-ml food treatment approximates normal food levels in selection.

To eliminate humidity as a factor in these experiments, we provided an agar base (at the same concentration as the agar in normal food) under the low-food treatments to make them up to the same total volume as the control (4 ml) treatment. Pupae from low-food treatments were therefore not exposed to desiccating conditions. Five vials were collected for each food level for each replicate population. Developmental time and viability data were collected in the same manner as the "standard" assay (see above), except that a postmortem analysis was not performed on the second (B, CB, CO) experiment; thus, no data on stage-specific survivorship of preadult phases were available in the second minimum-food experiment.

Body Size

Body size of B and O flies was estimated in two different ways: wet weight and thorax length. Adult wet-weight data were obtained by weighing flies at 24 h to 48 h from emergence, using offspring from parents at five ages (4, 9, 14, 19, 24 d from emergence). Eggs were collected from parents over an 8-h period, counted to exactly 60 per 8-dram vial, and reared in the standard manner at 25°C. After emergence, males and females were placed on standard banana food for approximately 24 h, ether-killed, and weighed in groups of five using a Cahn analytical balance. In total, 9897 flies were weighed. Although the holding period between emergence and weighing varied between 24 and 48 h, we interspersed selection treatments (B and O) in a regimented fashion to control against generating age-specific differences that might be misleading. This design does not address the issue of age-specific weight differences between selection treatments, which may have some role in the inconsistent results for weight found by Rose et al. (1984).

As another index of body size, we measured thorax lengths of B and O flies from the density experiment. The thorax was measured anterio-posteriorly along the dorsal side on flies preserved in 70% ethanol. Measurements were taken using a Wild M5 microscope fitted with a camera lucida aligned over a Summagraphics digitizing tablet connected to a Macintosh computer; the software was MacMeasure II (a highly modified version of W. Rasband's MacMeasure by A. R. Palmer, University of Alberta). This digital measuring system allowed high precision without the error associated with hand-recording of data. Three independent repeat measures of each thorax were made for 20 individuals of each sex for each food treatment; the mean of the three repeat measures was used as a single datum in the analysis. The total number of measurements was 4800 (4 food treatments × 2 selection treatments × 5 replicate populations × 2 sexes × 20 individuals × 3 repeat measures). Because measurements took a week to complete with three people working, sampling error was controlled by always pairing a B and an O treatment in any measurement session, with food levels randomized per measurement session separately for B and O replicates within each selection treatment.

Adult Starvation Resistance

Starvation resistance was measured on B and O flies at two adult ages: immediately upon emergence and after 4 d of feeding on standard Rose Lab banana-molasses medium. Flies were collected at 6-h intervals from the estimated peak of emergence and split into immediately starved (3 ± 3 h of adult life) and 4-d (96 ± 3 h) fed groups. The flies immediately starved were sexed and sorted into sealed and humidified vials (following Service et al. 1985) in groups of four, and mortality was scored at 6-h intervals. The flies treated with 4 d of feeding and mating were placed on standard banana medium in groups of eight (four males, four females) before removal to starvation vials. Forty flies of each sex were tested for each B or O population/treatment. The assay therefore involved 1600 flies in total.

Longevity

Longevity data reported here are a subset of data collected in 1990 (Nusbaum et al. in prep) in which longevity was assayed simultaneously on all laboratory selection treatments reported on in this communication, except the RU treatment. Briefly, longevity was measured by daily scoring of mortality of flies kept in mixed-sex vials (four male, four female). Flies were maintained on standard medium with transfer to fresh medium three times per week.

RESULTS

Long-Term Evolution of the O-B Difference in Developmental Time and Viability

The difference in egg-to-adult developmental time and viability between B and O selection treatments over the last five years (1988–1993) is given in table 1. The difference between selection treatments was analyzed in two different ways: factorial ANOVA and simple regression. In these analyses, we excluded the 1988 data because they were collected under a different ex-

TABLE 1. Long-term evolution of the B/O difference in developmental time and viability. Mean egg-to-adult developmental times and viabilities for the five B and five O populations from nine experimental assays over 5 yr of selection are shown. Os have shown consistently longer developmental times throughout the period of sampling. The earliest assay was conducted on an unconventional and lower nutrition culture medium, and this is reflected in lower viabilities; these data were excluded from long-term analyses. No temporal trends were detected in the difference between B and O, in either developmental time or viability, suggesting stability in both characters.

	Developmental time (h)						Egg-to-adult viability (%) (Males and females combined)		
	Males			Females					
Date	B	O	O-B	B	O	O-B	B	O	O-B
Jul 1988	221.71	233.92	12.21	220.01	230.68	10.67	65.5	65.1	−0.4
Dec 1990	234.60	242.60	8.00	233.68	241.26	7.57	NA	NA	NA
Mar 1991	232.06	247.32	15.26	227.26	242.62	15.36	80.7	84.3	3.6
Apr 1991	229.98	250.95	20.97	226.57	249.47	22.90	76.0	84.8	8.7
Jun 1991	229.80	237.40	7.60	224.90	232.60	7.70	80.3	83.3	3.0
Mar 1992	234.38	250.28	15.90	229.52	246.32	16.80	87.4	91.4	4.0
May 1992	238.61	250.73	12.12	235.04	247.48	12.44	85.8	90.8	5.1
Nov 1992	223.73	236.05	12.32	220.46	232.34	11.88	89.1	89.6	0.5
Feb 1993	230.06	248.31	18.25	225.73	243.74	18.01	85.0	91.4	6.4
Mean	230.55	244.17	13.63	227.02	240.72	13.70	81.2	85.1	3.9
SD	5.29	6.83	4.43	5.17	7.10	5.04	7.7	8.8	3.0
SE	1.76	2.28	1.48	1.72	2.37	1.68	2.7	3.1	1.0

perimental protocol and with lower replication levels (see Materials and Methods).

Analysis of variance on developmental time, with selection treatment, sex, and assay date as factors, incorporating replicates, revealed highly significant differences between B and O treatments (B < O; $P < 0.001$), sexes (male < female; $P < 0.001$), and assay ($P < 0.05$). No interaction terms were significant. A two-factor ANOVA on viability (selection × assay date) showed a significant difference between stocks (B < O; $P < 0.01$) and assay date had a significant effect ($P < 0.05$). The interaction term was not significant.

We tried to determine whether the significance of the assay date factor reflected a secular trend in the difference between the selection treatments over evolutionary time by linear-regression analysis. The difference values were regressed against time. Regression slopes for developmental time and viability were not significantly different from zero for either the averaged male/female developmental time data or egg to-adult viability. It appears that any trends in differentiation are more subtle than the between experiment variability, if they are present at all. A slight upward trend in overall viability of both selection lines likely has resulted from improvements in the assay technique, rather than any sort of adaptation.

The 1988 data were analyzed separately using the two-tailed t-distribution with $n - 2 = 6$ df. Using this test, the developmental time difference between B and O was statistically significant ($P < 0.01$), but the viability difference was not. The viabilities in this first experiment were considerably lower than those reported for later experiments (table 1). This probably reflects the lower nutritional content of the cornmeal medium used and that the stocks were not adapted to that medium. We do not invest much confidence in these results because of the differences in assay technique, but we have included them for sake of completeness. It is surprising, then, that the developmental time data compare very favorably, in both absolute magnitude and relative values, with the data from later experiments. Note, however, that the difference between O and B developmental times remains stable across a variety of food levels in the minimum food experiments and may be a generally more robust finding.

Developmental Time and Viability of Reverse-Selected (RU) Populations

Figure 1 summarizes the egg-to-pupa and egg-to-adult viability data from our experimental assay of B, O, and RU (reverse-selected) treatments. Figure 2 shows the frequency distributions for the emergence of females (the pattern for males is very similar). Developmental time has responded to reverse-selection by returning to values statistically indistinguishable from the Bs. The mean developmental times in this assay

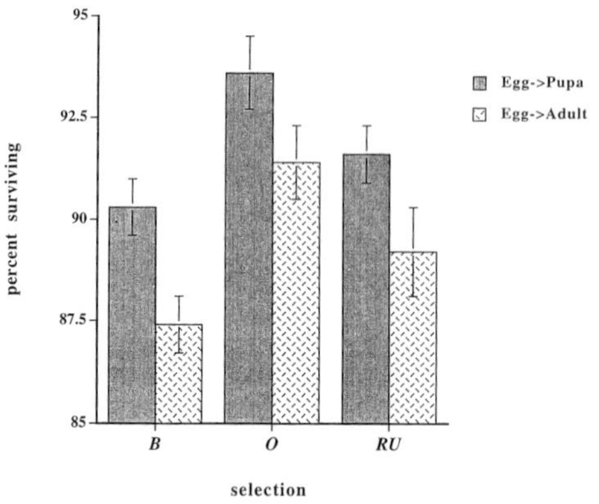

FIG. 1. Viabilities of B (early-reproduced), O (late-reproduced), and RU (reverse-selected from late-to early-reproduced) selection treatments. Each bar represents the mean of the replicate populations within a selection treatment (±SE). The O treatment displayed significantly higher egg-to-adult viability than either early-reproduced treatment. B and RU treatments were not significantly different.

were (male/female, ±SD): B = 234.8 ± 2.65/229.52 ± 3.33; O = 250.28 ± 4.26/246.32 ± 4.6; RU = 236.05 ± 1.58/231.33 ± 2.70. A one-tailed, unpaired t-test with $n - 2 = 8$ df showed a significant difference between B and O selection treatments (B < O; $P < 0.01$). A paired t-test on RU and O selection treatments (using only replicates 1–4, as there is no RU_5) also revealed a significant difference (RU < O; $P < 0.05$) in developmental time. B and RU treatments were not significantly different.

Both egg-to-pupa and egg-to-adult viabilities of the RUs were intermediate between B and O treatments (fig. 1). The t-tests showed significant differences ($P < 0.05$) for both egg-to-adult and egg-to-pupa viability between B and O lines. RUs, however, were not significantly different from either B or O for either viability character.

Stage-Specific Developmental Time and Pupation Height of B and O Lines

Table 2 combines data from our experimental assay of hatching time and hatchability of eggs with data from our assay of pupation time, time in pupa, time to adulthood, and pupation height for the B and O treatments.

The average times from oviposition to hatching were 20.48 h (B) and 21.12 h (O). The average hatchabilities were 95.4% (B) and 93.8% (O). One-factor ANOVA of population means shows that

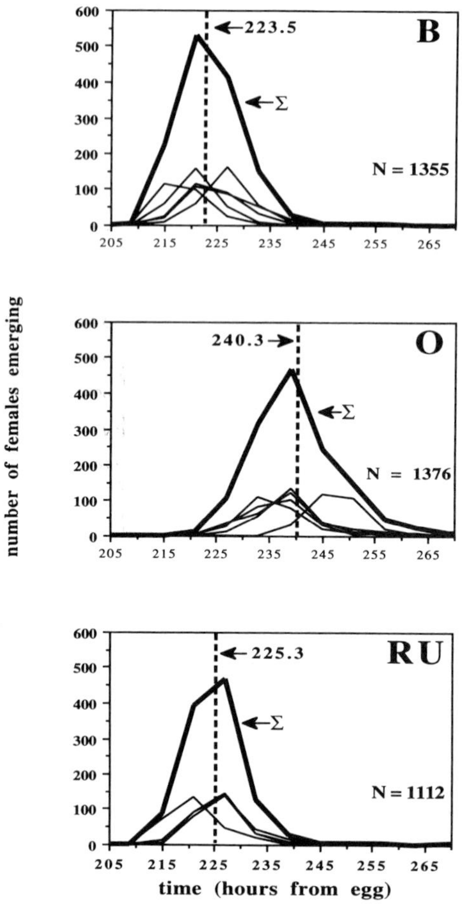

FIG. 2. Egg-to-adult developmental times of B, O, and RU selection treatments. The bold lines show the sum of the replicate populations, which are shown with fine lines. Laboratory selection for postponed senescence has significantly increased developmental time of the O treatment over their control B treatment, which is maintained in the same manner as the ancestral population. Reverse selection (from late- to early-reproduced) in the RU treatment has restored fast development to statistical identity with the Bs.

the two selection treatments do not differ significantly in either hatching time or hatchability ($P = 0.10$ and $P = 0.38$, respectively). Note that this measurement of the percentage of hatching does not distinguish between hatchability of fertilized eggs and the laying of infertile eggs. The purpose of the experiment was to test whether hatching percentage played a role in determining egg-to-adult viability differences; thus, in this us-

TABLE 2. Timing and survivorship of major preadult stages in the B (baseline) and O (old) treatments. The O-B difference is reported at the bottom. Data for egg-hatching time come from the "hatching-time" experiment. All other data are from the "pupation-time" experiment. Each cell represents ten observations of 60 larvae. Statistical significance of the O-B difference is based on the one-tailed t-distribution with $n = 5 - 1$ df.

	Egg to adult		Egg to larva		Egg to pupa		Pupa to adult		Pupation height (mm)
	Time (h)	Viability (%)	Time (h)	Viability (%)	Time (h)	Viability (%)	Time (h)	Viability (%)	
O									
1	228.52	86.8	20.52	96.0	121.88	90.3	106.64	96.5	38.38
2	227.52	87.3	20.87	95.0	122.05	90.5	105.47	96.8	39.86
3	226.31	88.2	19.37	97.0	121.12	91.3	105.19	96.9	29.56
4	227.31	80.7	20.45	94.0	122.68	88.0	104.63	92.7	29.84
5	229.81	82.0	21.21	95.0	124.13	88.3	105.68	93.7	36.38
Mean	227.89	85.0	20.48	95.4	122.37	89.7	105.52	95.3	34.80
SD	1.3	3.4	0.7	1.1	1.1	1.4	0.7	2.0	4.8
SE	0.6	1.5	0.3	0.5	0.5	0.6	0.3	0.9	2.2
B									
1	236.41	91.8	21.11	95.0	128.34	94.7	108.07	97.2	35.73
2	256.00	91.5	21.70	95.0	149.12	94.3	106.88	97.2	29.58
3	236.82	92.8	21.48	88.0	128.88	94.8	107.94	98.0	18.32
4	259.10	92.0	21.32	93.0	149.30	93.3	109.80	98.7	28.45
5	241.80	89.0	19.98	98.0	136.16	92.5	105.64	96.5	31.27
Mean	246.03	91.4	21.12	93.8	138.36	93.9	107.67	97.5	28.67
SD	10.8	1.5	0.7	3.7	10.4	1.0	1.5	0.9	6.4
SE	4.8	0.7	0.3	1.7	4.6	0.4	0.7	0.4	2.9
O-B difference	18.13 **	6.4 **	0.63 NS	−1.6 NS	15.99 **	4.2 **	2.15 *	2.2 *	−6.13 NS

* $P < 0.05$; ** $P < 0.01$.

age, the distinction between "percentage of hatching" and "hatchability" is not important. All other differences in the timing of major ontogenetic events were significant statistically when tested using the one-tailed t-distribution at 95% confidence, with O > B in all cases; the one-tailed test was considered appropriate because of the long-standing observation and a priori expectation of greater developmental time of the Os. The average times from egg to pupa were B = 122.37 h and O = 138.36 h; the difference (O − B) was 15.99 h ($P < 0.001$). The average times from pupa to adult were 105.52 h and 107.67 h (B and O, respectively; $P < 0.05$); the difference was 2.15 h.

Thus, the overall developmental time difference between B and O selection treatments is predominantly due to a difference in the duration of the larval phase: Time in the egg accounts for about 2% of the difference; time in pupa accounts for about 11% of the difference; time as larva accounts for about 87% of the difference in total developmental time from egg to adult.

Pupation heights of B and O selection lines did not differ significantly when compared using a two-tailed t-test (table 2). A failure to find a difference in pupation heights confirms the results of Mueller et al. (1993) for these selection stocks. It should be noted that, in the both studies, pupation heights were restricted to vials considerably shorter than the potential maximum height of pupation. Pupation height may reflect larval energetic expenditure on wandering and pupation site selection, but pupation in a relatively confined container may not allow differences to be observed.

Developmental Time, Viability, and Longevity in Intermediate Generation-Time Selection Treatments

Because all slower developing treatments display increased developmental times, higher preadult viabilities, and greater longevities, we have incorporated data from an assay (Chippindale et al. MS) that simultaneously surveyed developmental time and viability in B, CB, O, and CO treatments. Data from this assay are presented in figure 8. CBs were significantly slower developing than Bs ($P < 0.05$; two-tailed t-test) and showed significantly higher egg-to-adult viability ($P < 0.05$; two-tailed t-test).On the other hand, the CO treatment showed no significant

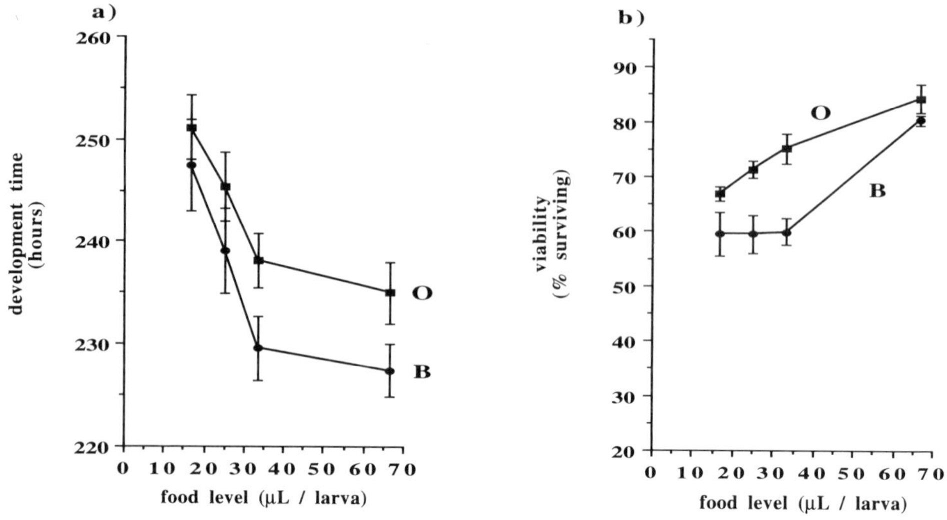

FIG. 3. Norms of reaction for (a) developmental time and (b) egg-to-adult viability of B and O selection treatments across a gradient of larval food levels. Each point is the mean (±SE) of the five replicate populations within a selection treatment. The O treatment displayed longer developmental time and higher egg-to-adult viability across the range of food levels employed.

evolution of either developmental time or viability from its O ancestor treatment. Longevity data are a subset of the data in Nusbaum et al. (in prep.). CB longevity had increased significantly in females (B = 40.97 d, CB = 45.84 d; $P < 0.01$; two-tailed t-test) but not males (B = 49.62 d, CB = 48.90 d), as previously reported by Rose et al. (1992). CO longevity had decreased significantly ($P < 0.01$) in both males (O = 86.83 d, CO = 63.18 d) and females (O = 74.34 d, CO = 58.62 d). Selection for intermediate generation time of CB from B and CO from O has therefore elicited some convergence in the two treatments but is asymmetrical with respect to the characters concerned: CBs have evolved longer development and greater viability without a change in longevity of both sexes; COs have evolved downward in longevity without any change in developmental time or viability from the Os. Apparently longevity can evolve without any impact on developmental characters, as shown in figures 8a and 8b.

Effects of Larval Food Manipulation on Developmental Time and Viability

Food-level manipulations produced a general pattern of reduced viability and increased developmental time with increased crowding in all selection treatments assayed. These results are summarized in figure 3 (B and O) and figure 4 (B, CB, CO). We have used some caution in equating these food-level manipulations with the terms "density" or "crowding" because some components of a "crowded" environment may be absent in our manipulations. L. D. Mueller (pers. comm. 1993) has argued that a truly crowded environment for *Drosophila* development would consist of intense scramble competition for food, as well as high waste-product build up, space limitations, and so on. Insofar as we have modulated the volume of food allowed per larva, we have performed a "density experiment," but we have also controlled humidity, surface area available for larval spiracular access and normal space for pupation, and perhaps provided an environment in which waste products would not build up to extreme levels in any of the food treatments. The present results do, however, closely resemble results from a true density experiment (in the sense of Mueller 1988) done on B and O lines (unpubl. data).

In the B versus O experiment, three-factor ANOVA with selection treatment, sex, and food level as fixed effects revealed highly significant effects of stock (B < O; $P < 0.01$), sex ($P < 0.05$) and food level ($P < 0.01$) for developmental time. No interaction terms were statistically significant. Both egg-to-adult and egg-to-pupa viabil-

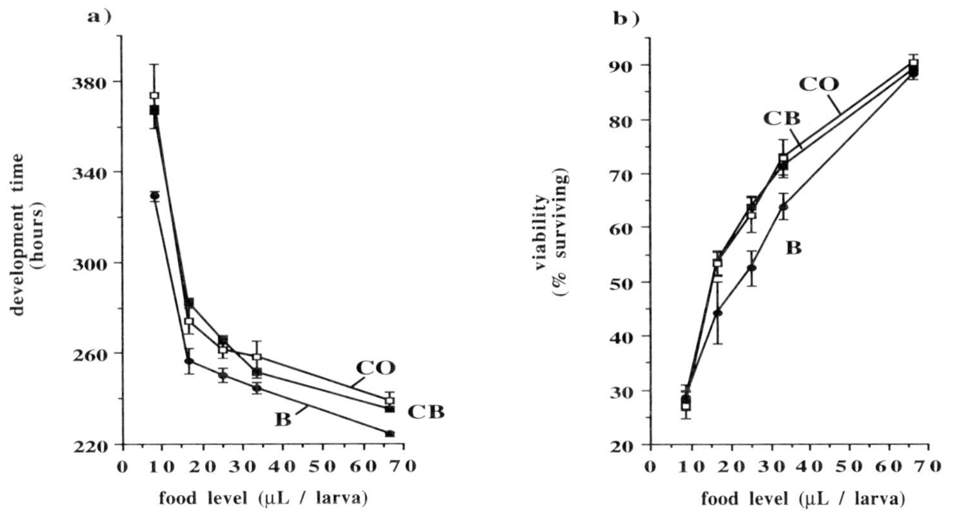

FIG. 4. Norms of reaction for (a) developmental time and (b) egg-to-adult viability of B, CB, and CO selection trteatments across a gradient of larval food levels. Each point is the mean (±SE) of the five replicate populations within a selection treatment. CB and CO treatments showed longer developmental time and higher egg-to-adult viability than the Bs and were indistinguishable from one another.

ities showed highly significant effects of selection (B < O; $P < 0.01$) and food level ($P < 0.01$) with no significant interaction effects.

In the B, CB, CO experiment, three-factor ANOVA on developmental time, with selection treatment, food level, and sex as factors, showed significant effects of selection ($P < 0.001$) and food ($P < 0.001$) and a significant ($P < 0.01$) interaction between these factors. The sex factor was not significant, in contrast to the B versus O experiment, and this was probably due to the addition of the 8.33μl/larva treatment. At this (lowest) food level, males developed faster than females in 12 of 15 populations. This reversal of relative developmental times between sexes is shown in figure 6. The three-way interaction of factors was also significant at $P < 0.05$. When selection treatments were paired to test differences between means, CB and CO were both significantly slower developing than B ($P < 0.001$; two-tailed t-distribution). CB and CO were not significantly different from each other in developmental time. A two-factor ANOVA on egg-to-adult viability revealed significant effects of selection treatment ($P = 0.0014$) and food ($P < 0.001$) with no significant interaction. The paired comparisons showed no significant differences between CB and CO, but both CB and CO were significantly higher than B in viability ($P < 0.05$).

Body Size of B and O Lines

Body size was estimated by both thorax length and wet-weight measures. Thorax lengths were measured over a gradient of larval food levels from the food-level experiment on B and O selection treatments. The norms of reaction for thorax length across a larval density gradient are shown in figure 5. Analysis of variance with selection treatment, sex, and food level as fixed effects revealed highly significant ($P < 0.01$) effects of sex (male < female) and food level (low < high). The sex × food interaction was significant ($P = 0.004$). No other interaction terms were significant. Selection treatment was not statistically significant (O > B; $P = 0.10$). When sexes were analyzed separately (ANOVA: stock × food), the O males were significantly larger than the B males ($P = 0.018$), though the females were not significantly different ($P = 0.77$). An important consideration is the difference between the two selection treatments at normal culture densities alone; at the 66.7μl per larva food level, there were no significant differences between B and O for either sex.

Wet weights were taken on young flies approximately 48 h from adult emergence. We weighed progeny from the same populations after the parents had been aged to 4, 9, 14, 19, and

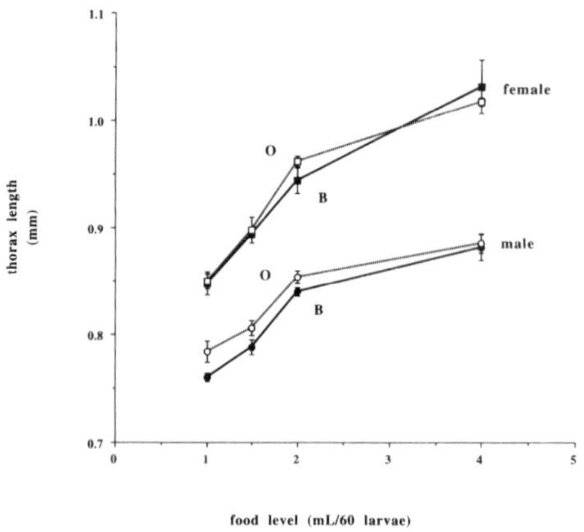

FIG. 5. Body size of the B and O selection treatments across a range of larval food levels. Thorax lengths were not significantly different between the two selection treatments, though sexes were highly significantly different.

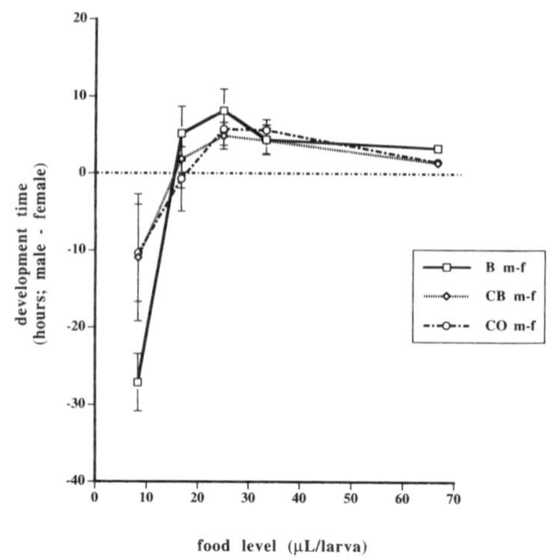

FIG. 6. The difference in developmental time between males and females in 15 divergent populations under larval food restriction. Female *Drosophila melanogaster* typically develop faster than do males. Under standard densities in our stocks this difference between sexes is highly consistent (e.g., table 4). The faster development of females, despite their size advantage, presents a fascinating puzzle. Here, however, we show that a reversal in relative developmental speed occurs under extreme food limitation in our stocks.

24 days of age; we wanted to make sure that parental age did not affect the data. Table 3 shows the results of this experiment, in which we weighed a total of 4 962 B flies and 4 961 O flies in groups of five. We found no significant difference in wet weight between the two selection treatments (ANOVA, with selection, sex, and parental age as factors). The difference between sexes was highly significant ($P < 0.01$) as was the parental age ($P < 0.01$). No interaction term was statistically significant, although the sex × age term was very close ($P = 0.059$).

Interestingly, the significance of the parental age factor in the ANOVA reflected a general decline in offspring size with parental age of the B treatment. Simple regression of the mean offspring weights against parental age for each replicate population and sex revealed a negative correlation for all tests. Regression-slope values were then used in a two-factor ANOVA, with selection treatment and sex as factors. Bs had a significantly more negative mean slope than Os ($p = 0.006$), and the interaction term was not statistically significant. When sexes were analyzed separately, the B male and female slopes were both significantly negative ($P < 0.05$), although both male and female slopes for the O treatment group were not significantly different from zero ($P = 0.34$ [males] and $P = 0.194$ [females]). Thus, it is apparent that parental age does affect offspring weight in these stocks: B offspring weight declines relative to O offspring weight with the age of the parents. This effect was not sufficient to generate an overall difference in weight between the Bs and Os and does not explain the inconsistent results of Rose et al. (1984), since the parents were of same age in their experiments.

Starvation Resistance of B and O adults

Starvation resistance was measured for both newly emergent and 4-d-old flies of the B and O treatments, as shown in figure 7. A three-factor ANOVA with selection, sex, and age as factors, showed significant effects of selection (O > B; $P < 0.05$), sex (f > m; $P < 0.01$), and age from emergence ($P < 0.05$). No interactions were significant. When the two different age classes were tested separately (two-tailed t-test with sexes pooled; df = 8), there were no significant differences between B and O at emergence. After 4 d of adult feeding, the B and O treatments differed significantly (B < O; $P < 0.01$).

Crosses within the O-Selection Treatment

Crosses between O populations were performed to test for mutation accumulation slowing their development relative to the Bs. These crosses failed to show any difference between

TABLE 3. Wet weights of B and O flies at 24 h from emergence. Means for the five replicates at five parental ages; "*N*" is the number of individuals weighed. Two-factor ANOVA with stock and sex as factors and populations as replicates reveals no significant difference between the stocks or interaction between stock and sex. The difference between the sexes is highly significant ($P < 0.01$). Regression of population means against parental age showed a significantly negative slope for the Bs but not the Os with age.

Parental	Male weight (mg)				Female weight (mg)			
age (d)	B	*n*	O	*n*	B	*n*	O	*n*
4	0.879	500	0.859	500	1.302	500	1.275	500
9	0.867	500	0.847	484	1.319	500	1.355	500
14	0.751	500	0.759	500	1.203	486	1.273	486
19	0.733	490	0.733	490	1.069	500	1.111	500
24	0.742	496	0.828	495	1.234	490	1.262	496
Mean	0.794	2486	0.805	2469	1.225	2476	1.255	2482
SD	0.072		0.056		0.100		0.089	
SE	0.032		0.025		0.045		0.040	

purebred and crossed groups of Os for egg-to-adult developmental time (table 4). The crossed samples had a mean developmental time of 234.9 ± 2.2/231.92 ± 2.3 (male/female, ±SE), although the pure samples had a mean developmental time of 233.4 ± 0.77/230.28 ± 0.84 (male/female, ±SE). An unpaired, two-tailed *t*-test, revealed no significant difference between crosses and purebred experimental treatments. Because it might be argued that deriving two samples of the purebred populations to balance the design might be pseudoreplication, we also tested the two groups of purebred populations seperately. This did no modify the conclusion that no differences existed for developmental time between pure and crossed groups.

DISCUSSION

Extended Developmental Time Has Evolved in Late-Reproduced Populations

The ten B and O lines are long-established, outbred populations of *Drosophila* selected for early- and late-life fertility, respectively. Despite extreme differences in the adult demography, both selection treatments have been handled identically with respect to the preadult conditions. Nonetheless, our assays of egg-to-adult developmental time have shown that the O line develops approximately 5% slower than the B line. This difference has been shown here for two different rearing media as well as with and without larval crowding or food limitation. The difference between O and B developmental times has also persisted throughout 5 yr of continued selection on the adult phase for the timing of reproduction, with no detectable directional trend.

The difference between B and O selection treatments in overall developmental time is mostly (87%) accounted for by the duration of the larval stage. The Os have a larval stage approximately 15 h longer than that of the Bs. It is probable that the O populations have increased in developmental time (rather than the Bs having decreased) for several reasons. First, the B-type selection treatment is identical to the ancestral maintenance regimen. The IV popu-

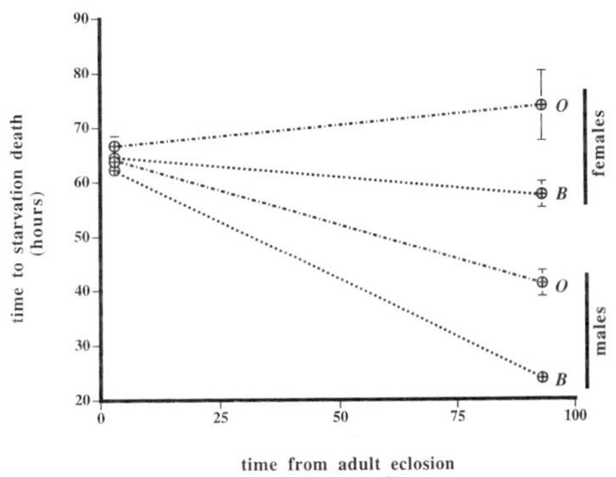

FIG. 7. Differentiation of starvation resistance in B and O selection treatments. Starvation resistance is strongly genetically correlated with longevity and widely evidenced to be a product of stored lipid reserves in *Drosophila melanogaster*. Adult O flies have repeatedly been shown to have higher levels of lipids and higher starvation resistance when assayed at 4+ d of age. Here we show that this starvation advantage is not present in the newly emerged imago. Differences between sexes and selection treatments occur as a function of adult age and are not present at eclosion. This result is consistent with the notion of antagonistic pleiotropy between early reproduction and aging related characters and inconsistent with a connection between extended development and postponed senescence.

TABLE 4. A comparison of egg-to-adult developmental time of purebred versus crossed O populations from an experiment run to test a mutation accumulation hypothesis for slower O development. Restoration of faster development in the crossed samples would point towards mutation accumulation or hybrid vigor (caused by inbreeding). The data indicate no such pattern.

Purebred group			Crossed group		
Cross	Male (h)	Female (h)	Cross	Male (h)	Female (h)
1 × 1 (1)	233.83	231.28	2 × 1	240.32	235.96
2 × 2 (1)	229.69	225.28	3 × 1	235.58	232.48
3 × 3 (1)	233.28	230.28	4 × 1	231.60	229.52
4 × 4 (1)	230.73	229.74	5 × 1	234.47	231.48
5 × 5 (1)	237.50	234.63	3 × 2	226.95	223.95
1 × 1 (2)	233.60	229.86	4 × 2	228.35	225.46
2 × 2 (2)	231.02	227.05	5 × 2	234.89	232.29
3 × 3 (2)	235.64	231.56	4 × 3	251.68	250.24
4 × 4 (2)	235.59	232.88	5 × 3	230.65	227.79
5 × 5 (2)	232.84	230.19	5 × 4	234.26	230.07
Mean	233.37	230.28	Mean	234.87	231.92
SD	2.44	2.67	SD	7.05	7.33
SE	0.77	0.84	SE	2.23	2.32

lation, from which both lines were derived, had gone through more than 120 generations of 2-wk propagation before selection was imposed on B and O, and should have been close to evolutionary equilibrium for those conditions. Second, reverse selection of the RU treatment from the Os to a B-type schedule returned developmental time to the faster pace of the Bs. Third, two other demographically selected treatments that have a lengthened adult period (CB, CO; 4–5-wk generations) have developmental times greater than the Bs, and statistically indistinguishable from the Os. Therefore, we conclude that the O stocks have a longer developmental time that is derived from that of the IV or B stocks, rather than the reverse.

Why Does Extended Development Evolve?

Inbreeding depression of the slower developing selection treatments is a highly unlikely explanation for the evolution of extended development in the O stocks because of the large O population sizes (in the range of 10^3). Inbreeding would be more likely to afflict the B populations than the Os, as the B treatment has undergone five times as many generations as the Os. Yet it is the Os that show slowed development. Inbreeding also should be accompanied by depression of other performance characters, which we have not observed here, or in other contemporaneous experiments on these lines (Chippindale et al. 1993; Leroi et al. 1994a). Furthermore, experimental hybridization between replicates within selection treatments has been performed periodically in the past (Hutchinson and Rose 1991) and in the present study for the O treatment, without any evidence of F_1 superiority, and thus hybrid vigor. Therefore, we conclude that inbreeding depression is not a plausible explanation of longer O development.

Another possible explanation is indirect selection. Early emergence will, all other things being equal, enhance early reproduction, because it provides more time for adults to mature sexually, both anatomically and physiologically, and more time to feed and mate in preparation for egg laying. Both B and O selection treatments have an absolute deadline for emergence from pupa at 2 wk, since any individuals that have not emerged by 2 wk are discarded. Bs are then immediately reproduced, whereas Os are transferred to cages for a further 8 wk before egg collection for the next generation. The 2-wk emergence deadline places a limit on the degree to which developmental time can evolve upward, and the emergence-distributions for adults of all populations examined in this study are entirely completed before the 2-wk deadline under normal culture densities. But the absence of direct (truncation) selection on developmental time does not preclude direct selection on early adult characters, which are correlated to developmental time for the aforementioned reasons. Thus, B stocks, for example, may be subject to indirect selection on developmental time because of selection on females in these stocks to lay fertile eggs at 14 d of age from their own oviposition as eggs, about 4 d from pupal eclosion. Later-reproduced selection treatments (O, CB, and CO) have, by contrast, experienced a relaxation of selection for early maturation and mating as a result of their extended adult maintenance, and thus they may be free of indirect selection on developmental time. They may have thereby evolved increased developmental time as a result of the removal of the long-standing selection pressure for earlier development imposed on the Bs. This conclusion does not, however, explain the evolutionary-genetic mechanism responsible for this increased developmental time.

When selection is relaxed, mutation accumulation is a reasonable expectation. Could this be the mechanism that produced the increased O, CB, and CO developmental times? The degree of mutation accumulation should depend on the number of generations elapsed since the relaxation of selection. However, selection treatments such as CB and CO have undergone different numbers of generations since derivation from 2-wk stocks (CB = 51, CO = 111 generations) but are statistically equivalent to the Os in developmental time. The strikingly uniform pattern of occurrence both between and within selection treatments is not consistent with random accumulation of selectively neutral mutations. In turn, the Os have been repeatedly assayed over the last 5 yr, yet have not shown a detectable upward trend in developmental time. These arguments do not preclude very limited action of mutation accumulation. It is perhaps most important, then, that crosses between O populations did not restore rapid development to the F_1 hybrids (table 4), as a mutation accumulation hypothesis would usually require.

Could the evolution of longer O developmental time reflect some general relationship between longevity and development? The "developmental theory of ageing" (Lints 1978, 1988) predicts the existence of a positive genetic correlation between developmental time and longevity. Superficial inspection of the present data seems to support the developmental theory: The longer lived O selection treatment does exhibit longer developmental time than its B control. At least three lines of reasoning suggest that this pattern has little to do with any profound connection between developmental time and aging. First, the continued upward selection response of O populations for longevity (Leroi et al. 1994a) is incongruent with the developmental theory of aging, since these populations have not displayed any detectable increase in developmental time over the last 5 yr or more. Second, the results of selection of lines with life-cycle times intermediate between B and O show no evidence for such a correlation (fig. 8a). CB populations, which have been demographically selected for reproduction at 2 to 3 wk from emergence have shown only a slight increase in longevity over their ancestor Bs, but are equal to the long-lived Os in developmental time. The COs, which have been back-selected demographically from the O populations (from 8 wk to 2 to 3 wk of adult selection before reproduction), have retained a developmental time equal to the Os, despite a substantial decline in longevity. Third, we failed to find any evidence for a direct contribution of extended development to either adult size or resistance to starvation stress. Rather, all the available evidence suggests that adults of demographically selected treatments (B, O, CB, CO) begin with essentially the same resources at emergence. The combination of these selection results is strong evidence that the larval phase has evolved independently from the aging phenotype. The developmental theory of aging is therefore refuted by our results, as it has been before (Zwaan et al. 1991, 1992; Zwaan 1993).

Another connection might be forged between preadult and adult survival, since all longer-lived treatments (CB, CO, O) also displayed higher preadult viabilities than the short-lived Bs. (This is the reverse of the pattern found by Partridge and Fowler [1992], who argued for a negative association between preadult and adult survival.). But such a hypothesis is easily refuted by consideration of intermediate generation-time selection treatments. Figure 8b illustrates the independence of longevity and egg-to-adult viability in these demographically selected populations. The CB treatment has evolved substantially higher viability with only a slight increase in average adult longevity, whereas the CO treatment has declined markedly in longevity from its O ancestor treatment but has not given up any preadult viability. Furthermore, in the second "minimum food" experiment, CB and CO treatments behaved identically in developmental time and viability across food levels despite considerable differences in longevity between the two treatments. We have failed to find any evidence for a connection between viability and longevity under conditions directly comparable to normal selection conditions, and with the stress of larval food limitation. It therefore appears that preadult viability and adult longevity can evolve independently, and we conclude that there is no simple connection between the two characters.

Over the 24 populations we have assayed, slower development is consistently associated with an increase in preadult viability. Although these differences seem small (e.g., O > B by 4.5% in egg-to-adult viability; table 1), preadult viability is a fitness character of enormous weight. The preadult period is subject to the full force of natural selection for survival (Charlesworth 1980). It is notable that there is an inverse relationship between preadult viability and developmental time across the stocks, over multiple experiments. This pattern suggests that there is antagonistic pleiotropy among loci affecting both developmental time and preadult viability. When O, CB, and CO stocks were removed from selection for early reproduction, they had relaxed indirect selection for early developmental time. This should have shifted the balance between selection on developmental time and preadult viability during the larval stage, causing a consistent shift in favor of alleles that slowed development, but enhanced viability. Because this effect is one of a shift in selective equilibrium during a life-cycle stage under strong selection, it is not surprising that stocks with very different periods of reduced selection on developmental time, such as the CBs and COs, should nonetheless have similar developmental times and viabilities. Since we have shown that developmental time was altered by selection without any change in body size or apparent provisioning of the imago, it is apparent that lower developmental *rate* confers higher preadult survivorship. Thus, despite the general perception that body size and adult fitness are critically associated in *Drosophila* (e.g., Robertson 1957; Tantawy and

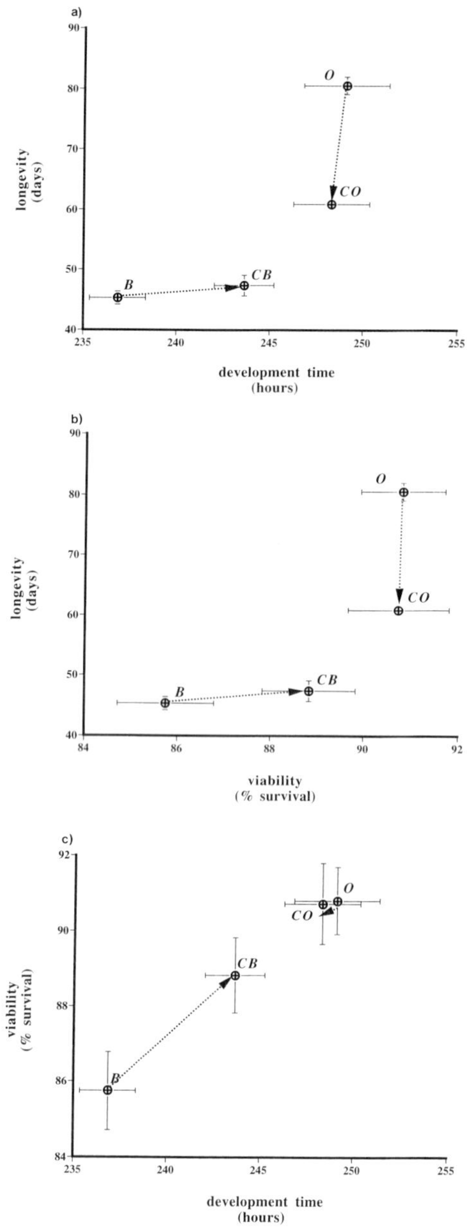

FIG. 8. Developmental time, viability, and longevity: What correlates with what? In this triptych, we summarize the results of demographic selection on B (2-wk), CB and CO (4–5-wk), and O (10-wk) treatments. Arrows emphasize the convergent selection applied to CB (2–5-wk generations) and CO (10–5-wk generations) from their B and O ancestor treatments. Part (A) shows a search for the "developmental theory of age- ←

ing", which posits that developmental time and aging are causally correlated. Clearly, however, developmental time and longevity can evolve quite independently, refuting this theory. In part (B), we show the independent evolution of viability and longevity. Although populations which display greater longevity also have higher preadult viabilities, the two characters are not genetically correlated. In part (C), the relationship between developmental time and viability is illustrated. Slower developing treatments (CB, CO, O) have higher egg-to-adult viabilities than fast developing (B) populations. The pattern supports the conclusion that antagonistic pleiotropy between developmental time and viability has driven the evolution of extended development in these stocks.

Vetukhiv 1960; Partridge and Fowler 1992), selection has, in this instance, acted on preadult viability rather than adult size. With the hypothesis of antagonistic pleiotropy between developmental time and viability, all the significant experimental results can be accounted for.

Phenotypic Manipulations of Larval Food Level Are Incongruent with Selection Results

Phenotypic manipulations have been presented as a good means of deducing evolutionary relationships among characters (e.g., Partridge 1992; Bell and Koufopanou 1986; but see Reznick 1992). But, unless genetic information is available in the form of selection or quantitative genetic results, this assertion is difficult to test. In an earlier paper (Chippindale et al. 1993), dietary manipulation of adults was compared with selection results for several fitness characters. Both methods produced a qualitatively similar trade-off between survival and reproduction, which may reflect a fundamental physiological relationship mediated by discrete allocative pathways. However, in that study, considerable differences between selection response and phenotypic manipulation were also identified. Leroi et al. (1994 b) extended this line of questioning and demonstrated that phenotypic trade-offs are easily altered by selection.

In the present context of larval growth and survival, we may also draw a comparison between the two methodologies. Among selection treatments, we have found a negative association between developmental rate and juvenile survival, whereas adult body size remained unaffected. Our manipulations of larval food level produced a strikingly different pattern among the same three traits. Food-limitation increases developmental time while at the same time leading to a smaller adult *and* reducing preadult viability. There is no correspondence between the selection result and phenotypic plasticity for these characters, suggesting that the physiological mechanisms producing longer-developing flies under selection are different from those highlighted by food limitation. Considering the evidence from controlled laboratory selection experiments, phenotypic manipulations seem to be poor predictors of evolutionary response.

Sexual Dimorphism in Developmental Time

A puzzle in the evolution of *D. melanogaster* is the coincidence of sexual dimorphism in body size (females are much larger than males) and developmental time (females develop faster than males). The timing of pupation may be related to attaining a critical minimum size for survival to adulthood. Males, which are about half the weight of females at emergence, should be expected to achieve this critical mass earlier, all other things being equal. Why do female larvae have such vastly higher growth rates? Our data offer little help in answering this question. However, in this context, an interesting feature of the crowding experiments was the reversal of relative developmental times of males and females under extreme food limitation in the second experiment (see fig. 6). In early-reproduced populations, it might be argued that the longer time to peak fertility after emergence in females (Ashburner 1989) would lead to greater selection for early emergence in females than males (Roper et al. 1993). In late-reproduced populations, such differential selection on the sexes should be removed. Sexual dimorphism in developmental time has persisted in later-reproduced populations in our laboratory nonetheless. The density dependence of this dimorphism is a curiosity that deserves further investigation.

Comparison with the Recent Literature

Partridge and Fowler (1992) showed extended development of their "old" lines compared with their "young" lines, which is echoed in our observation of longer developmental times of O (Rose "old") over B (Rose "young") lines. Partridge and Fowler speculated that longer development in their "old" flies leads to larger body size, reflecting increased investment in the "durability of the soma." The apparent penalties that the "olds" pay are exacted in preadult survivorship as well as later time of emergence. These results differ substantially from ours, as we find *higher* preadult survival rates in selection treatments with longer developmental times and *no difference* in body size. A slower *rate* of growth in the Os seems to be associated with greater preadult survival. Thus, the present study and that of Partridge and Fowler seem to be contradictory.

But the Partridge and Fowler (1992) study appears to have a number of problems that may preclude direct comparisons with our work. Roper et al. (1993, p. 455) discussed some of these problems, specifically, inbreeding depression and inadvertent selection: "Our results point to the need to avoid inadvertent direct selection on characters other than those intended. In the present context, standardization of larval density and avoidance of selection pressure for rapid development are clearly important." It is crucial to any claim for a correlated response that there be no direct selection on the character supposed to be correlated with the character under direct selection—in this case, no direct selection on larval competitive ability or developmental time when selecting on the adult. Yet, it is clear from Partridge and Fowler that their "young" and "old" lines differ substantially in the preadult conditions as well. Densities apparently were regulated by the duration of egg laying. The "young" lines, with more females laying at roughly peak fertility and before substantial mortality selection has acted, are likely to outlay the "olds" many times over. This difference in egg laying between "young" and "old" stocks probably leads to differences in effective population size and more rapid inbreeding depression of the "old" treatment (Roper et al. 1993). This difference in larval density is indicated in Partridge and Fowler: on average, about 69% (or about 3500 flies) of their "young" lines and about 83% (or about 2500) of "old" adults have emerged at day 17. In other words, the "old" lines developed *faster* than the "young" lines in normal culture and had higher survivorship under normal culture conditions. However, in their controlled developmental-time experiments, the mean time to emergence is only about 10.9 d ("young") and 11.8 d ("old"), both considerably less than 17 d. The two lines are reversed in rank order from the observations obtained from normal lab culture, for both speed and survival. Since developmental time is increased and viability is decreased by crowding of larval cultures (Bakker 1961; Roper et al. 1993; present data), it is probable that the Partridge and Fowler stocks are very crowded indeed. Moreover, direct truncation selection on the distribution of adult emergences will have acted more strongly on "young" than "old" treatments, since the former were normally more subject to crowding. Thus, the basic selection procedures used by Partridge and Fowler were not at all analogous to the usual late-fertility *Drosophila* selection procedures (e.g., Rose 1984).

In contrast to Partridge et al.'s results, Hillesheim and Stearns (1991, 1992) reported that flies selected for greater fresh weight at eclosion had reduced longevity. Inadvertent direct selection and differential inbreeding are likely to account for the two laboratories' disparate results as well. Indeed, in explaining the contradiction between their own results and those of Partridge and Fowler, Hillesheim and Stearns (1992, p. 750) stated: "The contradiction is apparent, not real. In selecting for heavier flies, we were also selecting for flies that had larger ovaries, laid more eggs at the start of life, and—perhaps therefore—died younger." Again, it appears that multiple types of selection were imposed on these stocks, making the distinction between direct and indirect responses to selection obscure. An additional problem with this study may be inbreeding depression. The authors used very small population sizes ($n_e \leq 10$) throughout selection on weight at emergence. Some attempt to counter inbreeding depression by crossing replicates was made at two points in selection, but selection was imposed again with small population size for about ten generations. Such crosses are likely to lead to linkage disequilibrium and also confound the replicate system for the purpose of statistical inference.

The problem of inbreeding depression is also raised by Roper et al. (1993). They presented evidence that both "young" and "old" selection treatments suffer from inbreeding depression, and that the "old" lines seemed to be more strongly afflicted. Here we look in greater detail at the data from Partridge and Fowler (1992) and contrast them with our selection treatments. Figure 9 shows a comparison of viability data from our B and O lines and the "young" and "old" stocks of Partridge and Fowler. The data on viability in the "young" and "old" Dahomey lines were calculated by combining their tabled data from the "competitive ability" experiment with their reported proportions of total flies emerging, to extract the wild-type viabilities. One thing that stands out in the comparison is the reversed ranking of B and O and "young" and "old" with respect to viability. The Partridge "old" lines had exceedingly low viabilities: even at high food levels, the olds have less than 50% average survivorship to adulthood (44.3%, 18.1%, and 82.0% for old populations 1–3, respectively). At the low food levels, it is particularly surprising to see the poor performance of the old treatment against a mutant stock, which might be expected to be highly inbred in its derivation. The low survivorship of old flies and the high variance between replicate old populations immediately raises the concern that inbreeding depression alone could account for the preadult differences between lines, since all developmental performance characters are depressed in the old lines.

Roper et al. (1993) attempted to recover the populations from inbreeding depression by crossing replicates within selection treatments. But it is questionable whether a single-generation cross is adequate to remove the confusion that inbreeding depression can engender for life-history investigation. Such crosses will lead to the arbitrary juxtaposition of components of inbred genomes, and are no substitute for crossing and then maintaining selection for a number of generations to reduce the frequency of deleterious alleles fixed by genetic drift in individual lines. The experiments of Roper et al. fail to support some of the earlier results (Partridge and Fowler 1992). In particular, incorporation of hybrid data and comparison with the base stock did not support the notion of a trade-off in performance between the two life stages.

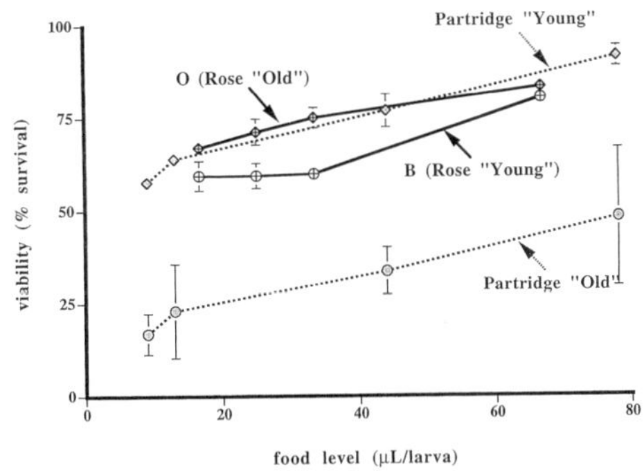

FIG. 9. Viability of Rose B and O selection treatments versus Partridge "young" and "old" treatments. Data for Partridge lines are from Partridge and Fowler (1992), minus the mutant stock used in their "competitive ability" experiment. Food levels are standardized for comparison. Note the reversed order of viabilities for early- and late-reproduced selection treatments between the two labs. The extremely low viabilities and heterogeneity among replicate populations in the Partridge and Fowler "old" treatment are probably indicative of inbreeding depression.

Conclusions

Although some of the present results seem to corroborate the studies of Partridge and Fowler (1992) and Roper et al. (1993), the evolutionary mechanisms involved in the extension of development in their and our late-reproduced stocks are quite different. In their studies, considerable inbreeding depression may have arisen, and some selection treatments appear to have experienced strong, inadvertent, direct selection on developmental rate. It is a routine finding that selection alters a character mean, and a mundane one that inbreeding depression can have pervasive effects on performance characters. Small differences in culture regime have apparently given rise to inadvertent selection and inbreeding in the work of Partridge and colleagues. Our selection system has evolved by different mechanisms. In our selection treatments, reduced developmental speed is apparently selectively favored because of beneficial effects on preadult viability. When selection for early reproduction has been relaxed in our late reproduced treatments, the primary target of selection has been preadult viability rather than increased body size or any other adult attribute. Thus, the antagonistic pleiotropy by which slower larval growth confers increased survivorship of the larval and pupal stages deserves attention, rather than any supposed trade-off between larval and adult survival.

The evolutionary relationship between development and adult life history is still poorly understood, despite some ambitious claims. For example, optimization models often assume a trade-off between developmental time and size of the adult (e.g., Werner 1986; Rowe and Ludwig 1991; Sibly et al. 1991), yet here we have found that adult body size and other early adult characters are conserved despite differences in larval developmental time. Differences in the adult life histories of our stocks have come about because of differences in adult selection, although preadult characters have evolved separately. There is no good evidence for the hypothesis of Partridge and Fowler (1992) that postponed senescence has evolved in laboratory-selected *Drosophila* by preadult investment or trade-offs between developmental rate and adult fertility and longevity. Preadult adaptation does not explain the laboratory evolution of postponed senescence in *D. melanogaster,* though it may be secondarily involved in such evolution.

ACKNOWLEDGMENTS

We thank A. R. Olvido for assistance in preliminary studies of developmental time. We also thank J. Shiotsugu, J. A. Alipaz, T. J. F. Chu, L. Phan, D. J. Borash, and C. Rosario, among others, for assistance in laboratory procedures. We thank A. M. Leroi, T. J. Nusbaum, L. D. Mueller, and T. J. Bradley for insightful comments throughout this project and in the preparation of the manuscript, and A.R. Palmer for the superb measurement software for the Mac. Special thanks to H. T. Van for many details of project support. This project was made possible by funding from U.S. Public Health Service grants AG06346 and AG09970 to M.R.R. from the National Institute on Aging.

Literature Cited

Ashburner, M. 1989. *Drosophila*: a laboratory handbook. Cold Spring Harbor Laboratory Press, Cold Spring Harbor, N.Y.

Bakker, K. 1961. An analysis of factors which determine success in competition for food among larvae of *Drosophila melanogaster*. Archives Néerlandaises de Zoologie 14:200–281.

Bell, G., and V. Koufopanou. 1986. The cost of reproduction. *In* R. Dawkins and M. Ridley, eds. Oxford Surveys in Evolutionary Biology, Vol. 3. Oxford University Press, Oxford.

Brett, W. J. 1955. Persistent diurnal rhythmicity in *Drosophila* emergence. Annals of the Entomological Society of America 48:119–131.

Charlesworth, B. 1980. Evolution in age-structured populations. Cambridge University Press, London.

Chippindale, A. K., A. M. Leroi, S. B. Kim, and M. R. Rose. 1993. Phenotypic plasticity and selection in *Drosophila* life history evolution. I. Nutrition and the cost of reproduction. Journal of Evolutionary Biology 6:171–193.

Chippindale, A. K., T. J. F. Chu, and M. R. Rose. 1995. Complex trade-offs and the evolution of starvation resistnace in *Drosophila melanogaster*. Evolution. *In press.*

Hillesheim, E., and S. C. Stearns. 1991. The responses of *Drosophila melanogaster* to artificial selection on body weight and its phenotypic plasticity in two larval food environments. Evolution 45:1909-1923.

———. 1992. Correlated responses in life history traits to artificial selection for body weight in *Drosophila melanogaster*. Evolution 46:745–752.

Hutchinson, E. W., and M. R. Rose. 1991. Quantitative genetics of postponed aging in *Drosophila melanogaster*. 1. Analysis of outbred populations. Genetics 127:719–727.

Ives, P. T. 1970. Further studies of the South Amherst population of *Drosophila melanogaster*. Evolution 38:507–518.

Leroi, A. M., A. K. Chippindale, and M. R. Rose. 1994a. Long-term laboratory evolution of a genetic life-history trade-off in *Drosophila melanogaster*. 1. The role of genotype-by-environment interaction. Evolution 48:1244–1257.

Leroi, A. M., S. B. Kim, and M. R. Rose. 1994b. The evolution of phenotypic life history trade-offs: an experimental study using *Drosophila melanogaster*. American Naturalist 144(4):661–676.

Lints, F. A. 1978. Genetics and ageing. Interdisciplinary topics in gerontology. Karger, Basil.

———. 1988. Genetics. Pp. 98–119 *in* F. A. Lints and M. H. Soliman, eds. *Drosophila* as a model organism for ageing studies. Blackie, London.

Lints, F. A., and C. V. Lints. 1969. Influence of preimaginal environment on fecundity and ageing in *Drosophila melanogaster* hybrids. I. Preimaginal population density. Experimental Gerontology 4:231–238.

Luckinbill, L. S., R. Arking, M. Clare, W. Cirocco, and S. Buck. 1984. Selection for delayed senescence in *Drosophila melanogaster*. Evolution 38:996–1003.

Luckinbill, L. S., J. L. Graves, Jr., A. Tomkiw, and O. Sowirka. 1988. A qualitative analysis of some life history correlates of longevity in *Drosophila melanogaster*. Evolutionary Ecology 2:85–94.

Mueller, L. D. 1988. Density-dependent population growth and natural selection in food-limited environments: the *Drosophila* model. American Naturalist 132:786-809.

Mueller, L. D., and V. F. Sweet. 1986. Density-dependent natural selection in *Drosophila*: evolution of pupation height. Evolution 40:1354–1356.

Mueller, L. D., J. L. Graves, and M. R. Rose. 1993. Interactions between density-dependent and age-specific selection in *Drosophila melanogaster*. Functional Ecology 7:469–479.

Partridge, L. 1992. Measuring reproductive costs. Trends in Ecology and Evolution 7:99–100.

Partridge, L., and K. Fowler. 1992. Direct and correlated responses to selection on age at reproduction in *Drosophila melanogaster*. Evolution 46(1):76–91.

Reznick, D. 1992. Measuring the costs of reproduction. Trends in Ecology and Evolution 7:42–45.

Robertson, F. W. 1957. Studies in quantitative inheritance. XI. Genetic and environmental correlations between body size and egg production in *Drosophila*. Journal of Genetics 55:428–443.

Roper, C., P. Pignatelli, and L. Partridge. 1993. Evolutionary effects of selection on age at reproduction in larval and adult *Drosophila melanogaster*. Evolution 47:445–455.

Rose, M. R. 1984. Laboratory evolution of postponed senescence in *Drosophila melanogaster*. Evolution 38:1004–1010.

Rose, M. R., and B. Charlesworth. 1980. A test of evolutionary theories of senescence. Nature 287: 141–142.

———. 1981a. Genetics of life history in *Drosophila melanogaster*. I. Exploratory selection experiments. Genetics 97:173–186.

———. 1981b. Genetics of life history in *Drosophila melanogaster*. II. Sib analysis of adult females. Genetics 97:187–196.

Rose, M. R., M. L. Dorey, A. M. Coyle, and P. M. Service. 1984. The morphology of postponed senescence in *Drosophila melanogaster*. Canadian Journal of Zoology 62:1576–1580.

Rose, M. R., L. N. Vu, S. U. Park, and J. L. Graves. 1992. Selection on stress resistance increases longevity in *Drosophila melanogaster*. Experimental Gerontology 27:241–250.

Rowe, L., and D. Ludwig. 1991. Size and timing of metamorphosis in complex life cycles: time constraints and variation. Ecology 72(2):413–427.

Service, P. M., E. W. Hutchinson, M. D. MacInley, and M. R. Rose. 1985. Resistance to environmental stress in *Drosophila melanogaster* selected for postponed senescence. Physiological Zoology 58: 380–389.

Service, P. M., E. W. Hutchinson, and M. R. Rose.

1988. Multiple genetic mechanisms for the evolution senescence in *Drosophila melanogaster*. Evolution 42:708–716.

Sibly, R. M., R. H. Smith, and H. Møller. 1991. Evolutionary demography of a bruchid beetle. IV. Genetic trade-off, stabilizing selection and a model optimal body size. Functional Ecology 5:594-601.

Stearns, S. C. 1992. The evolution of life histories. Oxford University Press, Oxford.

Tantawy, A. O., and M. O. Vetukhiv. 1960. Effects of size on fecundity, longevity, and viability in populations of *Drosophila pseudoobscura*. American Naturalist 94:395–403.

Tucić, N., D. Cvetković, and D. Milanović. 1988. The genetic variation and covariation among fitness components in *Drosophila melanogaster* females and males. Heredity 60:55–60.

Werner, E. E. 1986. Amphibian metamorphosis: growth rate, predation risk and the optimal size at transformation. American Naturalist 128:319–341.

Williams, G. C. 1957. Pleiotropy, natural selection, and the evolution of senescence. Evolution 11:398–411.

Zwaan, B. J. 1993. Genetical and environmental aspects of ageing in *Drosophila melanogaster*: an evolutionary perspective. Ph. D. diss. University of Groningen, The Netherlands.

Zwaan, B. J., R. Bijlsma, and R. F. Hoekstra. 1991. On the developmental theory of ageing. I. Starvation resistance and longevity in *Drosophila melanogaster* in relation to pre-adult breeding conditions. Heredity 66:29–39.

———. 1992. On the developmental theory of ageing. II. The effect of developmental temperature on longevity in relation to adult body size in *D. melanogaster*. Heredity 68:123–130.

Corresponding Editor: L. Chao

Evolution, 51(5), 1997, pp. 1536–1551

EXPERIMENTAL EVOLUTION OF ACCELERATED DEVELOPMENT IN *DROSOPHILA*. 1. DEVELOPMENTAL SPEED AND LARVAL SURVIVAL

Adam K. Chippindale,[1] Julie A. Alipaz,[2] Hsiao-Wei Chen,[3] and Michael R. Rose
Department of Ecology and Evolutionary Biology, University of California, Irvine, California 92717-2525

Abstract.—Developmental time is a trait of great relevance to fitness in all organisms. In holometabolous species that occupy ephemeral habitat, like *Drosophila melanogaster*, the impact of developmental time upon fitness is further exaggerated. We explored the trade-offs surrounding developmental time by selecting 10 independent populations from two distantly related selection treatments (CB_{1-5} and CO_{1-5}) for faster development. After 125 generations, the resulting accelerated populations (ACB_{1-5} and ACO_{1-5}) displayed net selection responses for development time of −33.4 hours (or 15%) for ACB and −38.6 hours (or 17%) for ACO. Since most of the change in egg-to-adult developmental time was accounted for by changes in larval duration, the "accelerated" larvae were estimated to develop 25–30% faster than their control/ancestor populations. The responses of ACB and ACO lines were remarkably parallel, despite being founded from populations evolved independently for more than 300 generations. On average, these "A" populations developed from egg to adult in less than eight days and produced fertile eggs less than 24 hours after emerging. Accelerated populations showed no change in larval feeding rate, but a reduction in pupation height, the latter being a trait relating to larval energetic expenditure in wandering prior to pupation. This experiment demonstrates the existence of a negative evolutionary correlation between preadult developmental time and viability, as accelerated populations experienced a severe cost in preadult survivorship. In the final assay generation, viability of accelerated treatments had declined by more than 10%, on average. A diallel cross demonstrated that the loss of viability in the ACO lines was not due to inbreeding depression. These results suggest the existence of a rapid development syndrome, in which the fitness benefits of fast development are balanced by fitness costs resulting from reduced preadult survivorship, marginal larval storage of metabolites, and reduced adult size.

Key words.—Development, developmental time, *Drosophila*, life-history evolution, selection, trade-offs, viability.

Received June 24, 1996. Accepted May 7, 1997.

Complex lifecycles, which typically involve dramatic shifts in morphology, foraging mode, and habitat as part of ontogeny, have evolved in a diverse range of taxa. At the level of metazoan phyla, complex lifecycles are the rule rather than the exception. The study of life-history evolution in organisms with such ontogenetic shifts presents the unusual challenge of connecting the dots between ecological and physiological processes occurring in separate stages, which may at first appear independent, into a coherent model of overall fitness. Holometabolous insects such as *Drosophila* are exemplary: the effectively aquatic larva is separated from the winged adult by a discrete transitional stage in puparium. The evolutionary-genetic trade-offs governing adult fitness components in this genus have been explored extensively using experimental techniques. In particular, a great deal is now known about the selective effects of early reproduction upon later reproduction and survival (e.g., Wattiaux 1968; Rose and Charlesworth 1981a,b; Luckinbill et al. 1984; Rose 1984; Service et al. 1985, 1988; Service 1987; Zwaan et al. 1991, 1995b; Partridge and Fowler 1992; Chippindale et. al. 1993; Roper et al. 1993; Leroi et al. 1994). But trade-offs governing fitness in the preadult stages have only recently received significant attention (Partridge and Fowler 1992; Chippindale et al. 1994, 1996; Zwaan et al. 1995a; Nunney 1996). The lack of information on genetic trade-offs within the juvenile stages belies their critical role in determining overall fitness.

Selection on the juvenile stages in a complex lifecycle can have multifarious effects on overall fitness. One axiom that immediately suggests itself is that, from the perspective of fitness in a later life stage, selection for survival will act at full force on earlier stages. The emphasis in the ecological literature has been upon the survival risks of juvenile foraging, where greater foraging time increases both growth rate and vulnerability to predation. From both theory and ecological observation, it is apparent that there are circumstances under which it is selectively favorable to accept a greater risk of mortality in favor of increased growth (Rowe and Ludwig 1991; Werner and Anholt 1993; Abrams 1996). But there may also be internal physiological risks associated with rapid development in the absence of predation. These risks would be manifest as lower levels of developmental stability or higher rates of juvenile mortality in faster developing genotypes. Populations selected for faster development may also display lower fitness due to reduced adult size and resource provisioning by the juvenile. These factors may, in turn, have impacts upon traits as diverse as environmental stress resistance, mating success, and fecundity.

In earlier work on laboratory-evolved populations of *D. melanogaster*, we presented evidence for a genetic trade-off between speed of growth and development on the one hand, and survival to adulthood on the other (Chippindale et al. 1994, 1996). These earlier investigations focused upon populations evolved for adult demographic and stress-resistance traits, where the evolution of developmental characters occurred despite identical rearing of selection treatments through to full adult maturity. But how general is the negative

[1] Corresponding author. Present address: Department of Biology, N. S. IV, University of California at Santa Cruz, Santa Cruz, California 95064; E-mail: adam@darwin.ucsc.edu.
[2] Present address: Committee on Evolutionary Biology, The University of Chicago, 1101 East 57th Street, Chicago, Illinois 60637; E-mail: jalipaz1@midway.uchicago.edu.
[3] Present address: Department of Biology, San Diego State University, 5500 Campanile Drive, San Diego, California 92182-4614; E-mail: wchen@sunstroke.sdsu.edu.

genetic correlation between developmental rate and viability in *D. melanogaster*? Given that the populations we have previously investigated were produced by laboratory natural selection on adult traits, and differ multifariously in adult life-history, we would like to confirm (or refute) the existence of the trade-off between viability and development time. Thus, having marshaled evidence from lines designed for the investigation of adult fitness trade-offs, we report on the evolution of 10 populations purpose-designed for the examination of trade-offs in the preadult period. These populations were directly selected for accelerated development and extremely early fertility.

Historically, it has not proven difficult to create lines divergent for developmental time through artificial selection. However, selection for such lines has generally involved the differential slowing of selection lines and controls (e.g., Sang and Clayton 1957; Bakker 1961, 1969; Clark et al. 1961). Attempts to select in both directions on development time have often led to asymmetrical responses and even cases in which fast-selected lines got slower during selection (Sang and Clayton 1957; Marien 1958; Clarke et al. 1961; Prout 1962; Tigerstadt 1969; Lints and Gruwez 1972; Marinkovic and Ayala 1986). Even Burnet et al. (1977) who succeeded in increasing larval feeding rate (sclerophapyngeal retraction rate) by nearly 50% at its third instar peak, did not record improvements in developmental rate. Falconer (1989; pp 213–215) recognized eight distinct causes for asymmetrical responses to selection. However, in the case of fitness characters, the list of suspects may be narrowed by a singular observation: In almost every case of an asymmetrical response for a fitness trait, it is easier to select down than up. Factors that act adirectionally, such as random genetic drift, genetic asymmetries, or genes with large effects are thus improbable explanations for the directional asymmetry of responses. Factors that depend upon alterations in the selection differential under selection due, for example, to differences in fertility are also extremely unlikely to account for this pattern, as fitter individuals will tend to have higher fertility. There is also little reason to suspect that selection protocols will tend to build in inadvertent or indirect selection in one direction more than the other. As Falconer (1989) observes, the main reasons we should *expect* directional asymmetry in the bidirectional responses to selection on fitness traits are: (1) if the character has been selected directionally in the past, then initial gene frequencies may be higher in the direction of higher fitness; and (2) fitness characters are generally susceptible to inbreeding depression.

Lack of genetic variation for faster development has been the general explanation for the limited success of artificial selection for faster development (e.g., Robertson 1963; Partridge and Fowler 1992). However, inbreeding and the accumulation of deleterious alleles in a population will tend to slow development, just as visible novel mutations do (e.g., Lindsley and Grell 1968; Tantawy and El-Helw 1970; Mukai and Yamazaki 1971). With traditional artificial selection designs, the minuscule population sizes employed are bound to lead to substantial levels of inbreeding. The population sizes used in traditional selection experiments on developmental time (e.g., Sang and Clayton 1957; Marien 1958; Clarke et al. 1961; Tigerstadt 1969; Lints and Gruwez 1972) are almost certain to engender inbreeding depression. With inbreeding, the experimenter is always fighting an uphill battle when selecting upward on components of fitness. And, even if a selection response is achieved, inbreeding depression will tend to give rise to broad-scale diminution of fitness and the observation of positive correlations among fitness components. While the effects of inbreeding may be controlled for if selected lines are compared to controls that experience matching levels of inbreeding depression, this procedure is rarely practiced and has little mitigating value where the concern is for the evolutionary correlations among traits within lineages.

The 10 populations that form the central focus of the present work were selected for faster development and very early fertility for 125 generations by using large populations ($n > 1000$) to avoid the confounding effects of inbreeding. A rapid and unwavering response to selection was observed. Unlike most earlier selection experiments on developmental time, we did not expend much effort evolving populations in the direction of slower development. Our design was intentionally asymmetrical because, given the preceding arguments, it is questionable what meaningful information is revealed by selection in the direction of lower fitness. In this paper, we examine the direct response of developmental time to selection for an accelerated lifecycle, and the correlated responses of a variety of preadult characters related to performance and survival.

Materials and Methods

The IV and Baseline Populations

All of the populations employed here descend from a single population of *D. melanogaster*, designated IV. The origin and subsequent maintenance of the IV population was described by Rose (1984). Briefly, the IV population was mass-sampled in 1975 from the wild population studied by Ives (e.g., Ives 1970). IV was subsequently maintained as an outbred population on two-week discrete generations under standard laboratory conditions. These conditions were: 24 hour light, 25°C temperature, high relative humidity, and low to moderate larval and adult densities with abundant banana/molasses food. Specifically, the IV population has been maintained at larval/adult densities of resulting from 50–150 eggs per 8-dram shell vials filled with 5 mL of medium. With 35 such vials being mixed together each generation, this population has had population sizes consistently in the thousands of breeding adults and is known to possess abundant genetic variation.

The B (baseline) selection treatment was derived from the IV population in 1980, as described by Rose (1984). This treatment, made up of five replicate populations (B_{1-5}), was maintained identically to IV in all respects. At the same time, five populations selected for postponed senescence, called O_{1-5}, were derived (Fig. 1). Despite selection for later life fertility, the early rearing period (two weeks, in vials) remained identical in the B and O selection treatments. At the larval densities employed during selection, a two-week rearing period is generally sufficient to allow the emergence of all normally developing adults. Occasionally, errors in the estimation of egg numbers at the beginning of a generation

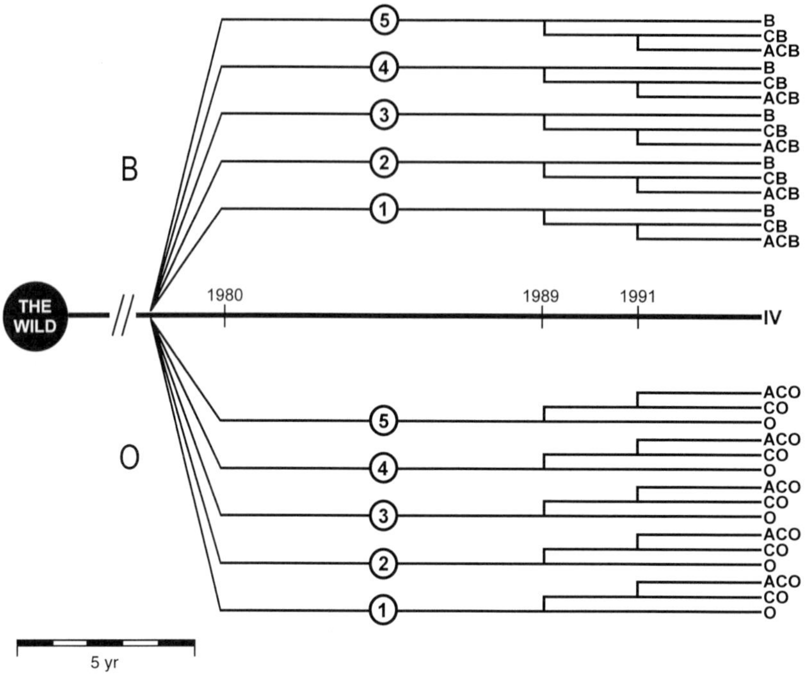

FIG. 1. The exact phylogeny (1975–1995) of major populations employed in the paper shows several important features of the experimental design: (1) relationships between B (and B-derived) and O (and O-derived) stocks of the same number have no special phylogenetic association because of the random assignment of replicate numbers in 1980; (2) accelerated (ACB and ACO) populations are most closely related to their same-numbered control/founder population (CB or CO); and (3) long periods of independent evolution separate the founding of each experimental population, both within and between selection treatments. B populations were used as a baseline because their handling was identical to IV, constituting 16–17 years (over 400 generations) of opportunity to approach evolutionary equilibrium for laboratory conditions. The B populations were used as controls for handling and general environmental effects, and to distinguish directional changes in the accelerated populations from directional changes in their controls.

led to crowding for a whole population, or for some replicate vials within a population. In crowded vials developmental time increases and it is likely that populations in the laboratory have experienced some weak and sporadic truncation selection. However, differences in density cannot explain developmental time differences between these selection treatments (Chippindale et al. 1994). Rather, the faster development of the B populations is believed to have resulted from continual selection for early maturation and mating, balanced by costs in preadult survival and size (Chippindale et al. 1994).

Ancestry of Accelerated Selection Treatments

On November 30, 1991, selection for accelerated development was initiated on derivatives of the CB and CO selection treatments, which are also control populations to other selected lines (see Rose et al. 1992). The 10 control/ancestor CB and CO populations were demographically selected on a generation time of four to five weeks beginning in 1989. These lines were given the standard two-week development period in vials before being placed in population cages with abundant banana/molasses medium for the remaining two to three weeks prior to oviposition of the next generation. The CB and CO selection treatments were chosen to found the accelerated populations for three main reasons. First, the CB and CO treatments had parallel recent histories, which involved selection for all of the general environmental and dietary conditions employed in selection for accelerated development. Second, these selection treatments had considerable phylogenetic separation, being derived from the B and O branches of the laboratory phylogeny more than 300 generations away from a common ancestor (Fig. 1). Third, the "C" treatments were expected to be stable evolutionarily because they are presumed to be under weak directional selection. Due to the generation time differences between accelerated and control populations in selection, it was impossible to maintain paired controls on the same schedule.

The developmental times and associated preadult characteristics of the CB and CO selection treatments relative to their ancestors were previously described by Chippindale et al. (1994, 1996). The derived selection treatments were designated ACB and ACO, where "A" stands for "accelerated." An accelerated population was derived from each of the $CB_{1\text{-}5}$ and $CO_{1\text{-}5}$ populations and the "A" population retained the same numbering.

Thus, a statistical pairing exists between same-numbered CB and ACB, or CO and ACO, populations (Fig. 1).

Selection

Selection for an accelerated lifecycle involved compression of all lifecycle stages. Development was selected upon directly, as only the fastest developing flies were transferred to population cages for breeding. Truncation selection was upon approximately the first 20% of the total eclosion distribution, estimated for each of 500 rearing vials individually. After pupae began to darken, heralding the onset of eclosion, the rearing vials were checked every three to eight hours. With each check, vials containing 15 (or more) emerged adults—irrespective of sex—were transferred to the appropriate cage and then discarded. Vials containing less than 15 adults were returned to the incubator. When all of the vials were gone, the food plates in all cages were supplemented with a live yeast and acetic acid paste to stimulate egg-laying. In total, approximately 1000 adults were put into each population cage, each generation. Eggs were collected as soon as possible, typically 24–36 hours after the last adults had been transferred to the cages. The total egg-to-egg generation time had improved from 12 days, in initial generations, to 10 days by the 50th generation, and eight days by the 100th generation of selection. This increase in speed was due to improvements in both development speed and early fertility, as both the time from egg to adult and from eclosion to egg-laying were reduced by selection.

Use of Controls

As primary controls, the CB and CO populations were presumed to be under weak selection and were not expected to show rapid change. However, for secondary control, the B selection treatment was run along with every experimental assay to provide a baseline for both accelerated and control populations. The B populations had been maintained for hundreds of generations on the same selection regimen as their ancestor IV population. The Bs were found to be at, or very close to, evolutionary equilibrium for developmental time and viability (Chippindale et al. 1994). The B treatment was therefore incorporated to verify that differentiation between the accelerated treatments and their ancestors resulted from increases in performance by the former, and not reductions by the latter. The B selection treatment also served as a control for environmental noise. By standardizing performance data for the other lines by the B treatment's performance, we were able to reduce such noise. Finally, prior to the present experiments, the B's were the fastest developing populations in the laboratory (Chippindale et al. 1994), thus establishing a performance benchmark for the other treatments.

Relaxed Selection on Emergence Time: The BG Treatment

An experiment was performed to examine the relationship between developmental time and viability in the B populations. As noted above, we have previously surmised that these populations experienced balancing selection on developmental time. To test this idea, an 11-generation selection experiment was performed in which selection for early emergence (one side of the balance) was relaxed on the B populations. Flies were allowed 18 days for emergence, rather than the customary two weeks, and then eggs were collected three days later (at 21 days from egg). This experiment was limited in duration because we were concerned with the genetic covariance of the preadult characters, not the long-term effects of drift and mutation accumulation. The BG selection treatment was experimentally assayed at generation 11 for developmental time and viability before being terminated.

Preparation of Selection Treatments for Experiments and General Handling

The selected lines and controls were assembled for the experiment by collecting 1000–2000 eggs from each of the 25 populations following routine selection. These samples were reared and held as adults in population cages to synchronize egg-laying for subsequent generations. Two full generations were then maintained without strong selection as follows: The flies were reared on medium from the same food batch using the standard two-week rearing period, then transferred to population cages for oviposition. The total generation time of these preexperimental generations was 2.5–3 weeks. Population sizes were 1200–2000 adults. Thus, a large unstressed sample of each population was established prior to experimentation, reducing parental and grand-parental non-genetic effects that may result from selection, and ensuring abundant eggs for the experiment.

Since several people participated in the set up of experiments and collection of data, efforts were made to distribute handling effects. For example, in the counting of eggs, each population sample was split between two counters, and counters were rotated across selection treatments. Similarly, selection treatments were physically and temporally interspersed in all parts of the experiments to reduce the potential for environmental noise to have directional effects (e.g., local temperature or light variation in an incubator affecting whole selection treatments differentially). Interspersion was carried out by grouping the populations by replicate number and handling these as blocks. Since each selection treatment consisted of five replicate populations, there were five blocks, each made up from the same-numbered representatives of each selection treatment. Blocks were randomly positioned and rotated at six-hour intervals to homogenize any differences resulting from position effects.

Development Time and Viability

Development time and viability were measured at generations 0, 5, 10, 20, 30, 36, 40, 45, 50, 75, 100, and 125 of "A" selection. In each experiment we measured the ACB, ACO, CB, and CO treatments, with the B treatment run in parallel as a baseline. Week-old adults from the second unselected generation were provided with fresh yeast paste ad libitum for three days prior to egg collection. The females were therefore at or close to their maximum daily fecundity (Chippindale et al. 1993), and were unlikely to be holding back advanced-stage embryos (Neyfakh and Hartl 1994). Eggs were collected by inserting petri dishes, partially filled with standard medium, into the population cages for four hours. At the end of this interval, the dishes were removed and labeled. Using fine brushes, exactly 60 eggs were counted into each of 10 vials filled with 5 ml (± 0.5 mL) of standard medium, except in generations 36 and 45, where five vials were used The vials were then incubated at 25°C until adults started emerging from the pupae. Checks on emerging adults were conducted at six-hour intervals, at 1:00, 7:00, 13:00, and 19:00. Flies were transferred at these hours to holding vials provisioned with food before counting.

In all experiments, viability was estimated in at least two ways: (1) egg to pupa; and (2) pupa to adult. (1) Pupae were counted by dotting them with a marker, and uneclosed adults were scored at the end of the experiment, after all viable adults had been counted. Earlier, pupae that had formed on the stopper were carefully removed and transplanted to the base of the vials, along with a piece of the stopper. (2) Measurements of survival to adulthood were drawn from the totals of adults counted from the development time estimates. After eight consecutive checks without emergence, the development time checks were terminated. However, for an additional week the vials were checked for extremely late adults on a daily basis. The vials were then left for another week before the tallying of pupae totals. In generations 36 and 100, egg hatchability was also measured. To measure hatchability (egg to larva viability), each person who counted eggs into the experimental vials for a given population also counted 100 eggs onto a petri dish filled with food-colored agar. The 50 petri dishes (two per population) prepared in this way were then incubated for 48 hours before being scored for unhatched eggs. The dishes were rechecked two days later. The unhatched egg total was used to compute hatchability for each population. In both the hatchability set-up and the larger development experiments, clear eggs were discarded because clear eggs are not fertile. However, the proportion of normal eggs that were infertile, as opposed to inviable, is unknown.

Stage-Specific Development Time

After 36 generations of selection, we measured both time from egg to pupa, and from pupa to adult. We assumed that little or none of the evolutionary response was due to modification of the egg duration based on earlier work (Chippindale et al. 1994) and based upon direct observation. Eggs were collected as in the standard development time assay (see above). Five vials for each population were then wrapped in clear plastic sleeves two days after egg collection. Pupae were distinguished from larvae according to the criteria of Ashburner (1989) for P_1 pupae: immobile, without visible larval segments, and with a smooth tanning of the pupal cuticle. At six-hour intervals, the plastic sleeves were scored with a different color of marking pen; one dot per pupal case. When all pupae had formed, the sleeves were removed from the vials and the colors were decoded into the times they represented. Adults emerging from this assay were sorted and sexed as in the standard assay (see above).

Larval Feeding Rate

Feeding rate was measured as the sclerophsryngeal retraction rate of the larvae approximately 96 hours after hatching from the egg, following the protocol of Mueller (1988). Eggs from the selection lines (at generation 50) and controls were collected after one hour of female oviposition onto non-nutritional agar plates. Twenty-four hours after egg-laying, hatchling larvae were brush-transferred to plates provisioned with a yeast suspension. Exactly 72 hours of feeding was allowed before feeding rate measurements were taken. This age was selected because the larvae reach their maximum feeding rate during the third instar, and sclerite retraction rates are known to stabilize somewhat at a plateau until pupation behavior is initiated (L. D. Mueller, pers. comm. 1993).

The assay of feeding rate was accomplished by counting the number of sclerite retractions in a one-minute period. Retraction rates were counted for 30 larvae of each replicate population by six different counters. For each group of counts, seven larvae were transferred to a fresh agar plate thinly coated with a relatively dilute yeast suspension for each round of counting. Larvae were then allowed three to five minutes to recover from transfer before counting began. Any larvae displaying obvious signs of damage or appetite loss were not counted. All counts were performed with a pairing of same-numbered replicates, with counters blind to selection treatment; populations were randomly assigned within each session. We encountered difficulty obtaining eggs from one population in the number five replicate, so data were collected for replicates 1–4 only.

Pupation Height

Pupation height is an energetic character that reflects larval wandering after the cessation of feeding. To estimate the time and energy expended in searching for a pupation site, the distance between the top of the food medium and the middle of each pupal case was measured. Experiments were conducted at generation 50 for the CB and ACB populations, and at generation 52 for the CO and ACO populations because only one set of tall vials was available. In each test, the B populations were run as a baseline. The 8-dram vials normally used for culture leave only a 3–5 cm gap between the food and the base of the sponge stopper. The available pupation space is probably insufficient to allow differences in prepupa migration distances (wandering) to be resolved by simple height measures. Vials with normal food volume and aperture size, but greater height (25 cm) were made to examine differences in pupation height. These vials were manufactured by cutting glass tubes, and were therefore open at both ends. To provide food for growth, we plugged the vials into petri dishes in cookie-cutter style. Five vials were measured for each population at standard density (60 eggs). As with the pupation time assay, clear plastic sleeves wrapped around the experimental vials were used to mark the locations of pupae and the food base. The sleeves were then unrolled and a computerized digitizing tablet (Summagraphics MacTablet) was used to record heights. Some difficulties were encountered in the first (generation 50) trial, as the seal between petri dish and vial was poor for some vials and drying of the food occurred. These vials were eliminated from the experiment, so actual sample sizes varied between two and five vials per population. In the second trial (at generation 52) this problem was reduced so that 68 of 75 vials were included for measurement.

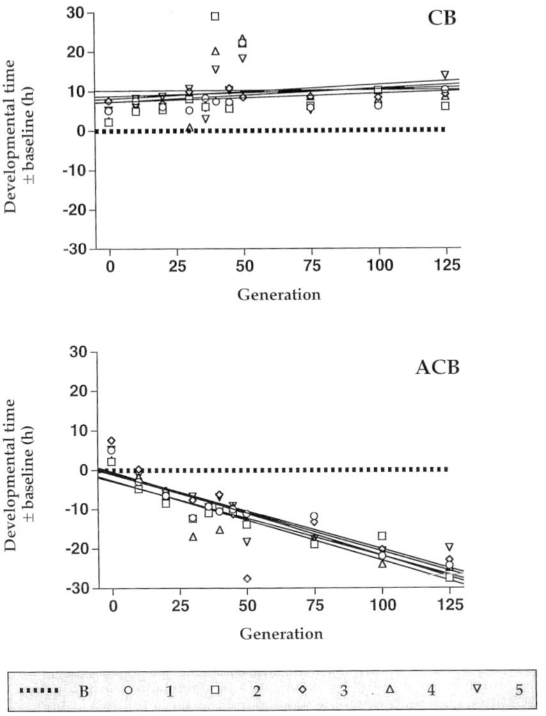

FIG. 2. The direct response to selection in the B-derived populations relative to the baseline. Individual populations are shown with least squares regression lines fitted to each. Each point represents the mean of the 600 individuals measured for each population. The control populations (CB; top) showed a slight trend toward increasing development time, gaining 0.02 hours per generation of accelerated selection, on average. The mean slope of the regressions on the accelerated (ACB) populations was −0.2 hours per generation, which was significantly negative ($P < 0.0001$). A total of 165,000 flies were measured for development time and viability in this experiment.

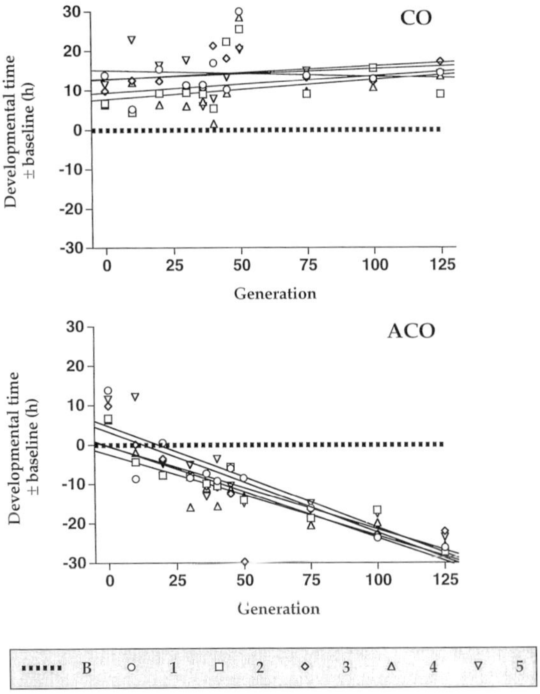

FIG. 3. The direct response to selection in the O-derived populations relative to the baseline. Individual populations are shown with least squares regression lines fitted to each. The control populations (CO; top) showed a slight but nonsignificant tendency toward increasing development time. The mean response of the accelerated (ACO) populations was −0.23 hours per generation which was significantly negative ($P < 0.0001$).

Crosses among Accelerated Populations

To test for inbreeding depression, we compared developmental time and viability of within-population-crossed ACO populations with F_1 hybrids among replicates. The experimental lines were drawn from generation 102 of accelerated selection. The full 5 × 5 matrix of reciprocal crosses was performed following the standard methods for development time and viability (see above). In such an experiment, the design is unbalanced because the true breeding group is outnumbered 4:1 by the crosses. Any handling error affecting a purebred population will therefore have a disproportionately greater effect upon the results of the experiment; at the limit, if a purebred population (one cell in the matrix) was entirely lost due to accident, nine of 25 cells in the design would be knocked out. For insurance, we chose to derive and handle two groups of each purebred (within cross) population. Thus, there were two $ACO_i \times ACO_i$ samples in the experiment. The two samples of each population were handled entirely independently. The CO treatment was also included in the experiment as a reference point. The full experimental design measured the development of 21,000 individuals.

RESULTS

Developmental Time

Figures 2 and 3 show the direct response to selection upon developmental time in the ACB and ACO selection lines, relative to controls (CB and CO). The B populations were used in each experimental assay as a baseline (zero line in Figs. 2, 3). The strategy behind this dual control system is described in the methods (see use of controls section).

The response to selection of the accelerated treatments has been remarkably steady, despite the fact that the selection intensity probably declined somewhat as the populations became more difficult to select (see methods, above). The accelerated selection treatments were significantly faster developing than their controls by generation 10, and significantly faster than the B selection treatment by generation 20 (ANOVAs not shown). As can be seen from Figures 2 and 3, the control populations were consistently slower devel-

TABLE 1. Slopes of linear regressions performed upon the long-term selection responses of accelerated populations and controls, relative to the B selection treatment. The t-values given are for a one-sample, two-tailed test against a hypothesized mean of zero. Significance of individual population regressions from the t-distribution with 9 or 10 df is given to the right of the slope value.

(a) Development time								
Replicate	CB		ACB		CO		ACO	
1	0.020	n.s.	−0.204	***	0.026	n.s.	−0.276	***
2	0.017	n.s.	−0.216	***	0.043	n.s.	−0.242	***
3	0.001	n.s.	−0.229	**	0.033	n.s.	−0.269	**
4	0.033	n.s.	−0.256	**	0.050	n.s.	−0.219	***
5	0.034	n.s.	−0.214	***	−0.015	n.s.	−0.203	*
Mean	0.021		−0.224		0.027		−0.242	
SE	0.006		0.009		0.011		0.014	
Treatment t-value	3.48	*	−24.52	***	2.41	†	−17.15	***
(b) Viability								
Replicate	CB		ACB		CO		ACO	
1	0.029%	n.s.	−0.125%	*	0.058%	†	−0.043%	n.s.
2	−0.008%	n.s.	−0.135%	**	0.031%	n.s.	−0.186%	*
3	−0.031%	n.s.	−0.133%	**	0.007%	n.s.	−0.054%	n.s.
4	−0.022%	n.s.	−0.185%	***	−0.040%	n.s.	−0.133%	**
5	0.003%	n.s.	−0.117%	**	−0.003%	n.s.	−0.092%	**
Mean	−0.006%		−0.139%		0.011%		−0.102%	
SE	0.010%		0.012%		0.016%		0.026%	
Treatment t-value	−0.57	n.s.	−11.71	***	0.65	n.s.	−3.84	*

† $0.05 < P < 0.1$; * $P < 0.05$; ** $P < 0.01$; *** $P < 0.001$.

oping than the baseline (B treatment) and appear to be increasing in development time slightly. Analysis of individual experiments revealed not only the rapid differentiation of the accelerated populations and controls, but also consistent differences between the sexes (M > F) and the significant effect of ancestry. The statistical significance of the last factor (ancestry) in every experiment resulted from the slower development of the CO treatment, and initially, its derivative ACO. As the ACO treatment caught up in development speed with the ACB treatment during selection, it is notable that the selection TS ancestry interaction became increasingly significant; this is believed to reflect the identity of the accelerated selection treatments later in the experiment and continued differences between CB and CO.

Table 1 gives the slopes for the selection responses of the accelerated populations and controls relative to the baseline (B). The accelerated treatments both showed significantly negative response slopes ($P < 0.0001$; two-tailed t-test with hypothesized mean = 0; 4 df). The rate of response for ACB was −0.20 hours per generation. The rate of response for ACO was −0.23 hours per generation. Surprisingly, both control treatments showed a slight positive trend: CB and CO increased in development time at an average rate of 0.02 hours per generation of accelerated selection. The slope was statistically significant for CB ($P < 0.05$) and marginally nonsignificant for CO (n.s., $P = 0.07$). It should be noted that the CB and CO controls experienced only one-quarter to one-third as many generations of selection as the ACBs and ACOs did because of the difference in generation time. The controls may therefore be increasing in development time at 0.06 to 0.08 hours per generation, on average. Such increases may reflect weak directional selection on developmental time.

The net response to selection after 125 generations was −33.4 hours (or 15%) for ACB and −38.6 hours (or 17%) for ACO (Fig. 2). The difference in net response resides entirely in preexisting differences in development time of the ancestral treatments (CB < CO). The CO-derived populations demonstrated a more rapid selection response in the first 30 generations. This difference in response may have resulted from stronger initial selection on the ACO selection treatment. This initial asymmetry in selection intensity was expected because both sets of lines were forced to lay eggs at the same age from egg, engendering stronger early-fertility selection on the ACO populations.

Developmental Time and Sexual Dimorphism in Accelerated Populations

At least initially, the intensity of selection was different upon males and females because of the difference in developmental time between sexes in the founder populations. Females develop slightly, but consistently, faster than males and therefore the sex ratio was skewed in the direction of more females per selection generation; this differential was approximately 1:2 (M:F). We expected selection to either alter the primary sex ratio in favor of males or reduce the difference between the sexes in developmental time. Neither of these expectations were met. Sex ratios of emerging adults from developmental time experiments never deviated significantly from 1:1 (analysis not shown). The difference in developmental time between sexes was highly conserved in all five selection treatments throughout the course of the experiment. The absolute difference between sexes actually expanded slightly, though nonsignificantly in any single generation, in the accelerated treatments relative to the other treatments. For example, at generation 125, the B males developed 2.2 ± 0.5 hours (1.0%) more slowly than the females, CB males developed 2.7 ± 0.6 hours (1.2%) more slowly than the females; for CO the M − F difference was +3.0 ±

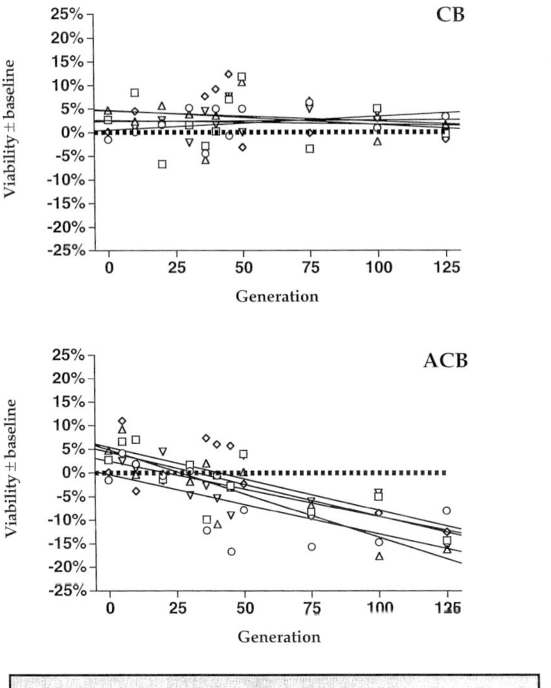

FIG. 4. The correlated response of egg-to-adult viability to selection on development rate in the B-derived populations relative to the baseline. Individual populations are shown with least squares regression lines fitted to each. The slower-developing control populations (CB; top) were generally more viable than the baseline (B) treatment. On average, the accelerated (ACB) populations declined by 0.13% per generation, which was significantly less than zero ($P < 0.0001$).

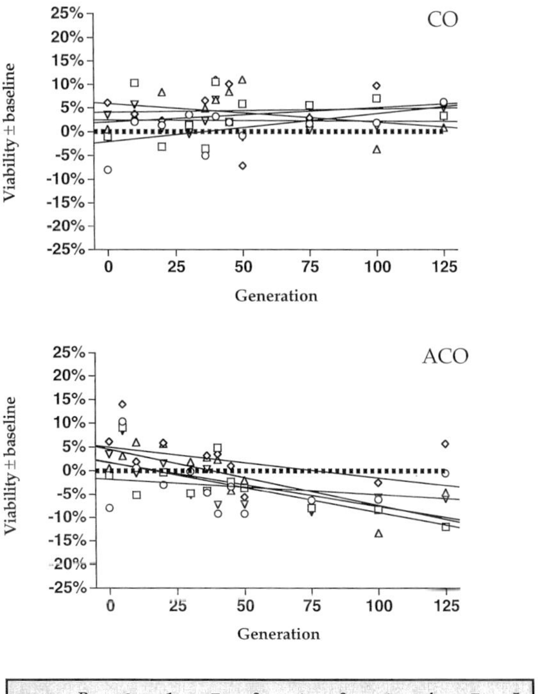

FIG. 5. The correlated response of egg-to-adult viability to selection on development rate in the O-derived populations relative to the baseline. Individual populations are shown with least squares regression lines fitted to each. The slower-developing control populations (CO; top) were generally more viable than the baseline (B) treatment. On average, the accelerated (ACO) populations declined by 0.06% per generation, which was significantly less than zero ($P < 0.05$). The responses of the ACO populations were more heterogeneous than their B-derived counterparts.

0.3 hours (1.3%); for ACB the M − F difference was + 4.5 ± 0.7 hours (2.4%); for ACO the M − F difference was +4.2 ± 0.6 (2.3%).

Viability of Accelerated Treatments

As Figures 4 and 5 show, there was substantial variability of individual populations throughout the experiment, even in the control lines. Greater levels of noise were apparent in generations 36 and 45, where smaller samples were used (300 flies per population instead of 600). Given the small differences in viability, the trait was, presumably, quite sensitive to experimental error. Despite this noise, a significant downward correlated response in viability was detected at generation 10 of selection, again at generation 30, and from generation 40 onward (individual analyses not shown). The linear regressions of viability over time (Figs. 4, 5; Table 1) give an estimated rate of decline of 0.139% per generation for the ACB selection treatment, significantly different from zero ($P < 0.0001$). For the ACO treatment, linear regression suggested a slower decline in viability (-0.102% per generation), which was significantly negative ($P < 0.05$).

The slope for the ACO treatment was affected by the anomolously high viabilities recorded late in the experiment for three populations (ACO_1 , ACO_3, and ACO_5; Fig. 5). We wanted to test whether these high values reflected a real diminution of the negative correlated response of viability at level of the selection treatment, or simply experimental noise in a subset of populations. The viability of the B selection treatment, used as a baseline, appeared normal in later generations and cannot be used to explain the change in ACO viability. We computed separate regressions for generations 0–50 (early generations) and 75–125 (late generations). For the ACB selection treatment, the slope for later generations was indistinguishable from the slope for earlier generations (early − late = 1.56 E-4; P = 0.84; two-tailed t-test; 4 df). For the ACO selection treatment there was a significant difference between early and late generations (early − late = −0.002; P = 0.03; two-tailed t-test; 4 df). The slope of the ACO regression on late generations was not significantly different

TABLE 2. Changes in the stage-specific timing of development after 36 generations of selection. Most of the total difference in developmental time observed in selection resulted from changes in the duration of the prepupal period, rather than the time spent in pupa. Values are the selection treatment means for the five replicate populations ± standard error.

Selection treatment	Egg to pupa (h)	Pupa to adult (h)	Egg to adult (h)
B	128.58	80.96	209.54
SE	0.61	0.28	0.81
CB	133.08	82.36	215.44
SE	0.39	0.32	0.23
ACB	120.20	79.31	199.51
SE	0.98	0.34	0.92
Difference (CB − ACB)	12.88	3.04	15.92
% change	9.68%	3.69%	7.39%
% of total change	80.89%	19.11%	n/a
CO	135.22	83.15	218.37
SE	0.57	0.35	0.46
ACO	119.66	79.51	199.17
SE	0.70	0.51	0.32
Difference (CO − ACO)	15.56	3.64	19.20
% change	11.51%	4.38%	8.79%
% of total change	81.03%	18.97%	n/a

from zero. It therefore appears that the initial downward response of viability in the ACO selection treatment reached a plateau late in the selection experiment. The flattening of the correlated response curve in the ACO selection treatment accounts for their more positive linear regression slope, compared to ACB over the entire experiment (as in Table 1).

Stage-Specific Developmental Time and Viability of Accelerated Treatments

Table 2 shows the stage-specific developmental time of the five selection treatments. Most of the overall developmental time difference between accelerated populations and controls (ACB = 80.9%; ACO = 81.0%) was accounted for by changes in the interval between egg-laying and pupation. Differences between the pupal duration of the 10 "A" populations and their controls accounted for the remaining 19% of the difference in overall developmental time. As previously described, viability of the selected lines declined steadily through the experiment. Over the entire experiment, 24.4% (CB) and 26.4% (CO) of the deaths occurring in the controls were pupal, while 29.3% (ACB) and 31.4% (ACO) of the mortality in the accelerated treatments was pupal. Analysis of treatment means over the whole selection experiment (paired t-test; 10 df) revealed that this difference was nonsignificant. Further inspection of the individual experimental assays supported the conclusion that, overall, the accelerated populations have declined in all stages of preadult viability approximately equally.

Developmental Time and Viability under Relaxed Selection: The BG Treatment

The BG treatment was created to test the hypothesis that the relatively low viability of the B treatment among demographic selection treatments was the result of sustained selection for rapid development because of their early-fertility selection treatment (Chippindale et al. 1994, 1996). When selection was relaxed by allowing 18 days for development (rather than 14 days in the Bs), the BG treatment rapidly increased in development time: In just 11 generations, the BGs increased by 13.6 hours, on average (Fig. 6). According to an ANOVA (factorial, with selection treatment and sex as factors), this result was highly significant (BG > B; $F_{1,16} = 50.75$; $P < 0.0001$). Sex was not a significant factor in the ANOVA and did not interact with selection treatment. In the same period of time, egg-to-adult viability increased by 3.7%, which was significant by one-tailed, paired t-test ($t_4 = -2.6$; $P = 0.03$). The major component of the viability response was found to be pupal survival, accounting for 3.0% of the observed increase in viability. Pupal viability was significantly higher in the BG treatment ($t_4 = -4.4$; $P = 0.005$).

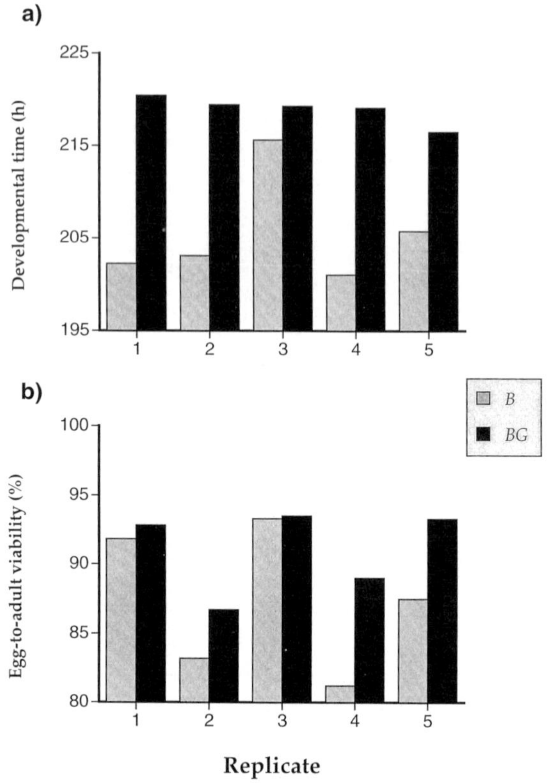

FIG. 6. Development time (a) and viability (b) of the BG selection treatment after 11 generations of selection. The BGs represent a relaxed selection treatment. Derived from the two-week (total generation) Bs, we increased the generation time to three weeks. This relaxation led to a rapid increase in development time and a correlated increase in viability, relative to the B ancestral treatment. The result is consistent with antagonistic pleiotropy between development rate and viability.

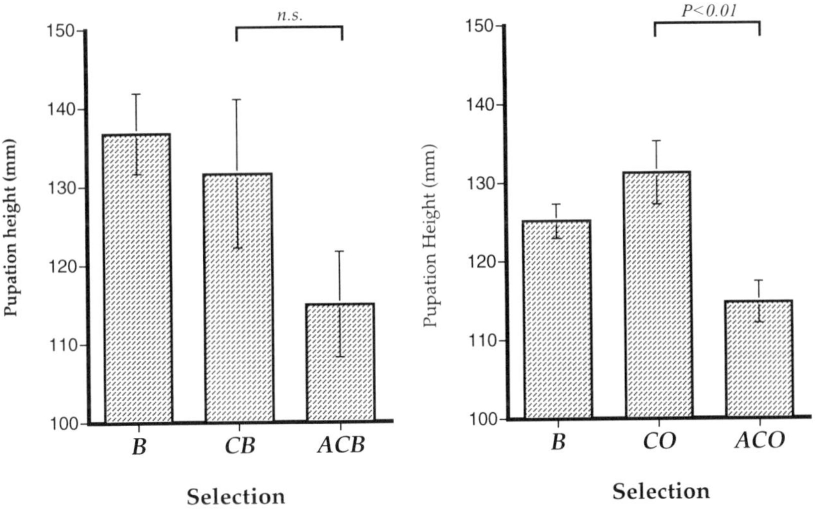

FIG. 7. Pupation heights of the accelerated selection treatments, controls, and baseline populations when reared in tall vials (maximum pupation height = 24 cm). The CB and ACB selection treatments were tested at generation 50 of "A" selection. The CO and ACO treatments were tested at generation 52 of "A" selection. In each case, the B treatment was run as a control. In general, the accelerated treatments showed reduced average pupation height relative to their controls, a result that was significant in the O-derived lines. Each bar is the mean of the five populations within a selection treatment (± standard error).

Larval Feeding Rate

Population means for feeding rate varied between 137.9 (B_1) and 162.1 (CO_3) sclerite retractions per minute. The selection treatment means were B = 145.0 (± 4.5), CB = 151.7 (± 3.5), CO = 147.3 (± 2.9), ACB = 147.1 (± 2.5), and ACO = 147.5 (± 2.5). Despite the substantial differences among selection treatments in larval development time, no significant differences were detected in feeding rate by a one-factor ANOVA on these data ($F_{4,15}$ = 0.53; P = 0.72). It appears, then, that there was no close connection between development rate and feeding rate, just as Burnet et al. (1977) found.

Pupation Height

The treatment means for pupation height are presented in Figure 7. The substantially smaller errors in the CO versus ACO comparison presumably reflect the greater sampling level and greater control over humidity in the second experiment. Surprisingly, the results for the control and accelerated populations were extremely uniform across experiments: CB and CO heights differed by less than 0.5 mm, and ACB and ACO by less than 0.25 mm (Fig. 7). The similarity in average height of the controls, on the one hand, and accelerated populations, on the other, was somewhat vexing because the Bs, examined in both experiments, differed by over 10 mm between experiments. If B is used as a baseline, the results look somewhat different: CB = −5.2 mm, ACB = −21.8 mm (generation 50), CO = +6.2 mm, ACO = −10.4 mm (generation 52). When standardized in this fashion no significant differences resulting from ancestry were apparent by unpaired, two-tailed, t-tests (CB vs. CO, P = 0.47; ACB vs. ACO, P = 0.07). Within experiments, ACB was not significantly different from CB (paired t-test; P = 0.09), but ACO was significantly lower in pupation height than CO (paired t-test; $P < 0.01$). Since differences in pupation height in this direction have been observed on a repeated basis for both ACB and ACO treatments relative to controls, it seems probable that the marginal significance of the CB versus ACB result was due to a lack of experimental power and not true equivalence of selection treatments.

Crosses among ACO Populations

Table 3 and Figure 8 shows the results from the diallel cross experiment on the ACO selection treatment. Of particular concern in this experiment was whether or not the F_1 hybrids would show hybrid vigor relative to the parental populations. Hybrid vigor would be suggested by significantly faster development or higher viability of the cross progeny than the better of the two parents (as in the overdominance tests below). We also looked for evidence of dominance, which would be suggested by a significant difference between the midparental and midcross values. The data were analysed by both paired t-test (parametric) and Wilcoxon signed-rank (nonparametric) test. The midhybrid mean viability 74.98% (± 1.26) did not differ significantly from the midparental mean of 72.59% (± 1.17) by t-test (9 df; t = −2.12; P = 0.09) The nonparametric test yielded a similar result (Z = −1.73; P = 0.08). There was no evidence of overdominance in either the parametric (9 df; t = −1.41; P = 0.19) or nonparametric (Z= −1.54; P = 0.12) tests for viability.

For development time, there was again a slight indication that the crosses performed better than the purebred popula-

TABLE 3. Results of the diallel cross between ACO populations. Part (a) shows the male/female mean values for developmental time. Part (b) shows the mean values for viability. The test of dominance compared the F_1 mean of the reciprocals to the midparental value, using a paired *t*-test (9 df) and the Wilcoxon signed rank test. The overdominance test compared the F_1 mean to the better parent, where "best" is determined in the direction of fitness, assuming that lower developmental time or higher viability are higher fitness.

Cross	Midparent mean (a)	Best parent (b)	Hybrid mean (c)	Dominance (c − a)	Overdominance (c − b)
(a) Developmental time (h)					
1&2	191.20	189.66	191.39	0.19	1.73
1&3	194.45	192.75	192.73	−1.72	−0.02
1&4	194.98	192.75	193.39	−1.58	0.65
1&5	192.77	192.75	188.24	−4.53	−4.51
2&3	192.91	189.66	192.20	−0.71	2.54
2&4	189.67	187.27	191.00	1.33	3.73
2&5	191.23	189.66	192.39	1.16	2.73
3&4	196.68	196.16	193.14	−3.55	−3.02
3&5	194.48	192.80	190.84	−3.64	−1.97
4&5	195.00	192.80	194.11	−0.90	1.30
Mean	193.34	191.62	191.94	−1.40	0.32
SE	0.68	0.80	0.53	0.64	0.85
$P_{(parametric)}$				0.06	0.72
$P_{(nonparametric)}$				0.08	0.74
(b) Egg-to-adult viability (percent survival)					
1&2	69.92%	74.17%	67.08%	−2.84%	−7.09%
1&3	75.00%	75.83%	75.84%	0.84%	0.01%
1&4	76.92%	79.67%	75.84%	−1.09%	−3.84%
1&5	69.87%	74.17%	72.92%	3.06%	−1.25%
2&3	70.75%	75.83%	76.59%	5.83%	0.75%
2&4	72.67%	79.67%	71.84%	−0.83%	−7.83%
2&5	65.62%	65.67%	72.92%	7.30%	7.25%
3&4	77.75%	79.67%	79.42%	1.67%	−0.25%
3&5	74.78%	84.00%	80.92%	6.14%	−3.08%
4&5	72.62%	79.67%	76.50%	3.89%	−3.17%
Mean	72.59%	76.84%	74.98%	2.40%	−1.85%
SE	1.17%	1.58%	1.26%	1.08%	1.36%
$P_{(parametric)}$				0.09	0.19
$P_{(nonparametric)}$				0.08	0.12

tions did: The average development time of the F_1s was 191.9 hours (± 0.5 h), whereas the average development time of the parental populations was 193.3 hours (± 0.07 h). The same battery of tests described for viability was performed upon development time. The dominance hypothesis produced a nearly significant result by *t*-test (9 df; $t = 2.18$; $P = 0.06$) as did the Wilcoxon signed-rank test ($Z = -1.78$; $P = 0.08$). There was no evidence for overdominance provided by either statistic ($P > 0.7$ in each case).

Thus, there was some evidence for directional dominance in each trait but no evidence of overdominance. In each case, the dominance was in the direction of greater fitness (Table 3), assuming that faster development or higher viability are higher fitness. These differences are extremely small, and none of the differences between hybrid and purebred populations proved statistically significant. The most significant result deriving from the cross experiment is that the reduction in viability of the ACO treatment cannot be explained by inbreeding depression.

DISCUSSION

Selecting for Faster Development

Development time is a fitness trait of considerable importance, particularly in species that typically mature in ephemeral habitats such as rotting fruit or necrotic cacti. The notion that *Drosophila* in the wild have undergone continuous directional selection for development speed has led some authors (e.g., Robertson 1963; Partridge and Fowler 1992) to suspect lack of additive genetic variance as an explanation for the failure of many artificial selection attempts at producing faster developing lines (e.g., Sang and Clayton 1957; Clarke et al. 1961; Burnet et al. 1977). The present results, along with the recent studies of Zwaan et al. (1995a) and Nunney (1996), demonstrate that developmental rate is not strictly maximized in this species. In other words, at least in certain laboratory strains, there is abundant, selectable, genetic variation for faster development.

Selection for faster development in the populations resulted in ACB and ACO (accelerated) populations that differed significantly from their controls after just 10 generations. The response continued, apparently unabated, until the last experimental assay at generation 125. There was no evidence of a plateau in the response, as often found in traditional artificial selection experiments (see, e.g., Falconer 1989). By the 125th generation the accelerated treatments developed 15–17% faster than their controls, completing egg to adult development in 7.6–7.7 days. As in our earlier study of postponed aging populations (Chippindale et al. 1994), this change in overall developmental time was principally

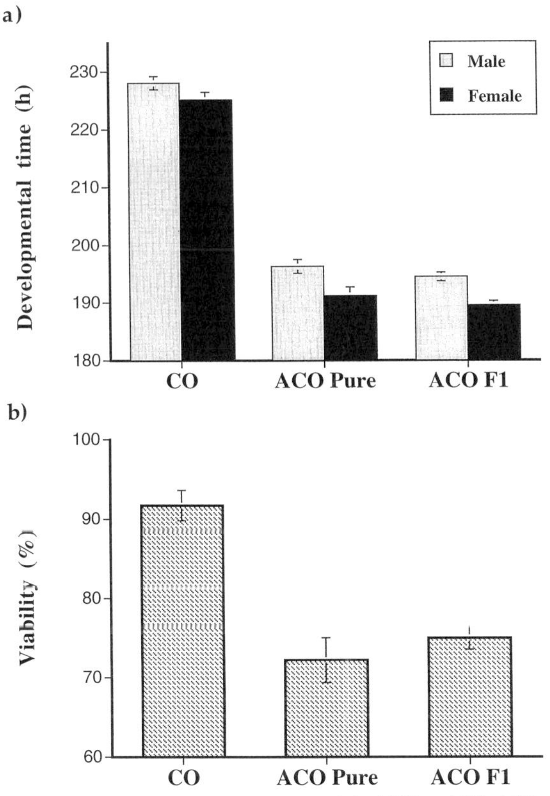

FIG. 8. Development time and egg-to-adult viability of CO, ACO, and F_1 hybrids of the ACO populations after 102 generations of selection. Crosses used a full diallel design to test for inbreeding depression of the ACOs as an explanation for their reduced viability. No evidence of inbreeding depression, dominance, or overdominance was obtained in this experiment, which involved 6000 individuals in the purebred treatment groups and 12,000 individuals in the hybrid groups. Error bars represent standard errors of population means.

due to a shortening of the larval stage. Although we did not assay hatching time of the accelerated populations, previous experiments (Chippindale et al. 1994) revealed no differences in hatching time between selection treatments differentiated for developmental time. Selection for embryogenic time also produces very small differences, on the order of an hour or less (Marinkovic and Ayala 1986; Neyfakh and Hartl 1994). It is therefore probable that most of the differentiation in time from egg-laying to pupa formation is the result of change in larval duration, rather than embryogenesis (egg duration). ACB and ACO populations pass through the larval stages approximately 25–30% faster than their controls. As might be expected, given their dramatically reduced opportunity for growth, the accelerated treatments are very small as adults. The multifarious effects of size differences upon adult life history are the subject of a sequel paper (Chippindale et al., unpubl. data). Here we will confine ourselves to the trade-offs that are internal to the juvenile stages. In particular, we discuss the costs of rapid development in terms of preadult mortality and pupation behavior.

Benchmarks for Performance

To our knowledge, the 10 accelerated (ACB and ACO) populations are the fastest developing and reproducing populations of *D. melanogaster* ever measured. On average, these "A" populations develop from egg to adult in less than eight days, and can produce fertile eggs less than 24 hours after emerging. The fastest individuals within a given population can go through a full (egg-to-egg) generation in less than eight days. Whether or not the accelerated populations break any absolute *Drosophila* speed records cannot be deduced with any certainty. The absolute performance level for any trait is bound to be tied to a perplexing array of environmental variables such as temperature, food quality, and density, as well as the degree of adaptedness of the population to the particular conditions used for measurement. While these factors limit comparison, we cautiously note that egg-to-adult developmental times for laboratory *D. melanogaster* populations at low density and 25°C typically fall into the 10–11 day range (e.g., Partridge and Fowler 1992; Zwaan et al. 1995a). Zwaan et al. were able to reduce development time to approximately 9.5 days after 16 generations of selection for faster development. Developmental times of wild populations reared in the laboratory at low density and 25°C are typically about nine days (204–223 h; mean = 212 h), based on data from 15 distinct European, Asian, and African populations (Louis et al. 1982; Draye et al. 1994; Bublii and Imasheva 1994). By these benchmarks, the B selection treatment, which had a mean developmental time of 216 hours (9 d) over the 11 experiments described in this paper, approximates the performance of natural populations of fruit flies measured in the lab, while the accelerated treatments are in a league of their own.

The Slower Development of Ancestral Populations

It would not be very interesting if we had simply evolved faster developing populations from mutation-ridden base stocks that had drifted toward slow development. Laboratory-adapted strains, maintained as well-nourished stocks with discrete generations before selection, have been employed in all three of the recent successful attempts at selection on developmental time (Zwaan et al. 1995a; Nunney 1996; present results). Under such conditions, selection for development time may have been relaxed in the base populations, leading to a relative acceleration through the elimination of alleles disrupting development when selection was reinstated. We know little about the opportunity for this factor to enter into the results of Zwaan et al. (1995a) or Nunney (1996), but a fraction of the response of the accelerated treatments considered here may be accounted for in this manner: Both founder selection treatments (CB and CO) exhibit slower development than the baseline (B) treatment. The extended lifecycle of these control/founder lines appears to foster a relaxation of selection for developmental time (Chippindale et al. 1994), and the long-term analysis presented here suggests that the control populations are slowly increasing in developmental time. The important question is: Did these founder populations lines evolve slower growth by drift or by selection?

Four lines of evidence suggest the action of selection rather than drift in the evolution of slower development in the ancestral treatments. First, CB and CO replicate populations showed considerable uniformity in their developmental times. Similarly all five populations in the BG selection treatment showed a rapid increase in developmental time when the generation time was slightly increased. The observation of a rapid, uniform response and the parallelism among independent replicates is suggestive of the action of direct selection rather than drift. Second, decreases in developmental rate were associated with gains in viability. Third, in previous work, crosses failed to show the restoration of faster development in hybrids of the O populations, which are slower developing than any of the selection treatments presently under consideration (Chippindale et al. 1994). Fourth, unlike traditional artificial selection protocols, the mass selection design used here was unlikely to engender inbreeding depression or drift. Population sizes were large (on the order of 10^3) and selection was allowed to act on the whole fertility phenotype. Thus, alleles with general fitness-depressing effects were constantly exposed to selection. In summary, the differences observed among the ancestral selection treatments probably resulted from inadvertent selection upon developmental time via changes in the relative intensity of selection for earlier emergence and balancing selection for other fitness components.

The Trade-Offs Surrounding Developmental Time

Faster development has not evolved without costs in other fitness-related characters. Most strikingly, the ACB and ACO populations have declined in excess of 10% in preadult viability. As discussed earlier, decrements in preadult survivorship have a 1:1 relationship with fitness. Whenever a key fitness character declines in the course of selection, one has to suspect inbreeding depression, which is an extreme case of genetic drift. In the evolution of the accelerated treatments, a significant role for inbreeding depression seems unlikely. Not only were population sizes very large, but the replicate populations generally displayed congruence in their selection responses. Moreover, after more than 100 generations of selection, F_1 hybrids among lines of the ACO treatment showed no improvement in viability, development time, or adult longevity.

An unusual feature of the correlated response to selection seen in egg-to-adult viability was the heterogeneity of the response after the 50th generation. While the ACB selection treatment showed a steady decline in viability over the entire experiment, the ACO treatment appeared to reach a plateau approximately halfway through the experiment. This significant change in the slope of viability in the ACO treatment explains the shallower linear regressions computed for the ACO populations relative to their ACB counterparts over the entire (125 generation) experiment (Table 1). Unfortunately, our experiments were not designed to probe the detailed responses of individual populations, having been conducted with approximate selection intensities and low sampling frequency. These factors, combined with the observed sensitivity of viability data to experimental noise, make us reluctant to interpret the heterogeneity among populations too strongly. However, the non-linear correlated response of ACO for viability is of interest for at least two reasons: First, it suggests a difference in the genetical structure of the two selection treatments, and, second, it suggests the decoupling of viability from the directly selected traits after extended selection.

The larval period in *Drosophila* normally includes a wandering phase at the end of the third instar. Feeding stops as the larva seeks out a suitable pupation site. Sokolowski and colleagues (e.g., Sokolowski and Bauer 1989; De Belle et al. 1993) have documented discrete variation in pupation behavior relating to crawling distance of the larvae. "Rovers" crawl much further than "sitters," and the determination of larval behavior appears to be under the control of a Mendelian genetic system. In the laboratory, pupation height of larvae above the medium in culture vessels provides an index of crawling distance. Mueller and Sweet (1986) and Mueller et al. (1993) demonstrated differences among laboratory lines, and significant effects of pupation height on survivorship, presumably due to humidity differences or other microenvironmental differences within bottles. It appears, then, that pupation height is a potentially sensitive index of time and energy spent by terminal-phase larvae in finding a suitable spot to pupate, and may have consequences for survival. The accelerated selection treatments had lower average pupation heights than their controls. Although the difference between the accelerated and the control treatment was highly significant in the comparison of ACO with CO, the ACB versus CB comparison was less convincing (P = 0.09). However, the experiment on the ACB and CB populations suffered from reduced power because of experimental problems (see methods section). In culture and experiment, the casual observation was frequently made that both sets of accelerated populations had lower pupation heights than their respective controls. We therefore believe that lower pupation height was a general by-product of selection for accelerated development.

Lower pupation height may help to account for the greater incidence of pupal mortality in the accelerated populations, however it is unlikely that within the narrow band of clear glass of the shell vials we used there will be much opportunity for humidity to vary substantially. We have noticed that when pupae form in contact with one another, the emergence of the imago is sometimes obstructed. But more probably, lower pupation is simply part of a syndrome involving marginal energetic stores available for crawling after the cessation of feeding. This idea is bolstered by the observation of lower pupation height in food-restricted larvae (Chippindale, unpubl. data) and the fact that the accelerated treatments displayed a shorter larval feeding period without an increase in feeding rate. We have also shown that starvation resistance of the newly emerged imago is significantly lower in accelerated populations than in controls (Chippindale et al., submitted) and that the dry weight, lipid and glycogen reserves of young adults are also reduced (Djawdan et al., unpubl. data). Thus, developmental time appears to be positively associated with growth and metabolite storage by larvae in these populations. The postfeeding wandering stage of the larva and the pupal stage together constitute a period of starvation in the development of a fly that lasts over 100 hours. It would be remarkable if the energetically expensive activities surrounding pupation and metamorphosis did not prove to be highly stressful. Thus, we suggest that the differences in pupation height and pupation survival we have observed indicate that selection for faster development has impinged upon larval storage of metabolites critical to successful pupation and metamorphosis.

Disagreements between Laboratories

In the experimental evolution system considered here—that derived from the IV population—a trade-off between developmental rate or growth rate and preadult viability has been supported by evidence from a wide range of selection treatments under different conditions (Chippindale et al. 1994, 1996; present results). We have shown that this trade-off manifests itself as changes in survivorship when development rate is directly selected up, or selection for faster development is relaxed. But other groups (e.g., Partridge and Fowler 1992; Zwaan et al. 1995a) found the reverse result: faster developing lines had higher egg-to-adult viability. Our concerns about the results of Partridge and Fowler (1992), who selected for postponed aging, were discussed in an earlier paper (Chippindale et al. 1994). Briefly, Partridge and Fowler found that their longer-lived OLD populations had both slower development and lower competitive ability than shorter-lived YOUNG populations. Roper et al. (1993) then showed evidence for both inbreeding depression and inadvertent direct selection on development rate operating in that selection system. With these confounding problems, untangling the relationships among preadult fitness characters would be extremely difficult.

The experiments of Zwaan et al. (1995a) pose a greater challenge to the generality of the trade-offs we propose. In the work of Zwaan et al., no differences in egg-to-adult viability were noted between the control and fast development selected lines. This result is not surprising given the sensitivity of viability estimates to experimental conditions demonstrated herein. We found that before generation 40 of selection, the "A" lines were not always significantly lower in viability than the controls were. Given the differences in the duration of the two experiments and replication levels (Zwaan et al. = 16 generations and four populations; present experiment = 125 generations and 20 populations) the failure of this group to detect the viability difference is not unexpected. But the slow selected lines in the experiment of Zwaan et al. evolved significantly *lower* viabilities over 16 generations. Some evidence that the slow lines might have been inbred during selection was provided by the authors, as the viability depression was still not recovered fully to control levels 12 generations after the cessation of selection, yet crosses among the slow lines produced comparable viability to the controls (i.e., hybrid vigor). One reason the slow lines in Zwaan et al.'s experiment experienced viability depression was due to their tendency to pupate in the medium, which is the opposite of the present results. It seems possible that the protocols employed by Zwaan et al. (1995a) selected for larvae that were simply sluggish owing to the selection of maladaptive genotypes. However, as the authors note, there was no evidence for inbreeding depression in other measured traits, such as body weight or longevity of adults, and the population sizes were the same in fast and slow treatments. Slow developing populations were also heavier, indicating that the slowing of development was not strictly a lower fitness phenomenon. We are therefore forced to consider the possibility that the Groningen 83 strain used by Zwaan et al. has a different genetic correlation structure for development time and viability than does the IV population and its derivatives. Given the dissimilarity between these two laboratories' results, it would be interesting to do parallel experiments on the IV and Groningen 83 populations under the same selective conditions.

Sexual Dimorphism in Developmental Rate

Males in *D. melanogaster* are typically slower developing than females, despite their substantially smaller size upon emergence. This sexual dimorphism in growth rate is among the mysteries of *Drosophila* biology. Presumably, for the multifarious advantages that larger size reportedly confers in sexual and natural selection there are counterbalancing fitness costs to being fast growing or big for males. In the present study, we selected accelerated populations by truncation selection upon adult emergence. As a result, we perturbed the sex ratio of the breeding populations in favor of females, thereby selecting more strongly upon males than females for developmental rate. However, over the 125 generations of selection, the accelerated populations showed no change in sexual dimorphism for developmental time; if any change was observed, it was in the direction of greater sexual dimorphism, not lesser. This absence of a significant change in the difference between the sexes suggests either that there was no sex-specific expression of heritable variation for developmental time or that there are strong canalizing mechanisms in growth maintaining the dimorphism. The stability of the relationship between the sexes for size and developmental time (i.e., the ratio of growth rates) defers to the more general evolutionary question: Why did the dimorphism of the sexes evolve in the first place? This question deserves detailed investigation.

A General Context

A refractory issue in laboratory studies, like this one, is the degree to which the findings may be applied to natural populations. All else being equal, we would have preferred to have done these experiments on natural populations in the wild. However, the alternatives to the laboratory-evolution approach to the study of adaptation each have drawbacks that can severely distort the interpretation of results (for a summary, see Rose et al. 1996). For example, studies of organisms recently introduced to the laboratory may suffer from confounding genotype-by-environment interactions associated with novel conditions imposed by the experimenter. In studies performed in the wild, it is impossible to control exact conditions or to replicate the design at the level of populations, let alone selection histories, with any certainty. These kinds of problems are particularly common and severe in the analysis of life-history traits, which are intimately bound up with overall performance and sensitive to the physiological condition of the organism. Our approach was to perform experiments with populations with realistic population structure, for which the selection history and phylogeny were known. These simplified, standardized laboratory conditions have been "the wild" for these populations for 15–20 years and hundreds of generations. For the most part, we observed parallel evolution in replicated populations derived from different selection histories, as well as from populations with similar selection histories that had been isolated for hundreds of generations (see also Chippindale et al. 1994, 1996). We are therefore confident that the general relationships among the major fitness traits are robust within this experimental system.

In the laboratory system we studied, there are substantial penalties associated with faster growth and development. This result is in accord with optimization theories designed around natural populations (e.g., Rowe and Ludwig 1991; Werner and Anholt 1993; Abrams et al. 1996). However, unlike those models, we demonstrate *intrinsic* costs to faster development, rather than costs mediated by extrinsic factors such as predation risk. Our experiments therefore help to extend the ecological models to situations where predation is not an operative factor. In holometabolous development the stakes are high: Not only do the usual absolute fitness costs of prereproductive mortality apply, but the adult experiences the broad-scale effects of differential larval success because it cannot grow to compensate for deficiencies. Moreover, the lifecycle of the fruit fly involves an abrupt transition from one environment to another, very different one. With this ontogenetic shift of environment, the conditions experienced by the adult may differ dramatically in quality to those experienced by the growing larva. Particularly when resources are ephemeral, the adult may be forced to disperse or resist a variety of environmental stresses the larva did not have to resist. For all of these reasons, larval growth is important to the entire age-specific fitness profile of *Drosophila*.

It is the cascade of negative consequences to later fitness resulting from relatively unsuccessful growth that leads to risk taking during growth. In this study and our previous work (Chippindale et al. 1994, 1996) we have probed these trade-offs in *Drosophila*. One general conclusion to be drawn from these studies is that trade-offs are not confined to single stages of a lifecycle, even when those stages are separated by radical morphological, behavioral, and ecological changes. We believe that this principle applies broadly to complex lifecycles, even if the specific characters involved do not.

Acknowledgments

This project would not have been possible without the dedicated hands of J. P. Mathot, L. Phan, and H. K. Brar, who helped maintain the stocks. The aforementioned, plus D. Coggeshall, K. J. Yee, Q. H. Nguyen, B. Lieu, S. Sen, F. Shazhadi, A. Ngo, and D. A. Templin assisted with particular experiments and counted many, many flies. L. D. Mueller showed us how to measure larval feeding rate, and he and T. J. Bradley both provided useful advice at various stages of the project. M. Ludwig kindly translated a reference from Russian. We thank an anonymous reviewer, and S. E. Via who reviewed and edited the paper in great detail, improving it substantially. Financial support came from US-PHS Grants AG06346 and AG09970 to MRR.

Literature Cited

Abrams, P. A., O. Leimar, S. Nylin, and C. Wiklund. 1996. The effect of flexible growth rates on optimal sizes and development times in a seasonal environment. Am. Nat. 147:381–395.

Ashburner, M. . 1989. *Drosophila*: a laboratory handbook. Cold Spring Harbor Laboratory Press, Cold Spring Harbor, NY.

Bakker, K. . 1961. An analysis of factors which determine success in competition for food among larvae of *Drosophila melanogaster*. Archs. Neerl. Zool. 14:200–281.

———. . 1969. Selection for rate of growth and its influence on competitive ability of larvae of *Drosophila melanogaster*. Neth. J. Zool. 19:541–569.

Bublii, O. A., and A. G. Imasheva. 1994. Genetic differentiation of three quantitative characters and natural populations of *Drosophila melanogaster* of Eastern Europe and Central Asia. Genetika 30:230–236.

Burnet, B., D. Sewell, and M. Bos. 1977. Genetic analysis of larval feeding behaviour in *Drosophila melanogaster*. II. Growth relations and competition between selected lines. Genet. Res. 30:149–161.

Chippindale, A. K., A. M. Leroi, S. B. Kim, and M. R. Rose. 1993. Phenotypic plasticity and selection in *Drosophila* life-history evolution. I. Nutrition and the cost of reproduction. J. Evol. Biol. 6:171–193.

Chippindale, A. K., D. T. Hoang, P. M. Service, and M. R. Rose. 1994. The evolution of development in *Drosophila melanogaster* selected for postponed senescence. Evolution 48:1880–1899.

Chippindale, A. K., T. J. F. Chu, and M. R. Rose. 1996. Complex trade-offs and the evolution of starvation resistance in *Drosophila*. Evolution 50:753–766.

Clarke, J. M., J. Maynard Smith, and K. C. Sondhi. 1961. Asymmetrical response to selection for rate of development in *Drosophila subobscura*. Genet. Res. 2:70–81.

De Belle, J. S., M. B. Sokolowski, and A. J. Hilliker. 1993. Genetic analysis of the foraging microregion of *Drosophila melanogaster*. Genome 36:94–101.

Draye, X., P. Bullens, and F. A. Lints. 1994. Geographic variations of lifehistory strategies in *Drosophila melanogaster* I. Analysis of wild-caught populations. Exp. Gerontol. 29:205–222.

Falconer, D. S. . 1989. Introduction to quantitative genetics. 3d ed. Longman Scientific and Technical, London.

Ives, P. T. . 1970. Further studies of the South Amherst population of *Drosophila melanogaster*. Evolution 38:507–518.

Leroi, A. M., A. K. Chippindale, and M. R. Rose. 1994. Long-term laboratory evolution of a genetic life-history trade-off in *Drosophila melanogaster*. 1. The role of genotype by environment interaction. Evolution 48:1244–1257.

Lindsley, D. L., and E. H. Grell. 1968. Genetic variations of *Drosophila melanogaster*. Carnegie Institute of Washington, Washington, DC.

Lints, F. A., and G. Gruwez. 1972. What determines the duration of development in *Drosophila melanogaster*? Mech. Age. Devel. 1:285–297.

Louis, J., J. David, J. Rouault, and P. Capy. 1982. Altitudinal variations of Afro-tropical *D. melanogaster* populations. *Drosophila* Inform. Serv. 58:100–101.

Luckinbill, L. S., M. J. Arking, M. J. Clare, W. C. Cirocco, and S. Buck. 1984. Selection for delayed senescence in *Drosophila melanogaster*. Evolution 38:996–1003.

Marien, D. . 1958. Selection for developmental rate of *Drosophila subobscura*. Genetics 43:3–15.

Marinkovic, D., and F. J. Ayala. . 1986. Selection for different rate of embryonic development in *Drosophila melanogaster* and *Drosophila simulans*. Genetika 18:205–219.

Mueller, L. D. . 1988. Density-dependent population growth and natural selection in food-limited environments: the *Drosophila* model. Am. Nat. 132:786–809.

MUELLER, L. D., AND V. F. SWEET. 1986. Density-dependent natural selection in *Drosophila*: evolution of pupation height. Evolution 40:1354–1356.

MUELLER, L. D., J. L. GRAVES, AND M. R. ROSE. 1993. Interactions between density-dependent and age-specific selection in *Drosophila melanogaster*. Funct. Ecol. 7:469–479.

MUKAI, T., AND T. YAMAZAKI. 1971. The genetic structure of natural populations of *Drosophila melanogaster*. X. Developmental time and viability. Genetics 69:385–398.

NEYFAKH, A. A., AND D. L. HARTL. 1994. Genetic control of the rate of embryonic development: selection for faster development at elevated temperatures. Evolution 47:1625–1631.

NUNNEY, L. . 1996. The response to selection for fast larval development in *Drosophila melanogaster* and its effect on adult weight: an example of a fitness trade-off. Evolution 50:1193–1204.

PARTRIDGE, L., AND K. FOWLER. 1992. Direct and correlated responses to selection on age at reproduction in *Drosophila melanogaster*. Evolution 46:76–91.

PROUT, T. . 1962. The effects of stabilizing selection on the time of development in *Drosophila melanogaster*. Genet. Res. 3:364–382.

ROBERTSON, F. W. . 1963. The ecological genetics of growth in *Drosophila melanogaster*. VI. The genetic correlation between the duration of the larval period and body size in relation to larval diet. Genet. Res. 4:74–92.

ROPER, C., P. PIGNATELLI, AND L. PARTRIDGE. 1993. Evolutionary effects of selection on age at reproduction in larval and adult *Drosophila melanogaster*. Evolution 47:445–455.

ROSE, M. R. . 1984. Laboratory evolution of postponed senescence in *Drosophila melanogaster*. Evolution 38:1004–1010.

ROSE, M. R., AND B. CHARLESWORTH. 1981a. Genetics of life-history in *Drosophila melanogaster*. I. Exploratory selection experiments. Genetics 97:173–186.

———. 1981b. Genetics of life-history in *Drosophila melanogaster*. II. Sib analysis of adult females. Genetics 97:187–196.

ROSE, M. R., L. N. VU, S. U. PARK, AND J. L. GRAVES. 1992. Selection on stress resistance increases longevity in *Drosophila melanogaster*. Expt. Gerontol. 27:241–250.

ROSE, M. R., T. J. NUSBAUM, AND A. K. CHIPPINDALE. 1996. Laboratory evolution: experimental wonderland and the cheshire cat syndrome. Pp. 221–241 *in* M. R. Rose and G. V. Lauder, eds. Adaptation. Academic Press, San Diego, CA.

ROWE, L., AND D. LUDWIG. 1991. Size and timing of metamorphosis in complex life cycles: time constraints and variation. Ecology 72:413–427.

SANG, J. H., AND G. A. CLAYTON. 1957. Selection for larval development time in *Drosophila*. J. Hered. 48:265–270.

SERVICE, P. M. . 1987. Physiological mechanisms of increased stress resistance in *Drosophila melanogaster* selected for postponed senescence. Physiol. Zool. 58:321–326.

SERVICE, P. M., AND M. R. ROSE. 1985. Genetic covariation among life-history components: the effect of novel environments. Evolution 39:943–944.

SERVICE, P. M., E. W. HUTCHINSON, M. D. MACINLEY, AND M. R. ROSE. 1985. Resistance to environmental stress in *Drosophila melanogaster* selected for postponed senescence. Physiol. Zool. 58:380—389.

SERVICE, P. M., E. W. HUTCHINSON, AND M. R. ROSE. 1988. Multiple genetic mechanisms for the evolution senescence in *Drosophila melanogaster*. Evolution 42:708–716.

SOKOLOWSKI, M. B., AND S. J. BAUER. 1989. Genetic analyses of pupation distance in *Drosophila melanogaster*. Heredity 62:177–184.

TANTAWY, A. O., AND M. R. EL-HELW. 1970. Studies on natural populations of *Drosophila*. IX. Some fitness components and their heritabilities in natural and mutant populations of *Drosophila melanogaster*. Genetics 64:79–91.

TIGERSTADT, P. M. A. . 1969. Experiments on selection for developmental rate in *Drosophila melanogaster*. Ann. Acad. Sci. Fenn. Ser. A IV Biol. 148:1–53.

WATTIAUX, J. M. . 1968. Cumulative parental effects in *Drosophila subobscura* Evolution 22:406–421.

WERNER, E. E., AND B. R. ANHOLT. 1993. Ecological consequences of the trade-off between growth and mortality rates mediated by foraging activity. Am. Nat. 142:242–272.

ZWAAN, B. J., R. BIJLSMA, AND R. F. HOEKSTRA. 1991. On the developmental theory of ageing. I. Starvation resistance and longevity in *Drosophila melanogaster* in relation to pre-adult breeding conditions. Heredity 66:29–39.

———. 1995a. Artificial selection for developmental time in *Drosophila melanogaster* in relation to the evolution of aging: direct and correlated responses. Evolution 49:635–648.

———. 1995b. Direct selection on life span in *Drosophila melanogaster*. Evolution 49:649–659.

Corresponding Editor: S. Via

Density-dependent natural selection in *Drosophila*: correlations between feeding rate, development time and viability

D. J. BORASH, H. TEOTÓNIO, M. R. ROSE & L. D. MUELLER

Department of Ecology and Evolutionary Biology, University of California, Irvine, California, USA

Keywords:

accelerated development;
ammonia;
density-dependent natural selection;
nitrogenous waste;
trade-off;
urea.

Abstract

We have previously hypothesized that density-dependent natural selection is responsible for a genetic polymorphism in crowded cultures of *Drosophila*. This genetic polymorphism entails two alternative phenotypes for dealing with crowded *Drosophila* larval cultures. The first phenotype is associated with rapid development, fast larval feeding rates but reduced absolute viability, especially in the presence of nitrogenous wastes like ammonia. The second phenotype has associated with it the opposite set of traits, slow development, slow feeding rates and higher viability. We suggested that these traits are associated due to genetic correlations and that an important selective agent in crowded larval cultures was high levels of ammonia. To test this hypothesis we have examined viability and larval feeding rates in populations kept at low larval densities but selected directly for (i) rapid egg-to-adult development, (ii) tolerance of ammonia in the larval environment and (iii) tolerance of urea in the larval environment. Consistent with our hypothesis we found that (i) larvae selected for rapid development exhibited increased feeding rates, and decreased viability in food laced with ammonia or urea relative to controls, and (ii) larvae selected to tolerate either ammonia or urea in their larval environment show reduced feeding rates but elevated survival in toxin-laced food relative to controls. It would appear that development time and larval feeding rate are important characters for larvae adapting to crowded cultures. The correlated fitness effects of these characters provide important insights into the nature of density-dependent natural selection.

Introduction

An important feature of early research on life history evolution, especially in relation to density-dependence, was the idea of trade-offs (MacArthur & Wilson, 1967). The early arguments were not genetic in nature but rather focused on the seemingly obvious point that individual organisms will ultimately have finite energy resources to divide among many competing needs (Cody, 1966). In our own work on density-dependent natural selection in *Drosophila* we have identified genetically based trade-offs that affect density-dependent rates of population growth (Mueller & Ayala, 1981; Mueller *et al.*, 1991) and efficiency of food utilization (Mueller, 1990; Joshi & Mueller, 1996).

Correspondence: L. D. Mueller, Department of Ecology and Evolutionary Biology, University of California, Irvine, Irvine, CA 92697–2525, USA. Tel./fax: +1 949 824 4744; e-mail: ldmuelle@uci.edu

More recently we found that there was a genetic polymorphism within populations that had adapted to crowded larval cultures (Borash *et al.*, 1998). Two phenotypic classes were identified that differed in the time they emerged from a crowded culture. The early emerging phenotype had a high larval feeding rate but low absolute viability especially in cultures in which the larval nitrogen waste product ammonia had been added. The later emerging phenotype had a slower feeding rate but higher absolute viability under several different conditions. These results were coupled with the observation that in crowded larval cultures the levels of ammonia increased exponentially during the course of the 20 days of larval growth.

These observations lead to the following hypothesis. The selection pressures in crowded *Drosophila* cultures change over the time it takes a larva to develop since the levels of food and levels of waste products are changing continuously and predictably over this period. Early in the life of the culture rapid development and high competitive ability are at a premium whereas late in the life of the culture the ability to tolerate waste products and low food levels are more important. The early phenotypic class has a high competitive ability due to their high feeding rate (Joshi & Mueller, 1988), but reduced efficiency of food utilization and tolerance of ammonia, presumably as a result of the genetic coupling of these traits with feeding rates. The late developing phenotype has the opposite set of traits that are most useful in the older, crowded *Drosophila* cultures.

If this hypothesis is true then the following predictions can be made concerning the evolution of these correlated sets of traits. Selection for rapid development should result in the appearance of the traits associated with the early phenotypes: high feeding rates and low viability especially in the presence of ammonia. Selection for the ability to tolerate ammonia in the larval environment should result in traits associated with the late phenotype: slow feeding rates but elevated viability in the presence of waste products. We test these predictions in this paper by studying *Drosophila* populations selected for rapid development and another set cultured for many generations in larval food with ammonia added. In addition we have tested a third set of populations that have been selected to tolerate urea in their larval food. These populations are interesting to address the question of whether the evolved response to larval ammonia is highly specific or whether it confers general resistance to other toxic nitrogenous compounds.

Methods and materials

Accelerated development populations

The derivation and maintenance of the accelerated development populations used in the present study are described by Chippindale *et al.* (1997). All of the populations are descended from a common ancestral population, the Ives population (Ives, 1970; Fig. 1). In 1980, replicate populations were derived from this ancestral population and subjected to two different demographic laboratory selection regimes as described in Rose (1984; Fig. 1). The B treatment (baseline) was maintained on the same discrete 2 week life cycle as the ancestral population. The O treatment (Old) was progressively selected for increased late fertility and longevity. Normally, the O populations reproduce at 8 weeks post-eclosion. Both B and O treatments are five-fold replicated. The B populations undergo their entire life-cycle in vials, including mating and egg laying, while the O populations are dumped into plexi-glass cages 2 weeks after egg collection and are maintained there until egg collection at 10 weeks. In 1989, intermediate selection generation treatments were derived from the B and O populations: the CB and CO populations. The CB populations were the direct descendents of the

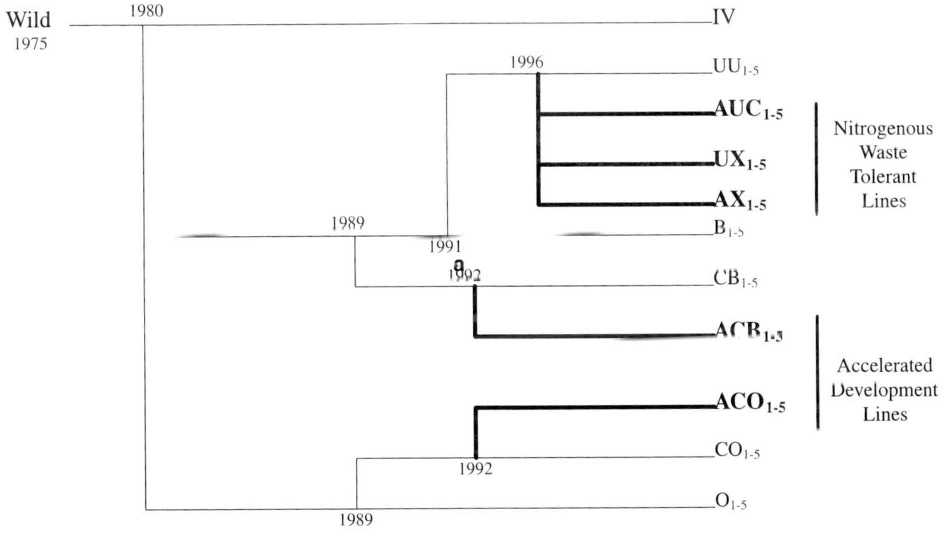

Fig. 1 The derivation of the accelerated development selection regime lines (ACB and ACO), their controls (CB and CO), and the nitrogenous waste tolerance selection regime lines (AUX, AX and UX). The graph is not chronologically to scale.

B populations, while the CO populations were derived from the O populations (see Rose *et al.*, 1992; Fig. 1). These 10 'C' populations spend the first 2 weeks in vials, reared in an identical manner to the B populations, after which the flies are dumped into plexi-glass cages and maintained until the eggs for the next generation are collected 2–3 weeks later.

In 1992, the accelerated development populations, ACB and ACO, were derived from the CB and CO populations, respectively (Fig. 1). Accelerated development selection was imposed on each of the derived populations by selecting for approximately the first 20% of the flies emerging from each of the uncrowded populations. The earliest emerging adults were dumped into plexi-glass cages and allowed to lay eggs for 24 h, after which eggs were collected for the next generation (for details see Chippindale *et al.*, 1997). By generation 100 of selection both ACB and ACO had generation times of 8 days. After approximately 110 generations, additional selection was suspended and all the populations were maintained on a fixed 9-day generation life-cycle. At the time of the present experiments approximately 210 generations had elapsed since the start of selection.

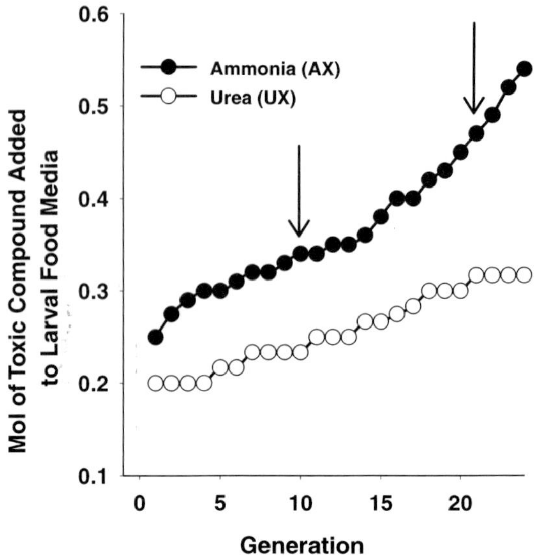

Fig. 2 The increase of supplemented levels of ammonium chloride (NH_4Cl) and urea added to the larval food medium of the AX and UX selection regimes, respectively. The arrows indicate the generations when assays of viability and feeding rates were performed.

Nitrogenous waste-tolerant populations

Fifteen nitrogenous waste-tolerant populations were derived from the five UU populations (Joshi & Mueller, 1996). The UU populations originally derived from the B populations had been maintained at low larval and adult densities, on a 3-week generation cycle, for over 100 generations, when these populations were created. The new populations created from the UU populations were named in accordance with the particular nitrogenous waste product they encounter as larvae. The AX selection regime line (Fig. 1) experiences ammonium chloride (pH about 5.5, equivalent to the food pH) as larvae. Each generation, for each of the five AX populations, 60 8-dram vials with an equal concentration of ammonium chloride (NH_4Cl) added to the plain banana-molasses food medium (Rose, 1984) were inoculated with approximately 60–80 eggs laid on non-nutritive agar. Plastic sleeves were inserted in the vials, so that pupae may be removed before eclosion, and hence only the larvae are exposed to ammonia. After >90% of the larvae had pupated, the sleeves were removed and placed into plexi-glass cages, with the plain banana-molasses food plate.

After a majority of adults had eclosed, yeasted food plates were inserted into the plexi-glass cages, to stimulate female oviposition, so that eggs may be collected and the next generation started. Initial levels of ammonia and urea were chosen to reduce viability but permit the collection of about 1000 breeding adults. It was expected that in the absence of ammonia or urea there would be around 3500 adults produced from all vials. While we did not precisely count eggs or adults we aimed to keep urea and ammonia levels high enough to result in only one-third of the eggs surviving. When it appeared that survival was significantly above this level the urea or ammonia levels were increased (Fig. 2). The UX selection regime was maintained in a similar fashion, with the exception that they received urea-supplemented food as larvae. The AUC regime served as a control, and did not receive any nitrogenous waste exposure as larvae. During the larval feeding phase, all populations were maintained in incubators under constant conditions (25 °C, 24 h light, and constant humidity). In natural environments *Drosophila* larvae would not be expected to encounter urea since ammonia is the natural nitrogen waste product (Borash *et al.*, 1998). In these experiments the final levels of ammonia in the AX populations were roughly 10 times greater than the levels reached in crowded *Drosophila* cultures (Borash *et al.*, 1998).

Pre-experimental conditioning

Before any experiments were performed, all populations were removed from selection for two full, 2-week generations and reared under identical larval and adult conditions, to remove environmental or maternal effects that may confound the inference of genetic differences between the populations. Assays were performed after 10 and 21 generations of selection, for the nitrogenous waste tolerance populations.

Experiments with the accelerated populations were performed separately.

Experimental assays

Viability

Eggs laid over a 4–6-h period from the CO/CB/ACO/ACB populations were placed in 8-dram vials at exact densities of 60 per vial. The following food treatments were used: plain food, plain food supplemented with 0.266 mol urea, or 0.350 mol NH_4Cl. Eight vials per food treatment were set up for each population. At the onset of eclosion, newly emerged adults were removed, using CO_2 anaesthesia, every 8 h, sexed and counted. Checks were stopped after 2 days had passed with no new adults emerging. The same process was repeated for the nitrogenous waste tolerance stocks.

Feeding rate

Eggs laid over a 3–4-h period were collected and placed onto a Petri-dish of non-nutritive agar, and allowed to hatch. After 24 h, larvae were collected and placed onto a new Petri-dish with a liberal amount of 37 g/100 mL live yeast-distilled water solution spread over the top of an agar base, so that all larvae would begin feeding at the same time. The larvae were allowed to feed on this medium for 48 h before measurements of sclerite retraction rates, a measure of larval feeding (Joshi & Mueller, 1988), were made. A single larvae was placed onto an agar plate with a thin layer of 10 g/100 mL live yeast-distilled water solution on top, and allowed to adjust to the environment for at least 30 s, which was a suitable length of time for acclimatization (Joshi & Mueller, 1996). Then the larvae were recorded for at least 1 min of feeding using a video camera (Hitachi CCD colour camera, model KP-C553) mounted on a dissecting microscope. The recording has a date and time camera titler connected to the VCR, so that the elapsed time of feeding can be followed.

Feeding rates were measured during a 1-week period, with the selection regimes broken up into blocks of populations bearing the same subscript being measured at the same time (i.e. AUCi, AXi, UXi or ACBi, ACOi, CBi, COi). See Borash *et al.* (1998) for more detailed descriptions of this process.

Our feeding rate protocol differs from previous techniques employed to measure this character (Joshi & Mueller, 1996; Chippindale *et al.*, 1997; Santos *et al.*, 1997). These earlier techniques involved a single person measuring one live feeding larva viewed through a microscope. This method does not allow individual feeding rate measures to be checked. Obtaining accurate feeding measures is particularly difficult for rapidly feeding larvae. Videotaping permits individual measurements to be recounted and slowed down if necessary. Videotape of each larva feeding was viewed by at least two people, and if separate measurements were off by more than 10% that larva was re-checked.

We would like to point out that the work of Chippindale *et al.* (1997) found no difference in feeding rates between the accelerated developmental populations (ACB and ACO) and their controls (CB and CO). Chippindale *et al.* (1997) used a large number of students to measure feeding rates of different larvae. Individuals with less experience are prone to giving biased results, where feeding rates of fast feeding larvae are underestimated. Because we videotaped the larvae, several students could repeatedly survey each larva and at speeds slower than real time if necessary. The increased accuracy our method affords may be sufficient to account for the different findings.

Statistical analysis

All statistical analysis used SAS for Windows 98. In the analysis of variance (ANOVA), selection regime was a fixed effect, and replicate was a random effect. For the viability assays, data were arcsin transformed before the ANOVA in order to normalize the data. Additionally, for the viability assays, food type was a fixed effect in our analysis. Differences between specific selection treatments were tested by Scheffe's test.

Results

Larval feeding rates in *Drosophila melanogaster* are highly correlated with competitive ability (Sewell *et al.*, 1975; Burnet *et al.*, 1977; Joshi & Mueller, 1988). The ANOVA indicated that populations which had undergone selection for accelerated development (ACB and ACO) had faster feeding rates relative to their controls (CB and CO). There were no differences between the accelerated development populations (ACB vs. ACO), nor were the control populations significantly different (CB vs. CO) in sclerite retractions measured per minute (Fig. 3).

Populations which had evolved with nitrogenous waste added to their larval food medium (AX and UX) displayed slower feeding rates than their controls (AUC) in both the generation 10 (not shown) and the generation 21 assays (Fig. 3). Scheffe's test to correct for multiple comparisons revealed that the mean feeding rates for the AX and UX were significantly lower than the AUC feeding rate. However, there was no detectable difference between the UX and AX populations (Fig. 3).

The ANOVA of the arcsin-transformed viability data revealed that both selection regime and the selection × food-type interaction were significant factors determining viability in the accelerated development populations and their controls. When reared on plain food, the CB, CO and ACO populations were significantly more viable than the ACB populations (Fig. 4A). When reared on food containing 0.350 mol ammonium chloride, all populations suffered a dramatic decline in viability. The CB and CO populations were significantly better at tolerating the damaging effects of ammonia in

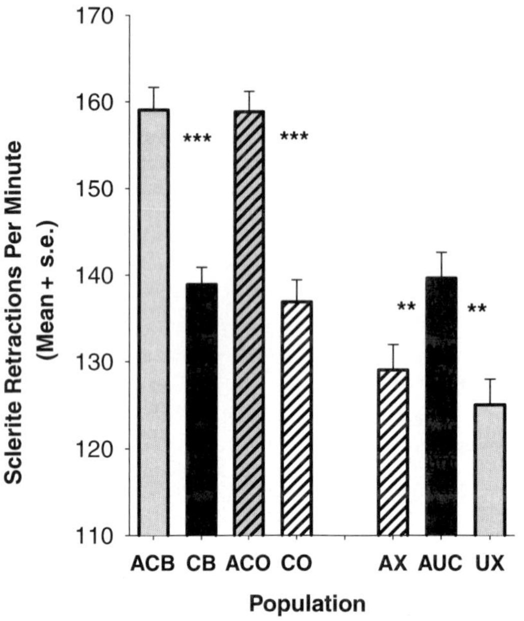

Fig. 3 Mean feeding rates. The mean feeding rate, measured as the number of sclerite retractions per minute, for the populations selected for accelerated development and the populations selected for ammonia and urea tolerance at generation 21. The error bars are standard errors around the mean of the five replicate populations of each selection regime. $^{**}P < 0.01$, $^{***}P < 0.005$.

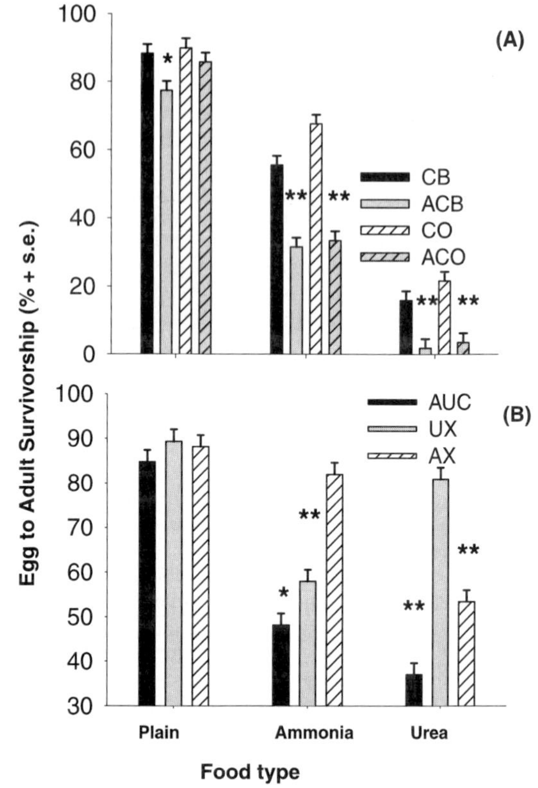

Fig. 4 Mean viability. (A) The mean viability of the populations selected for accelerated development and their controls, measured in three different food types: plain food, food supplemented with ammonium chloride, or food supplemented with urea. The error bars are standard errors around the mean of the five replicate populations of each selection regime. (B) The mean viability of the populations selected for ammonia and urea and their controls. The error bars are standard errors around the mean of the five replicate populations of each selection regime. $^{*}P < 0.05$, $^{**}P < 0.01$.

their larval food medium, compared to ACB and ACO populations. When reared on food containing 0.266 mol urea, all populations suffered an even more dramatic decline in viability, due to the toxic effects of urea. The CB and CO populations were significantly better at tolerating the damaging effects of urea in their larval food medium than were the ACB and ACO populations.

After 21 generations of selection for either urea or ammonia tolerance, there were no detectable viability differences among all populations when reared as larvae on plain food (Fig. 4B). When reared as larvae on food supplemented with 0.350 mol ammonium chloride, the AX populations displayed the greater viability, relative to the UX and AUC populations. The UX populations had a significantly greater viability than that of the AUC controls. When reared as larvae on food supplemented with 0.266 mol urea, the UX populations displayed the greatest viability relative to the AX and AUC populations. The AX populations had a significantly greater viability than that of the AUC controls. The generation 10 results were identical in terms of the direction of the differences shown above. The magnitudes of the differences were larger in the generation 21 assay.

Discussion

Borash *et al.* (1998) studied the conditions under which genetic polymorphisms may be maintained in a temporally varying environment in laboratory-adapted populations of *Drosophila melanogaster*. The theoretical work and experimental work of Borash *et al.* (1998) suggested that trade-offs in fitness components were a key to the maintenance of genetic polymorphisms. It was suggested that these trade-offs were reflected in a suite of genetic correlations between development time, feeding rate and tolerance of ammonia.

This study offers further evidence of causal genetic links between feeding rate, viability and sensitivity to nitrogenous waste. Populations which have been subjected to extreme demographic selection for early eclosion (ACB and ACO) demonstrate a reduced tolerance to both ammonia and urea. It had been previously found that the evolution of decreased developmental time is coupled with a reduction in viability on plain food (Chippindale *et al.*, 1997). In our experiments the differences in viability on plain food have disappeared in the ACO and CO populations and a small difference remains between the ACB and CB populations. When assayed in environments supplemented with nitrogenous compounds, however, the populations selected for rapid development show pronounced viability reductions relative to their controls. In ammonia-laced larval food the viability of the ACB populations is reduced to 57% of the CB level while the ACO viability is reduced to 49% of the CO.

There may be several explanations for the different observations of ACO and CO viability in this study and in Chippindale *et al.* (1997). (1) It might be that mutations with compensating effects on viability have arisen in the 85 generations since the Chippindale study (Lenski, 1988). However, in these relatively small populations (approximately 1000 adults) it is unlikely to have a repeated beneficial mutation event in five replicate populations in just 85 generations. (2) An alteration in the selection regime may have relaxed selection slightly, permitting viability to increase. Chppindale's study occurred during the active phase of selection in which the earliest developing adults were chosen for the next generation. That procedure was relaxed around generation 110 and adults were simple collected after 9 days compared with the peak of selection at 8 days. This sight relaxation of selection may have resulted in improved viability of the selected stocks since the relative fitness values of development time to viability had now been altered.

Populations that have adapted to high levels of nitrogenous waste show higher viabilities in the presence of these toxic compounds, while displaying lower feeding rates relative to that of their controls. Moreover, it appears that tolerance to nitrogenous substances was not specific, as populations which have evolved tolerance to one nitrogenous waste show some tolerance to a chemically distinct nitrogenous waste. There appears to be no cost associated with this response in environments not supplemented with nitrogenous substances, as the viabilities were not significantly different when the UX, AX and AUC populations were reared as larvae on plain food. This further indicates that genes conferring tolerance to one unique nitrogenous compound also provide a measurable degree of tolerance to a related, but chemically distinct, nitrogenous compound.

It appears that the ability to feed faster involves a reduction in the ability to tolerate nitrogenous waste. The mechanistic basis of this association is not completely understood. Since there is an increased influx of toxic food into a faster-feeding organism, it must have the ability to dispose of the nitrogenous substances or suffer the adverse effects created by their presence. In urea-selected populations, an increase in the excretion of urea may be the major mechanism responsible for tolerance (V. Pierce, personal communication). We have not ruled out that these populations have reduced the uptake of ingested nitrogen compounds through a decrease in gut permeability to that waste. These mechanisms, coupled or not, could have produced the observed patterns.

Acknowledgments

We would like to thank the countless number of students who have assisted in the general care and maintenance of the populations used in these assays, in particular S. Nguyen and A. Doan, whose tireless care of the nitrogenous waste tolerance stocks made much of this possible. We would like to thank A. Gibbs, V. Pierce, A. Chippindale, M. Shimada and two referees for their comments on the manuscript. H.T. is a PGDBM student supported by the Gulbenkian Foundation and Programa PRAXIS XXI (Portugal).

References

Borash, D.J., Gibbs, A.G., Joshi, A. & Mueller, L.D. 1998. A genetic polymorphism maintained by natural selection in a temporally varying environment. *Am. Nat.* **151**: 148–156.

Burnett, B., Sewell, D. & Bos, M. 1977. Genetic analysis of larval feeding behavior in *Drosophila melanogaster* II. *Gen. Res. Camb.* **30**: 139–161.

Chippindale, A.K., Alipaz, J.A., Chen, H.W. & Rose, M.R. 1997. Experimental evolution of accelerated development in *Drosophila*.1. Developmental speed and larval survival. *Evolution* **51**: 1536–1551.

Cody, M. 1966. A general theory of clutch size. *Evolution* **20**: 174–184.

Ives, P.T. 1970. Further studies of the South Amherst population of *Drosophila melanogaster*. *Evolution* **24**: 507–518.

Joshi, A. & Mueller, L.D. 1988. Evolution of higher feeding rate in *Drosophila* due to density dependent natural selection. *Evolution* **42**: 1090–1093.

Joshi, A. & Mueller, L.D. 1996. Density-dependent natural selection in *Drosophila*. Trade offs between larval food acquisition and food utilization. *Evol. Ecol.* **10**: 463–474.

Lenski, R.E. 1988. Experimental studies of pleiotropy and epistasis in *Escherichia coli*, II: Compensation for maladaptive effects associated with resistance to virus T4. *Evolution* **42**: 433–440.

MacArthur, R.H. & Wilson, E.O. 1967. *The Theory of Island Biogeography*. Princeton University Press, Princeton, NJ.

Mueller, L.D. 1990. Density-dependent natural selection does not increase efficiency. *Evol. Ecol.* **4**: 290–297.

Mueller, L.D. & Ayala, F.J. 1981. Trade-off between r-selection and K-selection in *Drosophila* populations. *Proc. Natl. Acad. Sci.* **78**: 1303–1305.

Mueller, L.D., Gonzalez-Candelas, F. & Sweet, V.F. 1991. Components of density-dependent population dynamics: models and tests with *Drosophila*. *Am. Nat.* **137**: 457–475.

Rose, M.R. 1984. Laboratory evolution of postponed senescence in *Drosophila melanogaster*. *Evolution* **38**: 1004–1010.
Rose, M.R., Vu, L.N., Park, S.U. & Graves, J.L. 1992. Selection for stress resistance increases longevity in *Drosophila melanogaster*. *Exp. Geront.* **27**: 241–250.
Santos, M., Borash, D.J., Joshi, A., Bounlutay, N. & Mueller, L.D. 1997. Density dependent natural selection in *Drosophila*: Evolution of growth rate and body size. *Evolution* **51**: 420–432.
Sewell, D., Burnet, B. & Connolly, K. 1975. Genetic analysis of larval feeding behavior in *Drosophila melanogaster*. *Genet. Res.* **24**: 163–173.

Received 10 August 1999; accepted 30 September 1999

Experimental Evolution of Accelerated Development in *Drosophila*. 2. Adult Fitness and the Fast Development Syndrome

Adam K. Chippindale[1], Julie A. Alipaz[2] and Michael R. Rose[3]

[1]*Department of Biology, Queen's University at Kingston, ON K7L 3N6, Canada, Tel: 613 533 6139*
chippind@biology.queensu.ca; also to whom correspondence should be sent

[2]*Stanford University, Hopkins Marine Station, Pacific Grove, CA 93950-3094*
jalipaz@stanford.edu

[3]*Department of Ecology and Evolutionary Biology, University of California, Irvine, CA 92697-2525, USA*
mrrose@uci.edu

Key words: Development rate, life-history, experimental evolution, trade-off, fecundity, body size, stress resistance, longevity, aging, growth, *Drosophila melanogaster*.

Abstract

Laboratory natural selection for rapid development and extremely early reproduction in *Drosophila melanogaster* was applied to ten independent populations from a known stock phylogeny. These populations, designated ACB and ACO, were contrasted with ten ancestor/control populations (CB and CO) and five long-established baseline (B) populations. After 100 generations of selection, these "accelerated" populations had evolved a total generation time of under eight days, compared with 12 days in their controls. Reduced pre-adult viability and pupation height were previously reported as costs of rapid development. Here we report that these populations also evolved substantially reduced body size (as measured by thorax length and dry weight), reduced early- and late-life female fecundity, reduced starvation resistance, and reduced longevity, while gaining only in the age of peak fecundity. These results suggest the strong and pervasive negative influence of selection for fast development and early fertility on fitness components expressed later in life. This syndrome of effects illuminates the direct connection of the pre-adult and adult stages through energetic trade-offs. We also discuss the potential role of antagonistic pleiotropy and mutation accumulation in the evolution of senescence in these populations, and the difficulty of dissecting the role of aging from the direct impact of miniaturization observed in the accelerated populations.

Introduction

Metamorphosis creates an ontogenetic bridge between two temporally separated niches. In holometabolous insects like *Drosophila,* this transition is between morphologies and lifestyles as distinct as between any two phyla: a burrowing semi-aquatic larva and a winged vagile adult. However, even with complete metamorphosis, trade-offs may connect life-stages. For example, a trade-off may occur between juvenile development rate and adult size if there is simultaneous selection for rapid development on the one hand, and body size, or larval stores to power the adult, on the other. This opposition of selective forces may be exacerbated when resources are limiting in space or time, as will frequently happen in crowded culture or in ephemeral habitats such as fruit rots.

In a classic study, Nicholson (1957) assayed genetic variation for autogeny in blowflies using an ingenious selection scheme. Autogenous insects are those able to produce viable eggs without a protein meal; conversely, mosquitoes and blowflies are typically anautogenous taxa. By excluding adults from the juvenile substrate and offering them only sucrose-agar plates, Nicholson was able to select blowflies for reproduction in the absence of protein in the adult diet. In his study, the carryover of reserves from larval feeding was responsible for the evolution of autogenous reproduction. Nicholson's work and later studies of *Lucilia* (Williams *et al.*, 1977) establish that selection influencing larval resource acquisition can originate in the adult stage. Similarly, *Drosophila* experimentally selected for adult stress resistance often adapt through adjustment of the juvenile life-history. In laboratory selection experiments, adult starvation and desiccation resistance are closely related to levels of metabolic reserves, particularly lipid and glycogen, accumulated prior to the stress event (e.g. Djawdan *et al.*, 1998; Harshman *et al.*, 1999). Selection for adult resistance leads to increased development time, growth rate, and increased larval storage of these metabolites (Chippindale *et al.*, 1996, 1998). In these experiments, reduced juvenile viability was one of the main costs to the evolution of increased resource acquisition. Such results underscore the comprehensive impact of selection on the life-cycle and the potential difficulty in seeking out trade-offs between single pairs of traits in isolation from the entire phenotype.

The genetic relationship between developmental time and body size in *Drosophila* has been intensively investigated using artificial selection. Adults in this genus are virtually amitotic, so larval size at pupation is likely to have a lasting impact on adult fitness. Unsurprisingly, it has proven straightforward to slow down development but difficult to move the trait in the direction of higher performance (faster development) with small artificially selected populations (e.g. Sang and Clayton, 1957; Marien, 1958; Clarke *et al.*, 1961; Prout, 1962; Tigerstadt, 1969; Lints and Gruwez, 1972; Burnet *et al.*, 1977; Marinkovic and Ayala, 1986). Selection for greater body size leads to increased development time, but selection for lesser body size does not necessarily decrease development time (Partridge and Fowler, 1993). These asymmetrical results tacitly suggest that the evolution of reduced vigor, perhaps due to inbreeding depression, often underpins the evolution of characters selected in the direction of lowered fitness, at least in small populations. In general, selection in the direction of reduced fitness, or with very small populations, is of limited value for the study of life-history evolution.

In contrast to earlier artificial selection experiments, most recent studies using laboratory natural selection and large populations (Partridge and Fowler, 1992; Chippindale *et al.*, 1994, 1996, 1997; Zwaan *et al.*, 1995; Nunney, 1996; Prasad *et al.*, 2000) have concluded that development rate and body size are negatively genetically correlated. The inference of a trade-off between faster development and adult fitness components is partly based on reports of various fitness advantages to greater body size in *Drosophila*, including male mating (e.g. Partridge and Farquar, 1983; Partridge et. al., 1987; Markow and Ricker, 1992), postcopulatory success (e.g. Bangham *et al.*, 2002), and female fecundity (e.g. Robertson, 1957; Mueller, 1985), to cite a non-exhaustive list of studies. However, a number of studies have failed to find adult fitness advantages accruing to greater body size in *Drosophila*, particularly in low to moderate density laboratory culture [Santos *et al.* 1994 (*D. melanogaster*), Zamudio *et al.* 1995 (*D. melanogaster*), Santos, 1996 (*D. buzzatti*); Joshi *et al.*, 1999; da Silva and Valente, 2001 (*D. willistoni*)]. These counter-results underscore the dubiousness of assuming that bigger is universally better. This in turn suggests that a number of growth and life-history strategies may evolve, depending upon the specific ecological circumstance encountered by a population (see discussion in Prasad *et al.*, 2001). Moreover, growth rate is highly labile evolutionarily in *D. melanogaster* (Chippindale *et al.*, 1994, 1996, 1997; Borash *et al.*, 2000; Prasad *et al.*, 2000), suggesting that any trade-off between final size and the speed of development may be complicated by changes in larval behavior or physiology. In the present study, we address this issue experimentally, asking whether strong, protracted selection for rapid development has a consistent impact on adult size and fitness.

Direct selection for faster development generally leads to reduced juvenile survival (Chippindale *et al.*, 1997; Prasad *et al.*, 2000, 2001), although this was not seen in the study of Zwaan *et al.* (1995). It is worth noting that the first two studies just mentioned were conducted using derivatives of the same original base population, while Zwaan *et al.* employed a different base population. However, the experiment by Zwaan *et al.* was also of much shorter duration (16 generations compared with 70+); significant reductions in viability were not seen by Chippindale *et al.* or Prasad *et al.* until substantially later in the selection process. Stage-specific mortality was predominantly larval in both studies, although the specific cause of death was not determined. Development is the coordination of growth with ontogeny, so it is possible that rapid evolution of development causes these processes to get out of step by, for example, executing an ontogenetic step with insufficient resources. Irrespective of the mechanism, increased juvenile mortality has unambiguous fitness consequences because the force of natural selection for survival is at its maximum before reproductive maturity (Hamilton, 1966).

Small differences in selection protocols may influence the kind and intensity of trade-offs between life stages. Two recent selection experiments in *Drosophila melanogaster*, initiated with descendants of the same original founding population, illustrate this point. The first was our own selection experiment (Chippindale *et al.*, 1997; the focus of the present paper as well) in which 10 populations were selected for rapid development and extremely early fertility (first available eggs). A sustained selection response was observed in all populations, with a mean net reduction in egg to adult development time of about 12% after 75 generations. In the second selection experiment, described in Prasad *et al.* (2000, 2001), collection of eggs was delayed until two and a half days after eclosion and adults were provided with abundant dietary yeast. Prasad *et al.* observed a 20% reduction in development time after 70 generations. Those authors suggested their stronger selection response might be due in part to less stringent selection for very early adult fertility. According to this hypothesis, less mature or poorly provisioned adults that emerge early could compensate by completing development or acquiring resources in the adult stage, allowing greater evolutionary lability in developmental time.

Here we extend the analysis of life-history evolution in populations selected for an accelerated lifecycle from juvenile to adult characters. In the course of selection for rapid development and very early fertility, total generation time (egg-to-adult-to-egg) was reduced from 12 days to less than 8 days (Chippindale *et al.*, 1997). Because these populations were strongly reverse-selected for age at reproduction, they allow us to test some predictions of the evolutionary theory of aging (Rose, 1991) as well as its genetic manifestation. The evolutionary theory of aging is based on the principle that the force of natural selection for survival declines after the onset of reproduction, predicting a decline in longevity for the accelerated populations over the course of selection. The tempo and mode of the response in later-life fitness and longevity may suggest something about the underlying genetics (e.g. the relative importance of antagonistic pleiotropy versus mutation accumulation in the evolution of aging). However, because of conjoint selection on development time and early fertility, we also address body size and resource storage effects on the adult life-history.

Materials and Methods

Selection Treatments and Controls

This study was conducted using a total of 25 *Drosophila melanogaster* populations. These populations and their controls are described fully in Chippindale *et al.* (1997; this volume). Briefly, the ten focal "accelerated" populations, selected for rapid development and early-fertility, were established in November 1991. These $ACB_{1\text{-}5}$ and $ACO_{1\text{-}5}$ populations were created with numbering matching their origin from each of the ten CB or CO control populations (described below). Accelerated selection was bipartite, consisting first of truncating the eclosion distribution so that only the first 20% (estimated) of emerging adults were transferred into population cages. Although density, temperature, and food quality were carefully regulated for larval growth, vials were individually censused and then transferred to adjust for microenvironmental variation. Approximately 1000 adults from a pool of 5000 juveniles growing up in vials were transferred to each cage. Second, once all adults had been moved to cages, eggs were collected as early as possible by adding yeast paste to the feeder petri dishes in the cages to stimulate oviposition. Eggs were always obtained within 36 h of the end of truncation selection. Selection therefore acted to compress all stages of maturation. The total (egg-to-adult-to-egg) generation time of the "A" treatments was reduced from 12 to less than 8 days over the first 100 generations of selection.

The ten ancestral / control CB and CO populations were maintained on a total generation time of 4–5 weeks. These controls were allowed a two week development period in vials before being placed in population cages with abundant banana/molasses medium for the 2–3 weeks prior to oviposition for the next generation. These populations experienced a relatively relaxed demographic selection regime, but shared all other aspects of basic culture with the accelerated populations: identical food, temperature, density, culture vessels;

and so on. An additional control treatment, designated B, was incorporated in most experiments. The five B (baseline) populations had been maintained by discrete generation vial culture for over 300 generations when accelerated selection was initiated, and were expected to be close to equilibrium for their selective regime. We therefore used the B selection treatment as a benchmark throughout the evolution and analysis of the accelerated treatments and their controls.

General Procedures

The growth medium employed throughout was a banana/yeast/barley malt/karo/agar/Tegosept medium that yields high juvenile survivorship. [The recipe can be obtained from the authors.] Populations were reared at regulated densities of 50–150 eggs per 25 × 95 mm vial, with approximately 5ml of food. Juveniles were reared at 25°C in 24 h light. After juvenile culture, adults were generally kept on the bench-top at ambient temperatures between 21°C and 24°C and were continuously lit. All experimental trait assays were conducted at 25°C in incubators.

Prior to each experiment, populations were placed under a standard selection regime with moderate density and relaxed selection on development time for two generations. This procedure helps to remove non-genetic factors resulting directly from selection that may influence the fitness of the measured subjects (e.g. maternal effects on egg quality) and synchronizes the populations prior to assay. In all the experimental assays, controls and selected treatments were interspersed and paired with respect to replicate number. For example, ACB_1, CB_1, B_1, CO_1, and ACO_1 were grouped in the same tray, with the second set of replicates in another tray, and so on. Where multiple rounds of measurement were made, selection treatments were again interspersed and the order of measurement reversed.

The general statistical model used in these experiments was a three-factor fixed effects ANOVA, with the factors being ancestry (B or O), selection treatment (accelerated or control), and sex (male or female). This model allowed us to examine the effects of the selection treatments in the context of differences in the prior selection histories of the control/ancestor populations: CB was derived from the early-fertility B selection treatment and CO from the late-fertility O selection treatment. Although the CB and CO populations had experienced convergent selection for almost three years (40–65 generations), they retained some differences from their ancestors (Chippindale *et al.*, 1996).

Body Size

Body size was measured by both thorax length (generations 20, 40, and 100 of accelerated selection) and dry body mass (generation 100 of accelerated selection). Thorax length was measured using a dissecting microscope (Wild M5 or Zeiss Stemi 2000c) and a digitizing tablet (MacTablet) and drawing tube connected to a Macintosh computer. Thorax length was taken as the anterior-posterior length of the dorsal surface of the thorax. For each population, 20 adults of each sex were each measured three times, in three rounds of measurement.

Dry weight was measured on adults flash frozen on dry ice within 6 h of eclosion. Because of development time differences, we collected adults from each population over a single 6 h interval representing the peak emergence period. This interval was directly measured via round-the-clock 6 h collections and counts of emerging adults. Adults were subsequently dried for at least 4 h in an oven at 60°C. Groups of ten same-sex flies were weighed on a Cahn C–33 microbalance to the nearest μg. Sixty adults of each sex were weighed.

Fecundity

Measurement of female fecundity was conducted for a 24 h egg production at generations 25 and 40 of the A-selection. In the generation 25 assay, both early- (immediately following eclosion) and late-life (at day 25 of adulthood) fecundity were measured for 5 successive days in egg-counting vials on a high yeast background. To make the yeast treatment, 100 μl of a suspension of 4 g yeast/ 20 ml distilled water/1 ml 1% acetic acid was pipetted into sucrose/yeast/agar medium tinted with charcoal to facilitate egg counts. One pair of flies was introduced into each vial at the beginning of the fecundity assay and these were transferred at 24 h intervals for five days. Fecundity of females typically peaks after 3–4 days of high yeast. Adults for the late-life fecundity assay were kept on banana medium without supplemental yeast (see above) for 25 days before being switched to egg-counting vials. This age treatment was chosen because it closely corresponds to the handling of CB and CO control populations, which lay eggs in middle age, and because relatively little mortality (i.e. demographic selection) was expected. For the early-fecundity assay, 76–80 pairs per population (1953 females in total) were measured; for the late-life assay, 60–69 pairs per population were measured (1658 females in total). The generation 40 assay only examined early-life fecundity for the first four days following adult emergence. 36–40 pairs were assayed (a total of 1904 females).

Starvation Resistance

Two experiments, at generations 20 and 40 of the A-selection, measured survival in humidified vials without food (see protocol in Service *et al.*, 1985). Starvation resistance was measured immediately following eclosion (3 ± 3 h) for adults from a 6 h collection at the peak of eclosion. The adults were immediately transferred to non-nutritional agar holding vials and allowed to mature before being sorted into the sealed and humidified test vials in groups of four same-sex flies. Ten vials (40 adults) of each sex were measured for each population; a total of 4000 adults were measured at each time point. The time to starvation death was measured by monitoring mortality at 1:00, 7:00, 13:00, and 19:00 daily until all the adults were declared dead on two successive checks. In each experiment, the adult starvation time was also determined after four days of adult feeding. In the generation 20 experiment, the adults were held in groups of four pairs on standard banana medium without any added live yeast slurry. In the generation 40 experiment, the high yeast treatment described in the fecundity methods (above) was employed.

Longevity

Longevity was measured at generations 40 and 100 of the ACB/ACO selection. In the first experiment, adults collected over a 12 h period were sorted into standard banana food vials with four pairs per vial and 20 vials per population (80 pairs/population; 4000 flies in total). Vials were examined daily for mortality and hand transfers without anesthesia were made three times per week to refresh the food. After exclusion of escapees and squashed flies, between 69 and 78 adults of each sex were successfully measured for each population (3689 in total). The experiment at generation 100 was identical except that only ten vials (40 pairs) were initiated for each population (1880 measured in total).

Results

Body Size

Thorax length declined steadily in the accelerated populations through selection, but consistent differentiation was not measurable until generation 100 (Fig. 1). Three-factor analysis of variance (ancestry × selection treatment × sex) detected no significant effect of ancestry (B or O) or selection (accelerated or control) after 20 generations and no interaction between selection and sex. Sex was highly statistically significant ($p < 0.001$), with the thorax of females being about 15% longer than males. Again at generation 40, only the sex factor was statistically

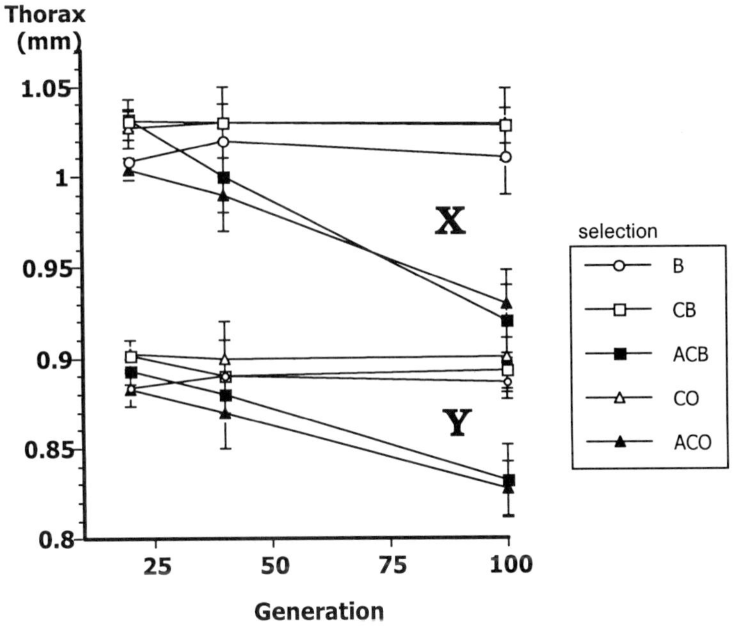

Fig. 1. The mean (± s.e.) thorax length of males (part Y) and females (part X) over the course of 100 generations of selection for accelerated development. The legend to the right gives the symbols for the respective selection treatments; each point is the grand mean of five population means.

significant by the $p < 0.05$ criterion ($F_{1,32} = 91.7$; $p < 0.0001$); however, the A-selected flies were nearly significantly smaller than their controls ($F_{1,32} = 3.9$; $p = 0.06$). Paired *t*-tests between accelerated treatments and their controls showed significant ($p < 0.05$) differences between the ACB and CB females and ACO and CO males only.

By generation 100, strong differentiation between controls and accelerated populations was apparent (Fig. 1). Females from accelerated populations had declined in thorax length by an average of 10.1% (ACB = –10.5%; ACO = –9.6%), and males by an average of 7.5% (ACB = –6.8%; ACO = –8.1%). These differences translated into high statistical significance. Three-factor ANOVA revealed the strong effect of selection treatment (A vs. C; $F_{1,32} = 26.0$; $p < 0.0001$), and sex ($F_{1,32} = 243.7$; $p < 0.0001$), and near significant interaction between the two factors ($F_{1,32} = 3.5$; $p = 0.07$). There was no indication that ancestry (B or O) had an effect on thorax size.

Dry weight measurements taken immediately post eclosion at generation 100 confirmed that the accelerated selection treatments had declined dramatically in size (Fig. 2). Accelerated females were 35.9% lighter than their controls on average, and A-selected males were 33.8% lighter than controls. Two-factor analysis of variance [selection × sex; no effect of ancestry (B or O)] confirmed the strong impact of accelerated development on size ($F_{1,32} = 544.0$; $p < 0.0001$) and the sex factor was also highly significant ($F_{1,32} = 172.2$; $p < 0.0001$). The highly significant interaction between sex and selection treatment ($F_{1,32} = 13.4$; $p < 0.001$) suggests a relatively greater negative response of females than males in body size.

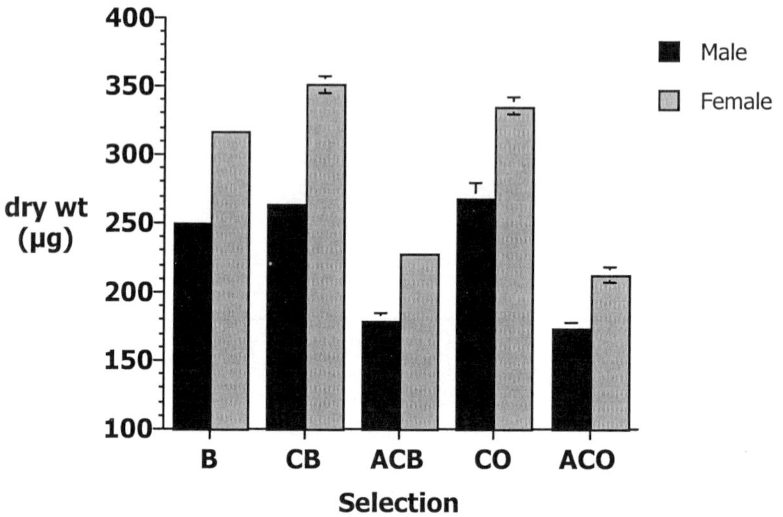

Fig. 2. Dry weight at eclosion (in micrograms; adult age = 3 ± 3 hours) of males (black bars) and females (grey bars) after 100 generations of selection for accelerated development. Error bars are standard errors based on the variance in means of the five replicate populations. Where not shown they are too small to be visualized.

Growth rate was estimated using the adult dry mass data divided by egg to adult development time at generation 100 (data from Chippindale, 1997). Overall, growth rate of A-selected populations was estimated at 1.00 ± 0.01 µg per hour of development and their ancestor/control populations had an average growth rate of 1.33±0.01 µg per hour. This 24.8% difference was highly statistically significant ($F_{1,32} = 308.5.0$; $p < 0.0001$). The ANOVA also indicated significant effects of sex (M<F; $F_{1,32} = 246.7$; $p < 0.0001$) and ancestry (O-derived < B-derived; $F_{1,32} = 8.4$; $p < 0.01$). The interaction between sex and selection factors was significant ($F_{1,32} = 8.9$; $p < 0.01$), reflecting the greater absolute reduction in growth rate of females from accelerated populations compared to males. However, this change was proportionally similar in both sexes (female = –25.7%, male = –23.8%). Accelerated selection reduced female growth rate to the equivalent of male growth rate in the controls; female controls were significantly faster growing and accelerated males significantly slower growing than this grouping (Tukey HSD post-hoc test).

Fecundity

The female fecundity data can be analyzed in several different ways as they were collected over successive days in each experiment. We have limited analysis to average fecundity (over all days) and peak fecundity (maximum 24h fecundity). In the generation 25 experimental assay, accelerated populations were found to have significantly lower average egg output, both early and late in life (Fig. 3). Early in life, ACB female average fecundity had declined by 17.1% and ACO fecundity had declined by 10.5%. Two-factor ANOVA [ancestry (B or O) and selection (A or C)] confirmed a significant effect of accelerated selection ($F_{1,32} = 6.0$; $p = 0.025$), with no effect of ancestry or interaction between these factors. Peak fecundity, however, was not significantly different between accelerated populations and controls ($F_{1,32}=2.6$; $p = 0.12$), nor was there any detectable effect of ancestry. Accelerated populations reached peak fecundity earlier and declined rapidly (on day 2 or 3 from the start of egg counts), while controls reached their peak on or later than day 3 of the assay and did not decline thereafter. Accelerated populations also began laying eggs earlier than any other selection treatment because of their faster development. Figure 3 is presented in terms of absolute age from the egg to underscore the earlier onset of reproductive maturity in these populations.

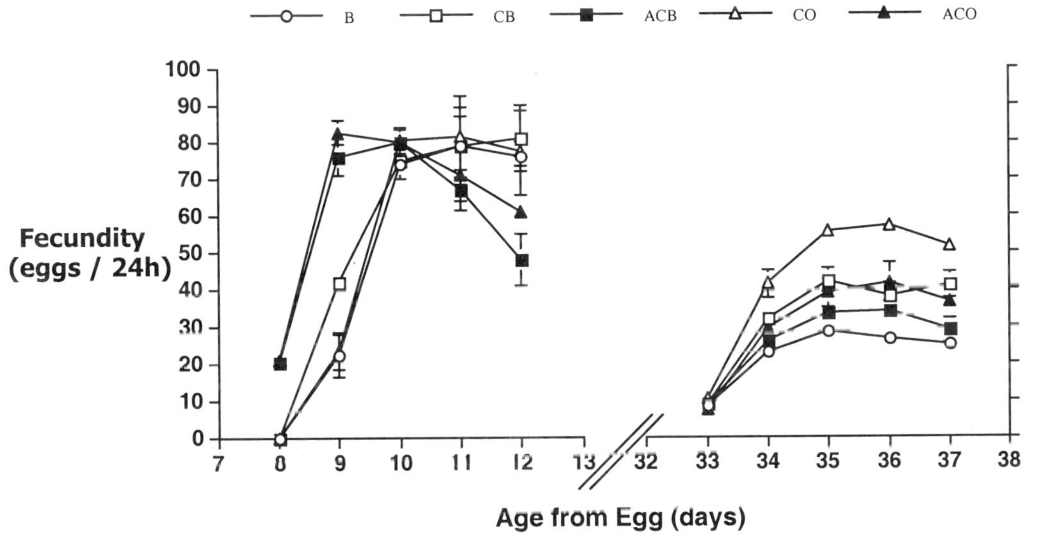

Fig. 3. Fecundity data from the 25th generation of accelerated selection. The abscissa gives the age (from egg) of the females laying in the experiment, beginning at 8 days for early fecundity and 33 days for late fecundity (25 days old as adults). Error bars are standard errors based on the variance in means of the five replicate populations. Where not shown they are too small to be visualized.

Late in life (days 25–30 from eclosion), ACB female average fecundity was 19.3% lower and ACO fecundity was 25.7% lower compared with their respective controls (Fig. 3). Two-factor ANOVA confirmed a significant effect of selection ($F_{1,32} = 15.6$; $p = 0.001$); a significant effect of ancestry (O > B; $F_{1,32} = 9.9$; $p = 0.006$); and no interaction between these factors. Peak fecundity was significantly different between accelerated populations and controls ($F_{1,32} = 16.1$; $p = 0.001$), and again there was a detectable effect of ancestry with the O-derived populations having higher late-life peak fecundity ($F_{1,32} = 11.9$; $p = 0.003$). Tukey's HSD test, applied post-hoc, showed that the ancestry effect was primarily driven by the CO selection treatment, which had significantly higher late-life peak and average fecundity than any other population ($p < 0.05$). The rank order of late fecundity was CO>CB>ACO>ACB>B for both measures of fecundity. The B selection treatment was significantly lower in fecundity (peak or average) than the control (CO or CB) populations but did not significantly differ from the accelerated populations (Tukey's HSD).

Average early fecundity of the accelerated females after 40 generations of selection was 18.1% lower than that of the control/ancestor treatment females (Fig. 4). The effect of selection was highly significant ($F_{1,32} = 24.5$; $p = 0.0001$), but the effect of ancestry was not ($F_{1,32} = 1.1$; $p = 0.3$). However, the interaction between these factors was significant ($F_{1,32} = 9.1$; $p < 0.01$) because of the much greater

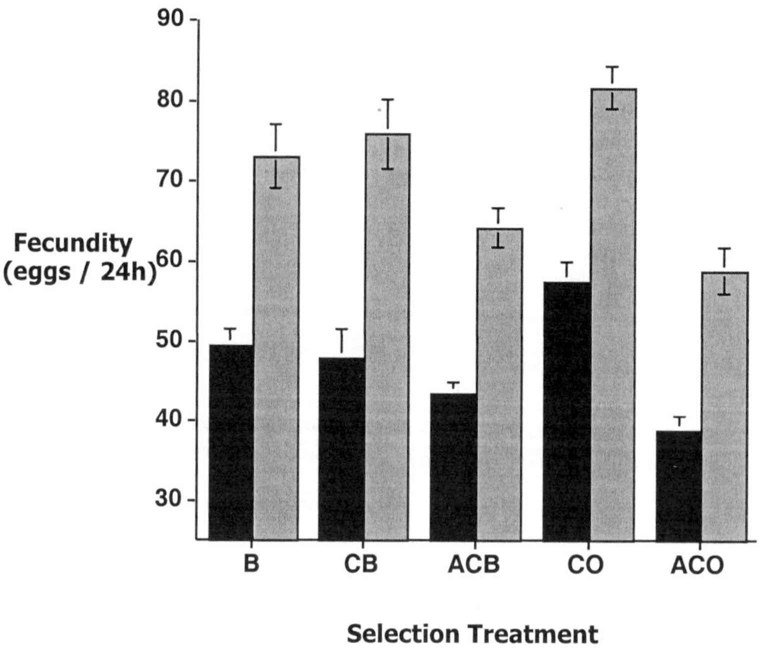

Fig. 4. Average (black bars) and peak (grey bars) 24 h fecundity of the five selection treatments from the 40th generation assay.

separation of the O-derived selection treatments than the B-derived treatments. Specifically, the CO populations had a significantly higher mean fecundity than any other selection treatment (Tukey's HSD, $p < 0.05$), and the ACO populations had the lowest fecundity; the difference between the CO and ACO was 18.6 eggs/day (ACO 33% lower) compared with 4.4 eggs/day difference (9.4% difference) between the CB and ACB females.

Starvation Resistance

Time to starvation death was measured on both sexes at two different ages and with two different adult nutrition treatments, generating a complicated set of main effects and interactions. Starvation resistance at eclosion, a measure of larval storage of metabolites, declined in response to selection for accelerated development (Fig. 5).

By the 20th generation of selection, starvation time of newly emerged males had dropped by 11.5% in the ACB treatment (from 60.8 ± 1.4 h to 53.8 ± 1.1 h), and 3.5% in the ACOs (from 56.8 ± 3.0 h to 54.8 ± 1.1 h), with the former being significant ($p < 0.01$ by paired, two-tailed *t*-test). In females, starvation time was reduced by 14.2% in the ACBs (from 61.9 ± 1.4 h to 53.1 ± 1.0 h), and 5.5% in the ACOs (from 60.1 ± 3.6 h to 56.8 ± 1.0 h). Again, only the comparison of CB and ACB was statistically significant ($p < 0.01$ by paired, two-tailed *t*-test). By the 40th generation, these differences had grown and all paired comparisons between accelerated treatments and controls were significant at $p < 0.01$; see Fig. 4. ACB males were 15.2% and females were 19.5% less starvation resistant than their ancestor/control CBs; ACO males/females were 14.8%/22% less resistant than their control/ancestors. Both CB and CO treatments had significantly higher starvation resistance at eclosion than the baseline (B) populations.

Mature adult starvation resistance (at day 4 from eclosion) was strongly influenced by the nutritional treatment (presence of live dietary yeast). The level of live yeast added to the medium has a positive and approximately linear effect on female fecundity over the range used in these experiments, and this has been shown to have an inverse effect on starvation time (e.g. Chippindale *et al.*, 1993). A three-factor analysis of variance (Selection, Age, and Sex as factors; results not shown) conducted on the generation 20 data revealed highly significant ($p < 0.0001$) effects of all main factors, as well as a strong interaction between age and sex ($p < 0.0001$). The interaction factor reflects the rapid decline in male starvation resistance with age in all selection treatments, while female starvation time increases slightly in the absence of live yeast. A three-factor analysis of variance (Selection, Age, and Sex as factors; results not shown) conducted on the generation 40 data revealed highly significant ($p < 0.0001$) effects of all main factors, significant interaction between age and sex ($p = 0.02$), selection and sex ($p = 0.03$), and selection and age ($p < 0.0001$). Under the yeasted conditions, both male and female starvation times fell with age, as predicted. The interaction between selection treatment and age was produced by the lower starting values of the accelerated treatments relative to controls (see above) but equivalent levels checked at day 4 (Fig. 5).

Separate analysis of the day 4 data (ANOVA; Ancestry, Selection Treatment, and Sex as factors) indicated that there were no significant effects of selection on the starvation time of mature adults in either assay generation. Sex (F > M) and ancestry (O-derived > B-derived) were both highly significant in both assays ($p < 0.0001$).

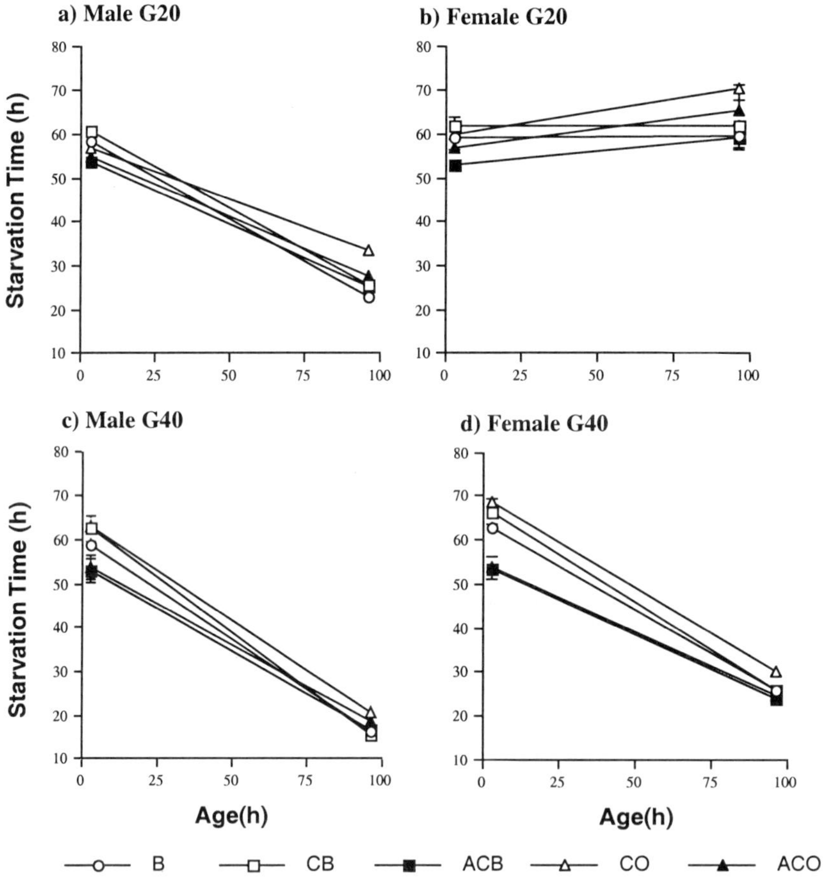

Fig. 5. Starvation time as a function of adult age. On the left side of each panel is the mean starvation time of each surveyed selection treatment at eclosion, and on the right is the starvation time after 4 days. The adult treatment was different at generations 20 and 40, with the latter experiencing higher levels of yeast and concomitantly reduced stress resistance. Error bars are standard errors based on the variance in means of the five replicate populations; where not shown they are too small to be visualized.

Longevity

Longevity of the accelerated populations had fallen significantly after 40 generations of selection (Fig. 6). From the three-factor ANOVA (factors: ancestry, selection, and sex), there was a strong effect of ancestry ($F_{1,32} = 28.1$; $p < 0.0001$) generated by the significantly higher longevity of the CO treatment compared with the CB treatment (CO–CB = 8.8 days/21%; intersexual average; Fisher PSLD, $p < 0.0001$). However, ACO had converged on ACB (ACO–ACB = 3.8d/9.5%; Fisher PSLD, $p = 0.03$), generating a significant interaction between ancestry and selection treatment ($F_{1,32} = 4.2$; $p = 0.047$). Averaged over the two sexes, the ACO populations were 7.7 days, or 15% shorter-lived than their controls; the ACB populations were 2.8 days, or 6.5% shorter-lived than their controls. The effect of selection treatment was significant ($F_{1,32} = 19.4$; $p = 0.61$) and the effect of sex was not ($F_{1,32} = 0.3$; $p = 0.047$).

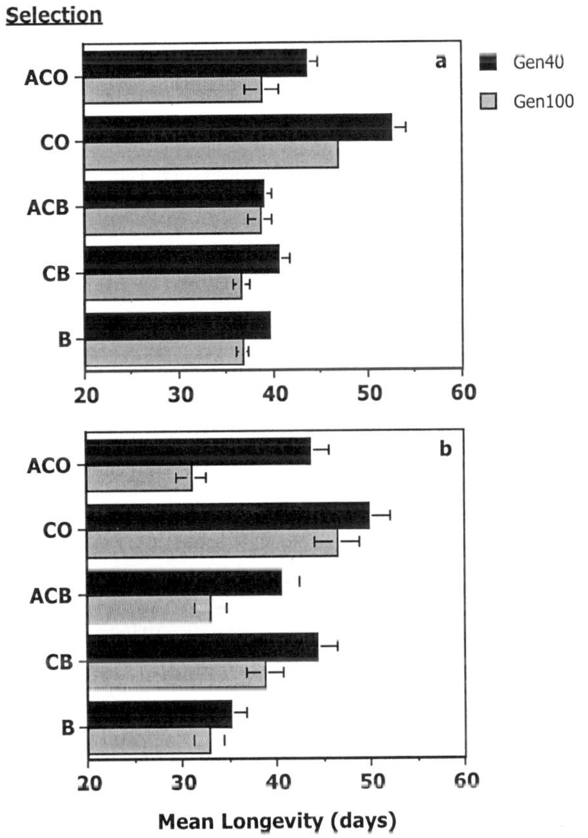

Fig. 6. Longevity (mean time of death) of the 5 treatments/25 populations at 2 points in the selection process. Part (a) gives male longevity and part (b) gives female longevity. Error bars are standard errors based on the variance in means of the five replicate populations; where not shown they are too small to be visualized.

Average longevity of the ACO selection treatment continued to drop relative to their controls between generations 40 and 100 (CO–ACO = 11.5d / 24.7%), but the ACB selection treatment did not change (CB–ACB = 1.9d/5.0%). Overall, these differences produced a highly significant effect of selection in the analysis ($F_{1,32} = 30.55$; $p < 0.0001$). Ancestry was again a significant factor, with O-derived populations longer-lived than B-derived populations ($F_{1,32} = 165.1$; $p < 0.0001$), and sex was also a significant factor (M > F; $F_{1,32} = 5.5$; $p = 0.02$). However, the average longevity of ACB and ACO selection treatments appears to have converged on the same absolute value of approximately 35 days, equivalent to the longevity of the early-fertility selected B selection treatment (mean longevity of 34.8 ± 1.0 days). Of the interaction terms, both the selection by sex interaction ($F_{1,32} = 9.5$; $p = 0.004$) and ancestry by selection ($F_{1,32} = 16.1$; $p = 0.0003$) were significant. The first of these interactions reflected the stronger longevity-reducing impact of accelerated selection on females compared with males. Females from the accelerated treatments were 24.8% shorter-lived than their controls, on average, while males were only 7.2% shorter-lived. This asymmetry was pronounced in both treatments: The ACO females were 33.3% and males were 17.0% shorter-lived than their controls; ACB females were 14.7% shorter-, while males were 5.0% longer-lived than the CB control (difference n.s. by Tukey's HSD). The ancestry by selection interaction reflects the greater decline in longevity of the ACO selection treatment, described above.

Discussion

The pursuit of trade-offs in life-history has employed both comparative and manipulative approaches, but the ultimate test of a trade-off is the observation of a negative genetic correlation between traits through selection. Experimental evolution in model organisms, most notably *Drosophila melanogaster*, has been extensively applied to life-history analysis. Yet despite the ease with which organisms like *Drosophila* are cultivated and made to evolve in the laboratory, generalities about the genetic architecture of fitness have proven more elusive than might be expected. A long list of potential confounds, including differential inbreeding depression between selection lines and controls, subtle adaptation to unintended sources of selection, and genotype-by-environment interactions in test environments (reviewed in Rose *et al.*, 1996), present a major challenge for the investigator. The present results illustrate another important problem for consideration: selection applied to the early life-history has an overriding impact on the later life-history, potentially obscuring later trait associations. Selection for accelerated development and extremely early fertility led to a comprehensive decline in components of fitness throughout the entire life-history. We refer to this cascade of deleterious fitness consequences as "the fast development syndrome."

First we consider the consequences of rapid development for the pre-adult stages. Our earlier work on these populations documented a substantial survival cost to rapid development expressed within the pre-adult period (Chippindale *et al.*, 1997). It was expected that the growth rate had increased to compensate for reduced development time, particularly because in other selection treatments within the same laboratory system, reduced viability was most clearly correlated with increased growth rate rather than absolute development time (Chippindale *et al.*, 1994, 1996, 1998). Our two measures of body size produce strikingly different answers to this question. If thorax length is applied, accelerated populations were seen to increase in growth rate. However, when calculated using dry body mass at eclosion, the growth rate declined strongly in the ACB and ACO selection treatments (approximately 25% in both sexes). The lack of congruence between these two growth rate estimates is not difficult to explain, considering that the first measure of body size is linear and the second is a cubic parameter. Here, we are concerned with the energetics of life-history, and mass is the appropriate measurement.

Prasad *et al.* (2000) report comparable reductions in growth rate in males of their fast development selection lines, and an even stronger reduction in females (about 32% below controls). While an apparent paradox, the evolution of reduced growth rate in response to selection for fast development can be understood when one considers that the pre-adult period consists of both growing and non-growing stages. For example, we found that 81% of the early selection response (measured at generation 36) was accounted for by changes in the larval period, with the remaining 19% being pupal (Chippindale *et al.*, 1997). The fly starves from the time the larva ceases to feed, from the third instar wandering phase to the time it has functional adult mouthparts. This period amounts to about four days of starvation, during which the energetically expensive process of metamorphosis occurs. Changes in larval feeding time may be disproportionately expressed as differences in adult weight because of an inability to reduce the pupal duration proportionately with the larval duration. In our previous study, we also observed a reduction in pupation height above the larval medium, suggesting that post-feeding motility is costly when there is strong selection for rapid development. Selection may target non-growing larval stages first, particularly if preserving body size is important to adult fitness or if there is a critical weight for successful pupation. As also reasoned by Prasad *et al.* (2000), development thus consists of phases during which growth or ontogeny takes on greater precedence, and the interplay between these processes can generate complicated outcomes for simple selection regimes.

The second major consideration is the impact of selection for an accelerated lifecycle on adult fitness. The evidence presented herein suggests that the accelerated fly emerges as a stripped-down version of its ancestors. In addition to being smaller, flies from accelerated populations displayed reduced starvation resistance at eclosion, a trait which is closely linked to total stored caloric reserves (Graves *et al.*, 1992; Djawdan *et al.*, 1998; Harshman *et al.*, 1999). Both body size and metabolic stores are likely to directly influence the early fertility of these populations. Accelerated populations were selected to produce eggs as early as possible after the addition of yeast-supplemented Petri dishes to their population cages. As would be expected from a functionally semelparous and r-type selection regime, accelerated females reached their peak fecundity earlier than control females. Interestingly, the accelerated females showed a sharp drop after peak fecundity while control females sustained high egg-output. Either the increased relative cost of each egg (perhaps due to small body size) intensified selection on the timing of reproduction to avoid gamete wastage, or the accelerated populations had become hypersenescent.

Having experienced no selection after about 36 hours of adult age, the evolution of reduced longevity and late-life fitness were predicted in the accelerated populations. In keeping with this prediction, late fecundity of females from the ACB and ACO selection treatments was 25% (peak) and 23% (cumulative) lower than their controls after 25 generations of selection. However, one cannot disentangle the effects of body size from those of senescence, particularly with a single point estimate of late fecundity in the selection experiment. Longevity provides a measure of the rate of senescence, and we found the predicted acceleration of mortality in the ACB and ACO populations. The early response was rapid and repeated in all ten replicate populations of the accelerated treatments, consistent with antagonistic pleiotropy between early- and late-life fitness components. However, the pattern of decline was far from monotonic later in selection. For example, the ACB selection treatment exhibited no measurable, additional reduction in lifespan between generations 40 and 100, and the changes in the ACO selection treatment after g40 were predominant only in females. The overall pattern is one of near perfect convergence on the baseline (B) mean longevity of approximately 35 days, with males being almost 20% longer-lived than females under the test conditions. The B populations had been selected for reproduction at 14 days (about 5 days old as adults) for over 350 generations at the time these experiments were conducted, and it remains an open question why natural selection maintains such extraordinary post-reproductive longevity under selection for semelparity. Assuming that there are genes with age-specific survival effects expressed in the first five weeks of adult life, we may speculate that the maintenance of an extended lifespan is the pleiotropic echo of selection for very early survival and fitness. Accordingly, building a high fitness young adult involves selection on loci with positive pleiotropic effects extending weeks beyond their immediate usefulness.

The longevity of the control (CB and CO) populations presents another curiosity. These ten populations were concurrently selected for reproduction at several weeks of adult age. Despite 40–60 generations of convergent selection for this intermediate age of reproduction, longevity and late-life fecundity remained highly differentiated between the CB and CO populations. The difference appears to be the result of a very slow downward response on the part of the CO populations, whose mean longevity did not change throughout the experiments documented in this study. It therefore appears that if antagonistic pleiotropy is involved in the evolution of aging in these populations, it is weak between middle and late ages.

Our study emphasizes the far-reaching impact of selection on early life fitness. Selection for fast development and early maturity are shown here to reduce performance in all other measured juvenile and adult traits. To add to this list, Teotónio *et al.* (2002) documented reduced early male and population fitness, and confirmed the reduced starvation tolerance and female fecundity of the ACO populations after 270 generations of accelerated selection (note: the protocol was simplified and direct truncation selection on development stopped after 125 generations). Teotónio *et al.* also demonstrated the resilience of these populations, as reverse evolution to a 14 day (B-type) cycle led to substantial recovery of the life-history characters of the ACO and ACB lines. Combined with the absence of hybrid vigor in the diallel crosses performed earlier with the ACO populations (Chippindale *et al.*, 1997), these results suggest that the depression of fitness components in the ACO populations was not the result of inbreeding.

When selection acts upon traits with far-reaching fitness impact, such as the body size trait, the relationships among the ancillary traits can be obscured. The fast development syndrome thus results in the appearance of pervasive positive relationships among fitness components because of their pleiotropic linkage to growth. The incredible shrinking fly is poor in a wide range of fitness-related categories, including fertility, survival and stress resistance, as documented by this study. Negative genetic correlations may still exist between such traits within these populations, but the overall path of selection is governed by the major traits, giving the appearance of uniformly positive genetic correlations among the populations. Similar problems will be encountered when the experimenter phenocopies these selective effects by, say, failing to control larval densities, or when other factors like differential inbreeding or novel environments are not taken into account. For example, experimenters often use density treatments to manipulate body size. Animals that are made smaller by increased competition during growth are likely to display reductions in a variety of adult fitness components because of both the size effect and the physiological stresses accompanying crowding. Similar artifacts may creep into selection experiments that were designed to address adult characters but fail to control the conditions of juvenile growth (e.g. Roper *et al.*, 1993, 1996).

The extraordinary speed and consistency with which populations of *Drosophila* adjust their pre-adult growth patterns to a variety of selection pressures is indicative of strong selection acting on these developmental patterns. In particular, our research suggests that both immediate survival and downstream fitness benefits promote slower growth and longer development in *Drosophila*. Despite the very different ecology and temporal separation of juveniles and adults in holometabolous insects, these stages are tightly connected by trade-offs affecting total fitness.

References

Bangham, J., T. Chapman, and L. Partridge. 2002. Effects of body size, accessory gland and testis size on pre- and postcopulatory success in *Drosophila melanogaster*. *Ani Behav* **64**: 915–92.

Borash D.J., H. Teotónio, M.R. Rose and L.D. Mueller. 2000. Density-dependent natural selection in *Drosophila*: Correlations between feeding rate, development time and viability. *J Evol Biol* **13**: 181–187.

Burnet, B., D. Sewell, and M. Bos. 1977. Genetic analysis of larval feeding behaviour in *Drosophila melanogaster*. II. Growth relations and competition between selected lines. *Gene Res* **30**:149–161.

Chippindale, A.K., A. Gibbs, M. Sheik, M. Djawdan, T.J. Bradley, and M.R. Rose. 1998. Resource acquisition and the evolution of stress resistance in *Drosophila melanogaster*. *Evolution* **52**: 1342–1352.

Chippindale, A.K., J.A. Alipaz, H-W. Chen, and M.R. Rose. 1997. Experimental evolution of accelerated development in *Drosophila*. 1. Larval development speed and survival. *Evolution* **51**: 1536–1551.

Chippindale, A.K., T.J.F. Chu, and M.R. Rose. 1996. Complex trade-offs and the evolution of starvation resistance in *Drosophila*. *Evolution* **50**: 753–766.

Chippindale, A.K., D.T. Hoang, P.M. Service, and M.R. Rose. 1994. The evolution of development in *Drosophila* selected for postponed senescence. *Evolution* **48**: 1880–1899.

Chippindale, A.K., A.M. Leroi, S.B. Kim, and M.R. Rose. 1993. Phenotypic plasticity and selection in *Drosophila* life-history evolution. 1. Nutrition and the cost of reproduction. *J Evol Biol* **6**: 171–193.

Clarke, J.M., J. Maynard Smith, and K.C. Sondhi. 1961. Asymmetrical response to selection for rate of development in *Drosophila subobscura. Gene Res* **2**: 70–81.

da Silva, L.B. and V.L.S.Valente. 2001. Body size and mating success in *Drosophila willistoni* are uncorrelated under laboratory conditions. *J Gene* **80**: 77–81.

Djawdan, M., A.K. Chippindale, M.R. Rose, and T.J. Bradley. 1998. Metabolic reserves and evolved stress resistance in *Drosophila melanogaster. Physiol Zool* **71**: 584–59.

Graves, J.L., E.C. Toolson, C. Jeong, L.N. Vu, and M.R. Rose. 1992. Desiccation, flight, glycogen and postponed senescence in *Drosophila melanogaster. Physiol Zool* **65**: 268–286.

Hamilton, W.D. 1966.The moulding of senescence by natural selection. *J Theor Biol* **12**: 12–45.

Harshman, L.G., K.M. Moore, M.A. Sty, and M.M. Magwire. 1999. Stress resistance and longevity in selected lines of *Drosophila melanogaster. Neurobiol Aging* **20**: 521–529.

Joshi, A., M.H. Do, and L.D. Mueller. 1999. Poisson distribution of male mating success in laboratory populations of *Drosophila melanogaster. Gene Res* **73**: 239–249.

Lints, F.A. and G. Gruwez. 1972. What determines the duration of development in *Drosophila melanogaster? Mech Ageing Dev* **1**: 285–297.

Marien, D. 1958. Selection for developmental rate of *Drosophila pseudoobscura. Genetics* **43**: 3–15.

Marinkovi´c, D. and F.J. Ayala. 1986. Selection for different rate of embryonic development in *Drosophila melanogaster* and *Drosophila simulans. Genetika* **18**: 205–219.

Markow, T.A. and J.P. Ricker. 1992. Male size, developmental stability and mating success in natural populations of three *Drosophila* species. *Heredity* **69**: 122–127.

Mueller, L.D. 1985. The evolutionary ecology of *Drosophila. Evol Biol* **19**: 37 98.

Nicholson, A.J. 1957. The self adjustment of populations to change. *Cold Spring Harbour Symposia in Quantitative Biology* **22**: 153–173.

Nunney, L. 1996. The response to selection for fast larval development in *Drosophila* and its effect on adult weight: An example of a fitness trade-off. *Evolution* **50**: 1193–1204.

Partridge, L. and M. Farquhar. 1983. Lifetime mating success of male fruitflies (*Drosophila melanogaster*) is related to their size. *Ani Behav* **31**: 871–877.

Partridge, L. and K. Fowler. 1992. Direct and correlated responses to selection on age at reproduction in *Drosophila melanogaster. Evolution* **46**: 76–91.

Partridge, L. and K. Fowler. 1993. Responses and correlated responses to artificial selection on thorax length in *Drosophila melanogaster. Evolution* **47**: 213–226.

Partridge, L., A.A. Hoffmann, and J.S. Jones. 1987. Male size and mating success in *Drosophila melanogaster* and Drosophila pseudoobscura under field conditions. *Ani Behav* **35**: 468–476.

Prasad N.G., M. Shakarad, D. Anitha, M. Rajamani, and A. Joshi. 2001. Correlated responses to selection for faster development and early reproduction in *Drosophila*: The evolution of larval traits. *Evolution* **55**: 1363–1372.

Prasad, N.G., M. Shakarad, V.M. Gohil, V. Sheeba, M. Rajamani, and A. Joshi. 2000. Evolution of reduced pre-adult viability and larval growth rate in laboratory populations of *Drosophila melanogaster* selected for shorter development time. *Gene Res* **76**: 249–259.

Prout, T. 1962. The effects of stabilizing selection on the time of development in *Drosophila melanogaster. Gene Res* **3**: 364–382.

Robertson, F.W. 1957. Studies in quantitative inheritance. XI. Genetic and environmental correlation between body size and egg production in *Drosophila melanogaster. J Gene* **55**: 428–443.

Roper C., P. Pignatelli, and L. Partridge. 1993. Evolutionary effects of selection on age at reproduction in larval and adult *Drosophila melanogaster. Evolution* **47**: 445–455.

Roper C., P. Pignatelli, and L. Partridge. 1996. Evolutionary responses of *Drosophila melanogaster* life history to differences in larval density. *J Evol Biol* **9**: 609–622.

Rose, M.R. 1991. *Evolutionary Biology of Aging*. Oxford University Press, New York.

Rose, M.R., T.J. Nusbaum, and A.K. Chippindale. 1996. Laboratory Selection: The Experimental Wonderland and the Cheshire Cat Syndrome. In: *Evolutionary Biology of Adaptation*, M.R. Rose and G.V. Lauder (eds.). Academic Press, New York, pp. 221–242.

Sang, J.H. and G.A. Clayton. 1957. Selection for larval development time in *Drosophila. J Heredity* **48**: 265–270.

Santos, M. 1996. Apparent directional selection of body size in *Drosophila buzzatti*: Larval crowding and male mating success. *Evolution* **50**: 2530–2535.

Santos, M., K. Fowler, and L. Partridge, 1994. Gene-environment interaction for body size and larval density in *Drosophila melanogaster*: An investigation of the effects of development time, thorax length and adult sex ratio. *Heredity* **72**: 515–521.

Service, P.M., E.W. Hutchinson, M.D. MacInley, and M.R. Rose. 1985. Resistance to environmental stress in *Drosophila melanogaster* selected for postponed senescence. *Physiol Zool* **58**: 380–389.

Teotónio, H., M. Matos, and M.R. Rose. 2002. Reverse evolution of fitness in *Drosophila melanogaster. J Evol Biol* **15**: 608–617.

Tigerstadt, P.M.A. 1969. Experiments on selection for developmental rate in *Drosophila melanogaster. Annales Academie Scientiarum Fennicae Series A IV Biologica* **148**: 1–53.

Williams, K.L., L. Barton-Browne, and A.C.M. Van Gerwan. 1977. Ovarian development in autogenous and anautogenous *Lucilia cuprina* in relation to protein storage in the larval fat body. *J Insect Physiol* **23**: 659–664.

Zamudio K. R., R.B. Huey, and W.D. Crill. 1995. Bigger isn't always better: Body size, developmental and parental temperature and male territorial success in *Drosophila melanogaster. Ani Behav* **49**: 671–677.

Zwaan, B.J., R. Bijlsma, and R.F. Hoekstra. 1995. Artificial selection for developmental time in *Drosophila melanogaster* in relation to the evolution of aging: Direct and correlated responses. *Evolution* **49**: 635–648.

Corpus Acknowledgments

We wish to thank the authors and the following publishers for their permission to reproduce the articles found in this volume.

1. M.R. Rose. 1984. Laboratory evolution of postponed senescence in *Drosophila melanogaster*. *Evolution* **38**: 1004–1010. *Publisher: SOCIETY FOR THE STUDY OF EVOLUTION.*
2. E.W. Hutchinson and M.R. Rose. 1991. Quantitative genetics of postponed aging in *Drosophila melanogaster*. I. Analysis of outbred populations. *Genetics* **127**: 719–727. *Publisher: GENETICS SOCIETY OF AMERICA.*
3. A.M. Leroi, A.K. Chippindale, and M.R. Rose. 1994. Long-term laboratory evolution of a genetic trade-off in *Drosophila melanogaster*. I. The role of genotype x environment interaction. *Evolution* **48**: 1244–1257. *Publisher: SOCIETY FOR THE STUDY OF EVOLUTION.*
4. T.J. Nusbaum, L.D. Mueller, and M.R. Rose. 1996. Evolutionary patterns among measures of aging. *Exper. Gerontol.* **31**: 507–516. *Publisher: ELSEVIER INC.*
5. P.M. Service, E.W. Hutchinson, M.D. MacKinley, and M.R. Rose. 1985. Resistance to environmental stress in *Drosophila melanogaster* selected for postponed senescence. *Physiol. Zool.* **58**: 380–389. *Publisher: UNIVERSITY OF CHICAGO PRESS.*
6. M.R. Rose, L.N. Vu, S.U. Park, and J.L. Graves. 1992. Selection for stress resistance increases longevity in *Drosophila melanogaster*. *Exper. Gerontol.* **27**: 241–250. *Publisher: ELSEVIER INC.*
7. M. Djawdan, A.K. Chippindale, M.R. Rose, and T.J. Bradley. 1998. Metabolic reserves and evolved stress resistance in *Drosophila melanogaster*. *Physiol. Zool.* **71**: 584–594. *Publisher: UNIVERSITY OF CHICAGO PRESS.*
8. A.G. Gibbs, A.K. Chippindale, and M.R. Rose. 1997. Physiological mechanisms of evolved desiccation resistance in *Drosophila melanogaster*. *J. Exper. Biol.* **200**: 1821–1832. *Publisher: THE COMPANY OF BIOLOGISTS LTD.*
9. D. Nghiem, A.G. Gibbs, M.R. Rose, and T.J. Bradley. 2000. Postponed aging and desiccation resistance in *Drosophila melanogaster*. *Exper. Gerontol.* **35**: 957–969. *Publisher: ELSEVIER INC.*

10. A.K. Chippindale, A.M. Leroi, S.B. Kim, and M.R. Rose. 1993. Phenotypic plasticity and selection in *Drosophila* life-history evolution. I. Nutrition and the cost of reproduction. *J. Evol. Biol.* **6**: 171–193. *Publisher: BLACKWELL SCIENTIFIC PUBLICATIONS ON BEHALF OF THE EUROPEAN SOCIETY FOR EVOLUTIONARY BIOLOGY.*
11. M. Djawdan, T. Sugiyama, L. Schlaeger, T.J. Bradley, and M.R. Rose. 1996. Metabolic aspects of the trade-off between fecundity and longevity in *Drosophila melanogaster. Physiol. Zool.* **69**: 1175–1195. *Publisher: UNIVERSITY OF CHICAGO PRESS.*
12. A.K. Chippindale, A.M. Leroi, H. Saing, D.J. Borash, and M.R. Rose. 1997. Phenotypic plasticity and selection in *Drosophila* life-history evolution. 2. Diet, mates and the cost of reproduction. *J. Evol. Biol.* **10**: 269–293. *Publisher: BLACKWELL SCIENTIFIC PUBLICATIONS ON BEHALF OF THE EUROPEAN SOCIETY FOR EVOLUTIONARY BIOLOGY.*
13. R.H. Tyler, H. Brar, M. Singh, A. Latorre, J.L. Graves, L.D. Mueller, M.R. Rose, and F.J. Ayala. 1993. The effect of superoxide dismutase alleles on aging in *Drosophila. Genetica* **91**: 143–149. *Publisher: KLUWER ACADEMIC PUBLISHING.*
14. J.E. Fleming, G.S. Spicer, R.C. Garrison, and M.R. Rose. 1993. Two dimensional protein electrophoretic analysis of postponed aging in *Drosophila. Genetica* **91**: 183–198. *Publisher: KLUWER ACADEMIC PUBLISHING.*
15. D.J. Deckert-Cruz, R.H. Tyler, J.E. Landmesser, and M.R. Rose. 1997. Allozymic differentiation in response to laboratory demographic selection of *Drosophila melanogaster. Evolution* **51**: 865–872. *Publisher: SOCIETY FOR THE STUDY OF EVOLUTION.*
16. R. Kurapati, H.B. Passananti, M.R. Rose, and J. Tower. 2000. Increased *hsp22* RNA levels in *Drosophila* lines genetically selected for increased longevity. *J. Gerontol.: Biol. Sci.* **55**: B552–B559. *Publisher: GERONTOLOGY SOCIETY OF AMERICA.*
17. P.M. Service, E.W. Hutchinson, and M.R. Rose. 1988. Multiple genetic mechanisms for the evolution of senescence in *Drosophila melanogaster. Evolution* **42**: 708–716. *Publisher: SOCIETY FOR THE STUDY OF EVOLUTION.*
18. J.L. Graves, E.C. Toolson, C. Jeong, L.N. Vu, and M.R. Rose. 1992. Desiccation, flight, glycogen, and postponed senescence in *Drosophila melanogaster. Physiol. Zool.* **65**: 268–286. *Publisher: UNIVERSITY OF CHICAGO PRESS.*
19. H. Teotónio and M.R. Rose. 2000. Variation in the reversibility of evolution. *Nature* **408**: 463–466. *Publisher: NATURE PUBLISHING GROUP.*

20. H. Teotónio, M. Matos, and M.R. Rose. 2002. Reverse evolution of fitness in *Drosophila melanogaster*. *J. Evol. Biol.* **15**: 608–617. *Publisher: BLACKWELL SCIENTIFIC PUBLICATIONS ON BEHALF OF THE EUROPEAN SOCIETY FOR EVOLUTIONARY BIOLOGY.*
21. L.D. Mueller, J.L. Graves, and M.R. Rose. 1993. Interactions between density-dependent and age-specific selection in *Drosophila melanogaster*. *Funct. Ecol.* **7**: 469–479. *Publisher: BLACKWELL SCIENTIFIC PUBLICATIONS ON BEHALF OF THE BRITISH ECOLOGY SOCIETY.*
22. A.K. Chippindale, D.T. Hoang, P.M. Service, and M.R. Rose. 1994. The evolution of development in *Drosophila melanogaster* selected for postponed senescence. *Evolution* **48**: 1880–1899. *Publisher: SOCIETY FOR THE STUDY OF EVOLUTION.*
23. A.K. Chippindale, J.A. Alipaz, H-W. Chen, and M.R. Rose. 1997. Experimental evolution of accelerated development in *Drosophila*. 1. Larval development speed and survival. *Evolution* **51**: 1536–1551. *Publisher: SOCIETY FOR THE STUDY OF EVOLUTION.*
24. D.J. Borash, H. Teotonio, M.R. Rose, and L.D. Mueller. 2000. Density-dependent natural selection in *Drosophila:* Correlations between feeding rate, development time, and viability. *J. Evol. Biol.* **13**: 181–187. *Publisher: BLACKWELL SCIENTIFIC PUBLICATIONS ON BEHALF OF THE EUROPEAN SOCIETY FOR EVOLUTIONARY BIOLOGY.*

Author Index

Subject Index